Advances in Optics and Optoelectronics

The Advances in Optics and Optoelectronics series focuses on the exciting new developments in the fast emerging fields of optics and optoelectronics. The volumes cover key theories, basic implementation methods, and practical applications in but not limited to the following subject areas:

AI Photonics
Laser Science and Technology
Quantum Optics and Information
Optoelectronic Devices and Applications
Fiber-Based Technologies and Applications
Near-infrared, Mid-infrared and Far-infrared Technologies and Applications
Biophotonics and Medical Optics
Optical Materials, Characterization Methods and Techniques
Spectroscopy Science and Applications
Microscopy and Adaptive Optics
Microwave Photonics and Wireless Convergence
Optical Communications and Networks

Within the scopes of the series are monographs, edited volumes and conference proceedings. We expect that scientists, engineers, and graduate students will find the books in this series useful in their research and development, teaching and studies.

Baojun Li · Yuchao Li · Hongbao Xin
Editors

Biophotonic Manipulation

Springer

Editors
Baojun Li
Guangdong Provincial Key Laboratory of Nanophotonic Manipulation
Institute of Nanophotonics
College of Physics and Optoelectronic Engineering
Jinan University
Guangzhou, China

Yuchao Li
Guangdong Provincial Key Laboratory of Nanophotonic Manipulation
Institute of Nanophotonics
College of Physics and Optoelectronic Engineering
Jinan University
Guangzhou, China

Hongbao Xin
Guangdong Provincial Key Laboratory of Nanophotonic Manipulation
Institute of Nanophotonics
College of Physics and Optoelectronic Engineering
Jinan University
Guangzhou, China

ISSN 2731-6009 ISSN 2731-6017 (electronic)
Advances in Optics and Optoelectronics
ISBN 978-981-96-4981-5 ISBN 978-981-96-4982-2 (eBook)
https://doi.org/10.1007/978-981-96-4982-2

This Springer imprint is published by the registered company Springer Nature Singapore Pte Ltd.
The registered company address is: 152 Beach Road, #21-01/04 Gateway East, Singapore 189721, Singapore

Preface

Since the first demonstration for particle trapping in 1986 by A. Ashkin, optical tweezers, based on a focused laser beam, have been widely used for the trapping and manipulation of different microscopic targets in a non-invasive and high-precision manner. This capability makes optical tweezers a powerful tool at the forefront of numerous studies in the interdisciplinary natural sciences, such as physics, material sciences, biological sciences and medicine. In particular, optical tweezers have shown its irreplaceable role for biophotonic manipulation and for the applications in biological systems. The great contribution of optical tweezers in biological systems was awarded the 2018 Nobel Prize in Physics.

The past half century has witnessed great advances in optical tweezers and biophotonic manipulation from the basic physical fundamentals to the practical biophotonic applications. The novel physics behind optical forces and light-matter interactions for optical trapping and manipulation have been revealed with new light beams such as structured or complexly shaped beams. Conventional optical tweezers (COTs) based on single focused laser beam has been widely used for the manipulation of different biological samples. In particular, the ability to manipulate single molecules with ultrahigh precision at a time via COTs reveals many new molecular insights of a specific molecule, which is distinguishable to the ensemble average over huge number of molecules. To meet the different requirements of biophotonic manipulation, and meanwhile to overcome the drawbacks of COTs, different biophotonic manipulation platforms have been developed. For example, surface plasmonic tweezers and photonic nanojet-based tweezers have been invented for the manipulation of nanoscale bio-particles, such as viruses and single molecules. Multi-physical coupling methods, such as optothermal tweezers, optoelectronic tweezers, opto-hydrodyanmic or optofluidic manipulation, etc., have been developed for high-throughput and multifunctional manipulation of different biological samples. In addition to the trapping and manipulation in the biological world, biophotonic manipulation techniques have also been used as a combination strategy to explore the biological world, and have gained many new insights at the molecular or subcellular level via precision imaging and sensing. When combined with machine learning, artificial intelligence, and spectral analysis, biophotonic manipulation has also shown huge potentials for identification of pathogens and diagnosis of diseases for precision medicine. In addition to the trapping and manipulation of cells, biophotonic manipulation techniques can also serve as a new tool for the modulation and regulation of cell functions. This optical modulation and regulation can turn a biological cell into a functional bio-micromotor or microrobot, with huge potentials to execute different biomedical tasks in different biomedical environments.

This book intends to provide a comprehensive review and perspectives of the rapidly expanding field of biophotonic manipulation, includes in-depth discussions on fundamental of optical forces, technologies of optical manipulation and their biological applications. Thus, this book is especially suitable for readers engaged in interdisciplinary research fields such as biophotonics, nanophotonics and biomedicine.

Part I of the book (Different Types of Optical Manipulation Techniques) mainly introduce the fundamentals of different types of optical manipulation techniques, including plasmonic tweezers (Chap. 1), optothermal tweezers (Chap. 2), holographic optical tweezers (Chap. 3), opto-hydrodynamic manipulation (Chap. 4), optofluidic manipulation (Chap. 5) and optoelectronic tweezers (Chap. 6). By enhancing the interaction between light and matter, as well as the coupling among multiple physical fields, the performance and functionality of optical manipulation have been greatly enriched.

Part II of the book (Biological Applications of Optical Manipulation Techniques) offers various biological and biomedical applications of optical manipulation techniques. Over the past few decades, scientists have developed diverse optical techniques to touch, to see, and to sense the biological world. The combination of optical manipulation, detection, and imaging technologies has greatly expanded the scope of their biological applications, including super-resolution imaging (Chap. 7), single-cell sensing (Chap. 8), nanoscale thermometry (Chap. 9), Raman detection (Chap. 10), single-molecule analysis (Chap. 11), light-driven microrobots (Chaps. 12 and 13), and nanoparticle assembly (Chaps. 14 and 15).

Therefore, from microscale to nanoscale, from single-function to multi-function, from cells to molecules, from physical mechanisms to biological applications, we are delighted to share with our readers the past, present, and future of biophotonic manipulation.

The idea of this book project was seeded in early 2023, we are very pleased to have received a positive response from our peers in this field, with a total of 15 book chapters submitted by nearly 50 authors. We would like to express our gratitude here to all the authors for their meticulous and rigorous work. Moreover, we would like to acknowledge the professional and effective assistance of the editors of Springer Nature. Finally, we hope the readers to enjoy the exciting journey into the varicolored landscape of the biophotonic manipulation.

Guangzhou, China
January 2025

Baojun Li
Yuchao Li
Hongbao Xin

Contents

Contributors

Jiajie Chen State Key Laboratory of Radio Frequency Heterogeneous Integration, Key Laboratory of Optoelectronic Devices, Systems of Ministry of Education and Guangdong Province, College of Physics and Optoelectronics Engineering, Shenzhen University, Shenzhen, China

Xiaoxia Chen Guangdong Provincial Key Laboratory of Nanophotonic Manipulation, College of Physics and Optoelectronic Engineering, Institute of Nanophotonics, Jinan University, Guangzhou, China

Xixi Chen Guangdong Provincial Key Laboratory of Nanophotonic Manipulation, Institute of Nanophotonics, College of Physics and Optoelectronic Engineering, Jinan University, Guangzhou, China

Xinbin Cheng Institute of Precision Optical Engineering, School of Physics Science and Engineering, Tongji University, Shanghai, China;
Technology Innovation Center of Mass Spectrometry for State Market Regulation, Center for Advanced Measurement Science, National Institute of Metrology, Beijing, China;
MOE Key Laboratory of Advanced Micro-Structured Materials, Shanghai, China;
Shanghai Institute of Intelligent Science and Technology, Tongji University, Shanghai, China

Ruping Deng Nanophotonics Research Center, Institute of Microscale Optoelectronics and State Key Laboratory of Radio Frequency Heterogeneous Integration, Shenzhen University, Shenzhen, China;
Institute of Modern Optics, Nankai University, Tianjin, China

Rongxin Fu School of Medical Technology, Beijing Institute of Technology, Beijing, China;
Zhengzhou Research Institute, Beijing Institute of Technology, Henan, China;
Chongqing Institute of Microelectronics and Microsystems, Beijing Institute of Technology, Chongqing, China

Zhiyong Gong School of Medicine, The Chinese University of Hong Kong, Shenzhen, China;
Guangdong Provincial Key Laboratory of Nanophotonic Manipulation, Institute of Nanophotonics, College of Physics and Optoelectronic Engineering, Jinan University, Guangzhou, China

Jinghui Guo School of Medicine, The Chinese University of Hong Kong, Shenzhen, China

Minru He MOE Key Laboratory for Nonequilibrium Synthesis and Modulation of Condensed Matter, Shaanxi Province Key Laboratory of Quantum Information and Quantum Optoelectronic Devices, School of Physics, Xi'an Jiaotong University, Xi'an, China

Tong He Guangdong Provincial Key Laboratory of Nanophotonic Manipulation, Institute of Nanophotonics, College of Physics and Optoelectronic Engineering, Jinan University, Guangzhou, China

Zhongpeng Huang Guangdong Provincial Key Laboratory of Nanophotonic Manipulation, Institute of Nanophotonics, College of Physics & Optoelectronic Engineering, Jinan University, Guangzhou, China

Ming Lei MOE Key Laboratory for Nonequilibrium Synthesis and Modulation of Condensed Matter, Shaanxi Province Key Laboratory of Quantum Information and Quantum Optoelectronic Devices, School of Physics, Xi'an Jiaotong University, Xi'an, China;
State Key Laboratory of Electrical Insulation and Power Equipment, Xi'an Jiaotong University, Xi'an, China

Yansheng Liang MOE Key Laboratory for Nonequilibrium Synthesis and Modulation of Condensed Matter, Shaanxi Province Key Laboratory of Quantum Information and Quantum Optoelectronic Devices, School of Physics, Xi'an Jiaotong University, Xi'an, China

Baojun Li Guangdong Provincial Key Laboratory of Nanophotonic Manipulation, Institute of Nanophotonics, College of Physics and Optoelectronic Engineering, Jinan University, Guangzhou, China

Gong Li School of Integrated Circuits and Electronics, Beijing Institute of Technology, Beijing, China

Hang Li School of Medical Technology, Beijing Institute of Technology, Beijing, China;
Zhengzhou Research Institute, Beijing Institute of Technology, Henan, China;
Chongqing Institute of Microelectronics and Microsystems, Beijing Institute of Technology, Chongqing, China

Heng Li Guangdong Provincial Key Laboratory of Nanophotonic Manipulation, Institute of Nanophotonics, College of Physics and Optoelectronic Engineering, Jinan University, Guangzhou, China

Yuchao Li Guangdong Provincial Key Laboratory of Nanophotonic Manipulation, Institute of Nanophotonics, College of Physics and Optoelectronic Engineering, Jinan University, Guangzhou, China

Zeran Li Guangdong Provincial Key Laboratory of Nanophotonic Manipulation, Institute of Nanophotonics, College of Physics & Optoelectronic Engineering, Jinan University, Guangzhou, China

Zonghao Li School of Integrated Circuits and Electronics, Beijing Institute of Technology, Beijing, China

Chenghong Lin Guangdong Provincial Key Laboratory of Nanophotonic Manipulation, Institute of Nanophotonics, College of Physics and Optoelectronic Engineering, Jinan University, Guangzhou, China

Chenchen Liu Guangdong Provincial Key Laboratory of Nanophotonic Manipulation, Institute of Nanophotonics, College of Physics & Optoelectronic Engineering, Jinan University, Guangzhou, China

Yao Lu Chongqing Institute of Microelectronics and Microsystems, Beijing Institute of Technology, Chongqing, China;
School of Integrated Circuits and Electronics, Beijing Institute of Technology, Beijing, China

Xingdi Luo Guangdong Provincial Key Laboratory of Nanophotonic Manipulation, Institute of Nanophotonics, College of Physics and Optoelectronic Engineering, Jinan University, Guangzhou, China

Changjun Min Nanophotonics Research Center, Institute of Microscale Optoelectronics and State Key Laboratory of Radio Frequency Heterogeneous Integration, Shenzhen University, Shenzhen, China

Ting Pan Guangdong Provincial Key Laboratory of Nanophotonic Manipulation, College of Physics and Optoelectronic Engineering, Institute of Nanophotonics, Jinan University, Guangzhou, P. R. China

Junle Qu State Key Laboratory of Radio Frequency Heterogeneous Integration, Key Laboratory of Optoelectronic Devices, Systems of Ministry of Education and Guangdong Province, College of Physics and Optoelectronics Engineering, Shenzhen University, Shenzhen, China

Yu-Xuan Ren Jinshan Hospital, and Institute for Translational Brain Research, Fudan University, Shanghai, China

Xuchen Shan School of Physics, Beihang University, Beijing, China

Yonghong Shao State Key Laboratory of Radio Frequency Heterogeneous Integration, Key Laboratory of Optoelectronic Devices, Systems of Ministry of Education and Guangdong Province, College of Physics and Optoelectronics Engineering, Shenzhen University, Shenzhen, China

Yang Shi Guangdong Provincial Key Laboratory of Nanophotonic Manipulation, Institute of Nanophotonics, College of Physics and Optoelectronic Engineering, Jinan University, Guangzhou, P. R. China

Yuzhi Shi Institute of Precision Optical Engineering, School of Physics Science and Engineering, Tongji University, Shanghai, China;
Technology Innovation Center of Mass Spectrometry for State Market Regulation, Center for Advanced Measurement Science, National Institute of Metrology, Beijing, China;
Shanghai Institute of Intelligent Science and Technology, Tongji University, Shanghai, China;
MOE Key Laboratory of Advanced Micro-Structured Materials, Shanghai, China

Dajing Wang School of Physics, Beihang University, Beijing, China

Fan Wang School of Physics, Beihang University, Beijing, China

Zhanshan Wang Institute of Precision Optical Engineering, School of Physics Science and Engineering, Tongji University, Shanghai, China;
Technology Innovation Center of Mass Spectrometry for State Market Regulation, Center for Advanced Measurement Science, National Institute of Metrology, Beijing, China;
MOE Key Laboratory of Advanced Micro-Structured Materials, Shanghai, China;
Shanghai Institute of Intelligent Science and Technology, Tongji University, Shanghai, China

Tianli Wu Institute of Physics, Henan Academy of Science, Zhengzhou, China;
Guangdong Provincial Key Laboratory of Nanophotonic Manipulation, Institute of Nanophotonics, College of Physics & Optoelectronic Engineering, Jinan University, Guangzhou, China

Hongbao Xin Guangdong Provincial Key Laboratory of Nanophotonic Manipulation, Institute of Nanophotonics, College of Physics and Optoelectronic Engineering, Jinan University, Guangzhou, P. R. China

Bingrui Xu Beijing Advanced Innovation Center for Intelligent Robots and Systems, School of Mechatronical Engineering, Beijing Institute of Technology, Beijing, China

Fan Yang Beijing Advanced Innovation Center for Intelligent Robots and Systems, School of Mechatronical Engineering, Beijing Institute of Technology, Beijing, China

Liliang Ye Guangdong Provincial Key Laboratory of Nanophotonic Manipulation, College of Physics and Optoelectronic Engineering, Institute of Nanophotonics, Jinan University, Guangzhou, P. R. China

Yongshi Ye Guangdong Provincial Key Laboratory of Nanophotonic Manipulation, Institute of Nanophotonics, College of Physics and Optoelectronic Engineering, Jinan University, Guangzhou, China

Xue Yun MOE Key Laboratory for Nonequilibrium Synthesis and Modulation of Condensed Matter, Shaanxi Province Key Laboratory of Quantum Information and Quantum Optoelectronic Devices, School of Physics, Xi'an Jiaotong University, Xi'an, China

Qian Zhang Guangdong Provincial Key Laboratory of Nanophotonic Manipulation, Institute of Nanophotonics, College of Physics and Optoelectronic Engineering, Jinan University, Guangzhou, China

Shuailong Zhang Zhengzhou Research Institute, Beijing Institute of Technology, Henan, China;
Chongqing Institute of Microelectronics and Microsystems, Beijing Institute of Technology, Chongqing, China;
Beijing Advanced Innovation Center for Intelligent Robots and Systems, School of Mechatronical Engineering, Beijing Institute of Technology, Beijing, China;
School of Integrated Circuits and Electronics, Beijing Institute of Technology, Beijing, China

Tiange Zhang School of Physics, Beihang University, Beijing, China

Xufeng Zhang Guangdong Provincial Key Laboratory of Nanophotonic Manipulation, Institute of Nanophotonics, College of Physics and Optoelectronic Engineering, Jinan University, Guangzhou, China

Yuquan Zhang Nanophotonics Research Center, Institute of Microscale Optoelectronics and State Key Laboratory of Radio Frequency Heterogeneous Integration, Shenzhen University, Shenzhen, China

Yanan Zhao Guangdong Provincial Key Laboratory of Nanophotonic Manipulation, College of Physics and Optoelectronic Engineering, Institute of Nanophotonics, Jinan University, Guangzhou, China

Jianxing Zhou State Key Laboratory of Radio Frequency Heterogeneous Integration, Key Laboratory of Optoelectronic Devices, Systems of Ministry of Education and Guangdong Province, College of Physics and Optoelectronics Engineering, Shenzhen University, Shenzhen, China

Part I

Different Types of Optical Manipulation Techniques

Surface Plasmon Optical Tweezers for Biomanipulation

1

Yuquan Zhang, Ruping Deng, and Changjun Min

Abstract

Surface Plasmon (SP) optical tweezers represent an innovative micro-nano manipulation technology that transcends the conventional optical diffraction limit. By exciting surface plasmon polaritons (SPPs) on metal surface or localized surface plasmons (LSPs) in metallic nanostructures, this technology generates optical hotspots at subwavelength scales, enabling the high-precision trapping and manipulation of microscopic objects. Leveraging their flexible all-optical modulation capability and significant local field enhancement properties, SP optical tweezers have demonstrated unique advantages in the non-contact manipulation of biological samples. This chapter introduces the fundamental principles of SP optical tweezers and their typical system designs, thoroughly exploring LSP systems based on structured substrates and SPP systems on smooth surfaces. Additionally, it outlines multimodal manipulation systems incorporating factors such as thermophoresis and electro-plasmonics. The chapter details the applications of SP optical tweezers at cellular, subcellular, and molecular scales, demonstrating their precise trapping capabilities across various scales and extensive application prospects. Finally, this chapter envisions the future development directions of SP optical tweezers and their potential impact on scientific research.

Keywords

Plasmonic tweezers · Electromagnetic field enhancement · Biomanipulation · Metallic nanostructures

1.1 Introduction

The radiation pressure of light acting on matter can be observed from the famous phenomenon of comet tails pointing away from the Sun. After the invention of laser, Ashkin trapped micron-sized particles in water using a laser, marking the beginning of the rapid development of optical tweezers technology [1–3]. Optical tweezers technology features non-contact and high precision properties, which make it widely employed in the biomedical field, such as trapping viruses, bacteria, and live cells, or indirectly used for measuring cellular mechanics. Optical tweezers manipulate target objects through the combined effects of gradient force and scattering force. When a high numerical aperture microscope is used in optical tweezers to focus a laser, creating a micron-scale focused light field at the focal point, which generates the gradient force. The strength of the gradient force is closely related to the gradient distribution of the light field intensity. Small objects can be trapped when the attractive gradient force overcomes the repulsive scattering force due to the scattering and absorption of objects. However, for traditional optical tweezers based on far-field optical field modulation, the spatial size of the focal spot is inevitably constrained by the optical diffraction limit [4], making it difficult to trap particles in deep sub-wavelength scale. To address this challenge, surface plasmon optical tweezers was proposed.

Surface plasmon (SP) is an electromagnetic wave that propagates only on the surface of metals, generated by the coupling and collective oscillation of free electrons on the

Y. Zhang · R. Deng · C. Min (✉)
Nanophotonics Research Center, Institute of Microscale Optoelectronics and State Key Laboratory of Radio Frequency Heterogeneous Integration, Shenzhen University, Shenzhen, China
e-mail: cjmin@szu.edu.cn

R. Deng
Institute of Modern Optics, Nankai University, Tianjin, China

B. Li et al. (eds.), *Biophotonic Manipulation*, Advances in Optics and Optoelectronics,
https://doi.org/10.1007/978-981-96-4982-2_1

metal surface with incident photons. It is a pure evanescent wave, exhibiting exponential decay away from the dielectric-metal interface, which strongly confines the electromagnetic field near the interface. SPs can be divided into two types: Surface Plasmon Polaritons (SPPs) that propagate on smooth metal surfaces, and Localized Surface Plasmons (LSPs) that exist in bounded geometries like metal particles. Due to the shortening of wavelength and the strong surface confinement effect, the electromagnetic field of SPs can be compressed in the deep sub-wavelength scale. This allows SP to "break through" the traditional diffraction limit and to generate electromagnetic hotspots at the nanoscale. This phenomenon also produces enhanced optical forces near the SP hotspots, which has stimulated research into plasmonic trapping and manipulation techniques.

In 1994, Kawata and others observed the motion of accelerated particles on a dielectric prism coated with a metallic film [5], attributing it to be a manifestation of surface plasmon resonance. In 2007, Quidant first conducted SP optical trapping experiments using a fabricated micron-sized metal disk structure [6]. Grigorenko et al. reported in 2008 the trap of 200 nm particles on metallic nanostructured substrates using optical tweezers [7], and managed to quench the Brownian motion of the trapped nanoparticles by nearly an order of magnitude with lower power, opening new avenues for the nanomanipulation of fragile biological objects. Soon afterwards, Quidant's group reported the trap of *Escherichia coli* [8], marking the first time that bacteria have been trapped on optical antenna.

The surface plasmon optical tweezers (also known as plasmonic tweezers) technology has attracted the interest of researchers studying LSPs near metallic nanostructures and SPPs on smooth metallic surfaces. Its fundamental operating mechanism and application potential have been studied in the fields of physics, chemistry, and biomedical research. So far, this technology has been used to trap hard structures (such as metals, dielectrics, semiconductors, and magnetic samples) as well as soft nanomaterials (such as proteins, polymer chains, and DNA). Today, this technology has been developed for many specific applications and has become an indispensable part of cutting-edge research in areas such as micro/nano-structured combinations, spectroscopy, and bio/medical detection.

In this chapter, we focus on SP optical tweezers technology, including its working principles and applications in biological manipulation. Initially, we present a concise overview of the excitation methods for LSP and SPP fields, followed by a theoretical derivation of the optical trapping forces. Subsequently, we introduce two distinct mechanisms for SP tweezers platforms: the LSP-induced trap system utilizing a structural substrate and the SPP-assisted manipulation system on a smooth surface. Additionally, we provide a brief overview of multi-modal manipulation systems that integrate factors such as thermophoresis and opto-thermoelectric forces. Then we introduce the applications of SP optical tweezers in biological field, categorized by the size of the targeted entities, including cellular, organelle, and molecular levels. Finally, we discuss the future development of this technology, its potential new applications, and its prospective impact on scientific research. Many representative studies have been introduced, which are significant and possess broad applicability, to illustrate the current state of the field. We believe that this chapter will furnish a comprehensive overview of this domain and offer insights and guidance for its advancement.

1.2 Fundamentals of SP Optical Tweezers

Due to the large wavevector of SPPs, their wavelength is shorter compared to the incident light, allowing SPPs to concentrate light energy in localized regions at sub-wavelength scales. This enables SPPs to achieve precise control over light-matter interactions at the nanoscale and significantly enhances electric field intensity and optical trapping capabilities, thereby improving the control of light-matter interactions in many applications. In this section, we outline two types of surface plasmon properties, corresponding to plasmonic trapping on smooth surfaces and trapping near microstructures, respectively.

1.2.1 Dispersion Properties of SPPs

SPs are the collective oscillations of free electrons occurring at the interface between a metal and a dielectric medium [9, 10]. This phenomenon arises from the discontinuity in the dielectric constant at the interface, where the dielectric constant transitions from a positive value in the dielectric medium to a negative value in the metal. SPs exhibit properties of surface propagation and near-field enhancement, and their propagation behavior is governed by the properties of the metal surface and the surrounding medium. Notably, even minor changes in the surrounding medium can lead to significant variations in the excitation of SPP field, rendering it highly sensitive and particularly suitable for applications in chemical and biological sensing. Figure 1.1a illustrates the classical model for the excitation of SPPs, consisting of an idealized half-infinite space formed by an isotropic dielectric with a dielectric constant of ε_1 ($z > 0$) and a metal with a dielectric function ε_2 ($z < 0$). Here ε_1 is a positive real number, while the relative permittivity of the metal, ε_2 is complex and expressed as $\varepsilon_2(\omega) = \varepsilon_r(\omega) + i\varepsilon_i(\omega)$, where $\varepsilon_r(\omega)$ is a negative real number and its absolute value is typically much larger than the imaginary part $|\varepsilon_r(\omega)| \gg |\varepsilon_i(\omega)|$.

The electromagnetic wave at the interface between metal and dielectric is a transverse magnetic (TM) mode, and its

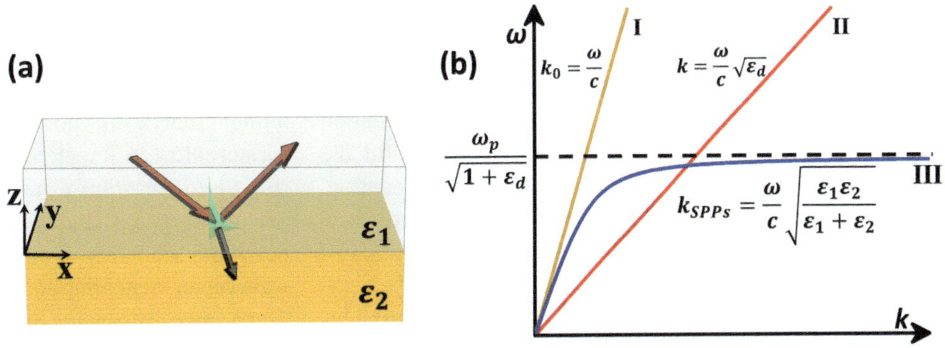

Fig. 1.1 **a** Surface plasmon polaritons excitation at a dielectric–metal interface. **b** Dispersion relationship curves of light in free space, dielectric, and SPPs. Here, ω_P is the plasma frequency of bulk longitudinal electron excitations

field distribution can be expressed as [11]: for the dielectric space where z > 0:

$$\begin{aligned} H_1(x, z, t) &= (0, A, 0)\exp(ikx - k_{1z}z - i\omega t) \\ E_1(x, z, t) &= \frac{iA}{\omega\varepsilon_0\varepsilon_1}(k_{1z}, A, ik) \\ &\quad \exp(ikx - k_{1z}z - i\omega t) \end{aligned} \tag{1.1}$$

for the metallic space where z < 0:

$$\begin{aligned} H_2(x, z, t) &= (0, B, 0)\exp(ikx - k_{2z}z - i\omega t) \\ E_2(x, z, t) &= \frac{iA}{\omega\varepsilon_0\varepsilon_2}(k_{2z}, B, ik) \\ &\quad \exp(ikx - k_{2z}z - i\omega t) \end{aligned} \tag{1.2}$$

In Eqs. (1.1) and (1.2), A and B are constants, and k is the wavevector along x direction. The wavevector along z direction $k_z = \sqrt{k^2 - \varepsilon(\omega/c)^2}$ is usually a pure imaginary number and determines that the electromagnetic field decays exponentially away from the interface z = 0. At the interface (z = 0), the tangential components of the electromagnetic field satisfy the continuity condition:

$$\begin{aligned} A &= B \\ Ak_{1z}/\varepsilon_1 &= Bk_{2z}/\varepsilon_2 \end{aligned} \tag{1.3}$$

Based on the relationship between frequency and wavevector k under TM polarization conditions, it can be further organized as follows.

$$\begin{aligned} k_{\text{SPPs}} &= k_0\sqrt{\varepsilon_1\varepsilon_2/(\varepsilon_1 + \varepsilon_2)} \\ &= \omega/c\sqrt{\varepsilon_1\varepsilon_2/(\varepsilon_1 + \varepsilon_2)} \end{aligned} \tag{1.4}$$

This is the dispersion relation for exciting SPPs at the dielectric-metal interface. Figure 1.1b shows the dispersion curves of light in free space, in a high refractive index medium, and for SPPs. Clearly, compared to the dispersion curve of light in free space (Curve I) and in high refractive index medium (Curve II), the wavevector of surface plasmon waves (Curve III) does not match the wavevector of incident light in free space. Therefore, surface plasmons are generally not directly excited by incident light from free space and require some wavevector enhancement to satisfy the wavevector matching condition for excitation.

Moreover, because dielectric constant of metals ε_2 is typically complex with a negative real part, the wavevector k_{SPPs} for SPPs is larger than the wavevector k_0 in free space. This results in a SPP wavelength $\lambda_{SPPs} = 2\pi/k_{SPPs}$ that is smaller than the wavelength λ_0 in free space for the same light frequency, and thus provides an unprecedented ability to compress light into subwavelength regions. Since the scale of surface plasmon excitation is much smaller than the wavelength of light, energy is confined to a subwavelength region near the metal surface. Within this small region, the light field is significantly enhanced compared to the incident light used for excitation. The maximum field enhancement of SPPs on a smooth surface can be described as follows:

$$A_{spp} = \left|\frac{E_{SPPs}}{E_0}\right|^2 = \frac{2}{\varepsilon_d}\frac{|\varepsilon_r|^2}{\varepsilon_i}\frac{a}{1+|\varepsilon_r|} \tag{1.5}$$

where $a = \sqrt{|\varepsilon_r|(\varepsilon_d - 1) - \varepsilon_d}$, ε_d is the dielectric constant of the substrate through which the metal film is illuminated to excite the SPP. Therefore, the field enhancement depends on the dielectric constants of the metal and the adjacent medium. Generally, the field enhancement of excited SPPs on gold/silver films is about one or two orders of magnitude in the visible light region.

1.2.2 Theory for SPP and LSP Excitation

SP lies in the collective oscillations of free electrons within a metal, driven by an external electromagnetic field (such as a light field), which occur along the metal surface or within nanostructures. According to the dispersion relationship in Fig. 1.1b, the wavevector of incident light in free space need to be increased to fit the wavevector of SP for excitation. In this section, we introduce several typical excitation methods for SPPs and LSPs.

1.2.2.1 Theory and Configurations for SPPs Excitation

As indicated by Eq. (1.4) and illustrated in Fig. 1.1b, the effective excitation of SPPs necessitates the compensation of the wavevector of incident light in order to fulfill the wavevector matching condition essential for SPP excitation. In the following section, we will provide a concise overview of several widely employed techniques for the excitation of SPPs.

The prism coupling method is recognized as a straightforward and efficient approach for the excitation of SPPs. Notably, Otto [12] and Kretschmann [13] independently introduced two distinct prism coupling techniques, as depicted in Fig. 1.2a, b, which are differentiated solely by the positioning of the metal film. Utilizing a prism with a high refractive index (where the refractive index $n_d = \sqrt{\varepsilon_d} > 1$), the wavevector of the incident light can be enhanced to match the wavevector of SPPs, as illustrated by the curve presented in Fig. 1.1b. When the condition for wavevector matching expressed as the below Eq. (1.6) is fulfilled, SPPs are generated at the upper interface between the metal and air.

$$Re[k_{SPPs}] = Re\left[\frac{\sqrt{\varepsilon_1\varepsilon_2/\varepsilon_1+\varepsilon_2}\omega_0}{c}\right] = Re\left[\frac{\sqrt{\varepsilon_d}\sin\theta_{sp}\omega_0}{c}\right] \quad (1.6)$$

where, ω_0 represents the angular frequency of light, and θ_{sp} indicates the incident angle for SPPs excitation. The prism coupling configuration presents several advantages, including minimal loss and high coupling precision, which contribute to its extensive application across various domains, including biology, photonics, and sensing.

Additionally, SPPs can be excited through near-field scattering from a tip positioned in proximity to the metal surface [15, 16], such as particles located on the surface (Fig. 1.2d) or from a nanoscale scatterer (Fig. 1.2e). The extensive spatial frequency spectrum associated with the near-field scatterers allows certain components to fulfill the wavevector matching condition for the excitation of SPPs. The primary advantage of excitation via near-field scattering is its compact configuration and the relative ease of implementation. However, it is important to note that only a limited fraction of the scattered light satisfies the wavevector matching condition, resulting in a low efficiency of SPPs excitation through this mechanism.

Furthermore, as illustrated in Fig. 1.2f [17], the incorporation of a periodic grating structure on the metallic surface facilitates the introduction of supplementary spatial frequencies, which in turn fulfills the wavevector matching condition for the excitation of SPPs. The equation governing wavevector matching is expressed as follows:

$$Re[k_{SPPs}] = Re[k_0 \sin\theta \pm mG] \quad (1.7)$$

wher θ represents the incident angle, G is the grating constant ($G = 2\pi/\Lambda$, where Λ is the grating period), m denotes the diffraction order (m is defined as a natural number). This technique demonstrates a compact on-chip excitation structure and a comparatively high efficiency in the excitation

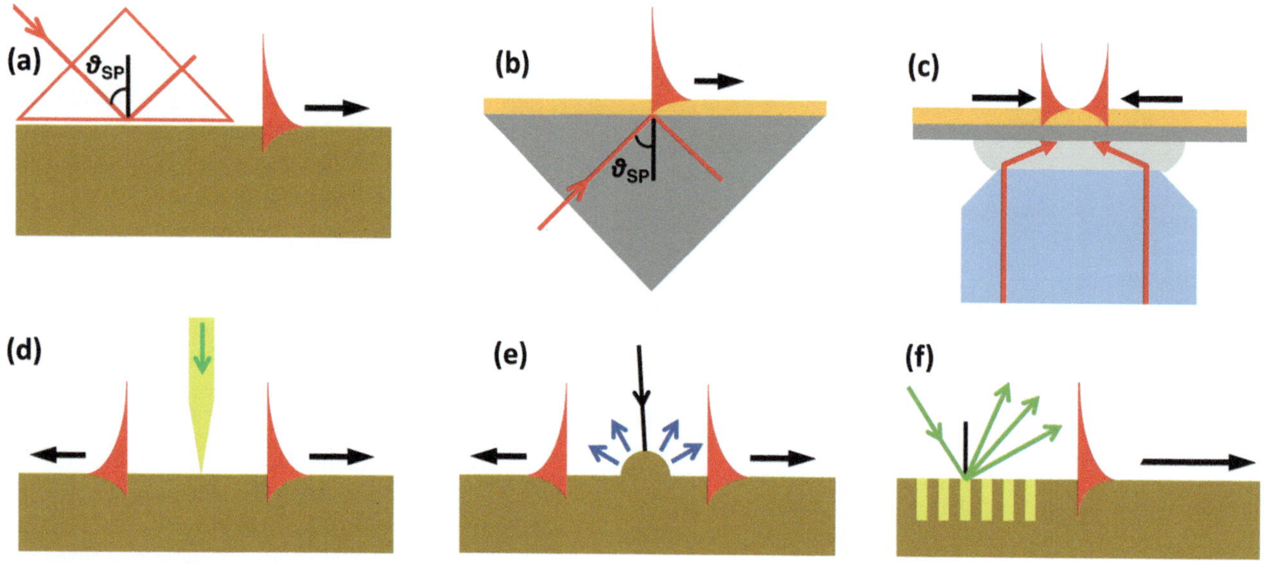

Fig. 1.2 Configuration of SPPs excitation (Adapted with permission from ref. [14], under Copyright license CC BY). **a** Otto geometry, **b** Kretschmann geometry, **c** focused excitation by a high NA objective lens, **d** excitation with a probe, **e** diffraction on surface structure, **f** diffraction on a grating

of SPPs. Nevertheless, it is associated with certain disadvantages, including elevated fabrication costs and restricted dynamic tunability.

Finally, high numerical aperture (NA) microscope objectives are also commonly used for exciting SPPs [18]. The utilization of a focused microscope objective enables the attainment of a sufficiently large incident angle, thereby fulfilling the wavevector matching condition for the excitation of SPPs on the upper surface of a metallic film. As illustrated in Fig. 1.2c, the incident light in this configuration exhibits a circularly symmetric distribution, which facilitates the excitation of an annular SPP that propagates inward. This process results in the formation of a highly focused SPP spot at the center of the optical axis, often referred to as a SPP virtual probe. This excitation technique can be seamlessly integrated with conventional microscopy systems, rendering it extensively applicable in the fields of biological imaging and sensing.

1.2.2.2 Theory and Configurations for LSPs Excitation

LSPs are defined as the collective oscillations of electrons that occur within metallic nanoparticles or other subwavelength metallic structures. These oscillations are triggered by the interaction of the metallic nanostructures and incident light, as illustrated in Fig. 1.3. Unlike propagating SPPs, LSPs are directly excited by the incoming light and restricted by the metallic nanostructures, rendering them non-propagating. Furthermore, since the energy of LSPs is restricted to the region of the metallic nanostructure, resulting in a substantially stronger electromagnetic field intensity at the LSP excitation point compared to that of the incident field. The resonance frequency of LSPs is intricately linked to the periodicity, morphology, and dimensions of the nanostructure and the refractive index of the surrounding medium. Consequently, the properties of the LSP field can be optimized through the design of different metallic nanostructures. By customizing metallic nanostructures, it is possible to achieve the maximum electromagnetic fields and electromagnetic field distributions better aligned with experimental objectives. These specific structures are capable of tightly confining the electromagnetic field within the narrow gap region between adjacent metallic structures, resulting in an electromagnetic "hotspot" phenomenon, which is characterized by an exceptionally high localized electric field intensity that exceeds the enhancement levels attainable with conventional metallic structures.

Figure 1.3 illustrates various types of configurations for LSPs excitation. Single metallic nanoparticle can achieve an electromagnetic field enhancement of LSP of approximately 10–100 times, where the resonance peak of such nanoparticles is influenced by their size and material composition. Gap plasmonic nanostructures can be categorized into two main types, both capable of achieving an electromagnetic field enhancement of around 10^{3-4}. The first category includes internal gaps within nanostructures that exclusively excite LSPs, as shown in Fig. 1.3b, d, e, which encompasses structures of nanogaps, nanoslits, and internal nanoparticle cavities. The second category is characterized by the gaps formed between a nanostructure and a planar metallic surface as shown in Fig. 1.3c, f. The local field enhancement of LSPs not only enables higher field gradients for optical trapping forces, but also open new possibilities for optical sensing based on the high electromagnetic field intensities.

1.2.3 Fundamental Theory of Optical Forces in SP Field

The optical forces utilized in optical trapping are dependent on the spatial properties of the electromagnetic field and can be classified into two main categories: scattering force and gradient force. The scattering force origins from

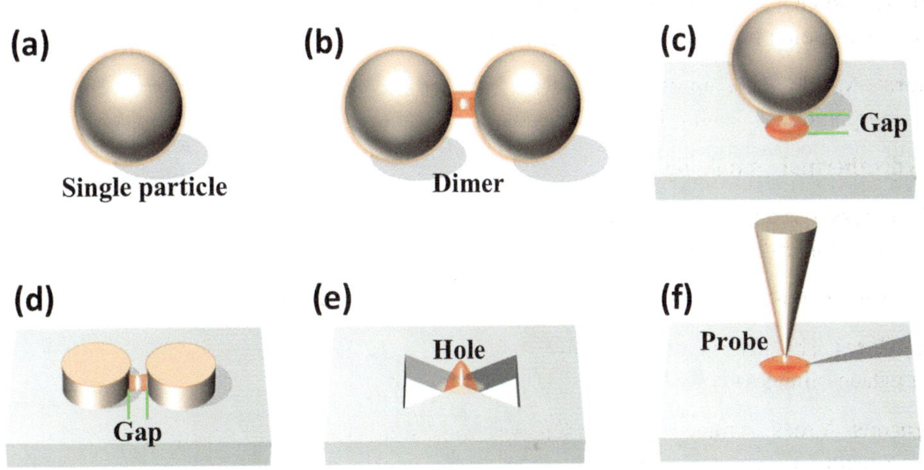

Fig. 1.3 Different scenarios of LSP excitation. a Single metallic nanoparticle. **b** Dimer structure of two nanoparticles. **c** Gap between a nanoparticle and a metal surface. **d** Nanostructure with surface protrusions. **e** Nanohole gap. **f** Tip of a probe and metal surface

the momentum transfer between the light and the object and usually push object along the direction of wavevector of light. This force arises when the path of light is altered due to a discontinuity in the refractive index. Conversely, the gradient force is primarily generated by the gradient of the light intensity and can attract object towards the high intensity region, and thus plays a crucial role in counteracting the scattering force and in the formation of optical traps.

In surface plasmon trapping systems, the magnitude of the optical forces relies on the degree of surface plasmon excitation, which in turn dictates the concentration of optical energy surrounding the structure, thereby affecting both the gradient and scattering forces. Since the late 1990s, many researchers have explored the forces present in optical near fields, revealing that surface plasmon also play a significant role in these interactions. In 2005, Quidant et al. examined the influence of radiative forces within optical near fields in their analysis of Rayleigh dielectric spheres, establishing a foundational framework for the systematic investigation of plasmonic traps [19].

To elucidate the properties of the interaction between a SPP field and a particle, the numerical Finite-Difference Time-Domain (FDTD) simulation can be employed to obtain the SPP field around the particle, and then the Maxwell Stress Tensor (MST) method can be used to calculate the optical forces induced by the SPP field. According to classical electromagnetic field theory [20], the total electromagnetic force (i.e., the Lorentz force) acting on a particle can be analyzed as follows:

$$f = \rho \boldsymbol{E} + \boldsymbol{j} \times \boldsymbol{B} \tag{1.8}$$

here, ρ represents the charge density, $\boldsymbol{j}$ is the current density, $\boldsymbol{E}$ is the electric field vector, and $\boldsymbol{B}$ is the magnetic flux density. By rewriting Eq. (1.8) according to the inhomogeneous Maxwell's equations, we can derive the following equation:

$$\begin{aligned} f = \nabla\left[\varepsilon(\boldsymbol{EE}) - \frac{\varepsilon}{2}\boldsymbol{I}E^2 + \frac{1}{\mu}(\boldsymbol{BB}) - \frac{1}{2\mu}\boldsymbol{I}B^2\right] \\ - \varepsilon\frac{\partial(\boldsymbol{E}\times\boldsymbol{B})}{\partial t} \end{aligned} \tag{1.9}$$

In the electrostatic and magnetostatic fields, the time-averaged total electromagnetic force acting on the particle can be expressed as:

$$\begin{aligned} \langle \boldsymbol{F} \rangle &= \langle \oint_s \boldsymbol{T} \cdot \boldsymbol{n} ds \rangle \\ &= \oint \left\{ \begin{array}{l} \frac{\varepsilon}{2} Re\left[(\boldsymbol{E}\cdot\boldsymbol{n})\boldsymbol{E}^*\right] - \frac{\varepsilon}{4}\left(\boldsymbol{E}\cdot\boldsymbol{E}^*\right)\boldsymbol{n} + \\ \frac{\mu}{2} Re\left[\mu(\boldsymbol{H}\cdot\boldsymbol{n})\boldsymbol{H}^*\right] - \frac{\mu}{4}\left(\boldsymbol{H}\cdot\boldsymbol{H}^*\right)\boldsymbol{n} \end{array} \right\} ds \end{aligned} \tag{1.10}$$

where, ε and μ denote the relative permittivity and relative permeability of the medium that envelops the particle, $\boldsymbol{n}$ signifies the unit normal vector that is orthogonal to the differential surface area ds, and $\boldsymbol{T}$ represents the Maxwell stress tensor. The MST algorithm represents a comprehensive analytical computational approach that calculates the total electromagnetic force exerted on a particle.

To elucidate the distinctions between SPP tweezers and traditional laser tweezers in their interactions with particles, it is essential to further decompose the total electromagnetic force into the gradient force and the scattering force. The gradient force is characterized as a dynamic Coulomb force, which is contingent upon the induced charge density within the geometric configuration. In contrast, the scattering force is indicative of the dynamic Laplace force, which is related to the intensity of the local magnetic field [21]. For Rayleigh particles, the force exerted upon them within an electromagnetic field can be approximated utilizing the electric dipole model, which is derived from the principles of Lorentz force law and Maxwell's equations. The time-averaged force can be expressed as follows, in accordance with the theoretical framework established by Spesyvtseva and Dholakia [22, 23]:

$$\begin{aligned} \langle \boldsymbol{F} \rangle = \frac{1}{4} Re\left[\alpha_p\right]\nabla|\boldsymbol{E}|^2 + \frac{\sigma(\alpha_p)}{2c} Re\left[\boldsymbol{E}\times\boldsymbol{H}^*\right] \\ + \sigma(\alpha_p)c\nabla\times\left(\frac{\varepsilon_0}{4\omega i}\boldsymbol{E}\times\boldsymbol{E}^*\right) \end{aligned} \tag{1.11}$$

here, the first term represents the optical gradient force [24], which is directly proportional to the gradient of the electric-field intensity. Particles with a high refractive index relative to the surrounding medium are pulled by the gradient force toward the region of maximum light intensity. The subsequent two terms correspond to the scattering forces [25, 26], which are proportional to the cross-sectional area of the particle. The second term represents the radiation pressure, which exerts a force on the particle following the direction of the Poynting vector. In contrast, the third term signifies the torque force, which is associated with the self-density of the light field and holds relevance in systems with tight focusing condition. Note that each term in Eq. (1.11) is related to α_p, which represents the effective polarizability of the particle. α_p is an adjusted form of the particle's intrinsic polarizability α_0, and it consider the scattering losses that arise from the interaction between the electromagnetic wave and the particle. α_p quantifies the material's propensity to polarize in response to the applied electric field $\boldsymbol{E}$. The relationship between α_p and α_0 is expressed as follows:

$$\alpha_0 = 4\pi n_m^2 \varepsilon_0 r^3\left(\frac{m^2-1}{m^2+2}\right) \tag{1.12}$$

$$\alpha_p = \frac{\alpha_0}{1 - i\alpha_0 k_0^3/6\pi\varepsilon_0} \tag{1.13}$$

herein, n_m represents the refractive index of the surrounding medium, ε_0 is the permittivity of the vacuum, r is the radius of the particle, m signifies the ratio of the particle's refractive index n_p to that of the surrounding medium n_m.

Stable trapping can be achieved when the gradient force is larger than the scattering force. This means that the potential well created by the trapping force exceeds the kinetic energy of the particle. Furthermore, the trapping efficiency of the optical tweezers system can be quantified by the stiffness of the optical trap, which is mathematically expressed as the ratio of the force applied to the particle by the optical trap to the displacement of the particle from the trap's center as:

$$U(r_0) = -\int_{\infty}^{r_0} \boldsymbol{F} \cdot dr \qquad (1.14)$$

The greater the optical trap stiffness, the stronger the optical trapping force acting on the particle for a given displacement. In experimental contexts, the careful selection of an appropriate force model is essential for accurate theoretical predictions. For instance, it is imperative to consider the effects of Brownian motion and the additional Stokes drag that influence the behavior of trapped particles. Consequently, to ensure stable trapping, the potential well must possess sufficient depth to effectively confine the particle, with typical values approximating $10k_BT$, where k_B represents the Boltzmann constant and T denotes the absolute temperature.

1.3 Typical SP Optical Tweezers Systems

The strong field enhancement effect of SP makes it highly applicable in the trapping, detection, and imaging of micro-nano objects. In the following section, we classify typical SP optical tweezer systems for two categories according to the structural properties of their underlying substrates, including the structural SPs optical tweezers system that incorporates metal nanostructures on the substrate, and the structureless all-optical modulated SP optical tweezers system that only utilizes a smooth metal film. Additionally, a concise evaluation of the respective advantages and disadvantages of these two systems is provided.

1.3.1 Plasmonic Traps Enabled by Metal Nanostructures

When an excitation laser irradiates a metal nanostructure, the scattering field of the nanostructure encompasses wavevectors thereby enabling SPs excitation. Furthermore, metal nanostructures possess the capability to concentrate electromagnetic field energy within specific gaps or hotspots of the structure, therefore enabling nanoscale trapping and detection that surpasses the diffraction limit. In structural SP optical tweezer systems, these metal nanostructures are generally fabricated on a dielectric substrate, which is subsequently overlaid with a fluid containing the target object intended for trap. Under the condition of surface plasmon resonance (SPR), the metal nanostructures on the substrate generate LSPs hotspots at the extremities of the structures, which typically exhibit electric field intensities several orders of magnitude greater than that of the incident light. According to Eq. (1.10), elevated field intensities can result in increased optical trap stiffness, thereby promoting the stable trapping of target objects in fluid.

1.3.1.1 Principle of Structural Plasmonic Tweezers

In plasmonic tweezers systems based on metallic nanostructures, the excitation efficiency of SPs is profoundly affected by the intrinsic properties of the metallic nanostructures, such as gaps and edges, as well as by the excitation conditions of the incident light, including wavevector matching and polarization orientation. In periodic metallic structures, efficient plasmon excitation typically requires satisfying the wavevector matching condition through coupling methods such as prism coupling or focused light coupling. In Sect. 1.2, we discussed the categorization of SPs excitation into three distinct methods: prism-based excitation, objective lens-based excitation, and grating structure excitation. The prism excitation technique, exemplified by the Otto and Kretschmann configurations, is characterized as a non-focusing excitation method. This approach utilizes prism coupling to excite a wide-field SPP field from the incident light, thereby can be used for large area trapping and rapid wide-field detection. Figure 1.4a illustrates a conventional plasmonic optical tweezers apparatus utilizing prism coupling excitation [6]. The substrate pattern of the system is resonantly excited by incident light that is coupled through the prism, thereby confining the electromagnetic field within the interstices of the nanostructures. This system produces a stable and robust method for trapping nanoscale targets within a solution.

In 2007, Quidant et al. reported the first demonstration of optical tweezers based on 2D surface plasmons, where single micrometer-sized dielectric beads were manipulated in a micron-sized gold disk [6]. By decreasing the size of the disk, the captured particle size can be reduced to approximately 1 μm. However, capturing smaller nanoscale particles remains limited. To address this challenge, researchers have developed various dimer and oligomer nanostructures, leveraging their enhanced electromagnetic coupling properties to achieve more precise control over plasmonic fields. For example, the Quidant group utilized gold optical nanoantennas to capture *Escherichia coli*. Prism-coupling structures are noted for their low loss and high excitation efficiency, rendering them prevalent in various domains, including biology, photonics, and sensing [27]. Nonetheless, the divergent nature of the incident light in prism-coupled excitation, which is advantageous for large-area trapping, results in relatively low manipulation precision. Moreover, it typically confines the size of trapped particles to the micron-to-submicron range, posing challenges for capturing smaller samples like molecules.

Utilizing an objective lens to concentrate the incident light beam results in the formation of a small focal spot that can be accurately directed onto the nanostructure, which greatly enhances the trap precision. In 1997, Nie and Emory demonstrated single-molecule detection using focused laser beams to excite LSPs on silver nanoparticles [28]. This is explained by the LSPs resonance excited by metallic nanoparticles, which significantly enhances the Raman scattering cross-section, allowing the originally weak signal to be detected. Ten years later, Grigorenko and his team first proposed subwavelength near-field optical tweezer system created near a nanopillars substrate [7], which optical near-fields of nanopillars provide a subwavelength trapping volume for the trapped nanoparticles. This research opens up exciting new possibilities across various scientific and nanotechnology fields by utilizing LSPs for optical trapping. In 2009, Quidant et al. considered the influence of the trapped particle itself on the local electric field. By generating a self-induced back-action (SIBA) optical trap near the cutoff resonance of nanopores in a metal film [29]. This technique significantly reduced the required minimum incident laser intensity. Additionally, SIBA can decrease the local intensity within the trap by an order of magnitude. This innovation has important implications for manipulating dielectric nanoparticles with lower energy consumption and less thermal damage.

Noteworthy, SP excitation system based on objective lens focusing, which not only enables precise manipulation of small targets but also facilitates real-time imaging. Figure 1.4b illustrates a simplified schematic diagram of the experimental configuration [30]. Although it is a challenge to directly image targets smaller than 50 nm, visualization can still be achieved by detecting the time evolution of the trapped particle's absorption and reflection within the SP field, and these recorded spectral data can be used to visualize the dynamics of the trapping events.

1.3.1.2 Design of Substrate Metal Nanostructures

Due to the ability of metallic nanostructures to confine SPs within exceedingly small dimensions, thereby generating a high-gradient electromagnetic fields that yield robust gradient force capable of counterbalancing the scattering force. In LSPs excitation systems, the design of metallic nanostructures is crucial, as their geometry and structural characteristics directly influence the excitation efficiency of LSPs. Particularly in nanostructures with gaps or edges, the orientation of the electric field plays a key role in determining the excitation efficiency. When the electric field direction is orthogonal to the gaps or edges of the structure, it facilitates more efficient plasmon excitation. This is because the electric field in this orientation better drives local charge separation, thereby enhancing the strength of the local electromagnetic field and ultimately improving the LSPs excitation efficiency. Through meticulous design of these nanostructures, researchers have effectively trapped a diverse array of micro- and nanoscale object, encompassing dielectric, semiconductor, metallic particles, and biological specimens.

Continuous advancements in nanofabrication technologies have driven the development of diverse and intricate nanostructure designs that employ in plasmonic traps. These include nanoholes, bowties, rings, waveguides, nanowires, disks, antennas, pyramids, cubes and cavities, as shown in Fig. 1.5. To establish a stable trapping potential, it is crucial to systematically design structural parameters, such as dimensions, morphology, orientation, and material composition. In addition, properties of the excitation beam and the requirement to generate hotspots at specific locations should also be considered. Researchers can design and simulate their models using simulation software such as MATLAB, COMSOL, and Finite Difference Time Domain (FDTD).

Presently, a variety of structural patterns have been extensively utilized for trapping diverse objects, through the fabrication of imprint pattern structures by selectively removing excess material from thin films deposited on dielectric substrates [7, 8, 31–35]. Plasmonic hotspots are generally excited at the edges of these structures, converging energy within the nanoscale gaps to produce localized electromagnetic field enhancement. To improve trapping stability, researchers have devised complex pattern designs that concentrate the energy into more confined volumes,

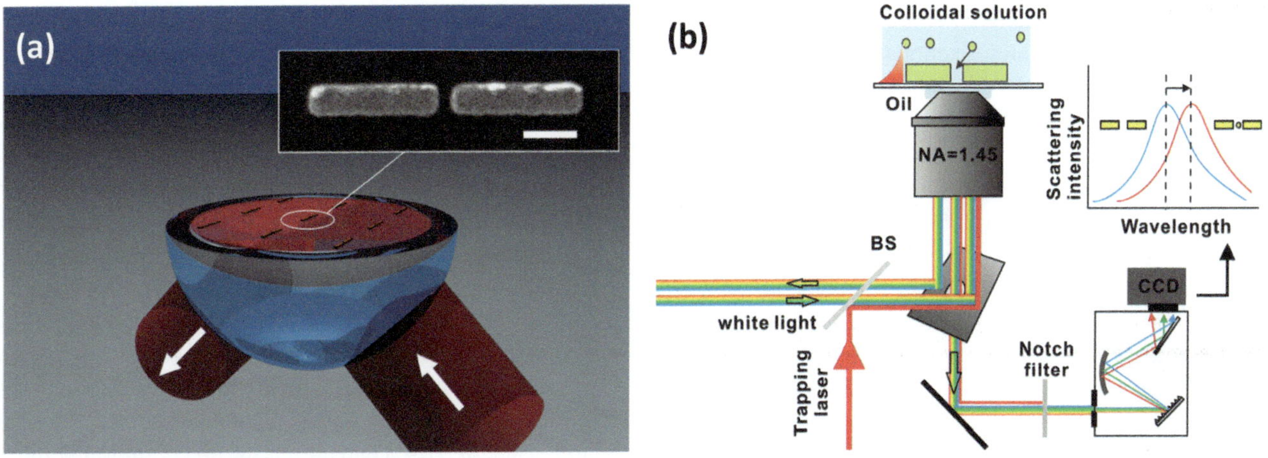

Fig. 1.4 **Principle of structural plasmonic tweezers. a** Surface plasmonic trapping configuration through the design of nanostructures, scale bar: 200 nm (Adapted with permission from ref. [8], Copyright © 2009, American Chemical Society). **b** Experimental configuration. The trapping events can be directly monitored using scattering spectra of the antennas (Adapted with permission from ref. [30], Copyright © 2010, American Chemical Society)

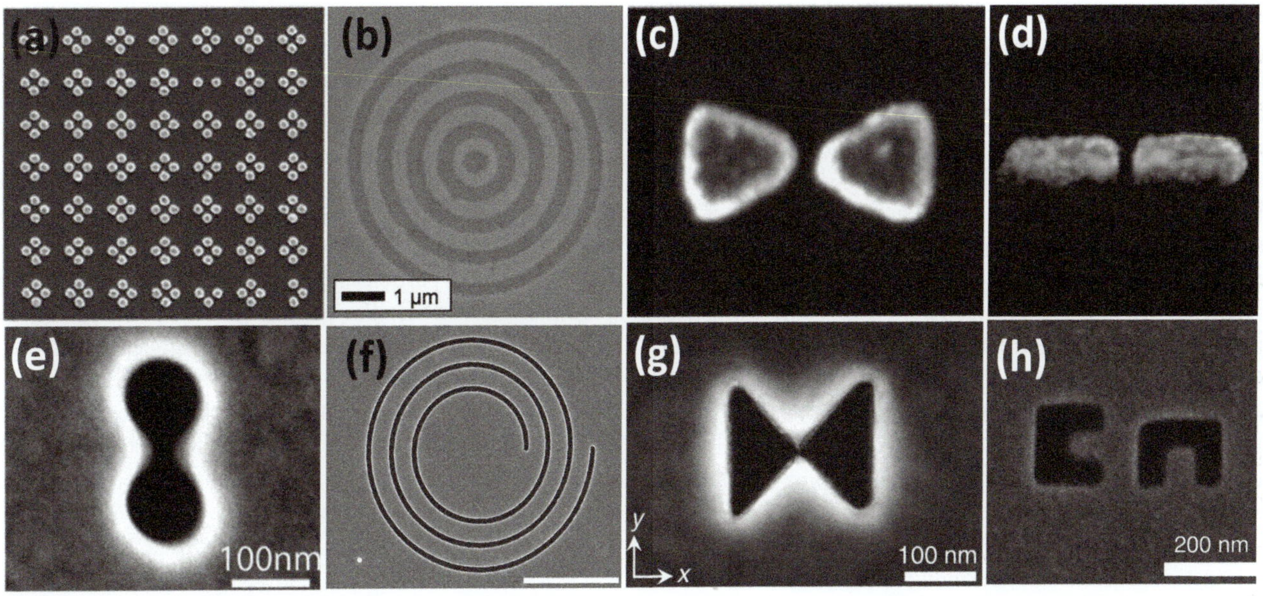

Fig. 1.5 **Schematic diagram of plasmonic nanostructures. a–d** Embossed patterns, **e–h** hollow aperture. **a** Nanodisks (Adapted with permission from ref. [36], Copyright © 2013, American Chemical Society); **b** Silver Bull's Eye nanostructure (Adapted with permission from ref. [35], Copyright © 2017, American Chemical Society); **c** Nanobowtie (Adapted with permission from ref. [37], under Copyright license CC BY); **d** Nanoantennas (Adapted with permission from ref. [30], Copyright © 2010, American Chemical Society). **e** Nanohole (Adapted with permission from ref. [38], Copyright © 2011, American Chemical Society; **f** plasmonic Archimedes spiral (Adapted with permission from ref. [39], Copyright © 2014, American Chemical Society; **g** Nanobowtie aperture (Adapted with permission from ref. [40], under Copyright license CC BY); **h** C-shaped engraving (Adapted with permission from ref. [33], Copyright © 2014, American Chemical Society)

thereby facilitating enhanced and more stable trapping effects. In addition to the previously mentioned embossed structures, the perforation or etching of metal films to generate nanoscopic cavity structures can also facilitate the localized of electromagnetic field [33, 35]. Figure 1.5e–h illustrates various designs for nanohole-based plasmonic trapping systems. These apertures have the capacity to enhance the interstitial field and create deeper potential wells, thereby increasing trapping efficiency. Moreover, the utilization of nanohole structures enables functionalities such as the trap of DNA molecules and DNA decoding.

1.3.1.3 Structural SP Optical Tweezers

Plasmonic metal nanostructures have been proved to enable stable optical manipulation [14, 41]. In 2007, Quidant and colleagues proposed SP optical tweezers based on the micro metal disk configurations. They engineered an array of gold disks on a quartz substrate and excited SP on the gold disks, which effectively trapped polystyrene microspheres suspended in a solution. Figure 1.6a illustrates the schematic diagram of the trapping system and the results. The next year, Grigorenko et al. introduced SP optical tweezers based on a nanostructured substrate (Fig. 1.6b) [7]. Their method involved the fabrication of a gold nanoparticle array on a glass substrate, resulting in a subwavelength plasmonic optical trap of 200 nm spheres. Furthermore, the steep field gradient in the central region of the plasmonic nanoantenna generates an optical force strong enough to overcome the thermal Brownian fluctuations and to trap quantum dot or a single molecule with low input power. Lee et al. proposed and demonstrated a non-fluorescent nanoscopic trapping and monitoring platform that can trap a single sub-5 nm particle and monitor it with a pair of floating nonlinear point sources, as show in Fig. 1.6e. In 2011, Kang et al. successfully demonstrated the trap and manipulation of nanoparticles utilizing a nanoantenna structure [32], by selectively exciting various locations on the nanoantenna with a light source, they effectively trapped and maneuvered 300 nm polystyrene spheres. Zheng et al. designed a C-shaped engraving (CSE) structure that resonates with different polarization directions [33], such as Fig. 1.6c. By adjusting the polarization direction of the incident light used for excitation, dynamic trapping and transport of targets can be achieved.

Because plasmonic patterns are mostly fixed on substrates, the structure limits the flexibility of SP optical tweezers. The optical trapping force is significantly influenced by the distribution of the SP field, and the SP field distribution is closely related to the properties of the incident light. Therefore, by changing the properties of the incident light, it is possible to achieve dynamic manipulation [42]. Lindquist et al. demonstrated the dynamic SP trap of nanoparticles using a computer-controlled spatial light modulator [35]. Crozier's group manually rotated the polarized angle of incidence light, causing the nanoparticles to rotate around the edge of a gold nanodisk [31]. Furthermore, by altering the polarization state, the spin and orbital angular momentum of the incoming light, it is possible to achieve dynamic control in SP tweezers [43–45]. Huang et al. demonstrated the rotation of optically isotropic dielectric particles within a single gold plasmonic Archimedean spiral. By employing circularly polarized laser excitation, they were able to reverse the spin direction, effectively changing the rotation direction [39].

Since plasmonic hotspots consistently form at specific locations near the fabricated structures, by designing multiple identical or different metal nanostructure arrays on the substrate, plasmonic hotspots can be generated simultaneously at multiple locations, enabling parallel or 3D trapping. In 2018, Yao et al. developed a flexible, label-free, and simple method to generate a three-dimensional plasmonic trapping array for the simultaneous separation and measurement of single-cell secretions [46]. However, due to the exponential decay of the plasmonic field around the fabricated structures, it is difficult to achieve complete three-dimensional control on a large scale. In 2014, Quidant was the first to showcase three-dimensional plasmonic manipulation of a single 50 nm dielectric particle by creating a bowtie plasmonic aperture at the end of a conical metal-coated optical fiber (see Fig. 1.6f) [47]. This technique allowed the object to be moved along with the fiber without needing to change the light source or the structure. Expanding on this idea, optical fibers have been connected to various plasmonic nanostructures, including apertures and conical tips, enabling the movement of the trapped sample several tens of micrometers in three-dimensional space.

However, in many nanostructure-based SP optical tweezers systems, the movement range of the trapped particles is often confined to relatively small area near the structures. Additionally, the nanoscale SP hotspots limit the size range of trapped particles, particularly not applicable to trapping mesoscopic metallic particles. Due to the high reflection and absorption of the mesoscopic metallic particles, there are strong scattering and repulsive forces when trapping such particles, making it difficult to be trapped. Therefore, there is a significant research value to explore SP tweezers with enhanced dynamics and broader trapping ranges, free from the constraints of substrate structures.

1.3.2 Structureless All-Optical Modulated SP Optical Tweezers

As mentioned in the previous section, LSPs are excited at the edges of nanostructures or within nanoscale gaps, which limits the trapping range to the areas near the hotspots. Unlike the structured SP optical tweezers, all-optically modulated SP optical tweezers overcome this obstacle, which are excited and propagate on smooth metallic surfaces without any structural assistance. This type of SP tweezers eliminates the complex fabrication processes for nanoscale metal structures, and can easily manipulate the excited SPP hotspots as well as the trapped particles by modulating the properties of excitation light, thus providing a significant advantage for dynamic manipulation.

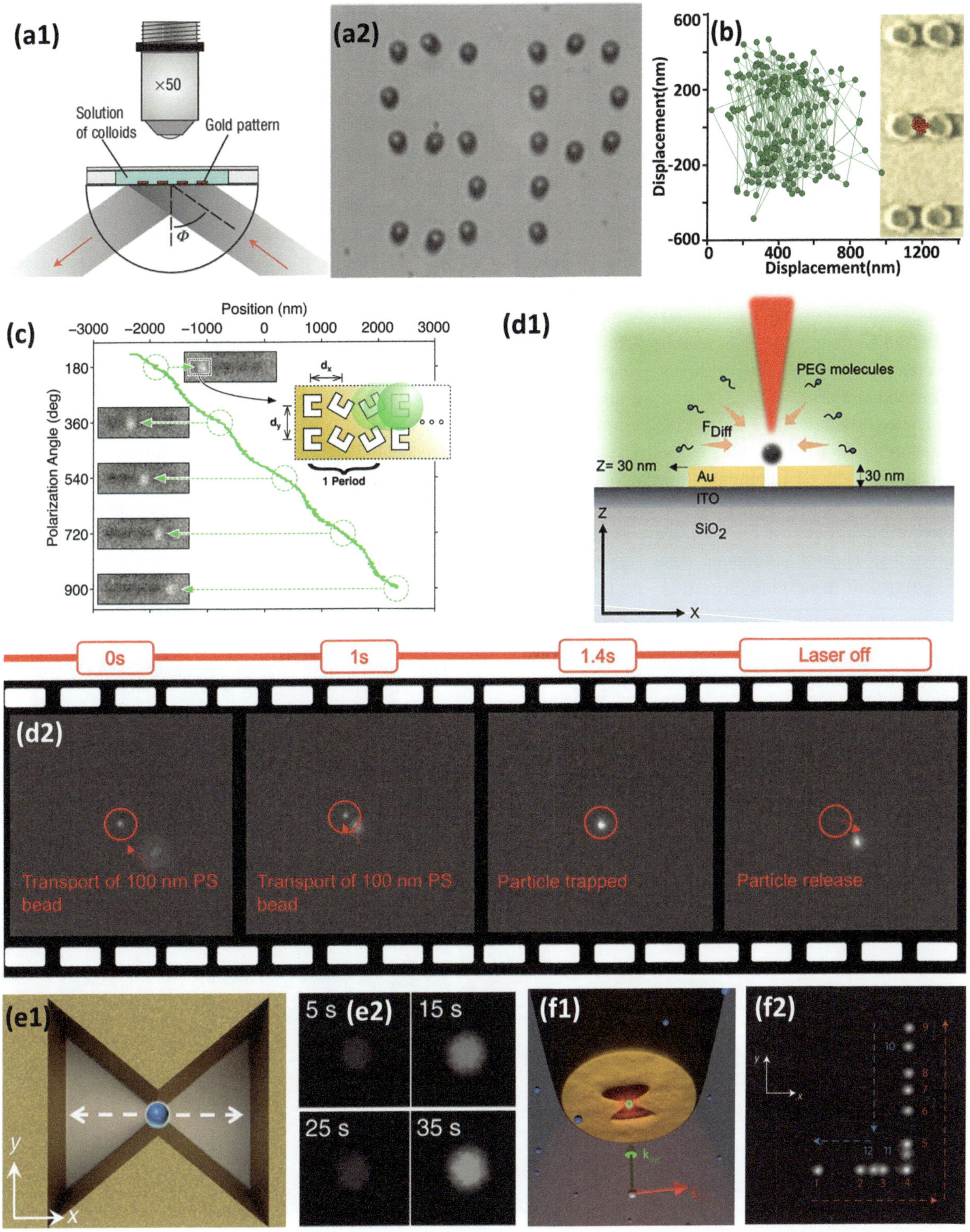

Fig. 1.6 The configuration for surface plasmon trapping using designed nanostructures. a Particle trapping and arrangement using SPPs metal disk structures (Adapted with permission from ref. [6], Copyright © 2007, Springer Nature Limited); **a1** Schematic diagram of the experimental setup, **a2** SP pattern formed by trapping PS microspheres with disk arrays. **b** Nanophotonic tweezers with subwavelength-sized optical traps and significantly enhanced trapping efficiency. (Adapted with permission from ref. [7], Copyright © 2008, Springer Nature Limited). **c** Plasmonic conveyor belt for remote transfer based on a C-shaped hole substrate (Adapted with permission from ref. [33], Copyright © 2014, American Chemical Society). **d** Plasmonic tweezers based on LSPR and photothermal depletion attraction forces. (Adapted with permission from ref. [37], under Copyright license CC BY), **d1** Schematic diagram of the experimental setup, where the PS microspheres are moved to the focus by thermal depletion force, **d2** Experimental snapshot of trapping 100 nm PS microspheres with a 5 mW laser. **e** non-fluorescent nanoscopic trapping and monitoring platform (Adapted with permission from ref. [40], under Copyright license CC BY), **e2** With the switching of the laser, quantum dots are captured in experimental fragments within a 5 nm gap. **f** Schematic diagram of scanning near-field optical tweezers and synthetic images of trapped particle (Adapted with permission from ref. [47], Copyright © 2014, Springer Nature Limited)

1.3.2.1 SPP Trapping Through a Prism

In 1996, Kawata et al. firstly observed that particle movement is accelerated on a dielectric prism coated with a metallic film [5]. In 2001, Chang et al. provided the first theoretical prediction of these plasmonic forces [48]. In 2006, Volpe et al. demonstrated that the optical forces generated by SPPs at the glass-metal and water-metal interfaces were approximately 40 times stronger for micron-sized dielectric particles compared to conventional optical forces [49]. Since SPPs are excited and propagate on smooth thin films, the resulting optical forces can act across the entire surface, facilitating large-scale particle trapping, moving and assembly. Dholakia reported a dynamic process for assembling dielectric microspheres on a gold nanofilm [50], utilizing TM-polarized light incident at the coupling angle. The study found that, as the power increased, temperature-induced variations could lead to the assembly of particles into elliptical ring structures. Kumar et al. introduced a comparable technique for assembling plasmonic nanoparticles [51, 52]. Figure 1.7a, c are the schematic diagrams of their device and some experimental results. Their experimental design allowed for the flexible control of the excitation position on the metal film by adjusting the relative positioning of the excitation source and the film itself. This methodology permitted the swift assembly and manipulation of nanoparticles across a uniform surface. Furthermore, the utilization of multiple laser beams for SPP excitation has the potential to enable concurrent multi-point assembly and distribution of the nanoparticles. In 2012, Yuan's group combined arbitrary beam theory (ABT) with generalized Lorenz–Mie scattering theory (GLMT) to numerically calculate the SPP forces exerted on micron-sized particles [53]. As shown in Fig. 1.7d, they experimentally verified this theory, offering deeper theoretical insights into SPP-based trapping and manipulation techniques. Darbari et al. proposed a novel optical tweezers system based on two counter-propagating SPPs [54, 55]. As illustrated in Fig. 1.7b, high-order counter-leaky SPPs were excited on a thin gold stripe, resulting in plasmon interference that balanced the opposing scattering forces to create a stable trap. In this configuration, particles with distinct intrinsic properties experience varying forces, making the system potentially valuable for advanced manipulation and sorting applications.

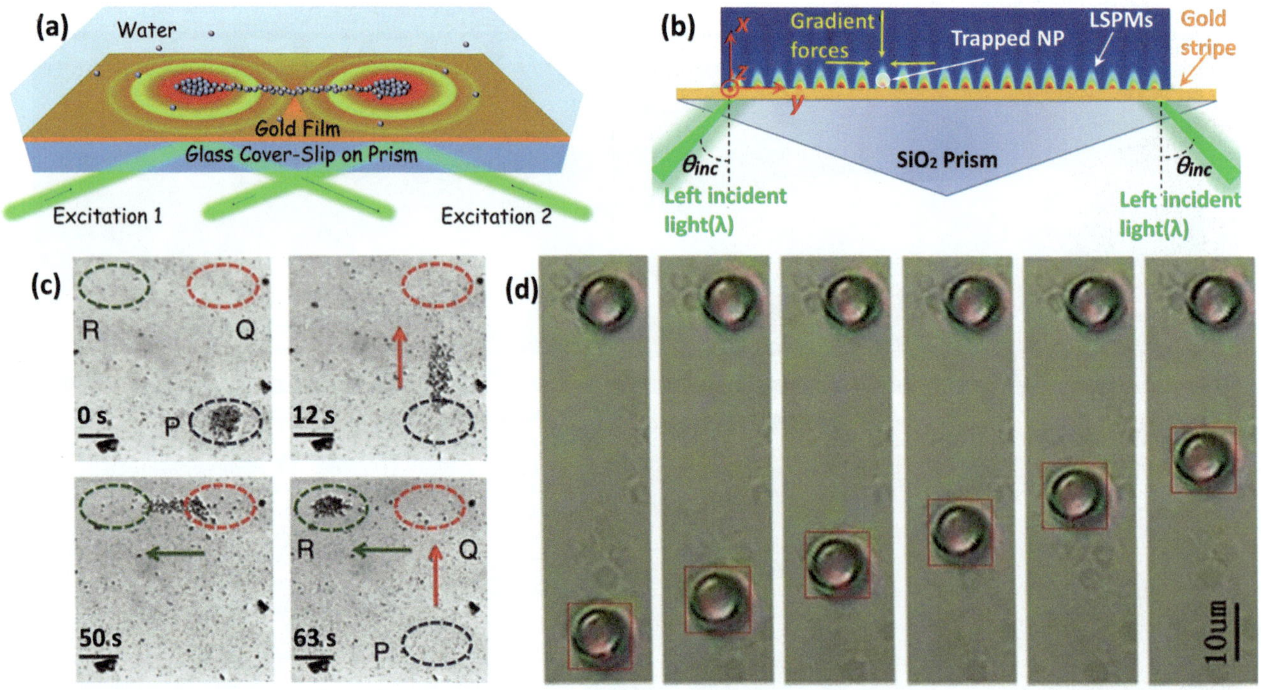

Fig. 1.7 All-optical modulation-based plasmonic tweezers system using prism excitation. a Schematic illustration of dual assembly of Ag nanoparticles on an Au thin film via prism coupling: green lines represent two 532 nm laser beams introduced at both ends of the prism, and black arrows indicate the direction of the laser beams (Adapted with permission from ref. [52], Copyright © 2016, The Royal Society of Chemistry). **b** Schematic representation of reconfigurable plasmonic trap arrays generated by the interference of total internal reflection surface plasmons at the edge of a gold waveguide, coupled through two counter-propagating LSPMs (Adapted with permission from ref. [54], Copyright © 2019 IEEE). **c** Plasmon-assisted assembly of Ag@Au nanoparticles at a metal-fluid interface (Adapted with permission from ref. [51], Copyright © 2014, Springer Nature Limited). **d** Trajectory of a single 8 μm PS microsphere under trapping (Adapted with permission from ref. [53], Copyright © 2012, Springer Science Business Media, LLC)

1.3.2.2 Focused Plasmonic Tweezers

The prism-excited SPPs tweezers are limited in their spatial resolution due to the use of a wide-field divergent laser for incident light, which excites a large area of SPPs on smooth metal surfaces. In 1998, Kawata et al. employed a focused laser beam to excite SPPs [18], generating interference-resistant plasmonic focus (also called virtual probe) with significant electromagnetic enhancement. Because the resulting plasmonic virtual probes are at deep-subwavelength scale and their electromagnetic field gradient are steep, these virtual probes are more effective for trapping. Based on this concept, in 2013, Min et al. proposed and implemented a novel focused plasmonic tweezers technology [56], achieving the first stable trapping of micron-sized metallic particles. As illustrated in Fig. 1.8a, a typical schematic of focused plasmonic tweezers based on a high-NA objective lens is presented, where an incident radially polarized light excites SPPs at a smooth metal-water interface. When particles enter the SPPs field, the plasmonic virtual probe attracts and traps the nearby particles. Additionally, they performed numerical analysis of the forces acting on gold nanoparticles as depicted in Fig. 1.8b, revealed that the net force points toward the center, enabling effective particle trap (Fig. 1.8c).

When the focused plasmonic virtual probe is compressed to dimensions smaller than the diffraction limit, it can produce a significant gradient force around the focus, thus enabling the trapping of particles with a large size range. Given that in the focusing process, SPPs propagate along the metal surface over a relatively extensive area, the gradient force is consequently dispersed across a larger spatial region. This principle allows for a focused configuration of surface plasmonic tweezers to effectively and stably trap metal objects within the size range from Rayleigh to Mie particles, thereby addressing the limitations associated with traditional plasmonic tweezers. The stiffness of plasmonic trapping is influenced by a multitude of factors, including Brownian motion and thermal convection. Due to SPPs only be excited by light beams that meet the wavevector matching condition, a considerable portion of

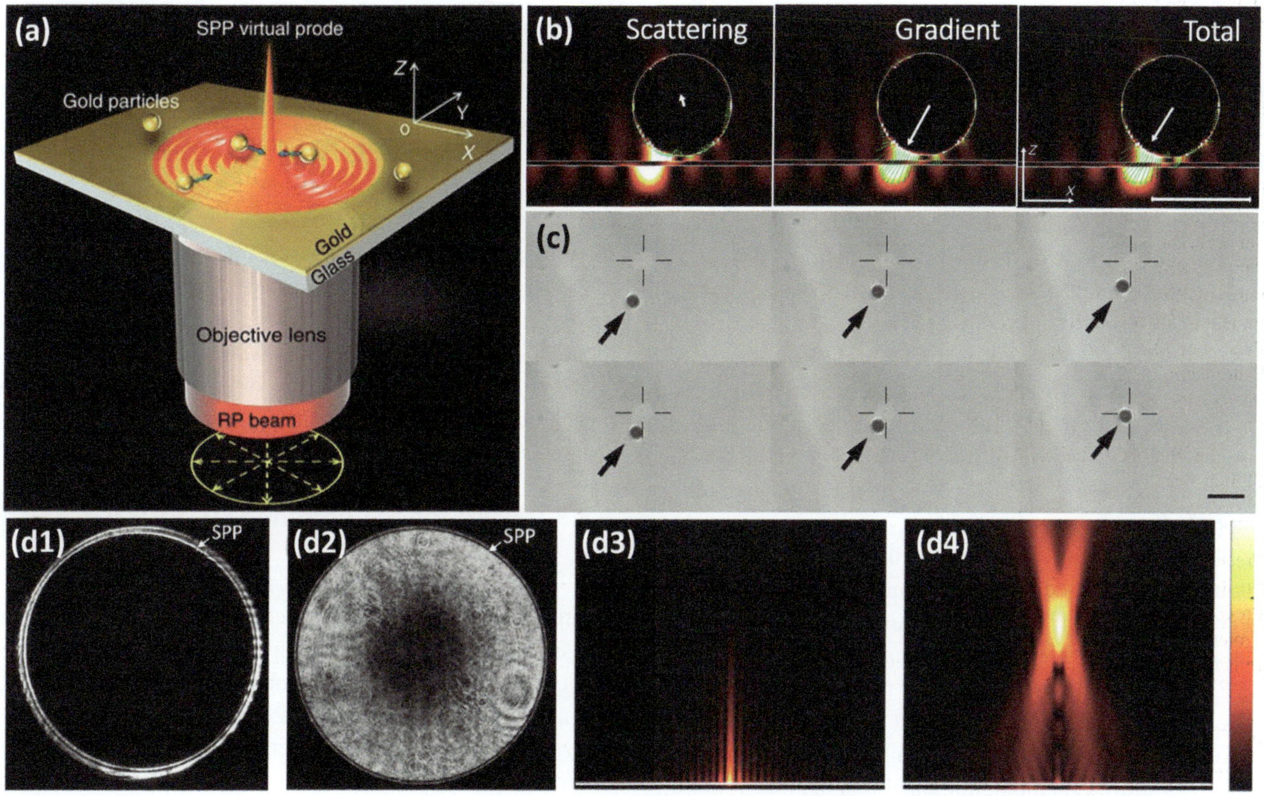

Fig. 1.8 Focused plasmonic trapping of metallic particles. (Adapted with permission from ref. [56], Copyright © 2013 Nature Publishing Group). **a** Schematic of trapping of metallic particles by an SPP virtual probe. Plasmons are excited on a smooth metallic film using a highly focused laser beam at the position that satisfies the coupling conditions, propagating to generate a stable plasmonic field through interference. **b** Plasmonic forces act on the particles located in the plasmonic field, **c** successive experimental results of gold particles (with a diameter of 1 ± 0.1 μm) trapped by the focused plasmonic tweezers. **d** Enhancing plasmonic trapping with a perfect radially polarized beam (Adapted with permission from ref. [57], Copyright © 2018, Chinese Laser Press). **d1**, **d2** The profile of the reflected light obtained at the back focal plane for PRPB and RPB. **d3**, **d4** Electric field intensities at the gold–water interface for PRPB and RPB

light transmits through the metal film, generating a substantial scattering force on the particle. By carefully designing the incident light beam that excites the focused plasmonic field, the trapping gradient force can be enhanced while suppressing the repulsive scattering force. In 2018, Wang et al. improved the efficiency of plasmon trapping by employing perfect radially polarized beam (PRPB) [57]. Their approach involved directing a ring-shaped incident beam onto a gold film, which allowed for the concentration of light exclusively within the range necessary for surface plasmon excitation. This method not only enhanced the SPP excitation efficiency of the incident light but also significantly improved the stability of plasmonic trapping by reducing the amount of scattered light and thermal motion compared to common Gaussian beams. In conclusion, focused plasmonic tweezers operating on smooth metal surfaces circumvent the need for structured substrates and the complexities they entail, and has great application value for dynamic manipulating a diverse array of nano- and microscale objects, encompassing traditional dielectric and metallic materials, quantum dots, and even biological molecules.

1.3.2.3 Dynamical Manipulation of SP Optical Tweezers

All-optical modulated plasmonic tweezers offer a platform for the manipulation of SPP field and trapped objects that does not rely on metallic structures. This approach not only simplifies the fabrication challenges associated with micro- and nanostructures, but also overcomes the limitations of traditional metallic structures with limited functionalities. By controlling the phase, amplitude, and polarization of the incident light beam, complex SPP fields can be generated, opening up new possibilities for the precise manipulation of objects in SPP tweezers. Whether through independent control of light properties, combinations of multiple properties, or multi-beam cooperative operations, all-optical control schemes provide enhanced flexibility and freedom in near-field control of trapped objects.

The polarization and phase of the incident laser beam, as independently adjustable parameters, can be used to generate optical torque, enabling particle trap, manipulation, and rotation, thus significantly enhancing the flexibility of SPP tweezers. The electromagnetic field enhancement properties of SPP effectively increase the optical torque, offering more opportunities for optical manipulation. By leveraging optical torque, the manipulation capability of optical tweezers is extended from planar control to particle rotation, greatly improving their operational capacity [58, 59]. In addition, non-uniform plasmonic fields generated by focused vector beams [60–62] produce unbalanced forces on non-isotropic objects. For example, as shown in Fig. 1.9a, linearly polarized beams can be used to excite two focal spots of SPP to manipulate the rotation of nanowires or nanorods [58, 63].

Compared to the polarization-based modulation, the phase of the excitation light in plasmonic excitation can be even more freely adjusted. Phase information can be conveyed into the plasmonic field through the excitation of SPPs, as exemplified by Airy plasmonic beams characterized by cubic phases [64] and plasmonic vortex fields characterized by spiral phases and orbital angular momentum (OAM) [65–67]. Small particles situated within these phase-modulated plasmonic fields are subjected to additional forces. Research has demonstrated that these particles follow circular trajectories within plasmonic vortex fields, thereby exhibiting rotational dynamics [68, 69]. The velocity of rotation is contingent upon the OAM (see Fig. 1.9b), thereby introducing a significant degree of freedom for targeted applications.

Through the manipulation of incident light field distribution, the versatility of SPP tweezers can be significantly enhanced to facilitate targeted trapping and manipulation of particles. Zhang et al. creating a volcano-shaped dual-wavelength plasmonic tweezers, achieved by exciting SPPs using coaxial beams of two wavelengths [71]. This innovative configuration enabled precise regulation of both the particle gap and the quantity of particles within the trapping potential well. Following this, they further developed a polarization-sensitive metalens, which permitted left and right handed circularly polarized light to be incident at distinct locations on a silver film [70]. This arrangement excited two independent plasmonic fields that produced both trapping force for target particles and anti-trapping force for the surrounding particles, thereby enabling the selective trapping of the target particles. Moreover, by integrating pre-designed holograms onto a spatial light modulator, the customized plasmonic field can be dynamically adjusted for the dynamic control of target particles, thereby obviating the necessity for mechanical modifications of the system [59].

1.3.3 Multi-mode Combined SP Optical Tweezers

To improving the stability of optical tweezers, recent advances suggest the integration of nanophotonics with other physical or chemical mechanisms for manipulation of samples. For instance, some recent researches verify that the proper utilization of photothermal effects can significantly improve the trapping capability and range. Thermophoretic refers to a technique that generates a local temperature gradient field through photothermal effects, and then uses thermophoresis to drive suspended particles along the temperature gradient toward warmer or cooler regions. The direction of particle movement depends on the Soret coefficient of the particles relative to the surrounding solution. Furthermore, by

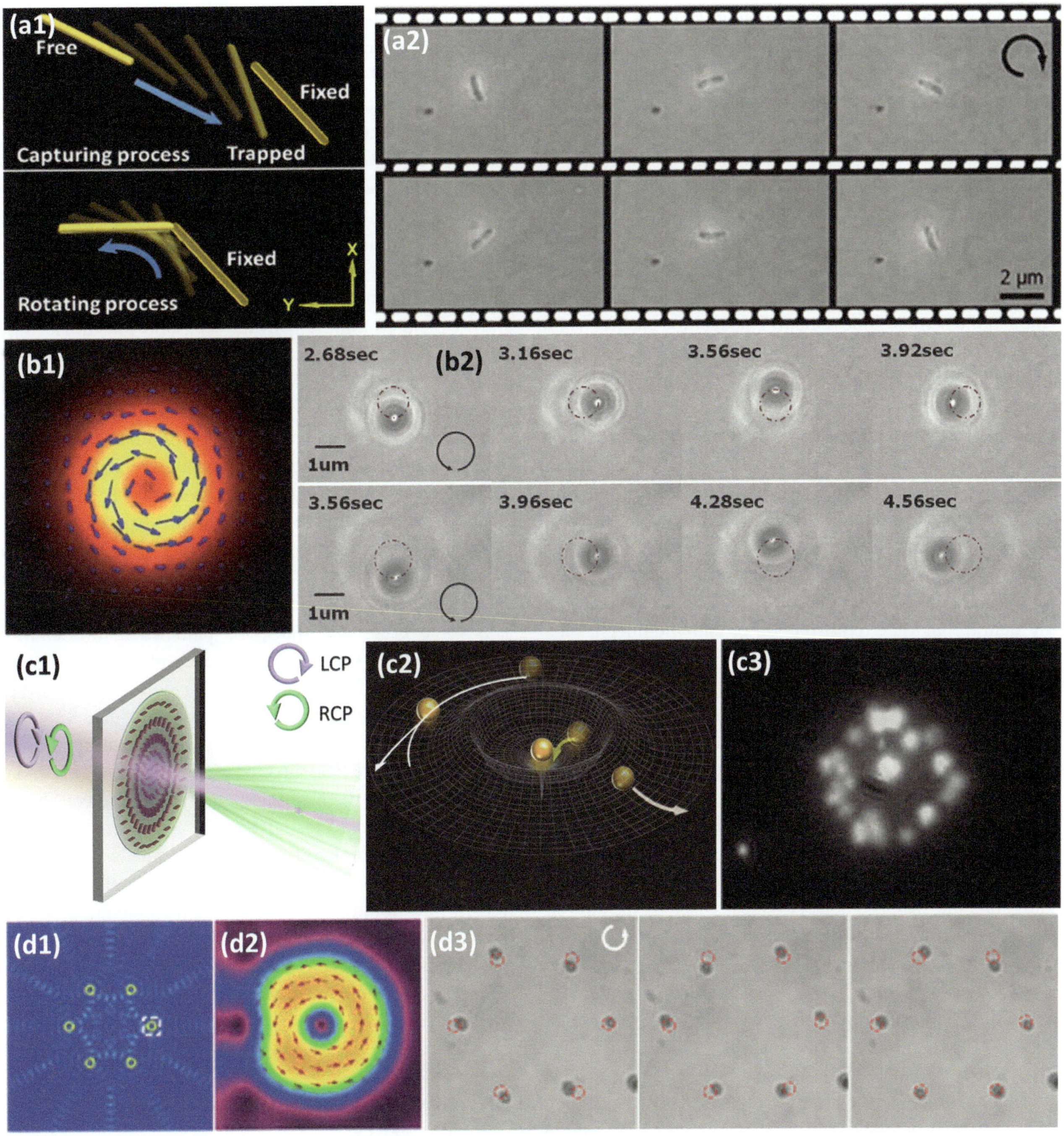

Fig. 1.9 All-optical plasmonic manipulation of micron and nano objects on smooth metal surfaces. **a** Schematic diagram of dynamic linearly polarized laser excitation plasmonic tweezers assembling a nanowire knot; the right silhouette image shows the continuous images of a single gold nanowire rotating clockwise by changing the incident polarization direction (Adapted with permission from ref. [63], Copyright © 2014, American Chemical Society). **b** Optical vortex excitation plasmonic tweezers for manipulating metal particles; **b1** The distribution of the plasmonic vortex field is shown on the left. (Adapted with permission from ref. [68], Copyright © 2015, Nature Publishing Group). **b2** The trajectory of metal particles in the OV optical tweezers (Adapted with permission from ref. [69], Copyright © 2012, Optical Society of America). **c** Customized plasmonic field for selective trap of metal nanoparticles. **c1** Schematic of a polarization-directed planar metal lens; **c2** Schematic of the force distribution (Adapted with permission from ref. [71] Copyright © 2017, The Royal Society of Chemistry); **c3** Experimental results of gold particles in the customized plasmonic field. **c1**, **c3** Adapted with permission from ref. [70], Copyright © 2018, American Chemical Society). **d1**–**d2** Normalized intensity distributions and energy flow distributions of the PV arrays, **d3** Experimental trapping states for gold particles in the PV array fields (Adapted with permission from ref. [59] Copyright © 2023, Optica Publishing Group)

utilizing the localized electromagnetic field enhancement induced by SPs and the increased ohmic losses resulting from metal absorption, this technology generates a highly concentrated heat spot with a large temperature gradient, which can be applied in biochemical experiments. It has shown great potential in fields such as photothermal therapy, material processing, and catalytic reactions [72–75]. In SP optical tweezers manipulation, by adjusting the local photothermal effect, the distribution and direction of the thermophoretic force acting on the trapped particles can be controlled to align with the desired trajectory, enabling precise manipulation of the particles' movement. This method enables stable trapping and handling of specific micro- and nano-structures, and opens up new application possibilities in biological and material sciences.

In 2017, Zheng's group demonstrated a light-induced thermophoresis tweezers, using a low-power light to heat a plasmonic substrate, which enabled the precise manipulation of single or multiple cells through thermophoresis [76, 77]. The following year, they developed Opto-thermoelectric nanotweezers (OTENT) technique based on a gold film. By adding surfactants to the solution, they created a thermoelectric field through the redistribution of cationic micelles and anions driven by the thermal field, to further enhance the trapping capabilities [78] as briefly presented in Fig. 1.10a, b. In 2022, Wang et al. proposed graphene-based opto-thermoelectric tweezers [79], which not only extends the operational wavelength to the visible-to-infrared range but also allows for multi-target trapping and related applications based on microstructural processing of the graphene substrate using laser direct writing. Additionally, the higher flatness of graphene resulted in a more uniform temperature gradient, mitigating the localized hotspot issues caused by the unevenness of gold films and thereby reducing the risk of thermal damage for bio-samples.

Utilizing using a special laser-cooling substrate and controlling the generation of a negative thermal gradient with the laser, thermophoresis-based trapping is achieved (Fig. 1.10c). Modulating the ambient temperature to achieve cooling has also emerged as an effective strategy for trapping. This approach not only enhances thermophoretic trapping forces under low temperature conditions but also significantly reduces thermal damage to target objects. These advancements broaden the application scope of opto-thermoelectric tweezers, offering greater flexibility and control in both trapping precision and sample preservation [72]. Furthermore, Wang and colleagues proposed a nanoscale dynamic surface-enhanced Raman scattering (SERS) mapping technique based on the two-dimensional nanoparticle scanning in a plasmonic-thermoelectric nano-tweezers system [75]. This system achieved stronger Raman signal compared to conventional plasmonic or thermoelectric tweezer platforms, and its Raman image resolution surpasses the diffraction limit and reaches ~ $\lambda/4.5$. Figure 1.10d illustrates the schematic diagram and the SERS scanning image of MOS_2. This breakthrough in SERS imaging technology holds significant promise for applications in biological detection.

Electro-plasmonic trapping (ETP) leverages the synergy between light-induced heating and an applied alternating electric field to rapidly activate near-field plasmonic traps. This enables the trap of nanoparticles located hundreds of micrometers away from the plasmonic nanocavity. In general, the generation process of the ETP mechanism proceeds as follows: when a plasmonic cavity is illuminated by a laser, light-induced heating creates a temperature gradient. Since the dielectric constant and conductivity of the fluid depend on the local temperature, their spatial distribution changes accordingly. Upon applying an external alternating electric field, the field interacts with the gradients in dielectric constant and conductivity, generating a body force within the fluid. This force drags and transports particles in the solution, enabling long-distance, high-efficiency particle transfer and trapping. Ndukaife and colleagues have developed multiple ETP-based plasmonic optical tweezers [74, 81, 82]. They proposed laser illumination in conjunction with an applied alternating electric field directed at a finite array of plasmonic nanoholes, thereby generating spatially varying electrohydrodynamic potentials (Fig. 1.10e) [74]. Such potentials facilitate the rapid trapping of biomolecules smaller than 10 nm, including small protein molecules, at femtomolar concentrations, while simultaneously reducing the likelihood of light-induced damage. Furthermore, the electric field-controlled mechanism in opto-thermo-electrohydrodynamic tweezers (OTET) permits the selective trapping of nanoparticles with different sizes (20–100 nm).

1.4 Biomanipulation of Plasmonic Tweezers

Optical tweezers have gained significant traction in the field of biological manipulation due to their capacity to accurately position biological specimens without inflicting damage. However, traditional optical tweezers are often constrained by the optical diffraction limit, which imposes certain limitations on their manipulation capabilities. SP optical tweezers have advanced its capability by concentrating energy into deep subwavelength volumes, thereby amplifying electromagnetic fields and breaking through the

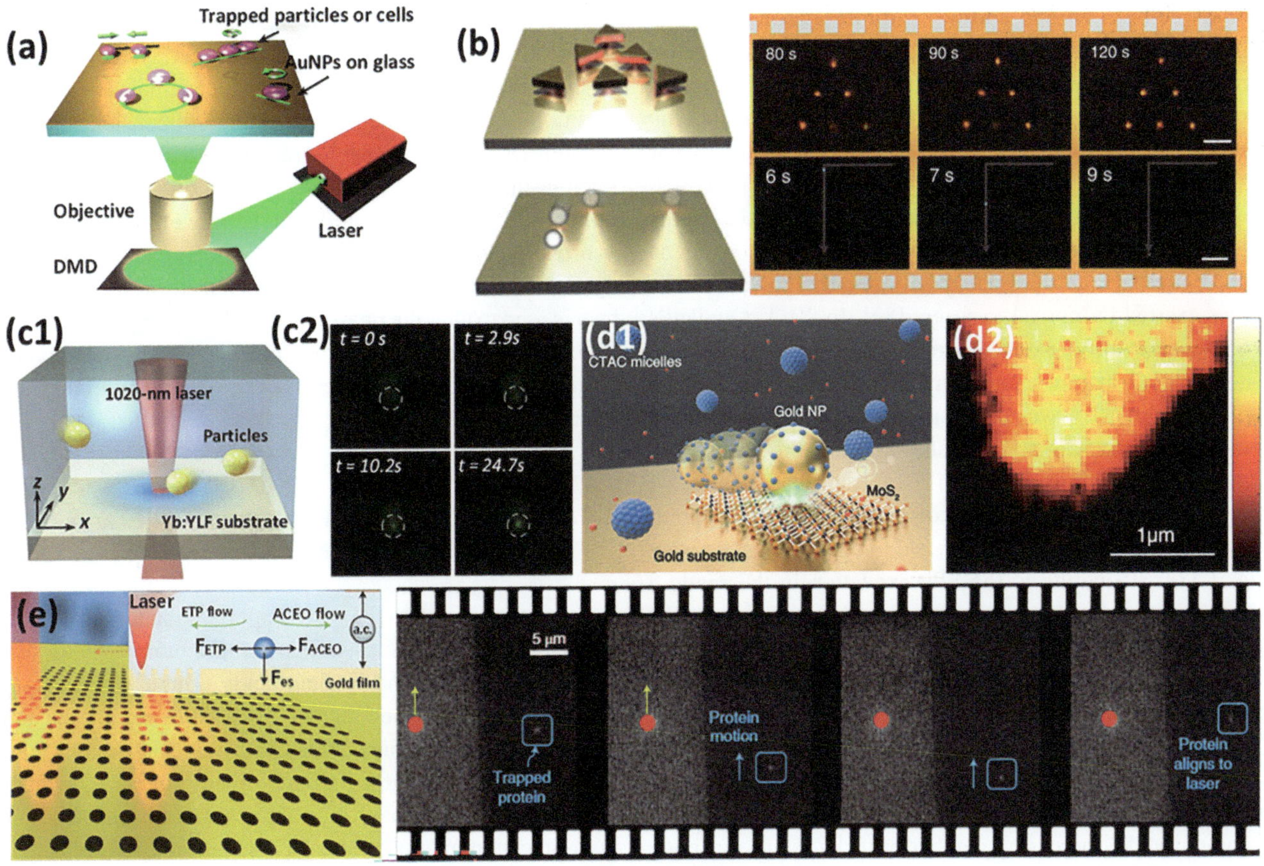

Fig. 1.10 **Photothermal tweezers based on plasmonic heating**. **a** chematic of a thermophoretic tweezers system designed for low-power and multifunctional manipulation of biological cells (Adapted with permission from ref. [76], Copyright © 2017, American Chemical Society). **b** Opto-thermoelectric nanotweezers (Adapted with permission from ref. [78], Copyright © 2018 Springer Nature). **c** Opto-refrigerative tweezers (Adapted with permission from ref. [80], Copyright © 2021, The American Association for the Advancement of Science). **c1** System schematic: excitation light at 1020 nm matches the Yb:YLF crystal transitions to produce anti-Stokes fluorescence, leading to internal cooling of the nanocrystals. The synergistic effect of optical cooling and thermophoresis enables dynamic manipulation of objects at the laser-induced cold spot. **c2** Experimental trap images of FITC-conjugated protein A/G. **d** Two-dimensional nanoscale SERS imaging achieved by dynamically scanning individual metal nanoparticles using a plasmonic thermoelectric nano-tweezers system (Adapted with permission from ref.[75], Copyright © 2022, American Chemical Society). **d1** Schematic of the SERS imaging process based on a plasmonic photothermal tweezers system, **d2** SERS scanning image of MOS_2. **e** Schematic of a photothermal electrohydrodynamic tweezers system, with sequential frames on the right showing the dynamic manipulation of a single BSA protein molecule. Right showing the dynamic manipulation of a single BSA protein molecule (Adapted with permission from ref. [74], Copyright © 2020, The Author(s), under exclusive licence to Springer Nature Limited)

diffraction limit. This technology facilitates the manipulation of a diverse range of targets, including cells, organelles, and small biomolecules, spanning the size spectrum from the microscale to the nanoscale. In addition, the interaction between SPs and metallic nanostructures not only enables manipulation at the nanoscale but also contain with analytical capabilities. Samples that are trapped can be subjected to high-sensitivity analysis through plasmon-enhanced techniques, such as fluorescence spectroscopy and Raman spectroscopy, thereby endowing the plasmonic tweezers with multifunctional attributes.

1.4.1 Applications of Plasmonic Tweezers in Cellular Scale

Cells represent the basic building blocks of all plant and animal life, and techniques for manipulating individual cells are essential in the field of biomedical research, which includes studies on cell interactions, fertilization, and transfection. Following Ashkin's pioneering achievement in 1987, which involved the trap and manipulation of bacteria and living cells, optical tweezers have undergone substantial advancements and have been widely applied in processes such as cell fusion and sorting, chromosome cutting, and cell transgenesis. These technological developments have significantly propelled research in cell biology, genetics, and biomedicine, providing innovative tools for investigating intracellular molecular mechanisms and intricate intercellular interactions.

In 2007, Lin et al. trapped a single yeast cell and used the dipole approximation model from Mie theory to calculate the light scattering field, providing an in-depth analysis of the forces exerted by optical tweezers [83]. In 2011, Quidant's group successfully achieved parallel trapping of yeast cells with an incident power of 20 mW by exciting SPs on the top of an optical waveguide using a simple fiber coupling setup [84]. This was accomplished on metallic disks integrated into a chip (Fig. 1.11b). In 2015, Kang et al. conducted experiments on random gold nanoislands and observed large-scale assembly of active *Escherichia coli* cells, as show in Fig. 1.11a, attributing to the SPP excitation in the random plasmonic nanostructures and the induced heat. This heat led to the formation of a combined effect of near-field plasmonic forces and long-range thermophoretic forces, enabling stable cell assembly after overcoming axial convective resistance [85].

The integration of photothermal effects with cell manipulation has generated significant interest in the research community. An increasing number of technologies utilizing photothermal principles for cell manipulation have been proposed [72–75, 78, 79, 86]. The integration of thermophoresis driven by plasmonic heating with plasmonic optical tweezers has emerged as a significant technique for cell manipulation. Inspired by the three-dimensional manipulation of optical fiber tweezers, Chen et al. proposed a trapping method of evaporating a 10 nm thick gold film on the end of a single-mode optical fiber in 2015. By leveraging the synergistic effect of natural convection and thermophoretic forces, they successfully trapped and accumulated PS microspheres and *Escherichia coli* on the end face of the 980 nm single-mode optical fiber. The experimental capture results are shown in Fig. 1.11c. In 2017, Zheng's group demonstrated the capability of light-guided thermophoretic trapping of cells, employing power levels that were 100–1000 times lower than those utilized by conventional optical tweezers (see Fig. 1.11e) [76].

Notably, the local electric field strength within the excited plasmonic region significantly exceeds that of the incident light, and thus serves to effectively enhance the optical spectral signals in that specific area. In 2021, Deng et al. demonstrated a graphene-based thermoelectric optical tweezers system (Fig. 1.11d) [87]. In their system, they accelerated metal nanoparticles using photothermal-electric forces, enabling the nanoparticles penetration into cells. By exploiting the LSP field enhancement effect of the metal nanoparticles, they successfully achieved in situ SERS detection within cells. However, such optical tweezers system relies on the substrate to generate electromagnetic fields or perform plasmonic heating, which limits its three-dimensional control capabilities. Recently, Zhao's group reported a 3D biocompatible plasmonic tweezers system as shown in Fig. 1.11f [88], utilizing SPP and the photothermal effect of thermosensitive hydrogels to manipulate individual cells. By exciting the propagation of SPPs through total internal reflection, the generated heat is used to alter the phase of poly(N-isopropylacrylamide) (PNIPAM) hydrogel, enabling the "on" and "off" control of light coupled with optical fibers, thereby trapping and releasing cells (Fig. 1.11g).

1.4.2 Applications of Plasmonic Tweezers at the Subcellular Scale

Extracellular vesicles (EVs) are important mediators of intercellular communication, containing key biomolecules such as proteins, lipids, and nucleic acids. These vesicles typically mature within the range of 30–100 nm and are released from the cell membrane. Plasmonic optical tweezers have become an ideal tool for the investigation of EVs, owing to their capacity to surpass the diffraction limit and facilitate high-precision manipulation of EVs. Currently, some studies have utilized optical tweezers to analyze the mechanical properties, internal components, and interactions between vesicles and their ligands or cells. These studies have provided new insights into intercellular signaling and have advanced developments in the field of biomedical research.

In 2018, Zheng's group successfully trapped anionic lipid vesicles using photothermal tweezers (see Fig. 1.12a), achieving the long-distance parallel trapping and dynamic manipulation over several hundred micrometers [91]. The following year, Kaufman et al. achieved the isolation, trapping, and positional control of vesicles using plasmonic wells, and detected the Raman spectral properties of single EV, providing a standard for the intrinsic molecular information of label-free EVs [92]. In the last two years, Ndukaife's group published several works on the plasmonic manipulation of EVs [37, 74, 81, 82, 93]. They proposed a geometry-induced electrohydrodynamic tweezers (GET), as schematized in Fig. 1.12b, which effectively overcame the low throughput and long-duration trapping limitations of existing plasmonic optical tweezers in trapping heterogeneous nanoscale EVs [81, 82]. GET achieves rapid parallel transport and trapping of individual EVs by generating multiple current dynamics potentials. Combined with nanoscale plasmonic cavities, it instantaneously enhances the trapping under laser illumination, avoiding thermal damage. This advancement has propelled the development of high-throughput, label-free trap

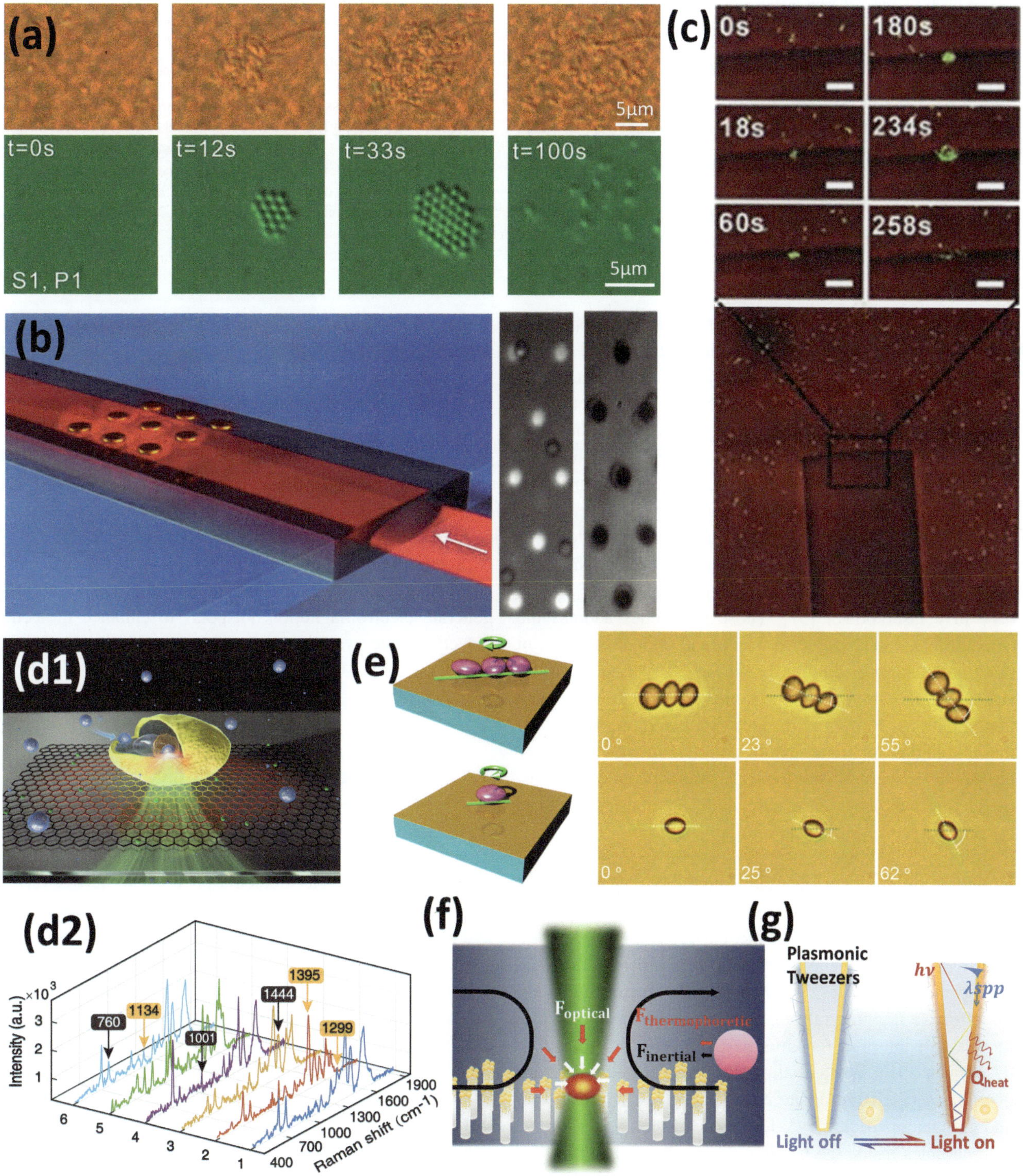
(a)
5μm
t=0s
t=12s
t=33s
t=100s
S1, P1
5μm
(b)
(c)
0s
180s
18s
234s
60s
258s
(d1)
(e)
0 °
23 °
55 °
0 °
25 °
62 °
(d2)
Intensity (a.u.)
×10^3
760
1134
1001
1444
1395
1299
Raman shift (cm-1)
(f)
F optical
F thermophoretic
F inertial
(g)
Plasmonic Tweezers
hv
λspp
Q heat
Light off
Light on

◀**Fig. 1.11 Plasmonic tweezers in cellular-scale applications. a** Low-power optical trapping and live cell assembly via random plasmonic nanostructures (Adapted with permission from ref. [85], under Copyright license CC BY). **b** SPs optical tweezers trap under low-power incident light via fiber-optic coupling on a dielectric channel waveguide, with arrayed metal discs on the waveguide (Adapted with permission from ref. [84], Copyright © 2011 American Institute of Physics). **c** Time-series side view images of *Escherichia coli* trap, where a 0.65 mW laser is turned on at 0 s and turned off immediately before the last trapped image (Adapted with permission from ref. [89], Copyright © 2015 Optical Society of America). **d** Graphene-based thermoelectric optical tweezers system for in situ Raman spectroscopy in cells (Adapted with permission from ref. [87], Copyright © 2022 Elsevier B.V. All rights reserved), where silver nanoparticles are introduced into cells using photothermal tweezers, enhancing the Raman signal around the particles through their LSPs effect; the upper image shows the experimental principle, and the lower image shows the SERS spectrum of yeast. **e** Dynamic manipulation of cell rotation using gold film thermophoretic tweezers (Adapted with permission from ref. [76], Copyright © 2017, American Chemical Society). **f** 3D plasmonic tweezers for single-cell manipulation using surface plasmon resonance and temperature-responsive hydrogels (Adapted with permission from ref. [88], Copyright © 2018, American Chemical Society). **g** Plasmonic tweezers system based on plasmonic heating effects (Adapted with permission from ref. [90], Copyright © 2023 Wiley-VCH GmbH), Mixed nanostructures are prepared by depositing AuNPs on zinc oxide nanorods (ZnONR), enhancing the trapping performance through plasmonic heating induced by local surface plasmon resonance

and spectral analysis of single EVs. In 2024, they proposed a trap method based on gold nano-bowtie, utilizing local surface plasmon combined with photothermal and depletion attraction forces. The thermophoretic depletion of polyethylene glycol (PEG) molecules in the vicinity of plasmonic hotspots generates pronounced concentration gradients, which facilitate the accurate localization of nanoscale particles through a synergistic interaction with diffusiophoretic forces, enabling stable trapping and manipulation of nanoscale particles such as extracellular vesicles and polystyrene beads. The platform demonstrated stable trap performance with as low as 0.78 mW laser power [37]. These researches address the constraints associated with conventional optical tweezers in the manipulation of diminutive biological specimens, while simultaneously improving the stability and accuracy of such manipulations. Furthermore, these innovations offer novel methodologies for investigating cellular interactions, diagnosing diseases, and developing therapeutic interventions, thereby propelling advancements in the technologies utilized for the analysis of biological samples.

Additionally, viruses are pathogens that cause a variety of infectious diseases. Studying viruses helps people understand their transmission mechanisms, immune responses, and mutation patterns, and is applied in vaccine development, antiviral drug research, and gene therapy, among other fields. With sizes comparable to vesicles (20–400 nm), viruses fall within the effective trapping range of plasmonic tweezers, making them a valuable tool for virus trap and related studies [94, 97, 98]. In 2018, Wang et al. proposed a particle-cluster-in-bowl array (PCIB) structure composed of a silver thin film and gold nanoparticles [94]. The cluster structure of the gold nanoparticles inside the bowl could be adjusted by controlling the voltage and deposition time, enabling the trap of targets of different sizes. These nanobowls also served as substrates for SERS, achieving the trap and Raman spectral detection of viruses and proteins (see Fig. 1.12c). In 2021, Yang et al. proposed an ACE2 protein-functionalized gold nanostructure "virus trap" SERS biosensor (Fig. 1.12d) [95]. This biosensor demonstrated exceptional sensitivity in detecting individual SARS-CoV-2 particles in water. It achieved over 100-fold virus enrichment and, combined with machine learning-based recognition standards, enabled virus detection within five minutes in complex environments. This advancement provides critical technical support for real-time monitoring and early warning systems. Recently, Luo's group proposed a multilayer plasmonic sensor system (Fig. 1.12e) [96], which is equipped with nanoscale trap capabilities and enables ultrasensitive and rapid detection of SARS-CoV-2 N protein. The system consists of multiple concentric cylinders coated with Ag–SiO_2–Ag layers, forming nanoholes that generate electromagnetic "hotspots" to enhance molecular detection. The system enables single-molecule-level detection of the SARS-CoV-2 N protein, demonstrating significant potential for early detection of infectious viruses.

1.4.3 Applications of Plasmonic Tweezers in Molecular Scale

The manipulation of individual molecules is essential for the investigation of single-molecule phenomena across various disciplines, including physics, chemistry, materials science, and life sciences, in addition to the development of nanoscale devices and molecular machinery. Nonetheless, due to the diffraction limit and Brownian motion, it is a challenge in the accurate trap of objects with size of a few nanometers such as molecules within a solution. Single-molecule manipulation techniques based on surface plasmon-enhanced electromagnetic fields present new opportunities to overcome the diffraction limit of traditional optical trapping. Due to their compact and deep subwavelength traps, plasmonic tweezers have found extensive applications in biological and medical research. They are particularly suited for integration with microfluidic systems, enabling functions such as the trap, detection, assembly, and sorting of molecules and proteins.

Since 2012, Gordon and colleagues have proposed nanopores in gold thin films to trap individual biomolecules, facilitating further biosensing research [99–101]. They

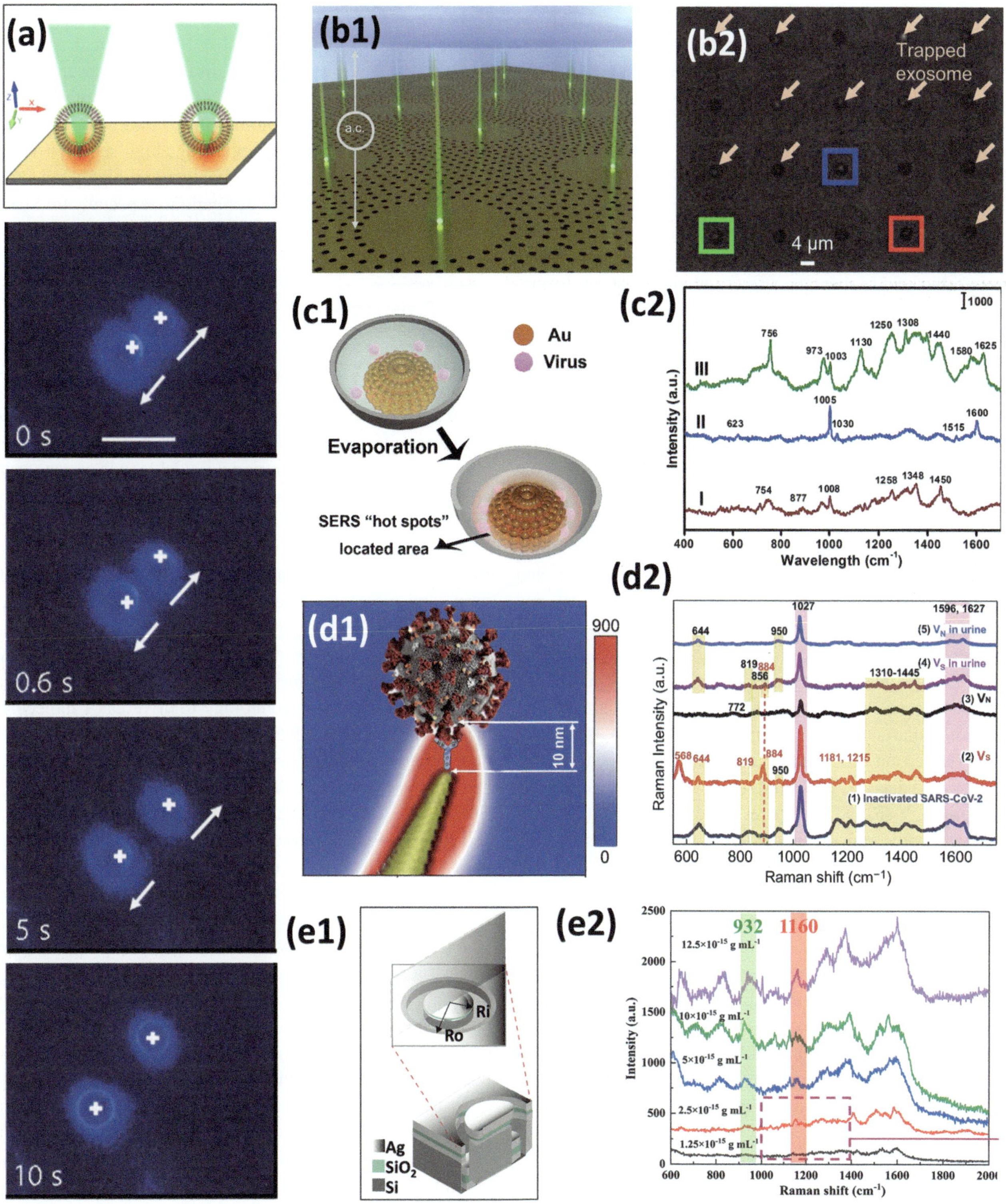

Fig. 1.12 Applications of Plasmonic Tweezers at the Subcellular Scale. a Trap method induced by temperature gradients based on plasmonic heating (Adapted with permission from ref. [91], Copyright © 2018, American Chemical Society). The left shows the schematic of the system principle, and the right presents experimental results. **b** Schematic of the GET system with the right image showing trapped single EVs of different sizes (Adapted with permission from ref. Adapted with permission from ref. [81], under Copyright license CC BY). **c** Demonstration of silver-covered Au PCIB structures (Adapted with permission from ref. [94], Copyright © 2018, American Chemical Society). On the right are the SERS spectra of enzyme bodies, bovine hemoglobin, and bovine serum albumin. **d** SERS detection system based on human ACE2-functionalized gold "virus trap" nanostructures (Adapted with permission from ref. [95], under Copyright license CC BY). **d1** Schematic of SARS-CoV-2 S in the 10 nm EM enhancement region, along with calculated intensity distribution, **d2** SERS spectra of inactivated SARS-CoV-2. **e** Multi-layer plasmonic nanotrapping (MPNT) device and measured SERS of different SARS-CoV-2N protein concentrations (Adapted with permission from ref. [96], Copyright © 2024 Published by Elsevier B.V.)

designed coaxial nanoaperture in the gold film, and then the spectral signal of the biomolecule trapped in the nanoaperture increased sharply (see Fig. 1.13b). Based on this structure, in 2012, they experimentally demonstrated the plasmonic trapping of a single bovine serum albumin (BSA) molecule and the conformational changes of its structure, with a hydrodynamic radius of 3.4 nm [100]. This breakthrough served as a foundation for further in-depth studies. More recently, Yousefi et al. utilized optical nanomanipulators with a double-nanopore structure to investigate the effects of ascorbic acid (a reducing agent) and pH on the conformational dynamics of individual ferritin molecules [102]. They revealed kinetics of a single ferritin molecule disassembling into its subunits. Figure 1.13a illustrated the fragmentation process of ferritin, where the "+" marks the corresponding states of the 24-mer, 22-mer, 16-mer, tetramer, or two dimer fragments. As researchers continue to explore protein manipulation, an increasing number of innovative protein detection techniques have been developed [100–107].

Nowadays a variety of single-molecule manipulation and detection techniques have been developed and applied, providing valuable insights into molecular behavior and enabling advancements in diagnostics, therapeutics, and personalized medicine [99, 103–105, 108–111]. In 2013, Dekker et al. demonstrated that nanopores could function as probes for investigating the thermal effects of individual plasmonic nanostructures. They designed various nanopore structures that successfully trapped single DNA molecules [112]. Based on this concept, they later developed a range of nanopore designs, including single and double nanopores, as well as V-shaped nanopores. These structures enabled the successful trap of single DNA molecules within the plasmonic gap (Fig. 1.13d) and facilitated the study of DNA translocation events with exceptional spatial resolution [113]. In 2020, Tatsuya and colleagues used gold nanopyramid dimers as substrates, employing plasmonic tweezers and thermophoresis to sort and permanently immobilize DNA molecules of different sizes [108]. Tian and his colleagues present a novel strategy for trapping, investigating, and releasing single molecules (2 nm) in solution using an adjustable plasmonic optical nanogap, addressing the challenge of Brownian motion that has previously hindered free-molecule trapping (see Fig. 1.13c).

In addition to trapping single molecules through substrate structures such as nanopores, intermolecular interactions can also be utilized to attach molecules to particles for indirect manipulation. Liu et al. developed a portable plasmonic-coupled SERS platform [106, 107], achieving single-molecule sensitivity, as shown in Fig. 1.13e. By controlling the distance between amino-functionalized Ag nanoparticles, they constructed tunable plasmonic nanogaps to enhance the SERS signal and detect analyte molecules trapped in the nanogaps.

In plasmonic tweezers systems, LSPs hotspots produced by the structures generate localized electric field enhancements, and thus significantly amplifies the spectroscopic signals of samples, establishing it as an essential platform for the spectroscopic study of biomolecules. Plasmonic tweezers, recognized for their robust trapping and high-sensitive detection capabilities, offer a cost-effective and highly sensitive platform for biomolecular investigations, especially for SERS and fluorescence spectroscopy.

1.5 Prospects

Over the past two decades, the plasmonic tweezers technology has undergone rapid development, significantly expanding the application of optical manipulation at the nanoscale. These advancements have provided opportunities for interdisciplinary research and integration with fields such as nanotechnology and biological sciences. Here, we analyze the future development of the plasmonic tweezers technology.

Plasmonic tweezers can be integrated into analytical platforms such as lab-on-a-chip systems for applications in cell, virus, and molecular detection. For instance, photothermal tweezers enhance detection sensitivity by concentrating the target species, making them suitable for detecting and purifying low-concentration samples. Chen et al. applied photothermal tweezers technology to drive DNA molecules in low-concentration solutions toward heated regions, thereby generating locally high-concentration DNA samples to meet the requirements of CRISPR technology [114].

Plasmonic tweezers can manipulate nanoparticles across cell membranes for intracellular delivery, with electromagnetic field enhancement enabling spectral signal detection of intracellular molecules, thereby achieving in vivo scanning imaging. Additionally, when combined with SERS technology, plasmonic tweezers can expand its applications in chemical detection. Furthermore, by integrating surface plasmon with thermophoresis, optical tweezers driven by photo-induced temperature fields can provide low-power, precise manipulation, large-area scanning, and wide-field imaging of target molecules [86, 115].

With the advancement of AI technology, deep learning has begun to optimize optical tweezers techniques [116, 117], such as improving the design, calibration, real-time control, and even tracking and analysis of trapped objects, thereby enhancing the stability, intelligent sorting, and path planning capabilities of optical tweezers. Due to its higher computational speed and versatility, deep learning can often achieve much better results. For instance, AI-driven calculations can design and optimize plasmonic substrate structures with higher SP excitation efficiency, optimize incident excitation light fields, and enable adaptive light field correction. Furthermore, portable applications are also a means for

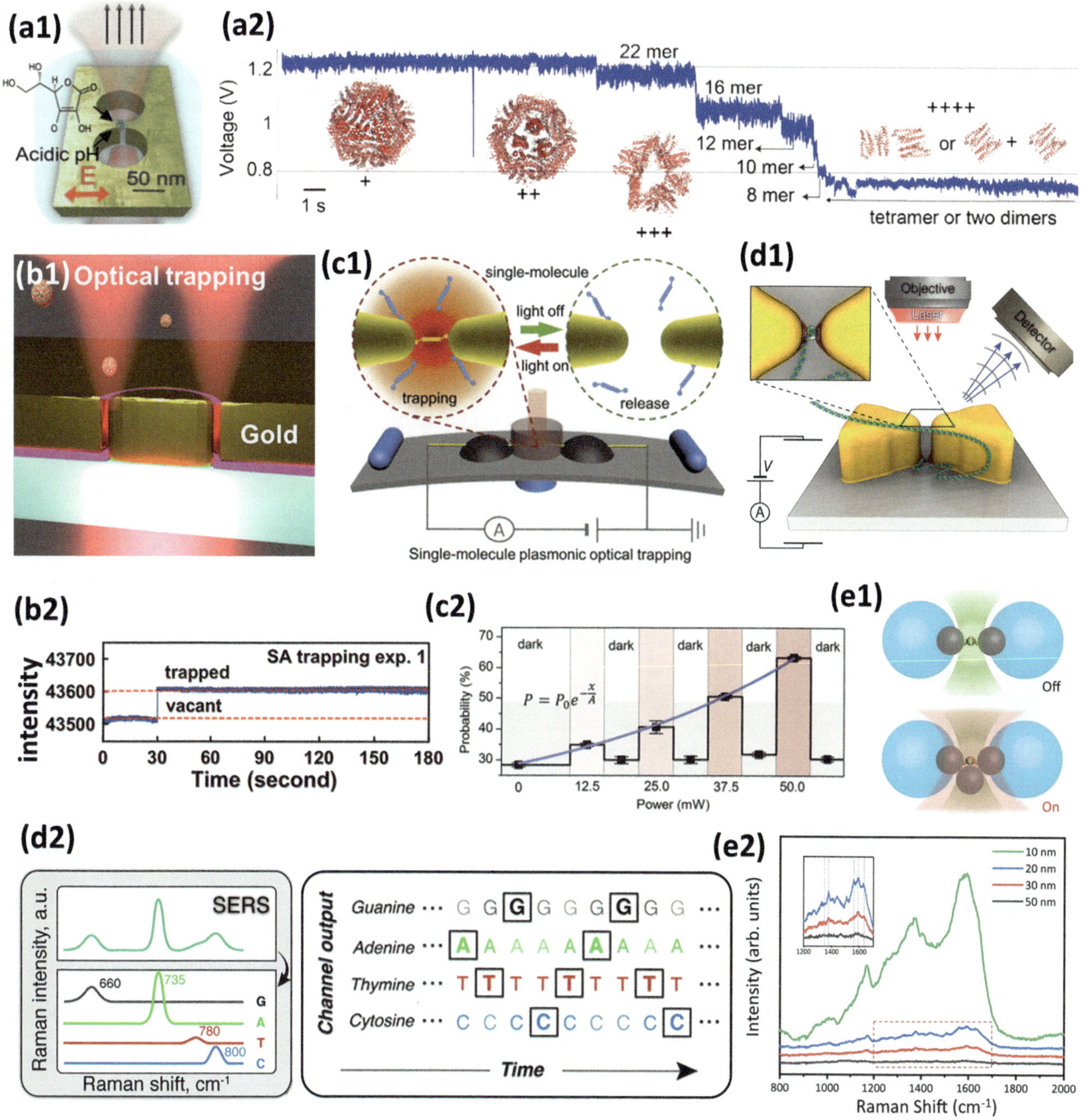

Fig. 1.13 **Plasmonic optical tweezers in molecular-scale manipulation and detection applications.** **a** Study of ferritin dynamics using optical tweezers and a dual nanopore structure (Adapted with permission from ref. [102], Copyright © 2024, American Chemical Society). **b1** Schematic diagram of protein trap using a 10 nm gap coaxial resonant nanopore (Adapted with permission from ref. [101], Copyright © 2018, American Chemical Society). **b2** The light power spectrum transmitted through a single coaxial nanopore in a 1 × PBS solution containing streptavidin (190 nM). **c** Direct trap, study, and release of 2 nm free single molecules in solution using plasmonic nanomaterials (Adapted with permission from ref. [109], Copyright © 2020 Published by Elsevier Inc.). **c2** The trap probability of single molecules as a function of laser intensity (691 nm). **d** Trap of DNA molecules and decoding of their nucleotide sequences using a plasmonic tweezers system combining gold butterfly knots and nanopores (Adapted with permission from ref. [113], Copyright © 2015, American Chemical Society). **e** Dynamic nanopore controlled by multiple beams (Adapted with permission from ref. [106], under Copyright license CC BY). **e2** SERS spectral features of hemoglobin in its native state when two AgNP-coated beads are trapped at different distances (50–10 nm). (Adapted with permission from ref. [107], under Copyright license CC BY)

optical tweezers technology to enter the market, for example, the nanoporous fiber aperture structures offer new development directions. Compared to traditional microscopes, this fiber structure is very compact and has been aligned during manufacturing, eliminating the need for complex adjustments.

1.6 Conclusions

In this chapter, we have thoroughly discussed the fundamental principles and mechanisms of SP optical tweezers, with a focus on LSPs-based nanostructure tweezers and all-optical modulated SPP tweezers systems. We also provided examples of the biological manipulation applications of the SP optical tweezers across different scales. SPs can overcome the diffraction limit of traditional optics, demonstrating high precision and broad potential in trapping micro- and nano-scale objects.

Metallic nanostructures offer an effective method for trapping small objects, while smooth metal surfaces further enhance the ability to manipulate them dynamically, extending the trapping range to mesoscopic and Mie metal particles. The selective generation of these plasmonic hotspots is achieved through structural design or modulation of the polarization and phase of the incident laser beams. Plasmonic tweezers based on nanostructures provide molecular-level precision, advancing fields such as life sciences, nanoscience, and materials science. Particularly, they hold vast promise in the development of novel functional materials, nanomedicines, and diagnostic tools.

Despite these advancements, expanding the applicability of this trapping technology still faces challenges, such as improving trapping efficiency and enhancing manipulation stability. However, we believe that with continued study, the applications of plasmonic tweezers will keep growing and paving the way for new applications.

Acknowledgements This work was supported by National Natural Science Foundation of China (62375177 and 62175157); Shenzhen Science and Technology Program (JCYJ20210324120403011 and RCJC20210609103232046); and Research Team Cultivation Program of Shenzhen University (2023QNT014).

References

1. Ashkin A, Dziedzic JM, Bjorkholm JE, Chu S. Observation of a single-beam gradient force optical trap for dielectric particles. Opt Lett. 1986;11(5):288–90. https://doi.org/10.1364/OL.11.000288.
2. Ashkin A. Acceleration and trapping of particles by radiation pressure. Phys Rev Lett. 1970;24(4):156–9. https://doi.org/10.1103/PhysRevLett.24.156.
3. Ashkin A. History of optical trapping and manipulation of small-neutral particle, atoms, and molecules. IEEE J Sel Top Quantum Electron. 2000;6(6):841–56. https://doi.org/10.1109/2944.902132.
4. Maragò OM, Jones PH, Gucciardi PG, Volpe G, Ferrari AC. Optical trapping and manipulation of nanostructures. Nat Nanotechnol. 2013;8(11):807–19. https://doi.org/10.1038/nnano.2013.208.
5. Kawata S, Tani T. Optically driven mie particles in an evanescent field along a channeled waveguide. Opt Lett. 1996;21(21):1768–70. https://doi.org/10.1364/OL.21.001768.
6. Righini M, Zelenina AS, Girard C, Quidant R. Parallel and selective trapping in a patterned plasmonic landscape. Nat Phys. 2007;3(7):477–80. https://doi.org/10.1038/nphys624.
7. Grigorenko AN, Roberts NW, Dickinson MR, Zhang Y. Nanometric optical tweezers based on nanostructured substrates. Nat Photon. 2008;2(6):365–70. https://doi.org/10.1038/nphoton.2008.78.
8. Righini M, Ghenuche P, Cherukulappurath S, Myroshnychenko V, García de Abajo FJ, Quidant R. Nano-optical Trapping of Rayleigh Particles and *Escherichia coli* Bacteria with Resonant Optical Antennas. Nano Lett. 2009;9(10):3387–91. https://doi.org/10.1021/nl803677x.
9. Stern EA, Ferrell RA. Surface plasma oscillations of a degenerate electron gas. Phys Rev. 1960;120(1):130–6. https://doi.org/10.1103/PhysRev.120.130.
10. Powell CJ, Swan JB. Origin of the characteristic electron energy losses in aluminum. Phys Rev. 1959;115(4):869–75. https://doi.org/10.1103/PhysRev.115.869.
11. Dykhne AM, Sarychev AK, Shalaev VM. Resonant transmittance through metal films with fabricated and light-induced modulation. Phys Rev B. 2003;67(19): 195402. https://doi.org/10.1103/PhysRevB.67.195402.
12. Otto A. Excitation of nonradiative surface plasma waves in silver by the method of frustrated total reflection. Z Physik. 1968;216(4):398–410. https://doi.org/10.1007/BF01391532.
13. Kretschmann E, Raether H. Notizen: Radiative decay of non radiative surface plasmons excited by light. Z Med Phys. 1968;23(12):2135–6. https://doi.org/10.1515/zna-1968-1247.
14. Zhang Y, Min C, Dou X, Wang X, Urbach HP, Somekh MG, Yuan X. Plasmonic tweezers: for nanoscale optical trapping and beyond. Light Sci Appl. 2021;10(1):59. https://doi.org/10.1038/s41377-021-00474-0.
15. Hornauer D, Kapitza H, Raether H. The dispersion relation of surface plasmons on rough surfaces. J Phys D: Appl Phys. 1974;7(9):L100. https://doi.org/10.1088/0022-3727/7/9/102.
16. Hecht B, Bielefeldt H, Novotny L, Inouye Y, Pohl DW. Local excitation, scattering, and interference of surface plasmons. Phys Rev Lett. 1996;77(9):1889–92. https://doi.org/10.1103/PhysRevLett.77.1889.
17. Nash DJ, Cotter NPK, Wood EL, Bradberry GW, Sambles JR. Examination of the +1, −1 surface plasmon mini-gap on a gold grating. J Mod Opt. 1995;42(1):243–8. https://doi.org/10.1080/09500349514550191.
18. Kano H, Mizuguchi S, Kawata S. Excitation of surface-plasmon polaritons by a focused laser beam. J Opt Soc Am B. 1998;15(4):1381.
19. Quidant R, Petrov D, Badenes G. Radiation forces on a rayleigh dielectric sphere in a patterned optical near field. Opt Lett. 2005;30(9):1009–11. https://doi.org/10.1364/OL.30.001009.
20. Irvine JM. Introduction to electrodynamics. Phys Bull. 1981;32(10):331. https://doi.org/10.1088/0031-9112/32/10/031.
21. Quidant R, Girard C. Surface-plasmon-based optical manipulation. Laser Photonics Rev. 2008;2(1–2):47–57. https://doi.org/10.1002/lpor.200710038.
22. Spesyvtseva SES, Dholakia K. Trapping in a material world. ACS Photon. 2016;3(5):719–36. https://doi.org/10.1021/acsphotonics.6b00023.

23. Albaladejo S, Marqués MI, Laroche M, Sáenz JJ. Scattering forces from the curl of the spin angular momentum of a light field. Phys Rev Lett. 2009;102(11):113602. https://doi.org/10.1103/PhysRevLett.102.113602.
24. Harada Y, Asakura T. Radiation forces on a dielectric sphere in the rayleigh scattering regime. Opt Commun. 1996;124(5):529–41. https://doi.org/10.1016/0030-4018(95)00753-9.
25. Pesce G, Volpe G, Luca ACD, Rusciano G, Volpe G. Quantitative assessment of non-conservative radiation forces in an optical trap. EPL. 2009;86(3):38002. https://doi.org/10.1209/0295-5075/86/38002.
26. Roichman Y, Sun B, Stolarski A, Grier DG. Influence of nonconservative optical forces on the dynamics of optically trapped colloidal spheres: the fountain of probability. Phys Rev Lett. 2008;101(12):128301. https://doi.org/10.1103/PhysRevLett.101.128301.
27. Shukla N, Chetri P, Boruah R, Gogoi A, Ahmed GA. Surface plasmon resonance biosensors based on Kretschmann configuration: basic instrumentation and applications. In: Biswas R, Mazumder N, editors. Recent advances in plasmonic probes: theory and practice. Cham: Springer International Publishing; 2022. p. 191–222.
28. Nie S, Emory SR. Probing single molecules and single nanoparticles by surface-enhanced Raman scattering. Science. 1997;275(5303):1102–6. https://doi.org/10.1126/science.275.5303.1102.
29. Juan ML, Gordon R, Pang Y, Eftekhari F, Quidant R. Self-induced back-action optical trapping of dielectric nanoparticles. Nat Phys. 2009;5(12):915–9. https://doi.org/10.1038/nphys1422.
30. Zhang W, Huang L, Santschi C, Martin OJF. Trapping and Sensing 10 nm Metal Nanoparticles Using Plasmonic Dipole Antennas. Nano Lett. 2010;10(3):1006–11. https://doi.org/10.1021/nl904168f.
31. Wang K, Schonbrun E, Steinvurzel P, Crozier KB. Trapping and rotating nanoparticles using a plasmonic nano-tweezer with an integrated heat sink. Nat Commun. 2011;2(1):469. https://doi.org/10.1038/ncomms1480.
32. Kang J-H, Kim K, Ee H-S, Lee Y-H, Yoon T-Y, Seo M-K, Park H-G. Low-power nano-optical vortex trapping via plasmonic diabolo nanoantennas. Nat Commun. 2011;2(1):582. https://doi.org/10.1038/ncomms1592.
33. Zheng Y, Ryan J, Hansen P, Cheng Y-T, Lu T-J, Hesselink L. Nano-optical conveyor belt, Part II: demonstration of Handoff between near-field optical traps. Nano Lett. 2014;14(6):2971–6. https://doi.org/10.1021/nl404045n.
34. Ghanashyam C, Sinha RK, Bankapur A. Surface-plasmon-polaritons for reversible assembly of gold nanoparticles, in situ nanogap tuning, and SERS. Small Methods. 2024;8(1):2301086. https://doi.org/10.1002/smtd.202301086.
35. Huft PR, Kolbow JD, Thweatt JT, Lindquist NC. Holographic plasmonic nanotweezers for dynamic trapping and manipulation. Nano Lett. 2017;17(12):7920–5. https://doi.org/10.1021/acs.nanolett.7b04289.
36. Chen K-Y, Lee A-T, Hung C-C, Huang J-S, Yang Y-T. Transport and trapping in two-dimensional nanoscale plasmonic optical lattice. Nano Lett. 2013;13(9):4118–22. https://doi.org/10.1021/nl4016254.
37. Anyika T, Hong I, Zhu G, Ndukaife JC. Application of resonant plasmonic bowtie nanoantennas for optically-assisted diffusiophoretic trapping of extracellular vesicles and nanoparticles. Laser Photon Rev. 2024;2400412. https://doi.org/10.1002/lpor.202400412
38. Pang Y, Gordon R. Optical trapping of 12 nm dielectric spheres using double-nanoholes in a gold film. Nano Lett. 2011;11(9):3763–7. https://doi.org/10.1021/nl201807z.
39. Tsai W-Y, Huang J-S, Huang C-B. Selective trapping or rotation of isotropic dielectric microparticles by optical near field in a plasmonic archimedes spiral. Nano Lett. 2014;14(2):547–52. https://doi.org/10.1021/nl403608a.
40. Yoon SJ, Lee J, Han S, Kim C-K, Ahn CW, Kim M-K, Lee Y-H. Non-fluorescent nanoscopic monitoring of a single trapped nanoparticle via nonlinear point sources. Nat Commun. 2018;9(1):2218. https://doi.org/10.1038/s41467-018-04689-5.
41. Kotsifaki DG, Chormaic SN. Plasmonic optical tweezers based on nanostructures: Fundamentals, advances and prospects. Nanophotonics. 2019;8(7):1227–45. https://doi.org/10.1515/nanoph-2019-0151.
42. Wang Y, Wei S, Ju Z, Min C, Somekh M, Yuan X. Dynamic optical tweezers for metallic particle manipulation via tunable plasmonic fields. Photon Res. 2024;12(9):1840–5. https://doi.org/10.1364/PRJ.528878.
43. Liaw J-W, Chien C-W, Liu K-C, Ku Y-C, Kuo M-K. 3D optical vortex trapping of plasmonic nanostructure. Sci Rep. 2018;8(1):12673. https://doi.org/10.1038/s41598-018-30948-y.
44. Bai Y, Yan J, Lv H, Yang Y. Plasmonic vortices: a review. J Opt. 2022;24(8):084004. https://doi.org/10.1088/2040-8986/ac7d5f.
45. Zheng Z, Yu Y. Controllable optical tweezer and spanner in evanescent fields via a single plasmonic metasurface. J Phys D: Appl Phys. 2024;57(36):365105. https://doi.org/10.1088/1361-6463/ad5600.
46. Yao Y, Ji J, Zhang H, Zhang K, Liu B, Yang P. Three-dimensional plasmonic trap array for ultrasensitive surface-enhanced Raman scattering analysis of single cells. Anal Chem. 2018;90(17):10394–9. https://doi.org/10.1021/acs.analchem.8b02252.
47. Berthelot J, Aćimović SS, Juan ML, Kreuzer MP, Renger J, Quidant R. Three-dimensional manipulation with scanning near-field optical nanotweezers. Nat Nanotechnol. 2014;9(4):295–9. https://doi.org/10.1038/nnano.2014.24.
48. Song YG, Han BM, Chang S. Force of surface plasmon-coupled evanescent fields on Mie particles. Optics Commun. 2001;198(1):7–19. https://doi.org/10.1016/S0030-4018(01)01484-5.
49. Volpe G, Quidant R, Badenes G, Petrov D. Surface plasmon radiation forces. Phys Rev Lett. 2006;96(23):238101. https://doi.org/10.1103/PhysRevLett.96.238101.
50. Garcés-Chávez V, Quidant R, Reece PJ, Badenes G, Torner L, Dholakia K. Extended organization of colloidal microparticles by surface plasmon polariton excitation. Phys Rev B. 2006;73(8):085417. https://doi.org/10.1103/PhysRevB.73.085417.
51. Patra PP, Chikkaraddy R, Tripathi RPN, Dasgupta A, Kumar GVP. Plasmofluidic single-molecule surface-enhanced Raman scattering from dynamic assembly of plasmonic nanoparticles. Nat Commun. 2014;5(1):4357. https://doi.org/10.1038/ncomms5357.
52. Pratim Patra P, Chikkaraddy R, Thampi S, Tripathi RPN, Pavan Kumar GV. Large-scale dynamic assembly of metal nanostructures in plasmofluidic field. Faraday Discuss. 2016;186:95–106. https://doi.org/10.1039/C5FD00127G.
53. Wang X, Xiao K, Min C, Zou Q, Hua Y, Yuan X-C. Theoretical and experimental study of surface plasmon radiation force on micrometer-sized spheres. Plasmonics. 2013;8(2):637–43. https://doi.org/10.1007/s11468-012-9449-y.
54. Ghorbanzadeh M, Darbari S. Efficient plasmonic 2D arrangement and manipulation system, suitable for controlling particle–particle interactions. J Lightwave Technol JLT. 2019;37(9):2058–64.
55. Ghorbanzadeh M, Moravvej-Farshi MK, Darbari S. Plasmonic optophoresis for manipulating, in situ position monitoring, sensing, and 3-D trapping of micro/nanoparticles. IEEE J Sel Top

Quantum Electron. 2017;23(2):185–92. https://doi.org/10.1109/JSTQE.2016.2593008.
56. Min C, Shen Z, Shen J, Zhang Y, Fang H, Yuan G, Du L, Zhu S, Lei T, Yuan X. Focused plasmonic trapping of metallic particles. Nat Commun. 2013;4(1):2891. https://doi.org/10.1038/ncomms3891.
57. Wang X, Zhang Y, Dai Y, Min C, Yuan X. Enhancing plasmonic trapping with a perfect radially polarized beam. Photon Res. 2018;6(9):847–52. https://doi.org/10.1364/PRJ.6.000847.
58. Zhang L, Dou X, Min C, Zhang Y, Du L, Xie Z, Shen J, Zeng Y, Yuan X. In-plane trapping and manipulation of ZnO nanowires by a hybrid plasmonic field. Nanoscale. 2016;8(18):9756–63. https://doi.org/10.1039/C5NR08940A.
59. Ju Z, Ma H, Zhang S, Xie X, Min C, Zhang Y, Yuan X. All-optically controlled holographic plasmonic vortex array for multiple metallic particles manipulation. Opt Lett. 2023;48(24):6577–80. https://doi.org/10.1364/OL.507098.
60. Man Z, Zhang S, Bai Z, Zhang Y, Ge X, Xing F, Sun Y-P, Fu S. All-optical and dynamic manipulation of surface plasmon polaritons by tailoring the polarization state of incident light. Laser Phys Lett. 2018;16(2):026001. https://doi.org/10.1088/1612-202X/aaf69d.
61. Dou X, Yang A, Min C, Du L, Zhang Y, Weng X, Yuan X. Polarization-controlled gap-mode surface-enhanced Raman scattering with a single nanoparticle. J Phys D: Appl Phys. 2017;50(25):255302. https://doi.org/10.1088/1361-6463/aa7114.
62. Shen Z, Su L. Plasmonic trapping and tuning of a gold nanoparticle dimer. Opt Express. 2016;24(5):4801–11. https://doi.org/10.1364/OE.24.004801.
63. Zhang Y, Wang J, Shen J, Man Z, Shi W, Min C, Yuan G, Zhu S, Urbach HP, Yuan X. Plasmonic hybridization induced trapping and manipulation of a single au nanowire on a metallic surface. Nano Lett. 2014;14(11):6430–6. https://doi.org/10.1021/nl502975k.
64. Minovich AE, Klein AE, Neshev DN, Pertsch T, Kivshar YS, Christodoulides DN. Airy plasmons: non-diffracting optical surface waves. Laser Photonics Rev. 2014;8(2):221–32. https://doi.org/10.1002/lpor.201300055.
65. Tan PS, Yuan X-C, Lin J, Wang Q, Mei T, Burge RE, Mu GG. Surface plasmon polaritons generated by optical vortex beams. Appl Phys Lett. 2008;92(11):111108. https://doi.org/10.1063/1.2890058.
66. Wei S, Wang D, Lin J, Yuan X. Demonstration of orbital angular momentum channel healing using a Fabry-Pérot cavity. Opto-electron Adv. 2018;1(5):180006. https://doi.org/10.29026/oea.2018.180006.
67. Man Z, Xi Z, Yuan X, Burge RE, Urbach HP. Dual coaxial longitudinal polarization vortex structures. Phys Rev Lett. 2020;124(10):103901. https://doi.org/10.1103/PhysRevLett.124.103901.
68. Zhang Y, Shi W, Shen Z, Man Z, Min C, Shen J, Zhu S, Urbach HP, Yuan X. A plasmonic spanner for metal particle manipulation. Sci Rep. 2015;5(1):15446. https://doi.org/10.1038/srep15446.
69. Shen Z, Hu ZJ, Yuan GH, Min CJ, Fang H, Yuan X-C. Visualizing orbital angular momentum of plasmonic vortices. Opt Lett. 2012;37(22):4627–9. https://doi.org/10.1364/OL.37.004627.
70. Wang X, Dai Y, Zhang Y, Min C, Yuan X. Plasmonic manipulation of targeted metallic particles by polarization-sensitive metalens. ACS Photonics. 2018;5(7):2945–50. https://doi.org/10.1021/acsphotonics.8b00282.
71. Zhang Y, Shen J, Xie Z, Dou X, Min C, Lei T, Liu J, Zhu S, Yuan X. Dynamic plasmonic nano-traps for single molecule surface-enhanced Raman scattering. Nanoscale. 2017;9(30):10694–700. https://doi.org/10.1039/C7NR02406A.
72. Kollipara PS, Li X, Li J, Chen Z, Ding H, Kim Y, Huang S, Qin Z, Zheng Y. Hypothermal opto-thermophoretic tweezers. Nat Commun. 2023;14(1):5133. https://doi.org/10.1038/s41467-023-40865-y.
73. Tiwari S, Khandelwal U, Sharma V, Kumar GVP. Single molecule surface enhanced Raman scattering in a single gold nanoparticle-driven thermoplasmonic tweezer. J Phys Chem Lett. 2021;12(49):11910–8. https://doi.org/10.1021/acs.jpclett.1c03450.
74. Hong C, Yang S, Ndukaife JC. Stand-off trapping and manipulation of sub-10 nm objects and biomolecules using opto-thermo-electrohydrodynamic tweezers. Nat Nanotechnol. 2020;15(11):908–13. https://doi.org/10.1038/s41565-020-0760-z.
75. Wang X, Zhang Y, Yu J, Xie X, Deng R, Min C, Yuan X. Plasmonic-thermoelectric nanotweezers for immersive SERS mapping. ACS Nano. 2022;16(11):18621–9. https://doi.org/10.1021/acsnano.2c07103.
76. Lin L, Peng X, Wei X, Mao Z, Xie C, Zheng Y. Thermophoretic tweezers for low-power and versatile manipulation of biological cells. ACS Nano. 2017;11(3):3147–54. https://doi.org/10.1021/acsnano.7b00207.
77. Kotnala A, Zheng Y. Opto-thermophoretic fiber tweezers. Nanophotonics. 2019;8(3):475–85. https://doi.org/10.1515/nanoph-2018-0226.
78. Lin L, Wang M, Peng X, Lissek EN, Mao Z, Scarabelli L, Adkins E, Coskun S, Unalan HE, Korgel BA, Liz-Marzán LM, Florin E-L, Zheng Y. Opto-thermoelectric nanotweezers. Nat Photonics. 2018;12(4):195–201. https://doi.org/10.1038/s41566-018-0134-3.
79. Wang X, Yuan Y, Xie X, Zhang Y, Min C, Yuan X. Graphene-based opto-thermoelectric tweezers. Adv Mater. 2022;34(8):2107691. https://doi.org/10.1002/adma.202107691.
80. Li J, Chen Z, Liu Y, Kollipara PS, Feng Y, Zhang Z, Zheng Y. Opto-refrigerative tweezers. Sci Adv. 2021. https://doi.org/10.1126/sciadv.abh1101.
81. Hong C, Ndukaife JC. Scalable trapping of single nano-sized extracellular vesicles using plasmonics. Nat Commun. 2023;14(1):4801. https://doi.org/10.1038/s41467-023-40549-7.
82. Hong C, Yang S, Ndukaife JC. Exosomes trapping, manipulation and size-based separation using opto-thermo-electrohydrodynamic tweezers. Nanoscale Adv. 2023;5(11):2973–8. https://doi.org/10.1039/D3NA00101F.
83. Miao X, Lin LY. Trapping and manipulation of biological particles through a plasmonic platform. IEEE J Sel Top Quantum Electron. 2007;13(6):1655–62. https://doi.org/10.1109/JSTQE.2007.910996.
84. Wong HMK, Righini M, Gates JC, Smith PGR, Pruneri V, Quidant R. On-a-chip surface plasmon tweezers. Appl Phys Lett. 2011;99(6):061107. https://doi.org/10.1063/1.3625936.
85. Kang Z, Chen J, Wu S-Y, Chen K, Kong S-K, Yong K-T, Ho H-P. Trapping and assembling of particles and live cells on large-scale random gold nano-island substrates. Sci Rep. 2015;5(1):9978. https://doi.org/10.1038/srep09978.
86. Zhong Y, Peng Y, Chen J, Zhou J, Dai X, Zhang H, Qu J, Shao Y. Optical temperature field-driven tweezers: principles and biomedical applications. Acta Opt Sin. 2023;43(14):1400001. https://doi.org/10.3788/AOS230530.
87. Deng R, Zhang Y, Wang X, Xie X, Song Y, Bu J, Min C, Yuan X. *In situ* intracellular Raman spectroscopic detection with graphene-based thermoelectric optical tweezers. Sens Actuat B Chem. 2022;361:131722. https://doi.org/10.1016/j.snb.2022.131722.
88. Kang S, Nisar MS, Lu Y, Chang N, Huang Y, Ni H, Novikov SM, Wang Y, Cui Q, Zhao X. A 3D biocompatible plasmonic tweezer for single cell manipulation. Small Methods. 2023;7(2):2201379. https://doi.org/10.1002/smtd.202201379.
89. Chen J, Kang Z, Kong SK, Ho H-P. Plasmonic random nanostructures on fiber tip for trapping live cells and colloidal particles. Opt Lett. 2015;40(17):3926–9. https://doi.org/10.1364/OL.40.003926.

90. Lee SY, Kim HM, Park J, Kim SK, Youn JR, Song YS. Enhanced plasmonic particle trapping using a hybrid structure of nanoparticles and nanorods. ACS Appl Mater Interfaces. 2018;10(48):41655–63. https://doi.org/10.1021/acsami.8b14787.
91. Hill EH, Li J, Lin L, Liu Y, Zheng Y. Opto-thermophoretic attraction, trapping, and dynamic manipulation of lipid vesicles. Langmuir. 2018;34(44):13252–62. https://doi.org/10.1021/acs.langmuir.8b01979.
92. Kaufman LH, Cooper T, Wallace GQ, Hawke DC, Betts D, Hess D, Lagugné-Labarthet F. Trapping and SERS identification of extracellular vesicles using nanohole arrays. Plasmon Biol Med. 2019;XVI(SPIE):9–18. https://doi.org/10.1117/12.2506633.
93. Hong C, Hong I, Jiang Y, Ndukaife JC. Plasmonic dielectric antennas for hybrid optical nanotweezing and optothermoelectric manipulation of single nanosized extracellular vesicles. Adv Opt Mater. 2024;12(12):2302603. https://doi.org/10.1002/adom.202302603.
94. Wang Y, Yu Y, Liu Y, Yang S. Template-confined site-specific electrodeposition of nanoparticle cluster-in-bowl arrays as surface enhanced Raman spectroscopy substrates. ACS Sens. 2018;3(11):2343–50. https://doi.org/10.1021/acssensors.8b00711.
95. Yang Y, Peng Y, Lin C, Long L, Hu J, He J, Zeng H, Huang Z, Li Z-Y, Tanemura M, Shi J, Lombardi JR, Luo X. Human ACE2-functionalized gold "virus-trap" nanostructures for accurate capture of SARS-CoV-2 and single-virus SERS detection. Nano-Micro Lett. 2021;13(1):109. https://doi.org/10.1007/s40820-021-00620-8.
96. Li D, Yue W, He Q, Gao P, Gong T, Luo Y, Wang C, Luo X. Single-molecule detection of SARS-CoV-2 N protein on multilayered plasmonic nanotraps with surface-enhanced Raman spectroscopy. Talanta. 2024;278:126494. https://doi.org/10.1016/j.talanta.2024.126494.
97. Durmanov NN, Guliev RR, Eremenko AV, Boginskaya IA, Ryzhikov IA, Trifonova EA, Putlyaev EV, Mukhin AN, Kalnov SL, Balandina MV, Tkachuk AP, Gushchin VA, Sarychev AK, Lagarkov AN, Rodionov IA, Gabidullin AR, Kurochkin IN. Non-labeled selective virus detection with novel SERS-active porous silver nanofilms fabricated by electron beam physical vapor deposition. Sens Actuat B Chem. 2018;257:37–47. https://doi.org/10.1016/j.snb.2017.10.022.
98. Sampad MJN, Saiduzzaman S, Walker ZJ, Wells TN, Wayment JX, Ong EM, Mdaki SD, Tamhankar MA, Yuzvinsky TD, Patterson JL, Hawkins AR, Schmidt H. Label-free and amplification-free viral RNA quantification from primate biofluids using a trapping-assisted optofluidic nanopore platform. PNAS. 2024;121(16):e2400203121. https://doi.org/10.1073/pnas.2400203121.
99. Gordon R. Future prospects for biomolecular trapping with nanostructured metals. ACS Photonics. 2022;9(4):1127–35. https://doi.org/10.1021/acsphotonics.2c00231.
100. Pang Y, Gordon R. Optical trapping of a single protein. Nano Lett. 2012;12(1):402–6. https://doi.org/10.1021/nl203719v.
101. Yoo D, Gurunatha KL, Choi H-K, Mohr DA, Ertsgaard CT, Gordon R, Oh S-H. Low-power optical trapping of nanoparticles and proteins with resonant coaxial nanoaperture using 10 nm Gap. Nano Lett. 2018;18(6):3637–42. https://doi.org/10.1021/acs.nanolett.8b00732.
102. Yousefi A, Zheng Z, Zargarbashi S, Assadipapari M, Hickman GJ, Parmenter CDJ, Bueno-Alejo CJ, Sanderson G, Craske D, Xu L, Perry CC, Rahmani M, Ying C. Structural flexibility and disassembly kinetics of single ferritin molecules using optical nanotweezers. ACS Nano. 2024;18(24):15617–26. https://doi.org/10.1021/acsnano.4c01221.
103. Nicolaï A, Delarue P, Senet P. Theoretical insights into subterahertz acoustic vibrations of proteins measured in single-molecule experiments. J Phys Chem Lett. 2016;7(24):5128–36. https://doi.org/10.1021/acs.jpclett.6b01812.
104. Chauhan N, Saxena K, Jain U. Single molecule detection; from microscopy to sensors. Int J Biol Macromol. 2022;209:1389–401. https://doi.org/10.1016/j.ijbiomac.2022.04.038.
105. Alfaro JA, Bohländer P, Dai M, Filius M, Howard CJ, van Kooten XF, Ohayon S, Pomorski A, Schmid S, Aksimentiev A, Anslyn EV, Bedran G, Cao C, Chinappi M, Coyaud E, Dekker C, Dittmar G, Drachman N, Eelkema R, Goodlett D, Hentz S, Kalathiya U, Kelleher NL, Kelly RT, Kelman Z, Kim SH, Kuster B, Rodriguez-Larrea D, Lindsay S, Maglia G, Marcotte EM, Marino JP, Masselon C, Mayer M, Samaras P, Sarthak K, Sepiashvili L, Stein D, Wanunu M, Wilhelm M, Yin P, Meller A, Joo C. The emerging landscape of single-molecule protein sequencing technologies. Nat Methods. 2021;18(6):604–17. https://doi.org/10.1038/s41592-021-01143-1.
106. Fu W, Chi H, Dai X, Zhu H, Mesias VSD, Liu W, Huang J. Efficient optical plasmonic tweezer-controlled single-molecule SERS characterization of pH-dependent amylin species in aqueous milieus. Nat Commun. 2023;14(1):6996. https://doi.org/10.1038/s41467-023-42812-3.
107. Dai X, Fu W, Chi H, Mesias VSD, Zhu H, Leung CW, Liu W, Huang J. Optical tweezers-controlled hotspot for sensitive and reproducible surface-enhanced Raman spectroscopy characterization of native protein structures. Nat Commun. 2021;12(1):1292. https://doi.org/10.1038/s41467-021-21543-3.
108. Shoji T, Itoh K, Saitoh J, Kitamura N, Yoshii T, Murakoshi K, Yamada Y, Yokoyama T, Ishihara H, Tsuboi Y. Plasmonic manipulation of DNA using a combination of optical and thermophoretic forces: separation of different-sized DNA from mixture solution. Sci Rep. 2020;10(1):3349. https://doi.org/10.1038/s41598-020-60165-5.
109. Zhan C, Wang G, Yi J, Wei J-Y, Li Z-H, Chen Z-B, Shi J, Yang Y, Hong W, Tian Z-Q. Single-molecule plasmonic optical trapping. Matter. 2020;3(4):1350–60. https://doi.org/10.1016/j.matt.2020.07.019.
110. Kabtiyal P, Robbins A, Jergens E, Castro CE, Winter JO, Poirier MG, Johnston-Halperin E. Localized plasmonic heating for single-molecule DNA rupture measurements in optical tweezers. Nano Lett. 2024;24(10):3097–103. https://doi.org/10.1021/acs.nanolett.3c04848.
111. Belkin M, Chao S-H, Jonsson MP, Dekker C, Aksimentiev A. Plasmonic nanopores for trapping, controlling displacement, and sequencing of DNA. ACS Nano. 2015;9(11):10598–611. https://doi.org/10.1021/acsnano.5b04173.
112. Jonsson MP, Dekker C. Plasmonic nanopore for electrical profiling of optical intensity landscapes. Nano Lett. 2013;13(3):1029–33. https://doi.org/10.1021/nl304213s.
113. Belkin M, Chao SH, Jonsson MP, Dekker C, Aksimentiev A. Plasmonic nanopores for trapping, controlling displacement, and sequencing of DNA. ACS Publications; 2015. Accessed 16 Oct 2024. https://doi.org/10.1021/acsnano.5b04173.
114. Chen J, Chen Z, Meng C, Zhou J, Peng Y, Dai X, Li J, Zhong Y, Chen X, Yuan W, Ho H-P, Gao BZ, Qu J, Zhang X, Zhang H, Shao Y. CRISPR-powered optothermal nanotweezers: diverse bio-nanoparticle manipulation and single nucleotide identification. Light Sci Appl. 2023;12(1):273. https://doi.org/10.1038/s41377-023-01326-9.
115. Yang J, Deng R, Wang X, Zhang Y, Yuan X, Min C. Research progress of opto-thermal tweezers and their applications. Chin J Lasers. 2024;51(18):62–83.

116. Ciarlo A, Ciriza DB, Selin M, Maragò OM, Sasso A, Pesce G, Volpe G, Goksör M. Deep learning for optical tweezers. Nanophotonics. 2024;13(17):3017–35. https://doi.org/10.1515/nanoph-2024-0013.

117. Deng R, Song Y, Yang J, Min C, Zhang Y, Yuan X, Liu W. AI-assisted cell identification and optical sorting [invited]. Chin Opt Lett. 2023;21(11):110009.

2 Optothermal Tweezers

Jianxing Zhou, Jiajie Chen, Yonghong Shao, and Junle Qu

Abstract

Optothermal tweezers (OTT) have emerged as a transformative technology in nanotechnology and life sciences, addressing the limitations of traditional optical tweezers, such as the diffraction barrier and thermal damage to samples. By coupling light and heat to create precise temperature gradients, OTT generate thermophoretic forces to manipulate particles with spatial resolution beyond the diffraction limit and at power densities significantly lower than conventional optical tweezers. This capability enables the trapping of delicate biological samples and nanoscale particles, including biomolecules, metallic nanoparticles, and polymeric materials, with minimal optical heating damage. OTT's versatility extends to large-scale particle assembly, facilitating the creation of complex nanostructures and biomimetic systems, and their applications span diverse fields, including targeted drug delivery, biosensing, synthetic biology, and single-cell manipulation. Recent advancements, such as the integration of low-temperature environments and opto-refrigerative designs, further enhance their precision and adaptability. This chapter systematically explores the principles of OTT, their biomedical and nanotechnological applications, and their transformative potential across disciplines. Additionally, we address the challenges in temperature gradient control, thermal effects on sensitive systems, and scalability for industrial applications. By overcoming these barriers through interdisciplinary efforts and innovative engineering, OTT are poised to revolutionize micro- and nanoscale manipulation, paving the way for unprecedented scientific and technological advancements.

Keywords

Optothermal tweezers · Optothermal effect · Temperature gradient · Thermophoresis · Thermo-osmotic flow · Convective flow · Large-scale particle assembly

2.1 Introduction

Since the 1970s, when Arthur Ashkin first demonstrated the manipulation of micro- and nanoparticles using optical forces and introduced the revolutionary concept of optical tweezers [1], this technology has evolved to become a cornerstone in the fields of biology [2], physics, and colloidal science [3]. Over the decades, traditional optical tweezers have enabled groundbreaking research, particularly in the manipulation of biological systems and colloidal particles. However, their application has faced two major challenges: the diffraction limit, which restricts the spatial precision of particle manipulation, and thermal damage, which limits the types of samples that can be safely handled.

To address these challenges, researchers have developed various advanced optical tweezers techniques. These innovations include near-field evanescent wave optical tweezers [4], fiber-optic tweezers [5], surface plasmon polariton tweezers [6], opto-thermoelectric tweezers [7], opto-refrigerative tweezers [8], and low-temperature optothermal tweezers [9]. Additionally, new methods have emerged that bypass traditional optical force manipulation entirely. Among these, optothermal tweezers have gained significant attention as a transformative tool in nanotechnology and life sciences.

Optothermal tweezers (OTT) leverage the coupling effects of light and heat to generate precise temperature fields, which in turn drive the trapping and manipulation of particles. Unlike conventional optical tweezers, which rely on direct

J. Zhou · J. Chen (✉) · Y. Shao · J. Qu
State Key Laboratory of Radio Frequency Heterogeneous Integration, Key Laboratory of Optoelectronic Devices, Systems of Ministry of Education and Guangdong Province, College of Physics and Optoelectronics Engineering, Shenzhen University, Shenzhen, China
e-mail: cjj@szu.edu.cn

B. Li et al. (eds.), *Biophotonic Manipulation*, Advances in Optics and Optoelectronics,
https://doi.org/10.1007/978-981-96-4982-2_2

optical forces to manipulate particles, OTT use lasers as localized heating sources. These create temperature gradients that induce thermophoretic forces [10], guiding particles to designated locations. This approach not only minimizes the power density required—typically three orders of magnitude lower than traditional optical tweezers—but also eliminates the limitations imposed by the diffraction barrier. The use of temperature fields instead of direct optical forces enables the trapping accuracy of OTT to exceed the spatial resolution constraints of traditional optical systems.

The benefits of OTT extend beyond their superior energy efficiency. By minimizing optical heating damage, they enable the manipulation of delicate biological samples and expand the range of particles that can be controlled, including nanoscale biomolecules, metallic nanoparticles, and polymeric materials. Moreover, OTT allow for large-scale particle assembly, enabling the development of complex nanostructures and biomimetic systems. These capabilities have established OTT as a versatile and essential research tool in fields such as nanotechnology, materials science, and biomedicine.

In this chapter, we provide a systematic overview of the fundamental principles underlying optothermal tweezers. We delve into their diverse biomedical applications, ranging from single-cell manipulation to the assembly of nanoscale devices, and highlight their potential to transform fields such as targeted drug delivery, biosensing, and synthetic biology. Furthermore, we explore the ongoing advancements in OTT technology, including the integration of low-temperature environments and opto-refrigerative designs [8, 9], which further enhance their precision and versatility.

Despite their many advantages, the development and application of OTT face several challenges. These include the precise control of temperature gradients, the potential for unintended thermal effects on sensitive biological systems, and the scalability of the technology for industrial applications. Addressing these challenges will require interdisciplinary collaboration and innovative engineering approaches.

Looking ahead, optothermal tweezers are poised to drive new discoveries and applications, cementing their role as a vital tool in both fundamental research and practical applications. By overcoming current limitations and leveraging emerging technologies, this tool could unlock unprecedented capabilities in the manipulation of matter at the micro- and nanoscale, heralding a new era of scientific innovation.

2.2 Principle of Optothermal Tweezers

2.2.1 Temperature Field

Various materials can absorb light energy at specific wavelengths under suitable conditions, converting it into heat. In laboratory settings, metals like gold, silver, and chromium, as well as a variety of two-dimensional materials, are frequently utilized for optical heating due to their efficient light-heat conversion properties. Among these, metal nanostructures are particularly notable under visible and near-infrared excitation due to a phenomenon known as plasmonic heating. This process occurs when incident light excites localized surface plasmon resonances on the surface of the nanostructure, leading to enhanced electron oscillations within the metal. The intensified electron-lattice collisions subsequently facilitate the conversion of photon energy into phonons, thereby generating heat. The resulting temperature field in the laser-irradiated region can typically be described by the heat transfer Eq. (2.1) [11]:

$$\rho C_p \frac{\partial T(r)}{\partial t} = \nabla \cdot [k \nabla T(r)] + Q \tag{2.1}$$

Here, ρ is the density of the metal substrate at constant pressure, C_p is the specific heat capacity of the metal substrate at constant pressure, T is the absolute temperature, t is time, k is the thermal conductivity of the metal substrate, and Q is the heat generated per unit volume per unit time by the nanostructure. Q can be expressed by Eq. (2.2) [12]:

$$Q = Re\langle J(r,t)\rangle \cdot E(r,t) \tag{2.2}$$

where J is the current density and E is the electric field in the metal nanostructure. When the temperature field reaches a steady state, it can be described by Eq. (2.3) [13]:

$$\nabla \cdot [k \nabla T(r)] = -Q = -Re\langle J \cdot E\rangle \tag{2.3}$$

2.2.2 Thermophoretic Force and Thermo-Osmotic Flow at Boundary

When a temperature gradient is applied to a fluid, suspended particles, molecules, and micelles migrate towards either the colder or hotter region, a phenomenon referred to as thermophoresis. This migration is driven by the thermophoretic force, which acts on the particles within the temperature gradient. The thermophoretic drift velocity, u, can be mathematically expressed as follows [10]:

$$u = -D_T \nabla T \quad (2.4)$$

where $D_T = DS_T$ is the thermophoretic mobility of the particle, D is the diffusion coefficient, and S_T is the Soret coefficient of the particle. The sign of S_T determines whether the particle is driven towards the cold or hot region. A positive S_T indicates movement towards the cold region, while a negative S_T indicates movement towards the hot region. ∇T represents the temperature gradient. The thermophoretic drift velocity depends on the surface properties of the particle and its interaction with the solvent.

The derivation of the thermophoretic drift velocity u is outlined as follows:

For micro- and nanoscale particles suspended in a liquid, the surrounding fluid is typically incompressible and characterized by a very low Reynolds number. Under these conditions, the velocity field around the particle can be described by the Stokes equation:

$$\eta \nabla^2 v - \nabla P + f = 0 \quad (2.5)$$

where η is the fluid viscosity, P is the hydrostatic pressure, and f is the force density exerted by the particle on the surrounding fluid. In the boundary layer approximation, within the region where the force density f is significant, the boundary layer thickness is small enough compared to the particle size that the particle surface can be considered flat. Near the particle interface, the vertical velocity component v_z is approximately zero, while the parallel component v_x varies more rapidly in the vertical direction z than in the parallel direction x. v_x can be approximated as a function of z only, reaching a steady value at a distance beyond the interaction range B. Thus, the parallel and vertical components of Eq. (2.5) can be separated and expressed as [14]:

$$0 = \frac{dP}{dz} - f_z \quad (2.6)$$

$$\eta \frac{d^2 v_x}{dz^2} = \frac{dP}{dx} - f_x \quad (2.7)$$

Given the boundary conditions $v_x|_{z=0} = 0$ and $dv_x/dz\big|_{z=\infty} = 0$, we obtain:

$$\overline{v_B} = v_x(\infty) - \frac{1}{\eta} \int_0^{\infty} z\left(\frac{dP}{dx} - f_x\right) dz \quad (2.8)$$

where η is the Viscosity of solution. For a uniform surface, dP/dx and f_x vary sinusoidally with the particle polar angle θ. Meanwhile, Eq. (2.8) can be generalized using local excess enthalpy and ultimately expressed in terms of the thermal creep coefficient as follows [15]:

$$\overline{v_B} = -\frac{1}{\eta} \int_0^{\infty} z h(z) dz \frac{\nabla_{=} T}{T} = \chi \frac{\nabla_{=} T}{T} \quad (2.9)$$

where $\nabla_{=}$T is the temperature gradient parallel to the particle surface, and χ is the thermo-osmotic coefficient. For spherical particles, the thermophoretic drift velocity can be given by the orientational average of the slip velocity at the boundary:

$$v_B = \overline{v_B} \sin \theta \quad (2.10)$$

$$u = -v_B(\theta) e_x = -\frac{2}{3} \overline{v_B} = -\frac{2}{3} \chi \frac{\nabla T}{T} = -D_T \nabla T \quad (2.11)$$

Here, $\overline{v_B}$ is the maximum slip velocity at $= \pi/2$, where the tangential velocity is parallel to the external temperature gradient. By determining the forms of P and f_x, the slip velocity v_B at the particle boundary can be established, and the drift velocity u can be further determined as the thermophoretic drift velocity generated by the temperature gradient.

Most particles exhibit a positive Soret coefficient S_T at room temperature, indicating thermophobic behaviour [16]. Typically, laser spots act as heat sources in experiments, as controlling heat sources is generally more convenient than controlling cold sources. Thus, to manipulate target particles using lasers, we need to reverse the sign of the target particle's Soret coefficient to convert them from thermophobic to thermophilic. The different methods to achieve this conversion define the various types of optothermal tweezers.

Thermo-osmotic flow exists at all solid–liquid interfaces with a tangential temperature gradient, as shown in Eqs. (2.8) and (2.9). The thermo-osmotic coefficient χ encapsulates all the information regarding liquid–solid interactions. When $\chi < 0$, the liquid is driven towards the cold region, and when $\chi > 0$, it is driven towards the hot region. It is noteworthy that thermophoretic force and thermal creep flow share the same physical origin; the thermophoretic force is derived from the thermo-osmotic flows at the particle's surface in the fluid.

In practical applications, optothermal tweezers operate within a complex environment that includes not only the particle–fluid interface but also the solid–liquid interface between the substrate and the fluid [17]. The substrate, often a heat-absorbing material such as a gold film, can act as an additional heat source due to the laser-induced optothermal effect. This dual heat source generates a tangential temperature gradient at the substrate-fluid interface, leading to thermo-osmotic flow along the interface. This thermo-osmotic flow can significantly impact particle trapping, either enhancing or interfering with the process, depending on the specific conditions. Despite its importance, this effect has been largely overlooked in many previous studies on optothermal tweezers. A deeper understanding of these interactions is crucial for optimizing trapping efficiency and expanding the applicability of optothermal tweezers in complex systems.

2.3 The Manipulation of Micro-nano Particles

2.3.1 The Manipulation of Solid Nanoparticles

In most optical tweezering systems, polystyrene (PS) particles are commonly employed as testing models. This is due to their size and morphological characteristics, which closely resemble those of biological surfaces, making them ideal for modeling complex biological substances. Additionally, PS particles have potential applications as drug delivery carriers. Their slight charge, hydrophobic nature, and surface roughness are key attributes relevant to biomedical research. A variety of nano- and microparticle manipulation techniques, such as opto-thermoelectric nanotweezers, opto-refrigerative tweezers, highly adaptable optothermal nanotweezers, natural convection optical tweezers, Marangoni convection optical tweezers, and thermo-electrohydrodynamic tweezers, have been developed to trap, manipulate, and enrich PS particles. However, these techniques differ in their levels of manipulation precision, applicability, and operational efficacy.

Here are some representative examples. In 2012, an optical tweezing system based on Marangoni convection was developed [18], utilizing optical heating to alter the surface tension at the liquid–gas interface. This change drives liquid flow toward regions of higher surface tension, thereby creating a trapping and enrichment effect. Marangoni convection optical tweezers offer significant advantages, including minimal constraints on sample properties and sizes and the ability to generate stronger trapping forces compared to natural convection. Furthermore, this method can trap particles at a distance from the heat source, mitigating thermal damage, though it offers lower precision for manipulating individual particles. However, it is particularly effective for cluster manipulation. More recently, Researchers from the Jinan University introduced a novel technique for particle manipulation and trapping using light-induced cold Marangoni convection (CMF) [19]. This approach enables collective particle migration and self-assembly, demonstrating significant potential for applications in drug delivery and precision medicine.

In 2015, as shown in Fig. 2.1, researchers from the Chinese University of Hong Kong developed a nano-optical tweezer system utilizing large-scale random gold nanoislands and gold film structures [20, 21]. This design leverages optothermal fields to manipulate polystyrene (PS) particles efficiently. The core mechanism relies on the combined effects of natural convection and thermophoretic forces. Optical heating creates a temperature gradient, inducing PS particles to migrate toward the high-temperature region where they are trapped and assembled. This method highlights the effective integration of optothermal principles for nanoscale particle manipulation. However, a notable limitation of this technique is its specificity to thermophilic particles (with negative Soret coefficients), which migrate toward higher temperatures. Since such particles are relatively rare in nature, the application scope of this method is inherently restricted. Despite this limitation, the development represented a significant milestone as one of the earliest uses of optothermal fields for targeted particle trapping and assembly.

As illustrated in Fig. 2.2, in 2018, researchers at the University of Texas at Austin introduced Opto-Thermoelectric Nanotweezers (OTENT) [7], enabling precise manipulation and in-situ spectral detection of metal nanospheres. The working principle involves the surfactant CTAC, which facilitates the trapping process. When optical heating generates a temperature gradient, ions with different Soret coefficients undergo thermophoresis, creating an internal electric field. CTAC adsorbs onto the surface of metal nanoparticles, rendering them positively charged and enabling their trapping within the electric field. Replacing the traditional optothermal-responsive substrate with graphene, a material offering superior thermal conductivity, further enhances the efficiency of this manipulation. OTENT operates with significantly lower laser power compared to traditional optical tweezers, making it particularly effective for manipulating inorganic materials. However, the approach has certain limitations. The reliance on surfactants like CTAC restricts its applicability for biological particles, as surfactants can alter or damage delicate biomolecular structures. Additionally, particles are trapped at the laser beam center, exposing them to localized heat and posing a risk of thermal damage, particularly for sensitive biological samples. These challenges highlight the need for further advancements to broaden OTENT' s application range while ensuring the safety of biological specimens.

In 2020, researchers developed Opto-Thermo-Electrohydrodynamic Tweezers (OTET) [23], a novel technique that combines optical, thermal, and electrohydrodynamic effects for precise particle manipulation. The principle relies on the coupling of a nanopore array with light to create a highly localized and enhanced hotspot, generating an optothermal-induced temperature gradient. This gradient drives a radially inward electrothermoplasmonic (ETP) flow. Simultaneously, an alternating current electro-osmotic (ACEO) flow, which moves radially outward, is induced. The interplay between these two opposing flows forms a stagnation zone with zero fluid velocity, where particles are effectively trapped. OTET offers several advantages,

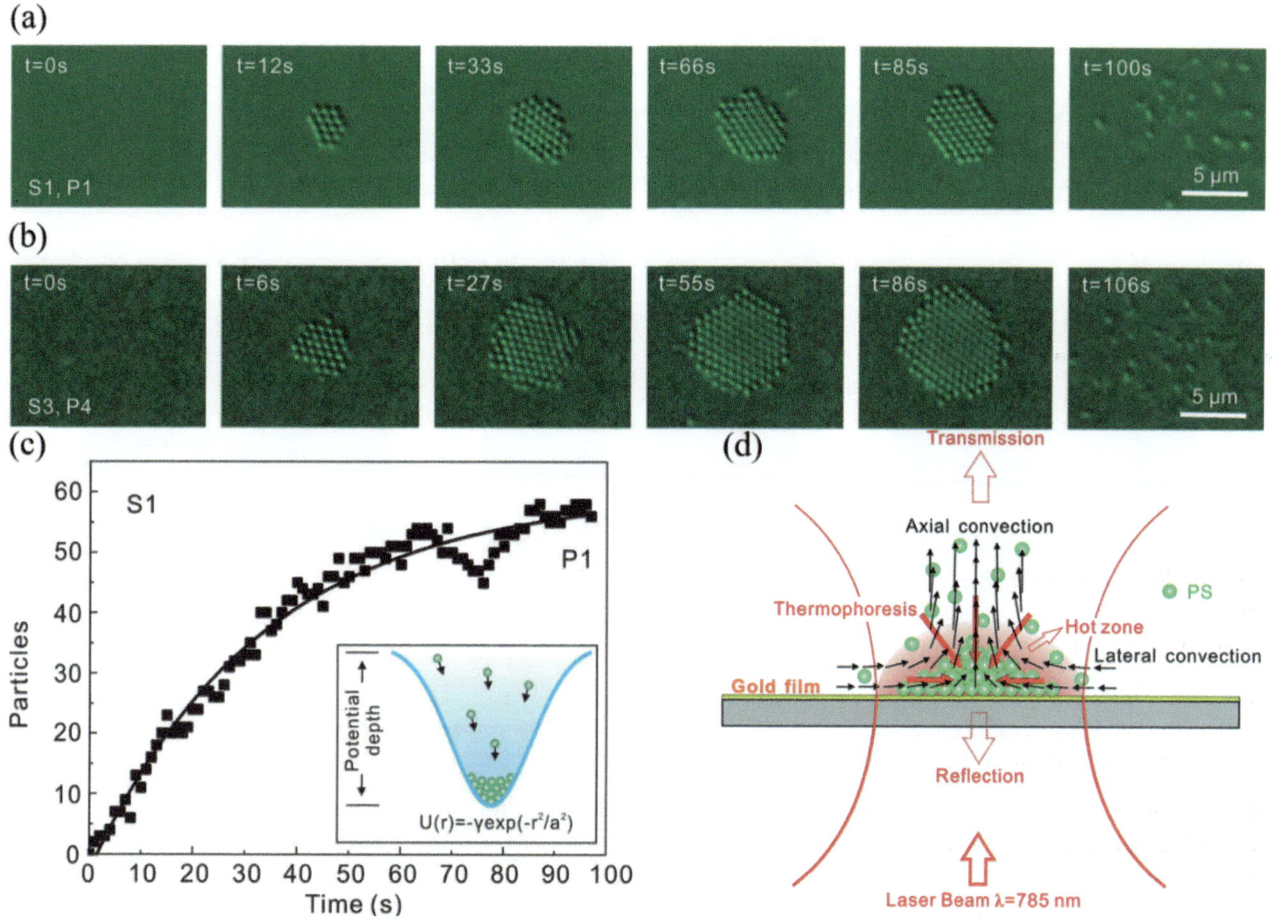

Fig. 2.1 Successive image frames obtained during the trapping of **a** 10 nm PS particles at 1 mW; **b** 30–140 nm particles at 14.5 mW; **c** Statistical analysis of the trapped PS as a function of time and laser power for 10 nm particles. The inset in **c** shows the schematic of the trapping potential well with a Gaussian profile, whose depth is described as U(r), where γ is a constant, r is the coordinate, and a is the width of the potential well [22] (Kang et al. [22], Copyright, with permission from Springer Nature); **d** Schematic illustration of the experiment. Thermal convection pushes the aqueous medium laterally into the hot zone and then axially out of the zone. Thermophoresis drives the PS from cold to hot region with a force magnitude larger than that of the axial convective drag force [21] (Kang et al. [21], Copyright, with permission from RSC Publishing)

including the ability to sort particles by size and reduce optothermal damage. Unlike traditional trapping methods, particles are held several micrometers away from the highly focused laser, minimizing exposure to intense heat while maintaining trapping efficiency.

In 2021, researchers introduced opto-refrigerative tweezers [8], a novel tool for particle manipulation that leverages the local cooling effect of an Yb:YLF crystal layer in heavy water (D_2O) solution to create a temperature gradient field for trapping thermophobic particles. The principle is akin to that of optical thermophoresis tweezers but with a significant distinction: it produces a temperature gradient field where the laser focus generates a cooler center. Thermophobic particles, characterized by a positive Soret coefficient, migrate toward and are trapped in the colder region. Since most biological particles also exhibit a positive Soret coefficient, opto-refrigerative tweezers hold great potential for biological applications, offering a non-invasive and efficient means of manipulation. However, challenges remain for this technology. The biotoxicity of nanocrystals used in the system and the heat generation from solvent molecules during laser operation need to be addressed to ensure compatibility with sensitive biological samples. Further developments in material selection and system optimization are essential to fully realize the potential of opto-refrigerative tweezers in biological and biomedical applications.

As illustrated in Fig. 2.3, researchers from Shenzhen University introduced a groundbreaking optothermal tweezers technology in 2024, known as Highly-Adaptable Optothermal Nanotweezers (HAONT) [24]. This innovative system is designed to accommodate a wide range of particle

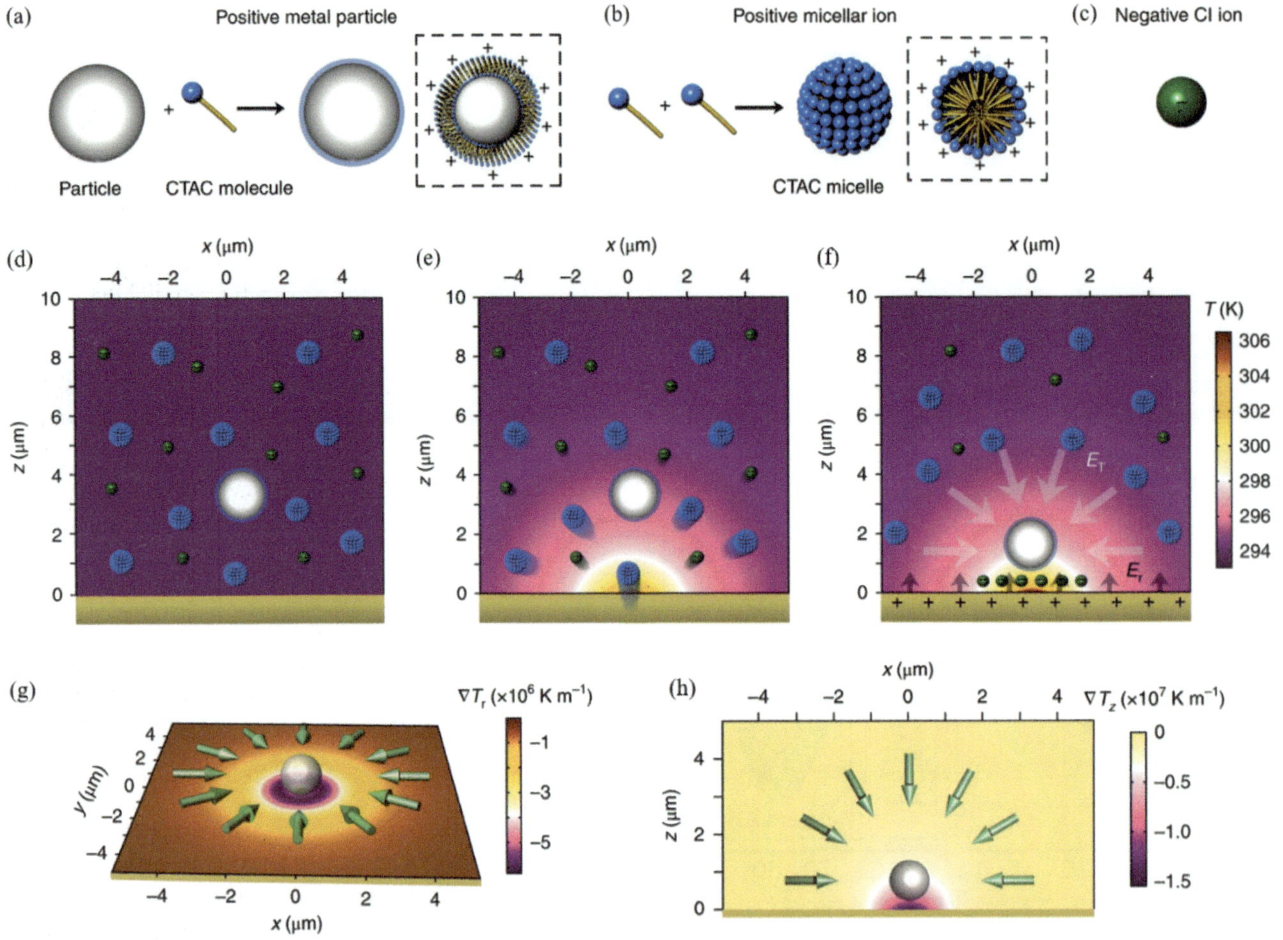

Fig. 2.2 **a** Surface charge modification of a metal nanoparticle by CTAC adsorption; **b** Formation of CTAC micelles; **c** Schematic view of a Cl^- ion; **d** Dispersion of a single metal particle and multiple ions in the solution without optical heating; **e** Thermophoretic migration of the ions under optical heating; **f** Steady ionic distribution under optical heating generates a thermoelectric field E_T for trapping the metal nanoparticle. The repulsive electric field E_r arises from the positive charges of the thermoplasmoic substrate and balances E_T; **g** Simulated in-plane temperature gradient ∇T_r and direction of the corresponding trapping force; **h** Simulated out-of-plane temperature gradient ∇T_z and direction of the corresponding trapping force [7] (Lin et al. [7], Copyright, with permission from Springer Nature)

sizes and types, making it a versatile tool for nanomanipulation. The key mechanism involves the addition of the polymer polyethylene glycol (PEG). The temperature gradient field induced by optical heating generates a concentration gradient field of PEG. This gradient significantly enhances particle transport, as PEG accumulates in the boundary layer around the particles, effectively driving their motion and facilitating trapping. HAONT demonstrates remarkable adaptability, enabling the trapping of diverse particles and biological macromolecules. Notably, it achieves efficient trapping of gold nanoparticles at the nanoscale—a feat that poses significant challenges for conventional optothermal tweezers. This advancement marks a significant step forward in the field of particle manipulation, expanding the capabilities of optothermal tweezers in both nanotechnology and biomedicine.

Similarly, as depicted in Fig. 2.4, the same group from Shenzhen University developed a novel type of low-temperature optothermal nanotweezers in 2024 [9]. These nanotweezers utilize a cooling mechanism to capture particles, leveraging the unique behavior of many particles that transition from thermophobic to thermophilic as the temperature decreases.

By lowering the environmental temperature, these low-temperature optothermal nanotweezers create conditions that facilitate the trapping of particles or biological macromolecules without requiring the addition of external solutes, such as polymers or surfactants. This approach not only simplifies the system but also broadens its applicability, particularly for sensitive biological samples that may be adversely affected by chemical additives.

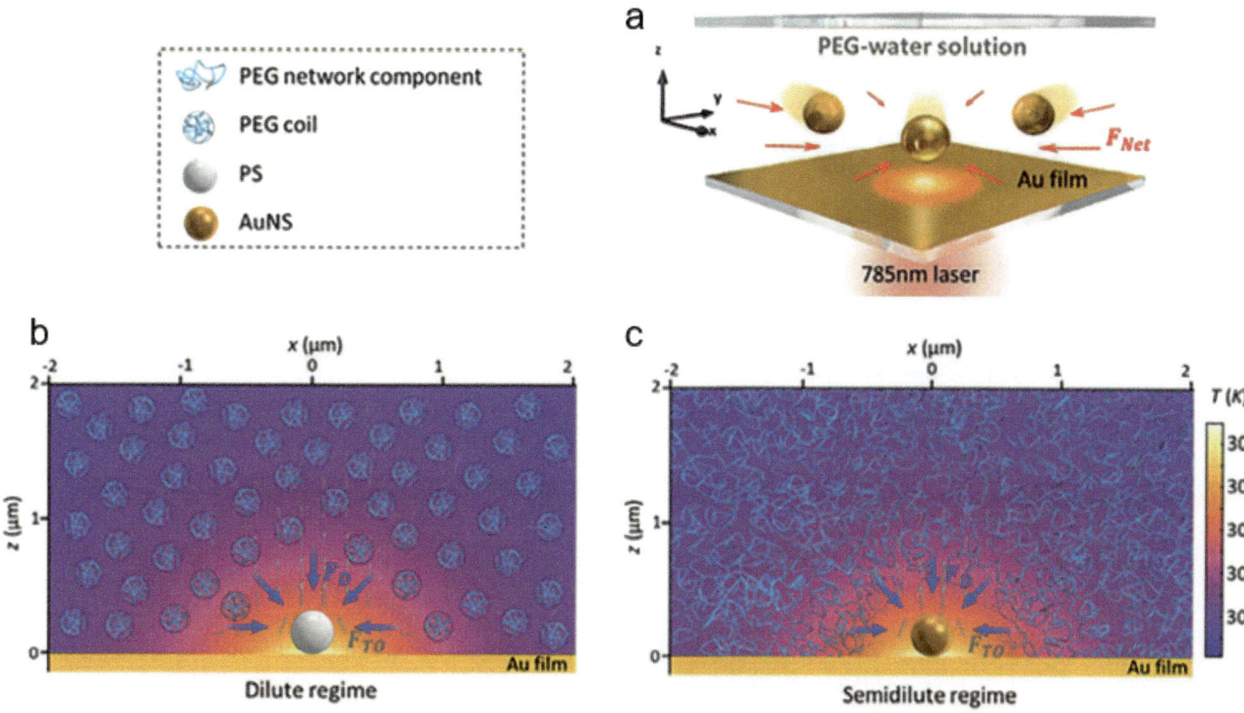

Fig. 2.3 **a** The microfluidic channel consists of two cover glass slides, and the bottom one is coated with 10 nm film and a 5 nm chromium adhesion layer. The height of the chamber is 30 μm. The F_{Net} is the net force for nanoparticle manipulation. **b** Polystyrene sphere (PS) trapping in dilute PEG solution. **c** Gold nanosphere (AuNS) trapping in semidilute PEG solution. The gray and blue arrows indicate the thermo-osmotic force (FTO) and diffusiophoretic force (F_D) respectively. The incident 785 nm laser spot diameter is 1 μm, and the incident optical power is 0.3 mW [24] (Chen et al. [24], Copyright, with permission from John Wiley and Sons)

2.3.2 The Manipulation of Proteins

Proteins are biological macromolecules composed of multiple amino acids, essential in living organisms on Earth. They play crucial roles in various biological processes. Studying the manipulation of single or multiple proteins is significant for biomedical research, such as antigen–antibody interactions, enzyme functions, and protein structure analysis. It is important to consider the impact of optical tweezers on protein structure during manipulation.

Proteins generally have four levels of structure. The primary structure is the linear sequence of amino acids in the polypeptide chain. The secondary structure is formed by hydrogen bonds between carbonyl and amino groups of different amino acids, mainly in the form of α-helices and β-sheets. The tertiary structure is the three-dimensional arrangement of secondary structures stabilized by hydrogen bonds and other interactions, determining a protein's basic function. The quaternary structure is a complex of multiple protein subunits with new functions. Thus, choosing an appropriate manipulation method that minimally affects protein structures is crucial.

In 2011, researchers from the University of Victoria used a dual nanopore structure to create a self-induced back-action (SIBA) optical trap, achieving enhanced local optical field trapping points [25]. This device stably captured 20 nm PS particles for over 5 min. Compared to 100 nm PS particles, the dual nanopore provided greater optical force for smaller particles at the same beam power. In 2012, the group used this device to capture bovine serum albumin (BSA) with a hydrodynamic radius of 3.4 nm, observing that BSA unfolded under optical forces [26]. This experiment enabled single-molecule protein studies.

In 2012, researchers from the Cornell University developed a novel photonic crystal nano-tweezer, using a silicon nitride photonic crystal (PhC) resonator to create an enhanced optical trap, achieving trapping and release of Wilson disease protein [27]. The resonator's temperature rise was less than 0.3 K, effectively avoiding optothermal effects on bioactivity. In 2013, researchers from the Hokkaido University used a resonance optical trapping (ROT) method with 1064 nm laser in an aqueous solution to effectively trapping myoglobin (Mb) [28].

However, current optical tweezers can only trap and fix proteins without precise manipulation. Precise manipulation of proteins remains challenging. Researchers have combined temperature fields with traditional optical forces, making temperature field optical tweezers a promising solution for

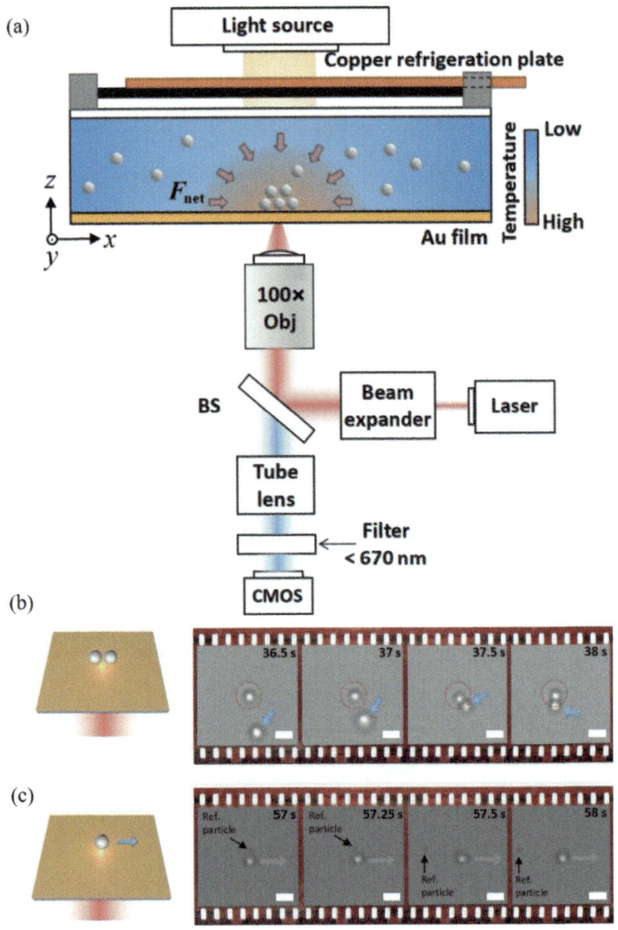

Fig. 2.4 **a** The device is composed of the optical heating generating part and the refrigerant chamber above the microfluidic cell The 785 nm laser is expanded and reflected by short wavelength pass BS before focusing at the Au flim. And a bright-field microscopy and CMOS device are used for the observation. **b** Image frames of two 500 nm-PS particles being trapped to the center of the laser spot. The laser spot is indicated by red dotted circle, and blue arrow shows the position of particles moving into the trap, the laser power is 0.1 mW. The corresponding movie is shown in Movie ESM1. **c** The translation manipulation of single 500 nm-PS particle near a reference (Ref) particle, the laser power is 0.167 mW. The scale bar is 1 μm. The room temperature is 25 °C [9] (Zhou et al. [9], Copyright, with permission from Springer Nature)

precise protein manipulation. As shown in Fig. 2.5, in 2019, researchers from the Leipzig University placed a chromium ring between two glass plates spaced 1 μm apart. They used an 808 nm focused laser to rotate and heat the chromium ring, creating a temperature field in the surrounding liquid film. This method captured amyloid fibrils using thermophoresis and thermal permeation flow [29]. It produced weak photobleaching, enabling long-term observation of amyloid fibrils (at least 10 min), providing a foundational tool for studying protein misfolding-related diseases.

Additionally, the use of an electric field can further enhance the manipulation and capture capabilities of temperature fields. In 2020, researchers from the Vanderbilt University developed Opto-thermo-electrohydrodynamic tweezers (OTET) capable of trapping, manipulating, and sorting particles several micrometers away from the laser focus, achieving the capture and manipulation of BSA particles smaller than 10 nm [23]. They diluted BSA to a concentration of 15 fM and focused a 932 nm laser on a gold nanohole array. By combining the laser with an AC electric field, they successfully trapped and manipulated BSA at approximately 8.6 μm from the edge of the gold nanohole array. Additionally, adjusting the frequency of the AC electric field changed the capture position relative to the edge of the gold nanohole array. As shown in Fig. 2.5, within a certain frequency range, the capture position moved closer to the edge of the gold nanohole array as the AC electric field frequency increased.

Therefore, as illustrated in Figs. 2.5 and 2.6, both methods successfully avoid trapping proteins in high-temperature regions, thereby minimizing the risk of thermal damage. However, at present, utilizing only temperature fields for the accurate and controlled manipulation of single proteins remains a challenge, requiring further advancements in technique and design.

2.3.3 The Manipulation of Nucleic Acid Molecules

DNA is a nucleic acid molecule with three levels of structure: primary (composed of various deoxynucleotides), secondary (a double helix formed by two DNA strands), and tertiary (further co-axial twisting of the secondary structure to form additional helices). Similarly, RNA has three structural levels: primary (composed of various nucleotides), secondary (local double helices), and tertiary (further folding of the secondary structure). Manipulating and trapping nucleic acid molecules is crucial for understanding their internal forces, structures, and interactions.

As shown in Fig. 2.7, in 2002, researchers from the Rockefeller University trapped plasmid-sized DNA through the interaction of convection and thermophoresis [30]. Using a 10 mW infrared laser to heat a glass chamber containing an ionic solution, they generated thermophoresis and circular convection that concentrated DNA at the lower central region of the chamber wall, enhancing its concentration by 13-fold.

In 2007, Researchers from the Tokyo University utilized thermophoresis to stretch single DNA molecules, generating an internal tension of approximately 0.1 pN on individual DNA molecules [31]. As illustrated in Fig. 2.8, a tethered DNA molecule was stretched along the temperature gradient, while a DNA molecule tethered at both ends formed an arc shape until the tension equilibrated with the induced force.

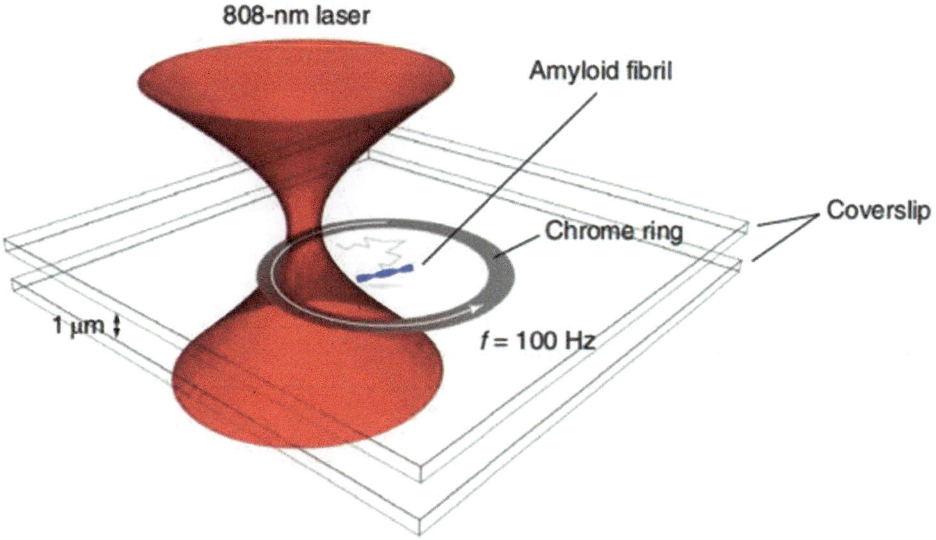

Fig. 2.5 Illustration of the trapping mechanism and the sample geometry. A chrome ring (10-μm inner diameter, 12-μm outer diameter, 50-nm height) is heated by a focused 808-nm laser rotating along the circumference of the ring at f = 100 Hz [29] (Fränzl et al. [29], Copyright, with permission from Springer Nature)

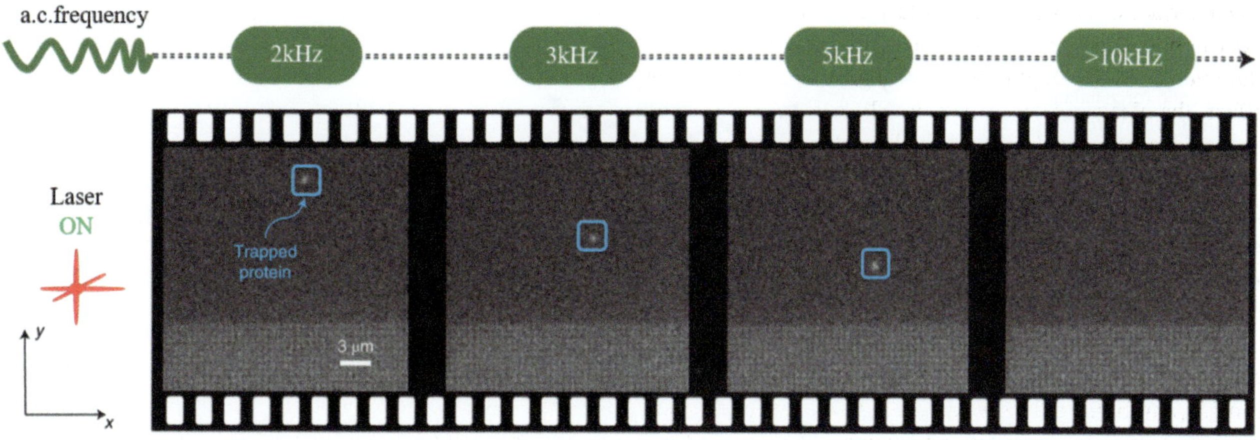

Fig. 2.6 Frame-by-frame demonstration of the trapped molecule's position under different a.c. frequencies. Under 2 kHz a.c. frequency, the single molecule is trapped farther away from the pattern's edge. When the a.c. frequency is tuned higher, from 2 to 5 kHz, the trapping position is shifted inward, closer to the pattern's edge. When the frequency increases to 10 kHz, the trapping fails [23] (Hong et al. [23], Copyright, with permission from Springer Nature)

In 2012, researchers from the Rockefeller University used thermophoresis and a PEG concentration gradient to manipulate and separate T4 DNA and RNA [32]. When the depletion force of PEG overcame thermophoresis, solutes were trapped in the heated area. In 2013, Maeda demonstrated the trapping of DNA in two-dimensional space using diffusiophoresis generated by optothermal effects [33]. The potential well force in diffusiophoresis is produced by interfacial osmotic pressure gradients, showing low material dependency. This method can trap DNA larger than 100 kbp and accumulate shorter DNA molecules, such as 4.3 kbp. Therefore, photo-induced temperature field optical tweezers expand the possibilities for light-driven macromolecular assembly. Studying the thermal motion, enrichment, and assembly of DNA and RNA is also significant for exploring the origins of life on Earth.

As shown in Fig. 2.9, in 2015, researchers from the University of Munich simulated the accumulation of nucleic acid molecules near evolving porous rocks [34]. By heating to generate convection and upward flow, asymmetric flow was

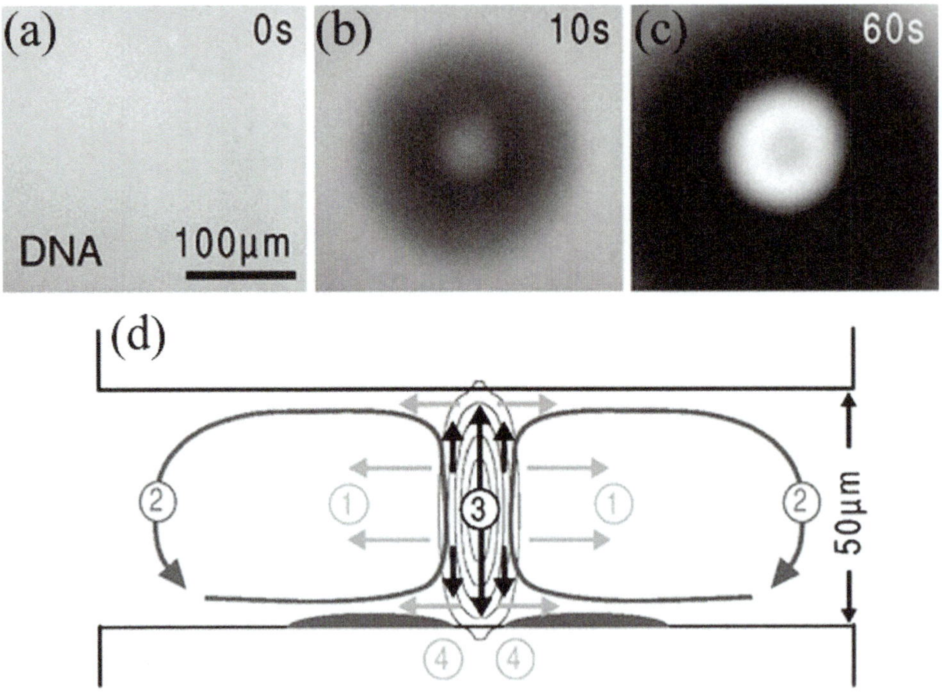

Fig. 2.7 Mechanism of thermophoretic trapping of DNA. **a** Image of stained DNA before heating; **b** Thermophoresis from central heating first repels DNA laterally; **c** Thermophoresis and toroidal convection trap DNA in the center toward the lower chamber wall in a ring geometry. The concentration enhancement is 13-fold; **d** The mechanism of thermophoretic trapping in the center is an interplay of lateral thermophoresis ①, ④ and axial thermophoresis ③ with convection ② [30] (Braun et al. [30], Copyright, with permission from American Physical Society)

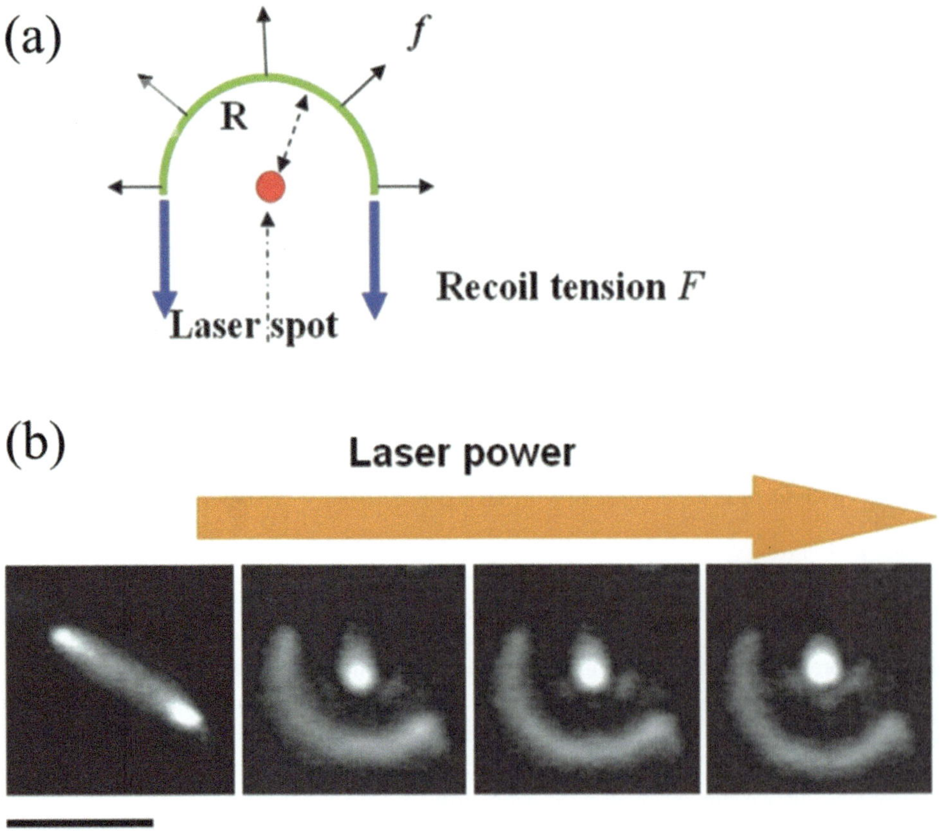

Fig. 2.8 **a** Stretching of two end tethered DNA by temperature gradient and force balance scheme; **b** DNA stretching for different laser powers. From left to right, no heating, 17.5, 19.5, and 21.5 mW. The scalar bar is 5 μm [31] (Jiang et al. [31], Copyright, with permission from AIP Publishing)

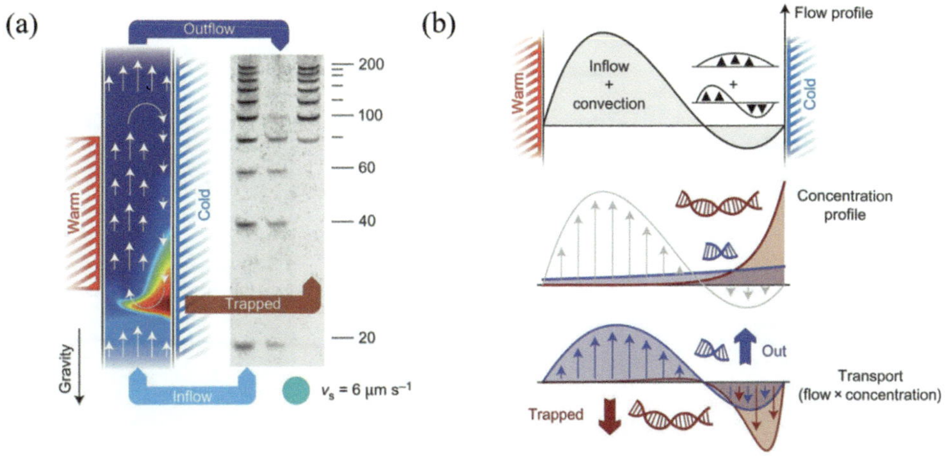

Fig. 2.9 Heat-driven fifilter selecting for strand length [34] (Kreysing et al. [34], Copyright, with permission from Springer Nature)

formed within the potential well, achieving the screening effect of trapping long DNA strands. The study suggests that in this ubiquitous non-equilibrium environment on Earth, the natural thermal enrichment of rare long-chain nucleic acids may have promoted the beginning of molecular evolution.

In 2015, Researchers from the Kyushu University focused infrared laser light into a chamber to create a PEG concentration gradient. Using a developed microfluidic device to control diffusiophoretic flow, they achieved sustained manipulation of DNA molecules and eukaryotic amoeba Dictyostelium discoideum [35]. The lower PEG concentration gradient guided the spontaneous movement of cells towards the hotspot. Since diffusiophoretic flow based on temperature and concentration gradients is less affected by material properties, this low material dependency is significant for manipulating soft biological materials. By 2020, researchers from the Kyushu University further summarized the manipulation of soft biological materials [36]. The relationship between intracellular phenomena (such as ATP hydrolysis coupled with the sodium–potassium pump for intracellular and extracellular material exchange) and physical transport remains unclear, making understanding biological transport mechanisms and their physical principles key to this research.

The examples above illustrate the combined effects of temperature fields and solute concentration gradients. Temperature gradients induce thermophoresis of different solutes, while large-scale low-concentration particles, such as nucleic acid molecules, are driven by a combination of thermophoresis and diffusiophoresis induced by the PEG solute concentration gradient. These techniques primarily exploit the synergistic effect of optothermal interactions to enrich nucleic acid molecules at a collective level. However, despite their effectiveness in particle manipulation at a larger scale, comprehensive studies and demonstrations of single-molecule manipulation and trapping of nucleic acids remain challenging.

2.3.4 Manipulation of Other Biological Particles

The manipulation and sorting of nanobioparticles such as bacteria, cells, vesicles, and viruses are crucial applications of optical tweezers. Typically, many virus studies focus on average characteristics, but optical tweezers enable the study of individual viruses. Current virus detection methods like PCR, ELISA, colloidal gold immunoassay, and chemiluminescent immunoassay do not allow for single-virus detection during the window period and can yield false negatives. Optical tweezers, however, can manipulate individual viruses, relocating them near specific cells or tissues, and combined with microscopy, they can observe the full invasion process and cell death caused by viruses.

In 1987, Ashkin demonstrated the optical trapping and manipulation of bacteria using 120-mW single-beam argon laser optical tweezers without significant thermal damage. Current particle sorting methods using microfluidic, acoustic forces, and optical forces focus on differences in mass, size, refractive index, or fluorescence staining. However, sorting label-free submicron particles with similar volume and refractive index but different shapes remain a major challenge.

Pathogenic bacteria in the bloodstream can produce harmful metabolites leading to infections, such as tuberculosis, scarlet fever, and bacterial dysentery, which severely affect human health. Conversely, beneficial bacteria like lactic acid bacteria and Bacillus subtilis improve the environment and daily life. Thus, single-bacterium detection and observation can enhance our understanding of bacterial biology and behavior, improving human living conditions.

As shown in Fig. 2.10a–f, in 2011, researchers from the Sun Yat Sen University developed a device for trapping and moving microparticles in solution based on subwavelength-diameter-fiber (SDF) induced negative photophoresis [37]. The group also trapped *E. coli* using 125 mW laser power

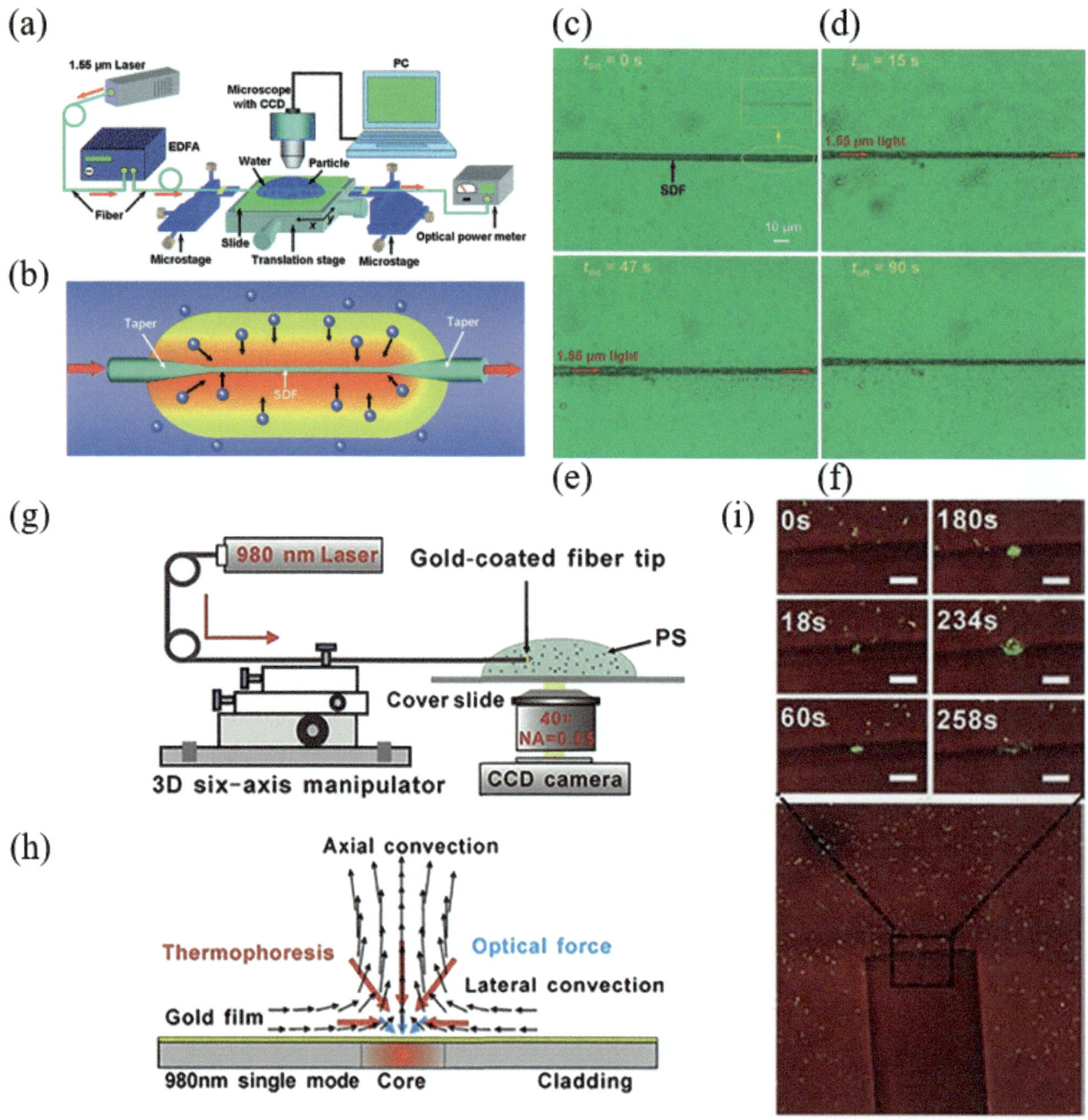

Fig. 2.10 **a** Experimental setup. The image was captured by the CCD, which shows the SDF immerged in the particle suspensions; **b** Schematic shows the propagating light along the taper–SDF–taper section and the negative photophoretic motions of particles radiated by the leaking light from the SDF; **c** For $t_{on} = 0$ s (without incident optical power); **d** For $t_{on} = 15$ s (with an incident optical power of 125 mW); **e** For $t_{on} = 47$ s (with an incident optical power of 125 mW); **f** Microscope image view after the laser off for $t_{off} = 90$ s [37] (Lei et al. [37], Copyright, with permission from Royal Society of Chemistry); **g** Experimental setup of optical fiber with precipitated gold film; **h** Force components in the trap region; **i** Side-view time-sequence images of trapping of *E. coli* cells. The 0.65 mW laser was switched on at 0 s and switched off immediately before capturing of the last image. The excitation laser was 488 nm, and the laser sweeping speed was 3 s per frame. Scale bar is 15 μm [20] (Chen et al. [20], Copyright, with permission from Optical Society of America)

and 1.2 μm diameter SDF, as shown in Fig. 2.10e–h. Considering human capillaries' diameters are 6–8 μm, this technology holds potential for vascular disease applications. The same year, they employed thermophoretic effects around tapered fibers to trap and transport silica dielectric particles in water [38]. Additionally, as shown in Fig. 2.10g–i, in 2015, researchers from the Chinese University of Hong Kong developed a method leveraging a 10 nm gold film deposited on the

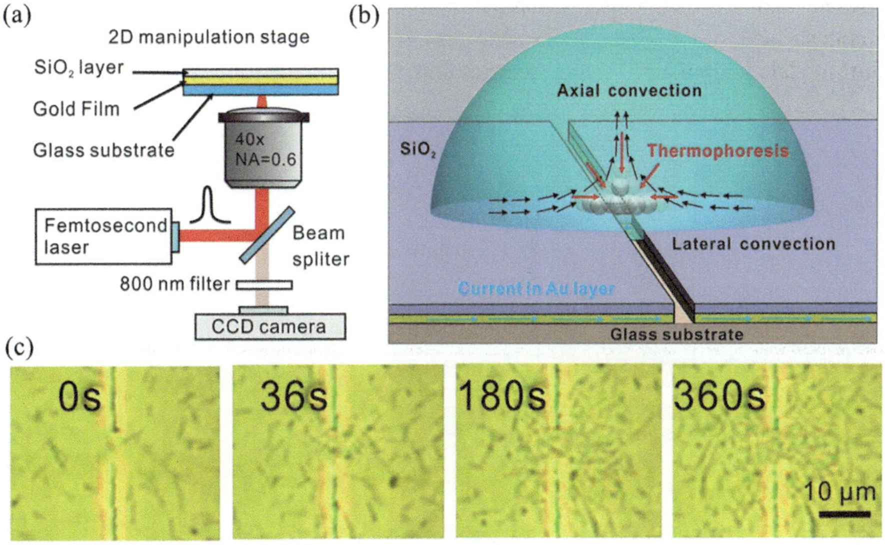

Fig. 2.11 **a** Setup of using a femtosecond laser to perform METH fabrication; **b** Force components in the trapping scheme; **c** Trapping of *E. coli* cells at 24 mA [39] (Chen et al. [39], Copyright, with permission from Springer Nature)

end of a single-mode fiber [20]. Utilizing natural convection and thermophoretic forces at the 980 nm single-mode fiber end, they achieved the trapping and aggregation of PS microspheres and *E. coli*. Notably, the minimum light power density reached 19 $\mu W/\mu m^2$, which is three orders of magnitude lower than traditional optical tweezers.

As shown in Fig. 2.11, in 2016, researchers from the Chinese University of Hong Kong introduced a photon-free trapping technique based on electrothermal forces [39]. Utilizing micro-scale electric thermal heater (METH), they achieved the trapping of dielectric PS particles and live *E. coli* through thermal convection and thermophoresis, with a threshold power density range of 10–100 $\mu W/\mu m^2$. Biological activity tests demonstrated that METH tweezers could capture live cells without causing damage or functional changes.

Additionally, in 2016, researchers from the Osaka Prefecture University developed a Photothermal assembling (PTA) method for bacterial counting [40]. They successfully trapped and counted PS particles and Pseudomonas aeruginosa. Localized thermal gradients induced by a laser create convective flows and bubbles in the liquid, aggregating floating objects into the bubbles for trapping. Compared to the cultivation method, this method offers speed, high precision, low cost, and suitability for bacteria of various shapes. It holds significant potential for healthcare and laboratory analysis.

In traditional optical tweezers systems, Rayleigh and Mie particles experience minimal shape-induced optical force differences, making shape-based bacterial sorting less effective. Pathogenic microorganisms often have distinct shapes, like cocci and bacilli. In 2019, researchers from the Nanyang Technological University developed an optofluidic nanophotonic sawtooth array (ONSA) based on nanophotonic Array [41]. This device generates a sawtooth-shaped optical field through optical coupling, interacting with submicron particles at the optical coupling hotspots. It applies different optical torques based on particle shape, selectively sorting bacteria of similar size and refractive index but different shapes (e.g., Staphylococcus aureus and *Escherichia coli*). As shown in Fig. 2.12, in 2019, researchers from the Chinese University of Hong Kong developed a method for 3D manipulation and lysis of single cells using gold-coated PDMS microchannels combined with laser-induced radial flow and optical scattering forces [42]. They integrated recombinase polymerase amplification (RPA) for nucleic acid amplification. This method enables single-cell detection within 30 min in a volume of 0.38 nL, providing a rapid approach for screening cancer cells and validating their drug resistance.

Due to the varying light absorption and refractive indices in different biological tissues, traditional optical tweezers face limitations in biomedical applications, including their use on vesicles. In 1989, Johnstone described the mechanism and function of exosome formation while studying the development of sheep reticulocytes, coining the term exosomes. Extracellular vesicles (EVs) released by cells, which include exosomes (40–100 nm), microvesicles (100–500 nm), and apoptotic bodies (400–3000 nm), contain various bioactive molecules from parental cells, such as genetic material and proteins. Detecting and sorting these vesicles can help diagnose and prevent various diseases due to their minimal invasiveness and repeatable sampling. However, due to the small size and complex composition of bodily fluids, isolating and enriching vesicles in samples like serum typically requires time-consuming, expensive, and complex ultracentrifugation, limiting their clinical application for disease diagnosis.

In 2019, researchers from the National Center for Nanoscience and Technology developed a thermophoretic

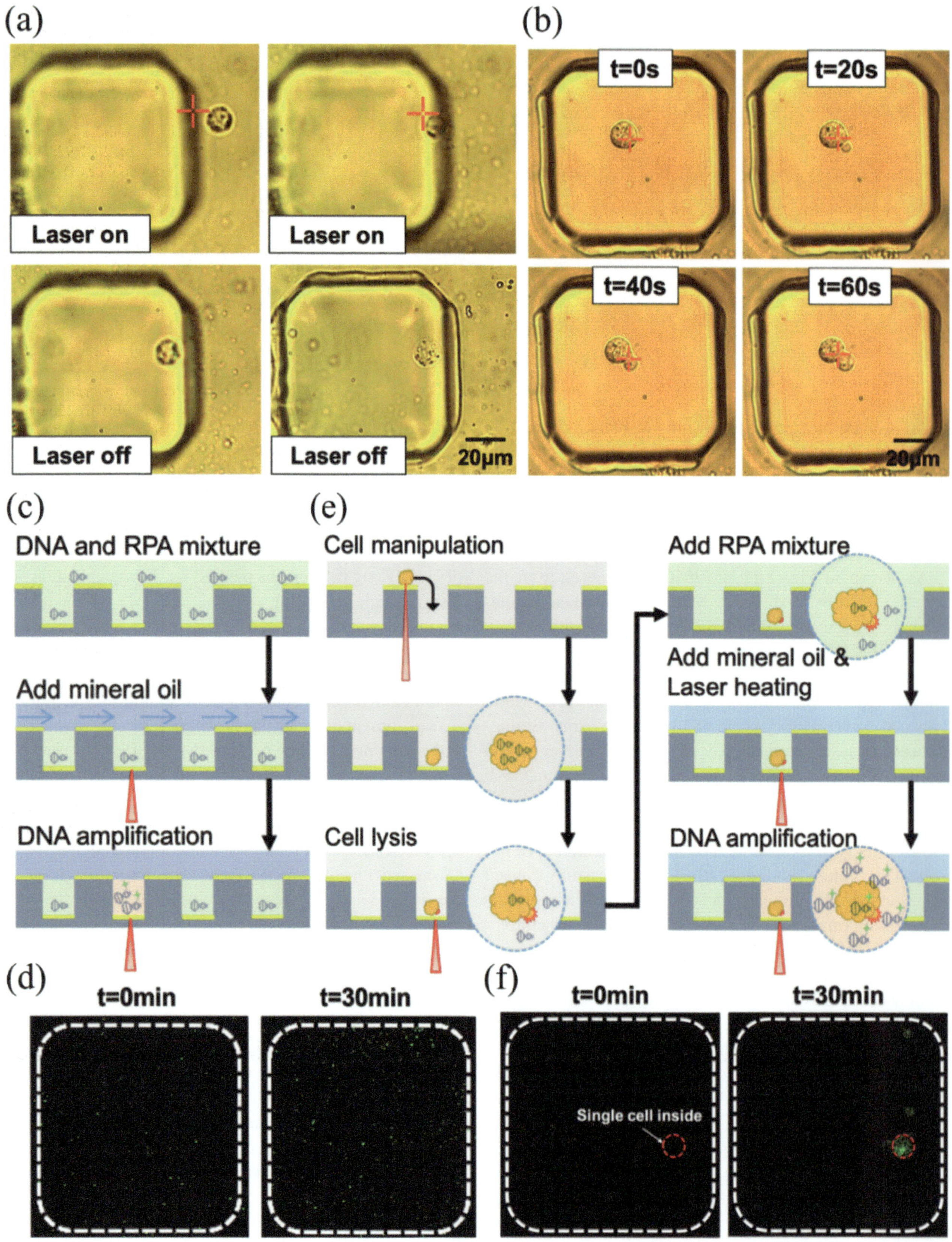
(a)
Laser on
Laser on
Laser off
Laser off
20μm
(b)
t=0s
t=20s
t=40s
t=60s
20μm
(c)
DNA and RPA mixture
Add mineral oil
DNA amplification
(e)
Cell manipulation
Cell lysis
Add RPA mixture
Add mineral oil &
Laser heating
DNA amplification
(d)
t=0min
t=30min
(f)
t=0min
t=30min
Single cell inside

◂Fig. 2.12 **a** Optical guiding of a free-flowing living single cancer cell into a micro-well; **b** Time-lapse images showing cell lysis during the first 60 s of laser illumination directed at the cell membrane; **c** Schematic diagram showing the procedures of DNA amplification in a particular micro-well. **d** Fluorescence image of the 0 and 30-min RPA reaction. **e** Schematic diagram showing the procedures for DNA amplification of a targeted cancer cell. **f** Fluorescence image showing amplifified DNA of a single cell inside the micro-well at 0 and 30-min RPA reaction [42] (Kong et al. [42], Copyright, with permission from Elsevier)

aptasensor (TAS) [43]. This sensor uses the interplay of thermophoresis, diffusion, and convection generated by local heating to enrich seven types of exosomes from blood, applicable for cancer detection and classification. TAS enriches EVs over 1,000-fold without prior serum sample pre-separation and purification, offering advantages like short processing time, high precision, and low cost. It detects seven EV protein markers in a serum sample in less than three hours, with sensitivity, specificity, and accuracy for stage I cancer detection at 95, 100, and 97%, respectively, comparable to in vivo tests.

In 2020, researchers from the Chinese University of Hong Kong developed a nanophotonic chip based on the previously mentioned ONSA [44]. This chip enables sorting of microspheres ranging from 100 to 500 nm by adjusting laser power and flow rates, demonstrating significant potential for sorting bacteria, viruses, and vesicles. However, capturing single viruses remains challenging due to their small size (typically 20–200 nm). In 2021, the team coated 500 nm PS particles with streptavidin, bound biotinylated adenovirus antibodies to streptavidin, and introduced these particles into a solution to trap and detect adenoviruses [45].

Additionally, as shown in Fig. 2.13, in 2022, researchers from the National Center for Nanoscience and Technology introduced a method named Tango for detecting EVs, leveraging the combined effects of temperature gradients and PEG concentration gradients induced by local heating [46]. Using infrared laser irradiation on microchambers containing serum samples, recognition probes, and PEG-8000, this method enriches EVs and facilitates their binding to recognition probes. Fluorescence signal detection enables the direct identification of prostate cancer EV markers (PSNAH and EpCAM) within 15 min, achieving precise diagnosis of prostate cancer from serum samples. Tango's application in EV detection promises to be a rapid, precise, and non-invasive cancer diagnostic method.

Additionally, one can use the particle size-selective capture capability of optothermal tweezers to sort and detect exosomes and microvesicles within extracellular vesicles (EVs). However, given the vast variety and extremely small size of vesicles, research on the capture and manipulation of vesicles using OTT is still in its early stages. Recently, as mentioned earlier, researchers from the Vanderbilt University developed the opto-thermo-electrohydrodynamic tweezers (OTET) technology, which achieved the trapping and manipulation of sub-10 nm BSA [23]. This technology also shows promise for the trapping and manipulation of individual vesicles.

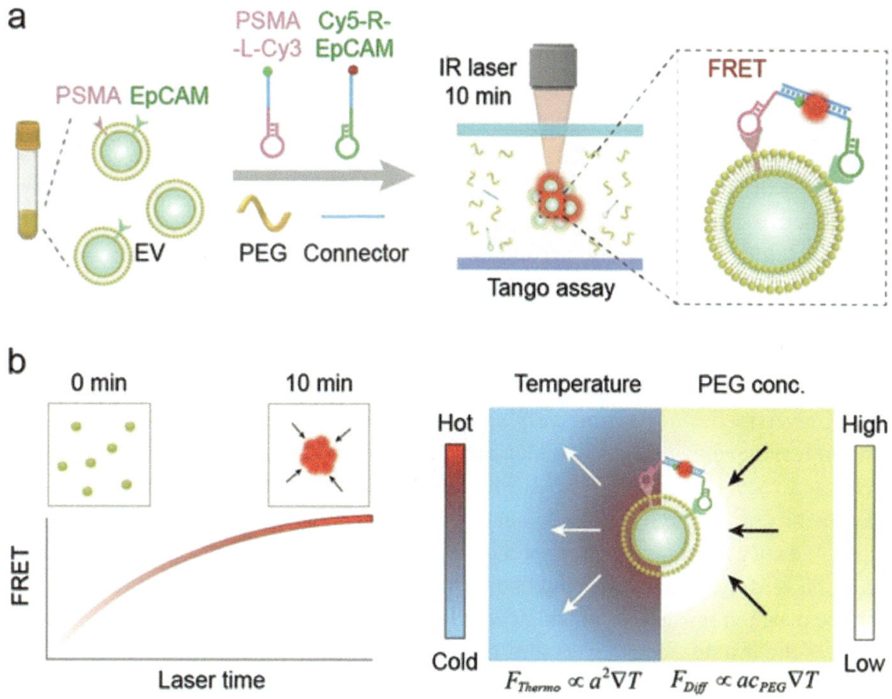

Fig. 2.13 Schematic of the one-step thermophoretic AND gate operation (Tango) assay. **a** Tango assay combining AND gate operation on EV membranes and PEG-enhanced thermophoretic accumulation of EVs. Two aptamer-based probes (PSMAL-Cy3 and Cy5-R-EpCAM) targeting PSMA and EpCAM and one connector strand (Connector) were designed; **b** Tango assay can directly detect tumor-derived EVs overexpression both PSMA and EpCAM within 15 min. A temperature gradient and a PEG concentration gradient were induced by local laser heating [46] (Deng et al. [46], Copyright, with permission from John Wiley and Sons)

2.3.5 Enhancing Biosensing Sensitivity

In many biosensing detection cases, the affinity interactions or biochemical reactions between biomolecules rely solely on the Brownian motion of sample particles in solution, leading to long detection times and low efficiency. This issue is exacerbated when the sample concentration is low or during single-molecule detection, often resulting in detection errors. To address this problem, Optothermal tweezers (OTT) can be used to enrich the detection samples, thereby significantly increasing detection sensitivity and reducing detection time. The driving mechanism of OTT for biomolecules primarily involves fluid perturbation induced by the thermal field.

In 2011, researchers from the University of Tennessee utilized AC electroosmosis (ACEO) and AC electrothermal (ACET) flows in an AC electrokinetics (ACEK) system to generate vortices for sample enrichment [47]. This allowed for rapid antigen–antibody binding detection, increasing speed by tenfold compared to traditional methods, with detection times around 3 min. The ACEO and ACET flows, driven by non-uniform electric heating, enabled a single-pump ACEK system. Their developed fluorescence immunoassay demonstrated higher sensitivity and faster detection than traditional methods, enhancing biosensing capabilities.

In 2018, researchers from the Barcelona Institute of Science and Technology combined localized surface plasmon resonance (LSPR) sensing with electrothermoplasmonic (ETP) flow effects, developing ETP-LSPR for IgG detection [48]. The ETP-assisted biosensor reduced IgG binding detection time from 50 to 17 min, improving speed by approximately 2.9 times. In 2020, researchers from the Osaka Prefecture University used the Marangoni convection effect for bacterial trapping and enrichment [49]. That same year, researchers from the University of Texas at Austin used the optothermal effect to induce Marangoni convection by generating bubbles at the liquid-substrate interface, accumulating particles near the bubble–liquid-substrate interface [50]. To avoid thermal damage to proteins, they used high-boiling point and low gas-solubility perfluoropentane for bubble generation. This method reduced single antigen–antibody detection time from 30 to 1 min, greatly enhancing sensing efficiency.

As shown in Fig. 2.14, in 2022, researchers from the Shenzhen University developed time-varying optothermal manipulation technique, significantly enhancing biosensing sensitivity and reducing sensing time at low molecular concentrations or during single-molecule analysis [51]. This technique, known as the optothermophoretic flipping (OTF), is an active biosensing enhancement method for wavelength-based surface plasmon resonance imaging (WSPRi). The principle involves the instantaneous reversal of thermophoretic forces near a gold film induced by laser switching, effectively driving the target proteins to bind with the gold film receptors, thereby enhancing biosensing. This method increased the sensitivity of traditional surface plasmon resonance imaging (SPRi) sensors by 23.6 times.

Furthermore, as shown in Fig. 2.15, in 2023, researchers from Shenzhen University developed a novel optothermal tweezer technology that integrates CRISPR biosensing technology [52], significantly enhancing the sensitivity of CRISPR for detecting target DNA. This innovation work, known as CRISPR-powered optothermal nanotweezers (CRONT), combines the precision of optothermal tweezers with the specificity of CRISPR-based DNA detection.

The process involves simultaneously introducing DNA-linked gold nanoparticles, CRISPR-associated proteins, and target DNA into a microfluidic channel. When all three components are trapped by the optothermal tweezers, the CRISPR-associated proteins are activated by the target DNA, leading to the cleavage of the DNA linked to the gold nanoparticles. By monitoring the separation of the gold nanoparticles at varying target DNA concentrations, quantitative detection of low target DNA concentrations can be achieved. This system allows for the detection of target DNA at much lower detection limits than traditional methods, making it a powerful tool for highly sensitive DNA analysis.

2.4 Summary and Outlook

Traditional optical tweezers (OTs) are extensively utilized in biomedicine, chemistry, and nanotechnology due to their remote, non-contact, and high-precision manipulation capabilities. However, their application in biological systems faces several challenges, including optical diffraction limits, high optical power density requirements, phototoxicity, and optothermal effects. Over the past decade, the advent of novel optothermal tweezers (OTT) has addressed some of these limitations through methods such as opto-thermoelectric nanotweezers and opto-refrigerative tweezers. These advancements retain the core benefits of traditional OTs while overcoming key drawbacks, positioning OTT as a vital tool in biomedical engineering and nanotechnology research.

This chapter first revisits the theoretical foundation of traditional optical tweezers and introduces OTT, focusing on the construction of temperature fields, thermodynamic principles, and particle manipulation techniques. Applications of OTT are explored, including the manipulation of nanoparticles, proteins, nucleic acids, viruses, bacteria, and other biological particles. The discussion highlights challenges and opportunities, emphasizing OTT's role in enhancing biosensing sensitivity through efficient conversion of light energy into mechanical energy. With power densities three orders of magnitude lower than those of traditional OTs, OTT can operate within the optical diffraction limit, enabling

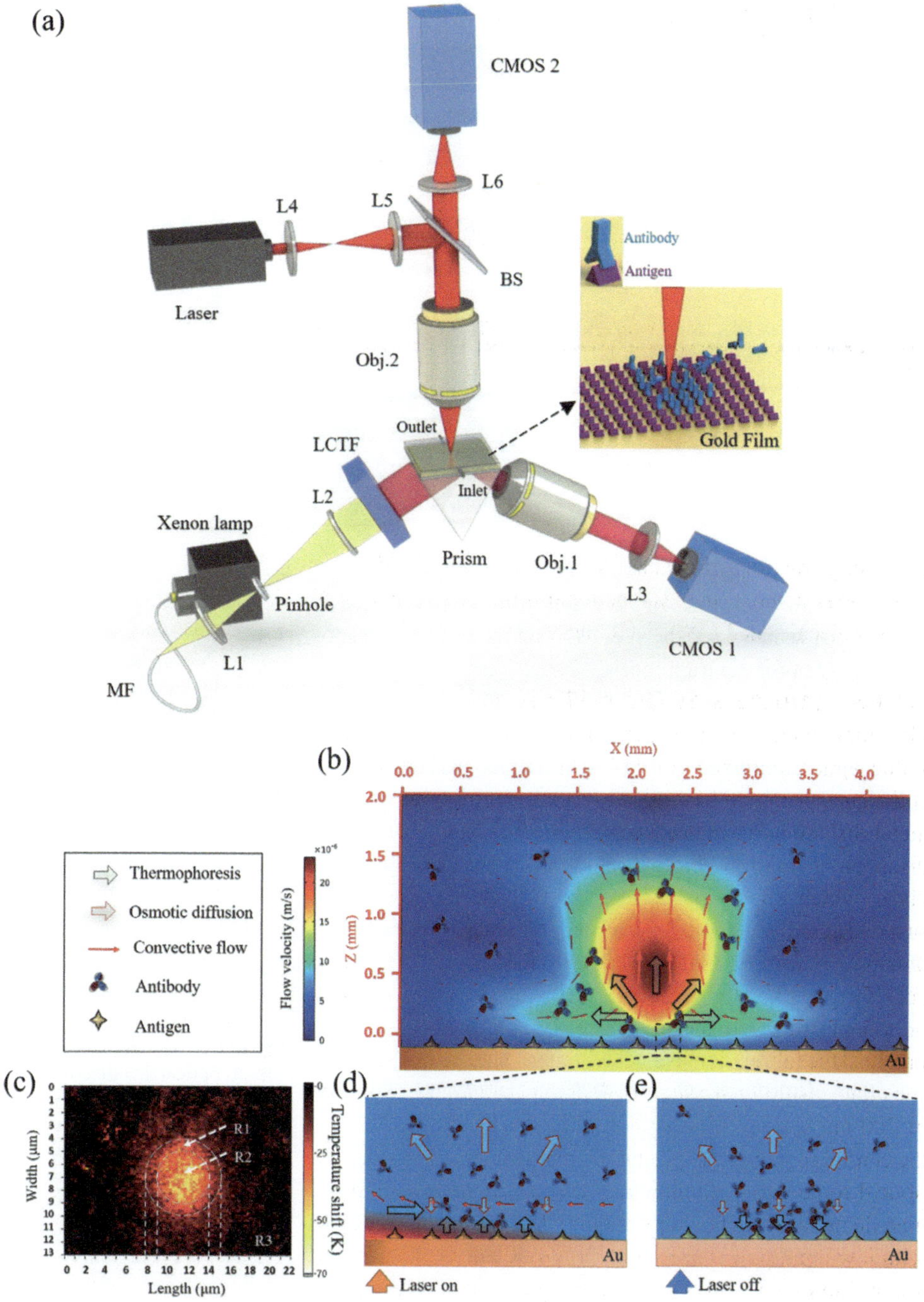

Fig. 2.14 **a** Optical setup. The WSPRi part and optothermal excitation part. Lens groups of L1, L2 and L4, L5: Beam expanding and collimating; L3, L6: Tube lens; Obj.1–2: Objective lenses; MF: Multimode optical fiber; BS: Beam splitter. Schematic illustration and theoretical analysis of OTF system; **b** Schematic side view of the entire system when the heating laser is switching on. The convective flow, which contains natural convective flow and thermo-osmotic flow, together with the thermophoresis, relocated and enriched the biomolecules from bulk solution to a ring shape region where the temperature is in relatively lower level; **c** Top view of the ring shape biomolecule aggregation region in the background of near-field temperature map. R1: ring-shape biomolecule enrichment and binding region; R2: heating laser focusing center region; R3: background region with no laser heating; **d** Zoom-in schematic picture of the biomolecule enrichment region when laser is switched on, the upward thermophoretic force induces inefficient binding of the biomolecules; **e** Zoom-in schematic picture of the biomolecule enrichment region when laser is just switched off, thermophoretic force induces efficient binding of the biomolecules [51] (Chen et al. [51], Copyright, with permission from Elsevier)

the trapping of opaque or similarly refractive particles with minimal laser power.

The development of OTT presents both significant opportunities and persistent challenges. Between 2002 and 2022, advancements in OTT have ranged from the use of optothermal tweezers to opto-thermoelectric nanotweezers, leveraging substrates from gold films to two-dimensional materials and constructing diverse spatial temperature fields. These innovations have enabled the manipulation of particles as small as sub-10 nm, such as bovine serum albumin (BSA) protein molecules. However, critical challenges remain, particularly in controlling biologically active particles like proteins, miRNA, exosomes, and other organelles. Major issues include:

1. **Temperature gradients** that may compromise the biological activity of manipulated targets.
2. **Dependence on surfactants** for particle modification, which may introduce chemical toxicity or alter protein structures, impacting their biophysical properties.

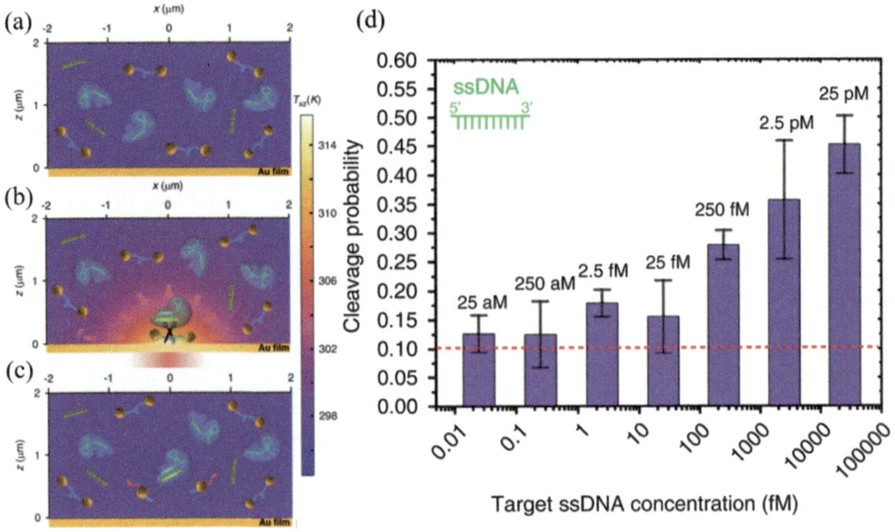

Fig. 2.15 **a** Dispersion of the three components in the solution without optical heating. **b** Optothermal net force (F_{Net}) induced migration and DNA@AuNS conjugate cleavage upon optical heating, the heating laser power is 0.5 mW. **c** Observation of the cleavage after the optical heating is switched off. **d** Cleavage probability of the DNA@AuNS at different target ssDNA (MP) concentrations [52] (Chen et al. [52], Copyright, with permission from Elsevier)

3. The current reliance on **2D potential wells** in OTT systems, which lack the precision and stability of traditional OTs for 3D trapping.

To advance OTT in biophotonics, these challenges must be urgently addressed. Potential solutions include developing biocompatible surfactants or harnessing environmental factors to control thermophoretic behavior, paving the way for more versatile and biologically compatible OTT systems.

Acknowledgements This work was supported by the National Key Research and Development Program of China (No.2022YFA1206300); National Natural Science Foundation of China (62275164, 62275168, 61775148, 61905145, 62204253, 22327802); Foundation from Guangdong Basic and Applied Basic Foundation (2021A1515011916, 2021QN02Y124, 2023A1515012250, 2022A1515011845); Foundation from Department of Education of Guangdong Province (2023ZDZX2052); Shenzhen Science and Technology R&D and Innovation Foundation (JCYJ20200109105608771); Shenzhen Key Laboratory of Photonics and Biophotonics (ZDSYS20210623092006020); Medical-Engineering Interdisciplinary Research Foundation of Shenzhen University, Medical-Engineering Interdisciplinary Research Foundation of ShenZhen University (2023YG033).

References

1. Ashkin A, Dziedzic JM, Yamane T. Optical trapping and manipulation of single cells using infrared laser beams. Nature. 1987;330:769–71.
2. Grier DG. A revolution in optical manipulation. Nature. 2003;424:810–6.
3. Gargiulo J, et al. Understanding and reducing photothermal forces for the fabrication of Au nanoparticle dimers by optical printing. Nano Lett. 2017;17:5747–55.
4. Garcés-Chávez V, Dholakia K, Spalding G. Extended-area optically induced organization of microparticles on a surface. Appl Phys Lett. 2005;86.
5. Zhao X, Zhao N, Shi Y, Xin H, Li B. Optical fiber tweezers: a versatile tool for optical trapping and manipulation. Micromachines. 2020;11:114.
6. Righini M, Zelenina AS, Girard C, Quidant R. Parallel and selective trapping in a patterned plasmonic landscape. Nat Phys. 2007;3:477–80.
7. Lin L, et al. Opto-thermoelectric nanotweezers. Nat Photon. 2018;12:195–201.
8. Li J, et al. Opto-refrigerative tweezers. Sci Adv. 2021;7:eabh1101.
9. Zhou J, et al. Low-temperature optothermal nanotweezers. Nano Res. 2023;16:7710–5.
10. Würger A. Thermal non-equilibrium transport in colloids. Rep Prog Phys. 2010;73:126601.
11. Landau LD, Lifshitz EM. Fluid mechanics, vol. 6. Elsevier; 1987.
12. Govorov AO, et al. Gold nanoparticle ensembles as heaters and actuators: melting and collective plasmon resonances. Nanoscale Res Lett. 2006;1:84–90.
13. Baffou G, Quidant R, García de Abajo FJ. Nanoscale control of optical heating in complex plasmonic systems. ACS Nano. 2010;4:709–16.
14. Anderson JL. Colloid transport by interfacial forces. Annu Rev Fluid Mech. 1989;21:61–99.
15. Bregulla AP, Würger A, Günther K, Mertig M, Cichos F. Thermo-osmotic flow in thin films. Phys Rev Lett. 2016;116:188303.
16. Braun M, Cichos F. Optically controlled thermophoretic trapping of single nano-objects. ACS Nano. 2013;7:11200–8.
17. Fränzl M, Cichos F. Hydrodynamic manipulation of nano-objects by optically induced thermo-osmotic flows. Nat Commun. 2022;13:656.
18. Hu W, Ishii KS, Fan Q, Ohta AT. Hydrogel microrobots actuated by optically generated vapour bubbles. Lab Chip. 2012;12:3821–6.
19. Shi Y, et al. Light-induced cold marangoni flow for microswarm actuation: from intelligent behaviors to collective drug delivery. Laser Photon Rev. 2022;16:2200533.
20. Chen J, Kang Z, Kong SK, Ho H-P. Plasmonic random nanostructures on fiber tip for trapping live cells and colloidal particles. Opt Lett. 2015;40:3926–9.
21. Kang Z, et al. Trapping and assembling of particles and live cells on large-scale random gold nano-island substrates. Sci Rep. 2015;5:1–8.
22. Kang Z, et al. Trapping and assembling of particles and live cells on large-scale random gold nano-island substrates. Sci Rep. 2015;5:9978.

23. Hong C, Yang S, Ndukaife JC. Stand-off trapping and manipulation of sub-10 nm objects and biomolecules using opto-thermo-electrohydrodynamic tweezers. Nat Nanotechnol. 2020;15:908–13.
24. Chen J, et al. Highly-adaptable optothermal nanotweezers for trapping, sorting, and assembling across diverse nanoparticles. Adv Mater. 2024;36:2309143.
25. Pang Y, Gordon R. Optical trapping of 12 nm dielectric spheres using double-nanoholes in a gold film. Nano Lett. 2011;11:3763–7.
26. Pang Y, Gordon R. Optical trapping of a single protein. Nano Lett. 2012;12:402–6.
27. Chen Y-F, Serey X, Sarkar R, Chen P, Erickson D. Controlled photonic manipulation of proteins and other nanomaterials. Nano Lett. 2012;12:1633–7.
28. Shoji T, Kitamura N, Tsuboi Y. Resonant excitation effect on optical trapping of myoglobin: the important role of a heme cofactor. J Phys Chem C. 2013;117:10691–7.
29. Fränzl M, et al. Thermophoretic trap for single amyloid fibril and protein aggregation studies. Nat Methods. 2019;16:611–4.
30. Braun D, Libchaber A. Trapping of DNA by thermophoretic depletion and convection. Phys Rev Lett. 2002;89:188103.
31. Jiang H-R, Sano M. Stretching single molecular DNA by temperature gradient. Appl Phys Lett. 2007;91.
32. Maeda YT, Tlusty T, Libchaber A. Effects of long DNA folding and small RNA stem–loop in thermophoresis. Proc Natl Acad Sci. 2012;109:17972–7.
33. Maeda YT. (2+ 1)-Dimensional manipulation of soft biological materials by opto-thermal diffusiophoresis. Appl Phys Lett. 2013;103.
34. Kreysing M, Keil L, Lanzmich S, Braun D. Heat flux across an open pore enables the continuous replication and selection of oligonucleotides towards increasing length. Nat Chem. 2015;7:203–8.
35. Fukuyama T, Fuke A, Mochizuki M, Kamei K-I, Maeda YT. Directing and boosting of cell migration by the entropic force gradient in polymer solution. Langmuir. 2015;31:12567–72.
36. Fukuyama T, Maeda YT. Opto-thermal diffusiophoresis of soft biological matter: from physical principle to molecular manipulation. Biophys Rev. 2020;12:309–15.
37. Lei H, Zhang Y, Li X, Li B. Photophoretic assembly and migration of dielectric particles and *Escherichia coli* in liquids using a subwavelength diameter optical fiber. Lab Chip. 2011;11:2241–6.
38. Xin H, Li X, Li B. Massive photothermal trapping and migration of particles by a tapered optical fiber. Opt Express. 2011;19:17065–74.
39. Chen J, et al. Thermal gradient induced tweezers for the manipulation of particles and cells. Sci Rep. 2016;6:35814.
40. Yamamoto Y, Shimizu E, Nishimura Y, Iida T, Tokonami S. Development of a rapid bacterial counting method based on photothermal assembling. Opt Mater Exp. 2016;6:1280–5.
41. Shi Y, et al. Nanophotonic array-induced dynamic behavior for label-free shape-selective bacteria sieving. ACS Nano. 2019;13:12070–80.
42. Cong H, et al. Target trapping and in situ single-cell genetic marker detection with a focused optical beam. Biosens Bioelectron. 2019;133:236–42.
43. Liu C, et al. Low-cost thermophoretic profiling of extracellular-vesicle surface proteins for the early detection and classification of cancers. Nat Biomed Eng. 2019;3:183–93.
44. Shi Y, et al. Optical potential-well array for high-selectivity, massive trapping and sorting at nanoscale. Nano Lett. 2020;20:5193–200.
45. Shi Y, et al. Trapping and detection of single viruses in an optofluidic chip. ACS Sens. 2021;6:3445–50.
46. Deng J, et al. One-step thermophoretic AND gate operation on extracellular vesicles improves diagnosis of prostate cancer. Angew Chem Int Ed. 2022;61:e202207037.
47. Liu X, et al. Development of an AC electrokinetics-based immunoassay system for on-site serodiagnosis of infectious diseases. Sens Actuat A. 2011;171:406–13.
48. Garcia-Guirado J, et al. Overcoming diffusion-limited biosensing by electrothermoplasmonics. ACS Photon. 2018;5:3673–9.
49. Tokonami S, et al. Light-induced assembly of living bacteria with honeycomb substrate. Sci Adv. 2020;6:eaaz5757.
50. Kim Y, Ding H, Zheng Y. Enhancing surface capture and sensing of proteins with low-power optothermal bubbles in a biphasic liquid. Nano Lett. 2020;20:7020–7.
51. Chen J, et al. Optothermophoretic flipping method for biomolecule interaction enhancement. Biosens Bioelectron. 2022;204:114084.
52. Chen J, et al. CRISPR-powered optothermal nanotweezers: diverse bio-nanoparticle manipulation and single nucleotide identification. Light Sci Appl. 2023;12:273.

Optical Micromanipulation with Structured Light Beams

3

Xue Yun, Minru He, Yansheng Liang, and Ming Lei

Abstract

Optical tweezers can trap and manipulate particles with size ranging from nanometers to micrometers with tightly focused laser beams, serving as critical tools in many fields, such as biomedicine, physical chemistry, and materials science. The continuous development of optical beam shaping technology has revolutionized the optical manipulation technique, evolving from stable capture of particles with a point trap to versatile manipulation of particles using novel beams with unique phase, amplitude, and polarization distribution. Nowadays, the elaborately structured light fields facilitate the driving of particles along more complex trajectories, enhancing optical forces, and providing more degrees of freedom for manipulating objects. In this chapter, we will focus on the generation of typical novel beams and their applications in particle manipulation, aiming to highlight the significant advancements made in optical tweezers technology in recent years and inspire further research in related fields.

Keywords

Optical tweezers · Optical manipulation · Self-similar beams · Non-diffracting beams · Vortex beams · Bessel beams · Cylindrical vector beams

X. Yun · M. He · Y. Liang (✉) · M. Lei (✉)
MOE Key Laboratory for Nonequilibrium Synthesis and Modulation of Condensed Matter, Shaanxi Province Key Laboratory of Quantum Information and Quantum Optoelectronic Devices, School of Physics, Xi'an Jiaotong University, Xi'an, China
e-mail: yansheng.liang@mail.xjtu.edu.cn

M. Lei
e-mail: ming.lei@mail.xjtu.edu.cn

M. Lei
State Key Laboratory of Electrical Insulation and Power Equipment, Xi'an Jiaotong University, Xi'an, China

3.1 Introduction

Optical tweezers (OTs), a groundbreaking tool that utilizes the radiation pressure and gradient force of light to manipulate small particles, have revolutionized numerous scientific fields since their invention by Ashkin in 1986 [1, 2]. These non-invasive, contactless manipulators are based on the principle that light exerts forces on objects, allowing for precise control of a huge variety of samples ranging from nanometers to micrometers in size. Initially, OTs relied on single-beam Gaussian light beams, but with the rapid evolution of laser technology and optical systems, the field has expanded to incorporate more advanced approaches. A significant breakthrough was the integration of holographic optical techniques into OTs. In 1998, Dufresne et al. proposed an innovative scheme by combining holographic optics with OTs [3], wherein diffraction optical elements were used to create multiple beam arrays from a single beam, allowing for the simultaneous manipulation of multiple objects. In 2000, Liesener et al. further developed holographic optical tweezers (HOTs) that use spatial light modulators (SLMs) to modify the phase of the incident light, enabling real-time dynamic manipulation of objects [4]. The multifunctionality of HOTs makes them invaluable tools in widespread applications, such as biophysics [5–7], atomic and molecular physics [8, 9], materials science [10, 11].

The demand for more complex and diverse particle manipulation has driven the further development of optical manipulation methods. The emergence of structured light has brought new opportunities to HOTs. Structured light refers to light beams where the amplitude, phase, and polarization are precisely modulated to create specific field patterns. Unlike traditional Gaussian beams, structured light beams have unique properties such as carrying orbital angular momentum (OAM), and can be designed to have specific intensity and phase gradients. These tailored light fields allow OTs to have richer manipulation capabilities. For instance, the well-known optical spanner that rotates particles is realized using

B. Li et al. (eds.), *Biophotonic Manipulation*, Advances in Optics and Optoelectronics,
https://doi.org/10.1007/978-981-96-4982-2_3

structured beams that carry OAM, such as Laguerre Gaussian beams [12, 13]. More complex trapping landscapes are capable of guiding particles around obstacles and along customized three-dimensional trajectories [14–16]. Additionally, vector beams with spatially varying polarization states have shown excellent performance in improving trapping efficiency [17–19]. Undoubtedly, the advent of structured light has driven the diversification of optical micro-manipulation applications and offered new degrees of freedom for the precision control of particles. This chapter focuses on typically structured light fields, including self-similar beams, non-diffracting beams, customized structured beams, and vector light fields, discussing their generation methods and applications in optical manipulation.

3.2 Optical Manipulation with Self-Similar Beams

Due to their diverse transverse intensity modes, self-similar beams occupy an important position in laser technology. During free-space propagation, the transverse shape of these beams remains unchanged and undergoes overall scaling due to divergence. Self-similar beams can be described as solutions to the paraxial Helmholtz equation, which consists of three complete families of exact and orthogonal solutions: Hermite-Gaussian beams, Laguerre-Gaussian beams, and Ince-Gaussian beams.

3.2.1 Hermite-Gaussian (HG) Beams

HG beams are solutions of the Helmholtz equation in Cartesian coordinates. The electric field of a HG beam can be expressed as:

$$\mathrm{HG}_{m,n}(x,y,z) = A_{m,n}\left[\frac{\omega_0}{\omega(z)}\right] H_m\left(\frac{\sqrt{2}x}{\omega(z)}\right) H_n\left(\frac{\sqrt{2}y}{\omega(z)}\right) \times \exp\left[ikz + ik\frac{x^2+y^2}{2R(z)} - i(m+n+1)\varsigma(z)\right], \quad (3.1)$$

where $A_{m,n}$ is a constant. ω_0 and $\omega(z)$ are the beam waist radius at distances $z = 0$ and z. H_m and H_n are known as the m-th and n-th order Hermite-Gaussian functions. k is the wave number, and $\varsigma(z) = \tan^{-1}(z/z_0)$. HG beams are characterized by different transverse modes, typically described by the parameters m and n, where m and n denote the mode number along the x- and the y-axis. Figure 3.1 shows the transverse intensity profiles of HG beams with several values of m and n. The fundamental mode corresponds to $(m, n) = (0, 0)$ and is the fundamental Gaussian beam.

The main method of generating HG beams is usually to fabricate lasers with special structures. For example, $\mathrm{HG}_{0,n}$ and $\mathrm{HG}_{n,0}$ modes can be directly generated in end-pumped solid-state lasers using the off-axis pumping scheme [20]. HG beams were first used in the field of optical micromanipulation in 1991, when Sato et al. used $\mathrm{HG}_{0,n}$ beams to capture polystyrene latex spheres and rotate erythrocytes [21]. They pointed out that compared with previous methods using fundamental Gaussian beams for trapping microparticles, higher-order mode laser beams can enhance the trapping force on high-refractive-index microparticles. By rotating the laser beam around its optical axis, the rotation of individual red blood cells was achieved, thereby verifying the potential application of higher-order mode lasers in noncontact and controlled rotation of microscopic biological cells. In addition, the use of the $\mathrm{HG}_{0,1}$ beam allows the creation of high-frequency optical trap with similar confinement strength to optical lattice for trapping atoms, advancing the possibility of making single highly confined lower dimensional condensates [22].

3.2.2 Laguerre-Gaussian (LG) Beams

LG beams are another fundamental class of laser beams that are solutions of the paraxial Helmholtz equation. Unlike HG beams, LG beams are expressed in cylindrical coordinates (ρ, φ, z), i.e.,

$$\mathrm{LG}_{l,p}(\rho,\varphi,z) = A_{l,p}\left[\frac{\omega_0}{\omega(z)}\right]\left(\frac{\sqrt{2}\rho}{\omega(z)}\right)^l L_p^l\left(\frac{2\rho^2}{\omega^2(z)}\right)\exp\left(-\frac{\rho^2}{\omega^2(z)}\right) \times \exp\left[-ik\frac{\rho^2}{2R(z)} \mp il\varphi + i(l+2p+1)\varsigma(z)\right], \quad (3.2)$$

where $L_p^l(\cdot)$ is the generalized Laguerre polynomial. The radial index p is related to the number of concentric rings in the intensity profile of the beam. LG beams with $p = 0$ have a single doughnut-shaped ring, while beams with higher p values have multiple rings. The azimuthal index l determines the number of twists in the phase structure of the beam, also known as the topological charge. Figure 3.2 shows the transverse intensity profiles of a set of LG beams with $p = 0, 1, 2$ and $l = 0, 1, 2, 3$. Solving in polar coordinates leads to radially symmetric transverse profiles. The lowest-order LG mode (LG_{00}) is the same as that of the lowest-order HG mode (HG_{00}), both representing the fundamental Gaussian beams. LG beams carry OAM of $l\hbar$ for per photon, which brings them massive applications [23]. The on-axial phase singularity of the LG vortex beams results in the donut-shaped intensity distribution with null intensity at the center.

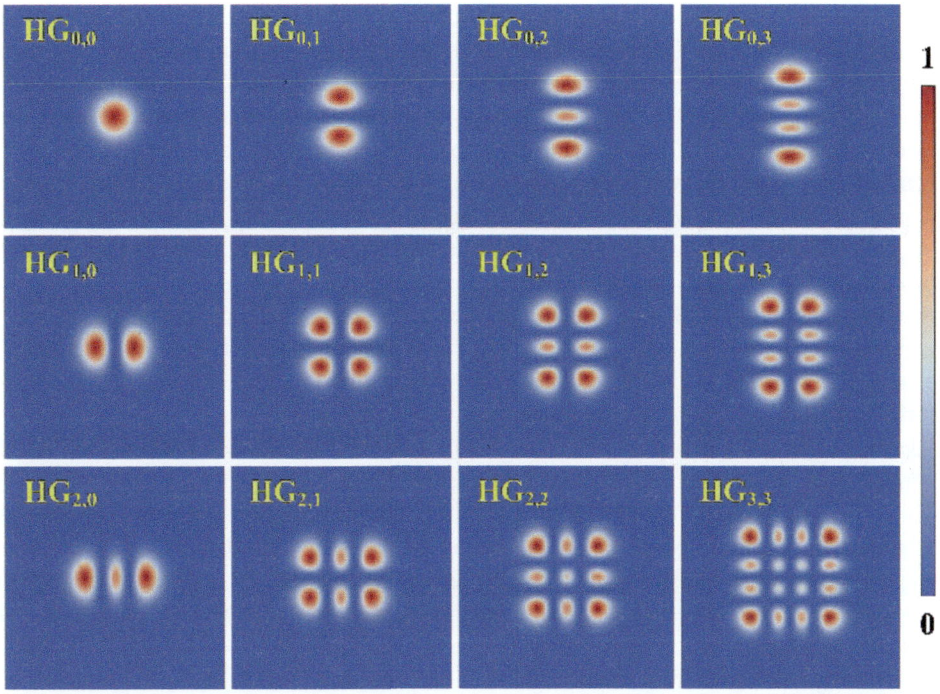

Fig. 3.1 Transverse intensity profiles of several low-order HG beams

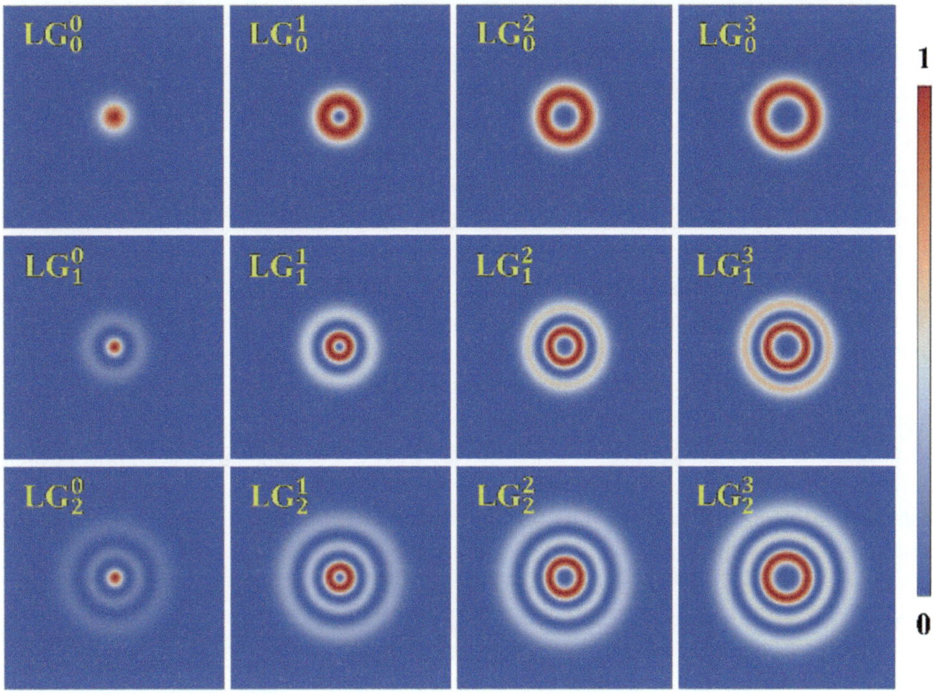

Fig. 3.2 Transverse intensity profiles of several low-order LG beams

Due to the extensive applications of LG beams, the generation of such beams has been a big concern. As high-order eigenmodes, LG beams can be output from the laser directly. By utilizing power-dependent thermal lensing of the laser material to create a limiting aperture within the gain medium, high-power LG modes can also be output from a side-pumped bounce amplifier [24]. Other direct methods include introducing loss-controlling elements (such as plano-convex lens, spot-defect mirror, etc.) into the resonator cavity [25–27], shaping the end-pump beams into the annular beams [28–30], as well as off-axis pumping [31]. Given the lack of flexibility for these methods, production outside the laser cavity is often preferred. A series of phase modulation elements can be used to transfer the fundamental Gaussian beams into LG beams, including spiral phase plates, fork-shaped grating, spiral zone plates, and SLMs [32–37].

LG beams have received widespread attention in the field of optical micromanipulation. Simpson et al. experimentally confirmed that, compared to Gaussian beams, using LG modes as trapping beams can enhance the axial trapping efficiency of particles [38]. This means that stable confinement can be achieved under lower laser power, reducing the photo damage to living cells, which is crucial for applications of optical tweezers in biological fields. Another notable advantage of higher-order LG beams in optical manipulation is that the presence of OAM allows the beams to exert mechanical torque on particles, enabling the orbital rotation of particles [23]. H. Rubinsztein-Dunlop and collaborators used forked diffraction grating phase holograms to efficiently generate focused higher-order LG beams and successfully applied them in optical manipulation experiments [39, 40]. This experiment directly observed the transfer of OAM from a laser beam to absorptive particles and confirmed that the angular momentum transfer is the primary reason for the particles' rotation. Subsequently, J. E. Curtis et al. further utilized the dynamic holographic optical tweezers technique to generate topological ring traps (Fig. 3.3a) and revealed the dependence of such vortex beams' structure and angular momentum on helicity [12]. Dielectric particles are attracted to the ring trap and rotate along the ring orbit due to OAM. These toroidal torque-exerting traps are known as the "optical spanners" [41–43] and have been used in a variety of applications, including measuring the viscous resistance of fluid [44–46], diagnosing the cell's viability [47, 48], measuring the torsional stiffness of DNA [49, 50], and optically driven micromachines [51–54].

Compared to the manipulation of dielectric particles immersed in liquid, optical trapping of aerosol particles in gas media presents greater challenges, primarily due to the low viscosity of gases and the strong influence of photophoretic forces [55, 56]. Nevertheless, research on aerosols is of significant importance in fields such as combustion science and atmospheric chemistry. In 2010, Shvedov et al. successfully achieved three-dimensional trapping and guiding of light-absorbing agglomerates of carbon nanoclusters suspended in the air using two counter-propagating LG beams, with trapping distances of up to several millimeters [57]. Figure 3.3b illustrates a schematic of two counter-propagating beams with topological charges of $\pm$ 1, while Fig. 3.3c shows an actual image of the particle trapped in air.

The optical trapping structures created by the self-interference of LG modes or interference with other light waves provide a powerful and versatile method for precise rotational control of particles. The interference pattern formed by the LG beam and a plane wave results in a spiral pattern with the number of spiral arms corresponding to the azimuthal index l of the LG beam [58]. The trapped objects are located within the spiral arms of this pattern. By introducing phase delay through altering the optical path length in the interferometer, the spiral pattern rotates, causing the trapped particles to rotate as well. Another interference pattern formed by two LG beams with opposite helicity (one positive and one negative) creates $2l$ bright spots on a radially symmetric ring structure [59]. Each bright spot provides a localized trapping point for the particles, which are trapped at these spots and arranged into a three-dimensional array. The ability to create and manipulate 3D structures at the microscale has potential applications in colloid physics, micromanipulation, and self-assembly of crystalline structures.

3.2.3 Ince-Gaussian (IG) Beams

The exact analytical solutions of the paraxial wave equation found in the elliptical cylindrical coordinates are referred to IG modes, which together with the HG modes and the LG modes form three complete families of exact solutions. Mathematically, the IG modes are described as [60]

$$\begin{aligned} \mathrm{IG}^{e}_{p,m}(\boldsymbol{r}) &= \frac{C\omega_0}{\omega(z)} C^m_p(i\xi,\varepsilon) C^m_p(\eta,\varepsilon) \exp\left[\frac{-r^2}{\omega^2(z)}\right] \\ &\quad \times \exp\left[ikz + i\frac{kr^2}{2R(z)} - i(p+1)\varsigma(z)\right], \end{aligned} \tag{3.3}$$

$$\begin{aligned} \mathrm{IG}^{o}_{p,m}(\boldsymbol{r}) &= \frac{S\omega_0}{\omega(z)} S^m_p(i\xi,\varepsilon) S^m_p(\eta,\varepsilon) \exp\left[\frac{-r^2}{\omega^2(z)}\right] \\ &\quad \times \exp\left[ikz + i\frac{kr^2}{2R(z)} - i(p+1)\varsigma(z)\right]. \end{aligned} \tag{3.4}$$

The elliptic coordinates (ξ, η, z) in a transverse z plane are defined as $x = f(z)\cosh\xi\ \cos\eta$, $y = f(z)\sinh\xi\ \sin\eta$, and $z = z$, where $f(z) = f_0\omega(z)/\omega_0$ and f_0 is the semifocal separation at $z = 0$. $\xi \in [0, \infty)$ and $\eta \in [0, 2\pi)$ represent the radial and angular elliptic variables. C and S are the normalization constants. The superscripts e and o are even and odd modes, and the C^m_p and S^m_p are the even and odd Ince polynomials of order p and degree m. $\varepsilon = 2{f_0}^2/{\omega_0}^2$ denotes the ellipticity parameter. Intensity profiles of even and odd IG modes with $\varepsilon = 4$ are shown in Fig. 3.4. When the ellipse parameter ε is 0 and ∞, the IG mode tends to the LG mode and the IG mode, respectively. Since $C^0_0(i\xi, \varepsilon) = 1$, the IG mode with $p = 0$ and $m = 0$ is the simple Gaussian mode.

Methods for generating IG modes can be divided into two main categories. One is the direct generation of IG beams by breaking the symmetry of the resonant cavity, using methods such as shifting the output coupler sideways [61, 62], off-axis pumping by tilting the microchip laser cavity [63], mode-gain control [64], and mode-loss control [65]. The second type involves using optical elements like SLM [66, 67], digital micromirror device (DMD) [68], and

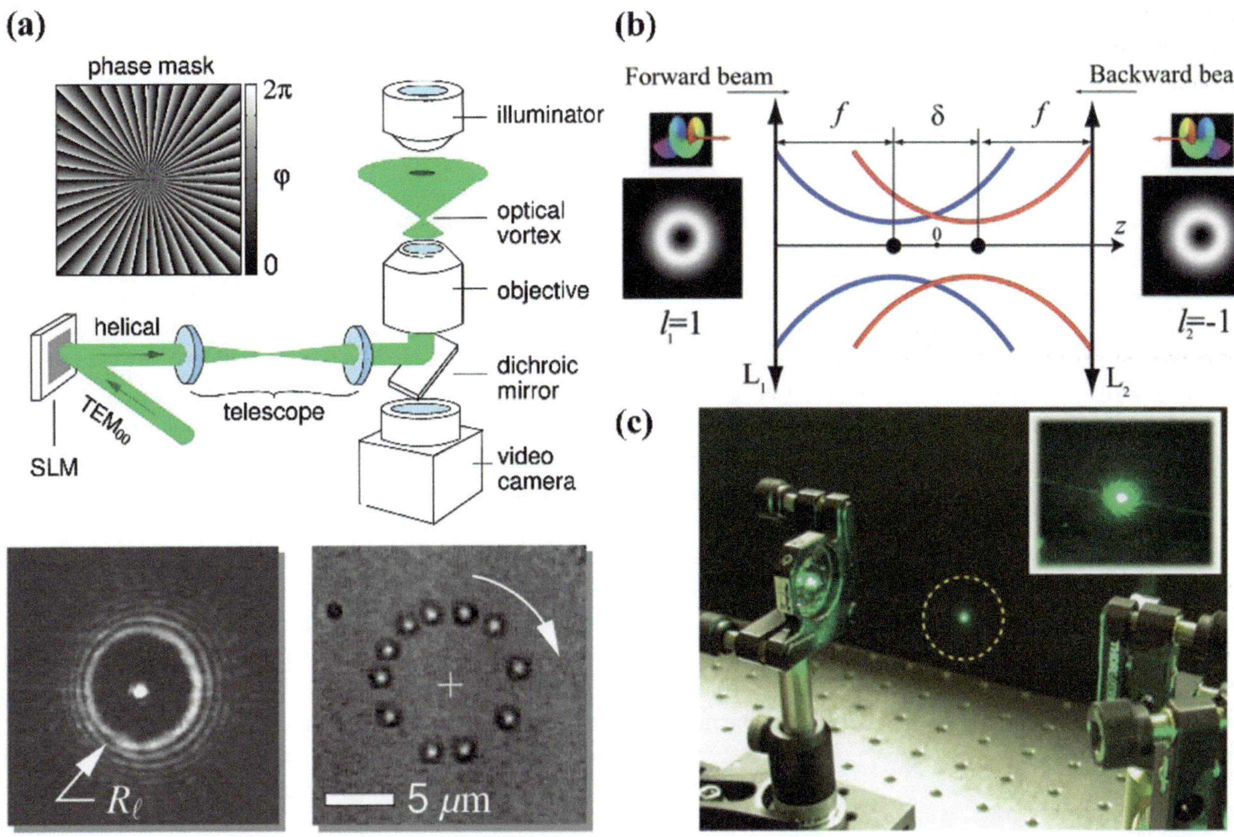

Fig. 3.3 Application of LG beams in optical micromanipulation. **a** Using holographic optical tweezers to generate higher-order LG beams and rotate microparticles (Reprinted figure with permission from Curtis et al. [12], Copyright (2024) by the American Physical Society). **b**, **c** The schematic and experimental result of trapping light-absorbing microparticles in air using two counter-propagating LG beams (Reprinted by permission from Springer Nature: Applied Physics A: Materials Science & Processing [57])

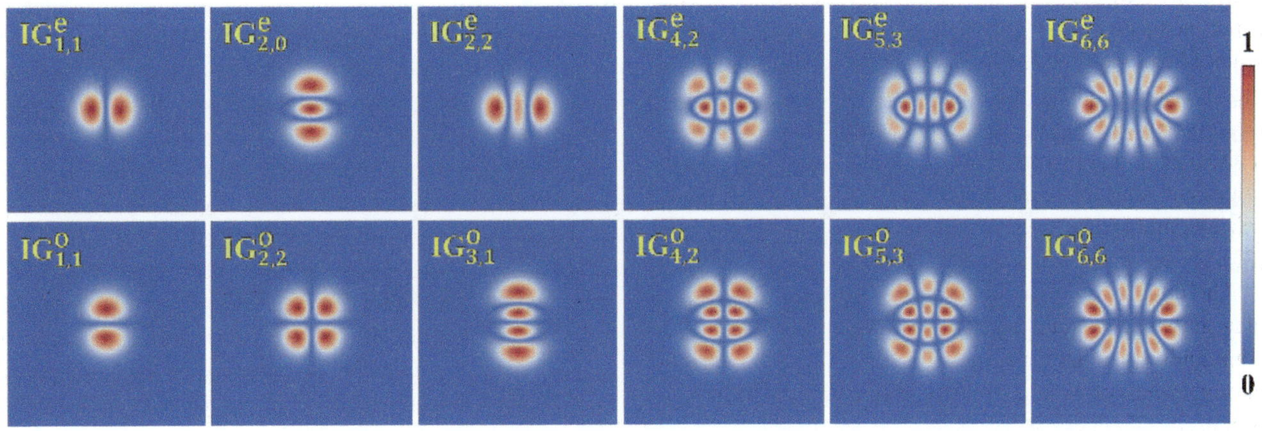

Fig. 3.4 Transverse intensity profiles of even (first row) and odd (second row) IG modes. The ellipse parameter ε is set to 4

meta-surface [69] to modulate the incident beam into IG modes.

Since IG modes not only have both odd and even modes but also exhibit different transverse intensity distributions related to the ellipticity parameters, their field distribution is more diverse compared to LG and HG modes. The greater variety of modes makes IG beams suitable for trapping particles, and the significant potential of IG beams in optical manipulation have been demonstrated. An example of trapping microparticles using IG beams with different modes is shown in Fig. 3.5

[66]. Figure 3.5a is the schematic of the experimental setup. The modulated IG beams are focused onto the sample stage by a microscope objective, and multiple silica microspheres with a diameter of 1.5 μm are simultaneously trapped at each bright spot of the focused IG beams (Fig. 3.5b).

3.3 Optical Manipulation of Nondiffracting Beam Beams

All electromagnetic waves, including self-similar beams, are affected by diffraction and inevitably scatter during propagation. However, for nondiffracting beams, which are also specific solutions of the Helmholtz equation, this physical characteristic can be ignored as they will maintain their shape and size of the transverse intensity distribution over longer propagation distances. More interestingly, these beams possess remarkable self-healing capability, meaning that after encountering an obstacle or disturbance, they can restore their original intensity distribution after propagating a certain distance. This section will introduce some well-known nondiffracting beams and their key applications in the field of optical micromanipulation.

3.3.1 Bessel Beams

In cylindrical coordinates (ρ, φ, z), the electric field distribution of an ideal Bessel beam takes the form [70–72]

$$E(\rho, \varphi, z) = A_0 \exp(ik_z z) J_l(k_\rho \rho) \exp(\pm il\varphi), \tag{3.5}$$

where A_0 is a constant and k_z and k_ρ are the longitudinal and radial wavevectors. $J_l(\cdot)$ represents l-th order Bessel function. The transverse intensity profiles of Bessel beams of different orders are given in Fig. 3.6a. Figure 3.6b shows the correspondingly normalized energy curves along the horizontal direction. The zero-order mode displays a central bright spot surrounded by a series of concentric rings with equal power, whereas for the higher-order Bessel beam, phase singularity on the beam axis causes the center of the beam to exhibit a bright ring, and the diameter of the bright ring increases with increasing order. All higher-order Bessel beams have helical phase $\exp(il\varphi)$ with respect to the azimuthal angle, indicating that they carry OAM $l\hbar$ per photon.

Ideally, Bessel beams contain an infinite number of central rings and carry infinite energy, allowing them to indefinitely counteract diffraction. However, this is not practically achievable due to aperture limitations in all experimental setups, which result in the truncation of the beam by the aperture. Nevertheless, Bessel beams with finite energy can be produced experimentally, although this results in a shorter distance of diffraction-free propagation. So far, various methods have been proposed to generate Bessel beams. The angular spectrum of a Bessel beam is a ring. Therefore, the most direct method to generate Bessel beams is to use a circular slit, which is also the first experimental realization of Bessel beams [70]. Figure 3.7a illustrates the schematic diagram of this method. A plane wave is incident perpendicularly on a circular slit with a diameter d and a width Δd, which is placed in the focal plane of a lens with radius R and focal length f. 0th-order Bessel beams are formed in the conical region behind the lens, achieving the maximum nondiffracting distance $Z_{\max} = R/\tan\theta$ and $\theta = \tan^{-1}(d/2f)$. Obviously, the drawback of this method is that a significant portion of the beam energy is blocked while passing through the slit, resulting in a very low generation efficiency. The method of using axicons offers the advantages of simple setup and high efficiency, making it a commonly used approach in current research for generating Bessel beams [73]. The experimental setup is shown in Fig. 3.7b. After passing through an aperture with radius size L, a plane wave illuminates an axicon with axicon angle ϕ. The axicon modulates the phase and deflects the beam into a conical wave, which interferes and superimposes during propagation, finally generating a 0th-order Bessel beam in the rhombic region, where the long axis of the rhombic region is the maximum non-diffracting range of the Bessel beam, i.e., $Z_{\max} = L/\tan(\phi(n-1))$ and n is the refractive index of the axicon material. Higher-order Bessel beams can be produced by replacing the incident plane wave with higher-order LG beams [74]. It is noteworthy that the axicon angle ϕ cannot be made infinitely small due to manufacturing constraints, which typically limits the $Z_{\max}$ in the axicon method to the centimeter range. Recently, methods based on annular lens elements and free lens modulation have been proposed to successfully achieve long-range Bessel beams that propagate without diffraction over distances of several meters [75, 76]. These advancements may play a significant role in applications such as stable long-range communication through turbulent media [77]. The schematic for generating Bessel beams using free lens modulation is shown in Fig. 3.7c. The digital free lens is formed by the superposition of the lens phase and vortex phase (Fig. 3.7d). By loading the hologram of the free lens phase onto an SLM, the angular spectrum of the Bessel beam can be directly shaped. After the angular beam undergoes a Fourier transform through a spherical lens, a long-distance Bessel beam is generated. The high flexibility of freeform lens modulation allows for the generation of diffraction-free beams with diverse transverse intensity distributions by customizing the shape of the free lens (Fig. 3.7e). Furthermore, a range of novel customized optical elements can be used to generate high-quality Bessel beams, such as meta-surfaces [78, 79], integrated silicon photonic chips [80], and Fresnel zone plates [81]. These methods provide greater flexibility and precise control in the generation and application of Bessel beams.

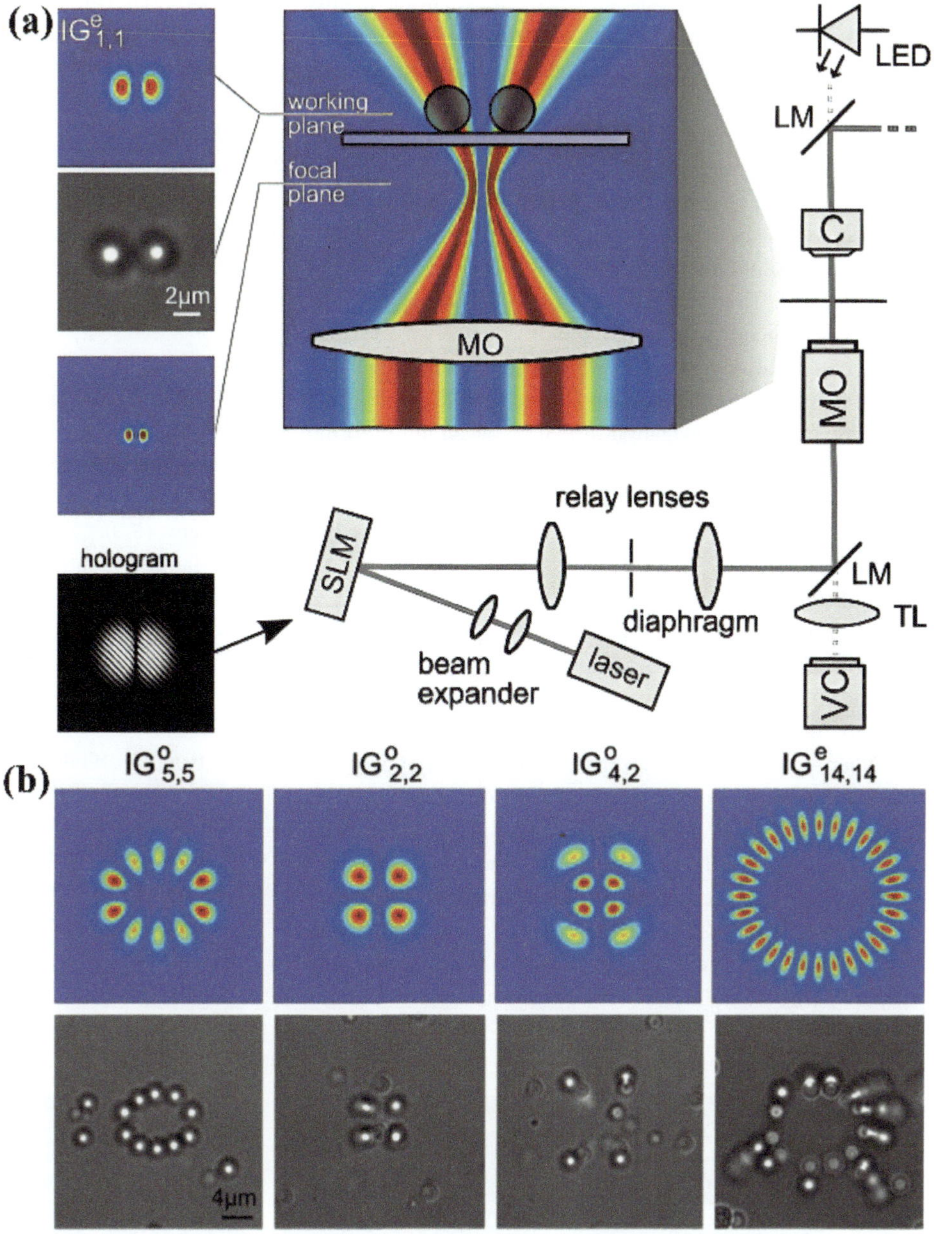

Fig. 3.5 Optical micromanipulation of particles using IG beams. **a** Diagram of experimental equipment for generating IG beams and manipulating particles based on SLM modulation. **b** IG intensity patterns (top) and the corresponding microstructures formed by optically assembling silica spheres (bottom) (Adapted with permission from Woerdemann et al. [66] with the permission of AIP Publishing)

Bessel beams address a significant limitation of conventional optical tweezers that use Gaussian beams, which cannot trap particles over large axial separations due to beam divergence and distortion when interacting with the particles. In contrast, Bessel beams possess unique properties that allow diffraction-free propagation and self-reconstruction, enabling the manipulation of microscopic particles in multiple, spatially separated locations [82, 83]. As shown in Fig. 3.8a, a zeroth-order Bessel beam's central maximum spot acts as a rod of light, simultaneously confining different types of microparticles in the two sample cells [83]. In this experiment, the Bessel beam has a central radius of 5 μm, with a diffraction-free propagation distance of approximately 4 mm. Sample cells I and II are separated by 3 mm, with each cell being 100 μm deep. Cell I contains hollow spheres (diameter ~ 5 μm) with a lower refractive index relative to the suspending medium, while Cell II contains silica spheres with a diameter of 5 μm. In Cell I, the hollow sphere is trapped between the central spot and the first ring of the Bessel beam. After passing through Cell I, the beam experiences distortion due to the presence of particles but reconstructs itself after a certain distance. Upon entering Cell II, the reconstructed beam successfully captures and stacks three silica spheres, forming a one-dimensional chain of particles.

Conventional OTs systems are limited to observing regions in the lateral plane (X–Y plane) parallel to the focal plane of the trapping objective, as shown in the imaging results in Fig. 3.8a. However, for diffraction-free optical fields such as Bessel beams, which can propagate over a certain distance along the beam propagation direction (Z direction), researchers would like to directly observe the optical trapping process in other planes, particularly in the axial plane (X–Z or Y–Z plane). For this purpose, Liang et al. developed an OTs system based on a right-angle microreflector (Fig. 3.8b), which allows simultaneous trapping and imaging in the axial plane [84, 85]. The system's effectiveness was further validated by investigating the trapping performance with various optical fields, including conventional Bessel beams (Fig. 3.8c, d) propagating along the optical axis and tilted Bessel beams (Fig. 3.8e, f) with propagation directions at an angle to the optical axis. Figure 3.8c, e shows the three-dimensional intensity profiles of the beams, and Fig. 3.8d, f shows the experimental results of the optical guiding of particles observed in the axial plane.

Similar to LG beams, the OAM of high-order Bessel beams can also be mechanically transferred to trapped microparticles, effectively driving their rotation [86, 87]. Going beyond trapping and rotating, one notable optical manipulation application of Bessel beams is optical traction. It is well known that when light interacts with matter, the objects experience radiation pressure along the propagation direction of light. However, there is a counter-intuitive result that in some special cases, particles are dragged in the light source's direction, so-called "optical traction" [88–91]. The gradientless zeroth-order Bessel beam has been demonstrated as a good candidate for tractor beam [92]. By properly manipulating the nonparaxiality of the Bessel beam, backward pulling forces can be generated along both the radial and longitudinal directions, thus enabling the traction of particles with different sizes and materials. Moreover, by suitably superimposing co-propagating Bessel beams, a special diffraction-free structured beam, known as frozen waves, can be produced [93, 94]. The intensity envelope of these wave fields remains static, and along the propagation direction, these beams contain multiple high-intensity peaks. Experimentally, frozen waves have been generated [95–97] and used for optical trapping of microparticles [98].

3.3.2 Mathieu Beams

Another solution to the Helmholtz equation in elliptic coordinates (ξ, η, z) is Mathieu beams. Specifically, in elliptic coordinates (ξ, η, z), the Helmholtz equation can be reduced to the Mathieu equation by the substitution $\eta = i\xi$, which can be written as [99, 100]

$$\frac{\mathrm{d}^2\omega}{\mathrm{d}\xi^2} + (a - 2q\cos(2\xi))\omega = 0, \tag{3.6}$$

where a is the separation constant. The ellipticity parameter q is proportional to the squares of the eccentricity of the elliptical coordinates f and wave vector k, satisfying the relation $q = f^2k^2/4$. Consequently, the field distribution of nondiffracting Mathieu beams is

$$\mathrm{M}_m^e(\xi, \eta, z, q) = C_m \mathrm{Je}_m(\xi, q)\mathrm{ce}_m(\eta, q)\exp(ik_z z), \tag{3.7}$$

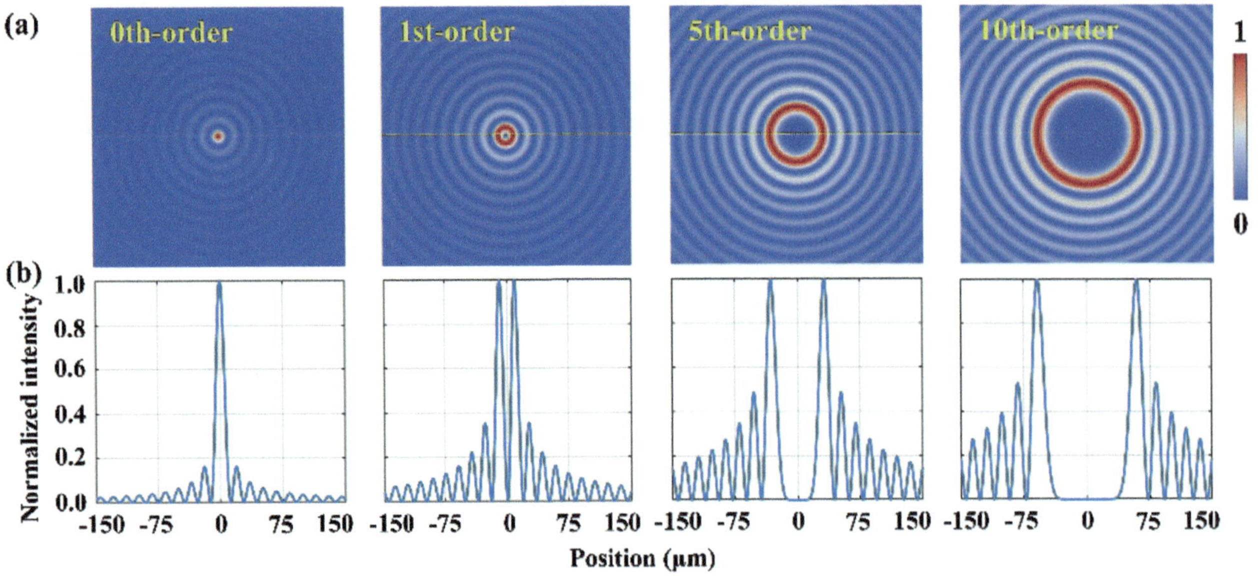

Fig. 3.6 Transverse intensity profiles of Bessel beams with different order (**a**) and the correspondingly normalized energy distributions along horizontal direction (**b**)

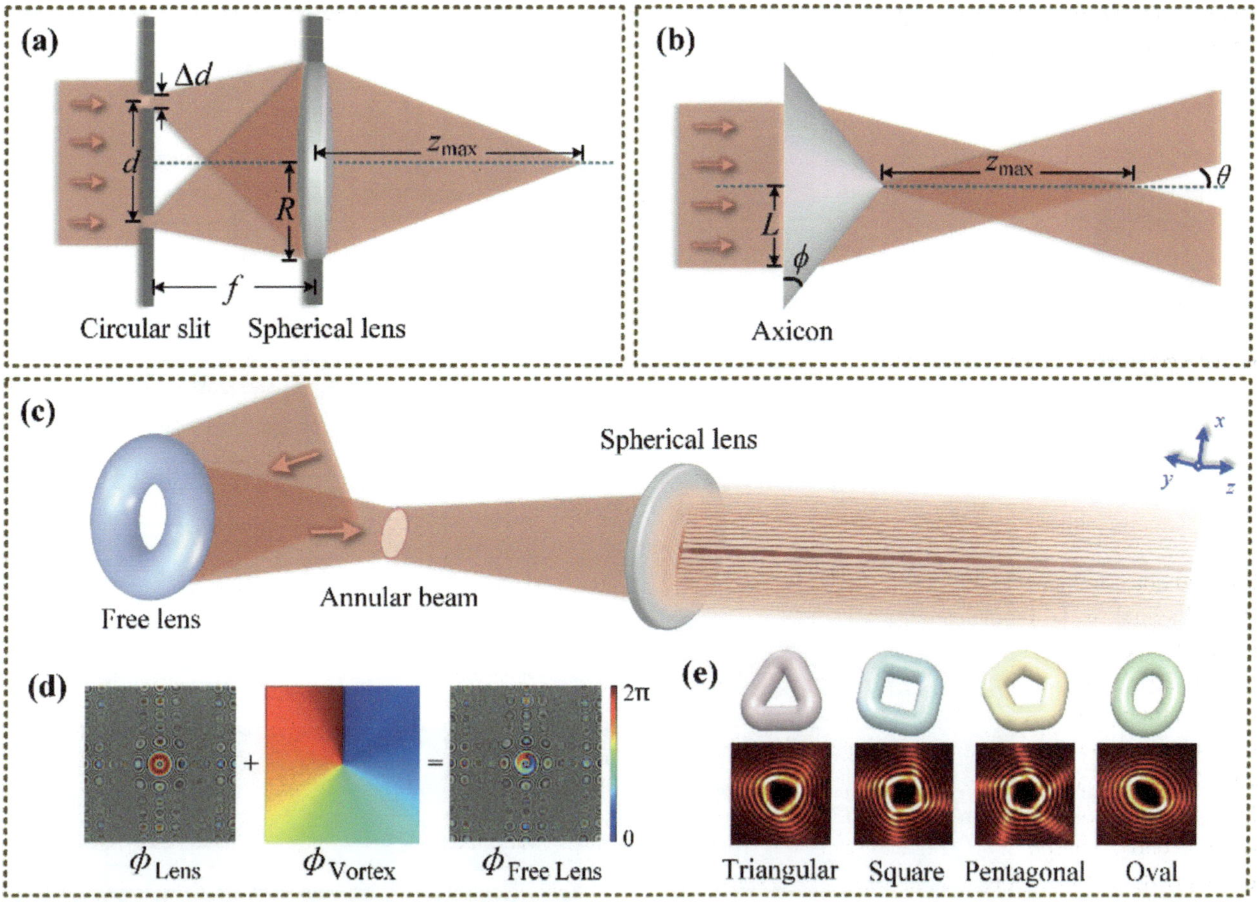

Fig. 3.7 Schematic diagram of two methods for producing Bessel beams. **a** Methods of using circular slit. **b** Methods of using axicon. **c–e** Methods of using free lens modulation to generate long-distance and tunable nondiffracting beams. **c** Schematic of the method. **d** Phase hologram of digital free lens. **e** Different shapes of free lenses produce non-diffracted beams with diverse transverse intensity distributions (Image **c–e** adapted with permission from [76] © Optical Society of America)

$$\mathrm{M}_m^o(\xi, \eta, z, q) = S_m \mathrm{Jo}_m(\xi, q)\mathrm{se}_m(\eta, q)\exp(ik_z z), \quad (3.8)$$

where C_m and S_m are the normalization constants. Je_m and Jo_m respectively are the even and odd radial Mathieu functions, whereas ce_m and se_m represent the even and odd angular Mathieu functions. m is the order of Mathieu functions, for even modes m takes 0, 1, 2, 3 …, and for odd modes, m takes 1, 2, 3 … The intensity distribution of even Mathieu beams with different parameters is shown in Fig. 3.9a.

Mathieu beams are generated by methods similar to Bessel beams, including circular slit-lens system [101], computer-generated phase hologram [102], axicon-based method [103, 104], and spatial light modulators [105, 106]. Very recently, a method for producing second-order even Mathieu beams using single-layer substrate transmissive metasurface has been proposed [107]. The metasurface unit is composed of a pair of metallic I-shaped resonators with half-period dislocation printed on a single-layer dielectric substrate (Fig. 3.9b). By controlling the rotation angle of the I-shaped resonators, continuous amplitude modulation from 0 to 1 and phase shift from 0 to π can be achieved, along with polarization conversion of the incident beam. This provides a novel and efficient approach for generating Mathieu beams.

Mathieu beams provide diverse field distributions with azimuthal symmetry, which are associated with the beam order m and the ellipticity parameter q. These non-diffracting 3D light fields can be tailored as versatile light moulds for the fabrication of structured materials at the microscopic scale [106]. As shown in Fig. 3.10a, b, due to the self-healing properties of Mathieu beams, multiple particles can be stacked along the propagation direction of the light, with each particle arranged in the same transverse direction as the preceding one. The light moulds formed by several intensity maxima distributed of Mathieu beam along the radial direction can capture and orient elongated particles (Fig. 3.10c). Dynamic rotation of structured materials is achieved through rotating the direction of the light moulds (Fig. 3.10d–g), which may find applications in microfluidics.

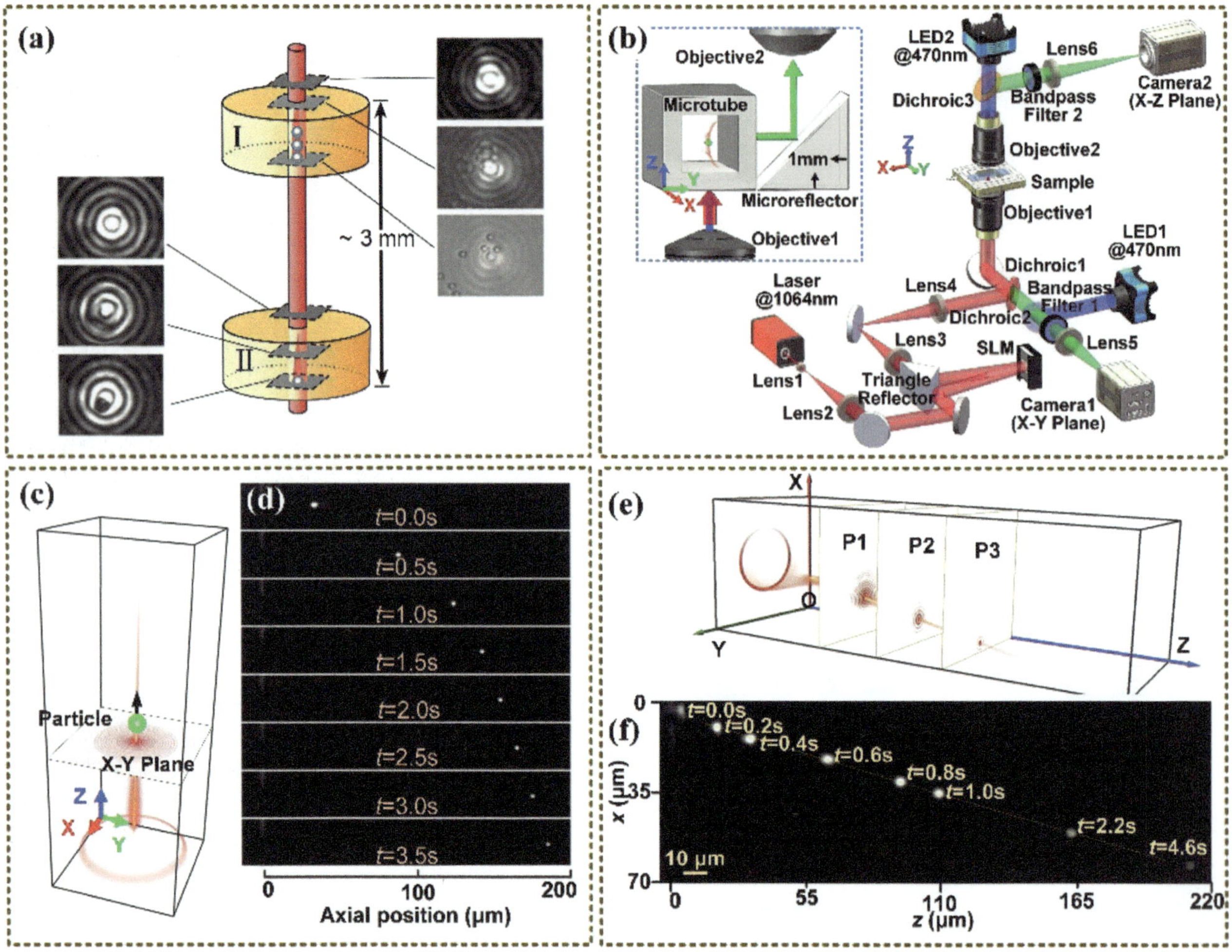

Fig. 3.8 Optical manipulation with Bessel beams. **a** Zeroth-order Bessel beam simultaneously manipulates particles in different sample cells (Reprinted by permission from Springer Nature: Nature [83]). **b–f** Simultaneous optical trapping and imaging in the axial plane. **b** Schematic diagram of the axial-plane imaging based on a right-angle microreflector. **c**, **e** respectively are the 3D intensity profiles of a Bessel beam propagating along the optical axis and a tilted Bessel beam. **d**, **f** are the corresponding images of optical guiding of microbead. **c**, **e**, **f** Adapted with permission from [84] © Optical Society of America). **d** Adapted with permission of Reports on Progress in Physics, from Liang et al. [85], permission conveyed through Copyright Clearance Center, Inc)

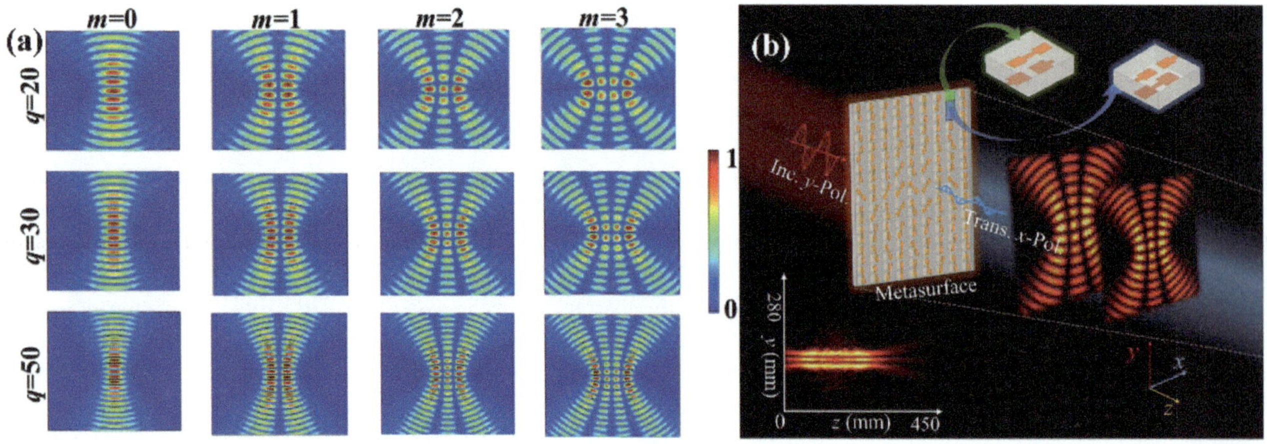

Fig. 3.9 **a** Simulated intensity distribution of even Mathieu beams with different order m and ellipticity parameter q. **b** Generating Mathieu beams with single-layer substrate transmissive metasurface (Reprinted from Yuan et al. [107], Copyright (2024), with permission from Elsevier)

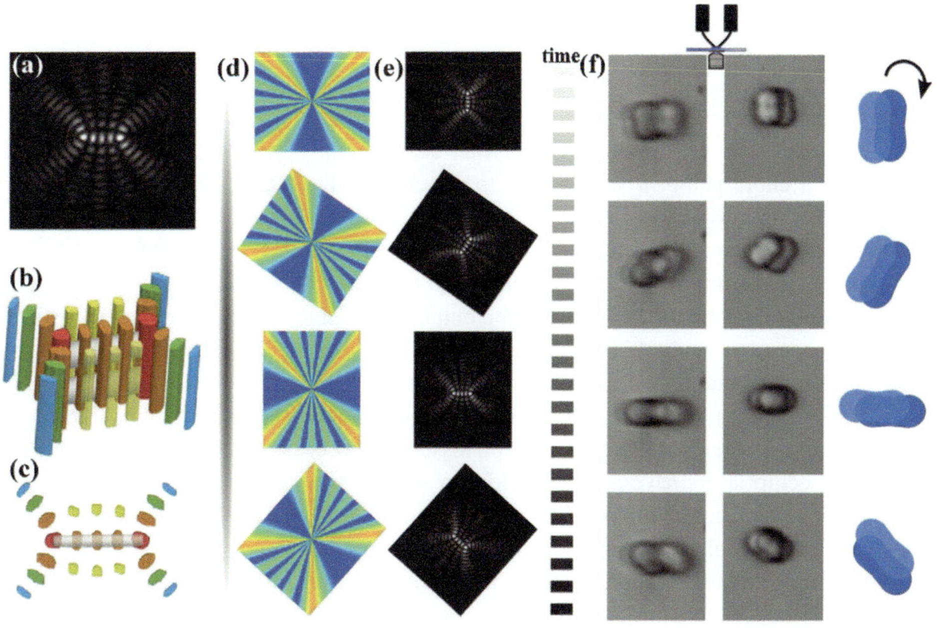

Fig. 3.10 Trapping and rotating non-spherical particles using Mathieu beams. **a, b** 2D and 3D intensity distribution of Mathieu beams with $m = 4$. **c** Elongated particle is captured and oriented by several intensity maxima distributed of Mathieu beam. **d** Rotating hologram. **e** Rotating Mathieu beam. **f, g** Screenshots of non-spherical particles rotating and corresponding model diagrams at different times (Adapted with permission from [106] © Optical Society of America)

Additionally, the superposition of odd and even modes of Mathieu beams can generate helical Mathieu beams carrying OAM, characterized by the transverse intensity distribution of a set of concentric elliptical rings with varying intensities [102, 108, 109]. Consequently, these beams can also transfer OAM to particles, thereby inducing particle rotation [108, 110].

3.3.3 Airy Beams

Under paraxial approximation, the transmission and evolution of one-dimensional Airy wave packets in free space follow the potential-free Schrödinger equation [111, 112]:

$$i\frac{\partial \phi}{\partial \zeta} + \frac{1}{2}\frac{\partial^2 \phi}{\partial s^2} = 0. \tag{3.9}$$

In this function, ϕ is the electric field envelope of Airy beams. $\zeta = z/(kx_0^2)$ denotes a normalized propagation distance, where x_0 is an arbitrary transverse scale and k is the wavenumber of the beam. $s = x/x_0$ represents a dimensionless transverse coordinate. Solving Eq. 3.1.9 yields a diffraction-free Airy wave solution, written as:

$$A(\zeta, s) = \mathrm{Ai}\left(s - \left(\frac{\zeta}{2}\right)^2\right)\exp\left(i\left(s\frac{\zeta}{2} - \frac{\zeta^3}{12}\right)\right), \tag{3.10}$$

where $Ai(\cdot)$ is Airy function. The term $(\zeta/2)^2$ confirms that the non-diffracting Airy beam experiences self-bending during propagation in free space, similar to the parabolic trajectory of objects under the influence of gravity [113]. This property is also referred to as self-acceleration. This unusual characteristic of Airy wave packets distinguishes them from other non-diffracting field families, although they share the common feature of carrying infinite energy and propagating invariantly. To experimentally generate Airy beams with finite energy, a feasible approach is to introduce an exponential aperture function, i.e.:

$$A(\zeta, s) = \mathrm{Ai}\left[s - \left(\frac{\zeta}{2}\right)^2 + ia\zeta\right] \exp\left[a\left(s - \frac{\zeta^2}{2}\right) + i\left(s\frac{\zeta}{2} + \frac{a^2\zeta}{2} - \frac{\zeta^3}{12}\right)\right]. \tag{3.11}$$

The positive parameter a is a decay factor that satisfies $0 < a < 1$. Further, the electric field envelope of two-dimensional (2D) Airy beam is given as

$$\mathrm{A}(\zeta, s_x, s_y) = \prod_{k=x,y} \mathrm{Ai}(\zeta_k, s_k). \tag{3.12}$$

Figure 3.11 illustrates the propagation dynamics of ideal Airy beam (Fig. 3.11a) and finite-energy Airy beam with a decay factor set to 0.05 (Fig. 3.11b). where the first column shows the parabolic trajectory of the beam and the second column shows the transverse intensity profile at $\zeta = 7$. The transverse intensity profile of the 2D Airy beam is shown in the third column. It can be observed that the finite-energy Airy beam also exhibits the distinctive features of ideal Airy beam, i.e., the beam remains diffraction-free during propagation, and the transverse intensity tends to move along the parabolic trajectory. Such special beams with curved propagation properties in free space are known as self-accelerating beams [114, 115].

Based on the theoretical foundation that the Fourier transform of a truncated Airy beam is a Gaussian beam modulated with a cubic phase, a simple way to experimentally generate Airy beams is to irradiate a Gaussian beam onto spatial light modulator loaded with a cubic phase mask and then Fourier-transform the modulated beam with a spherical lens [112]. Using a diffraction grating for output coupling in a laser can also generate Airy mode. In this case, the diffraction grating imposes a transverse cubic phase on the diffracted light [116]. However, a drawback of this method is its relatively low efficiency due to diffraction losses. In recent years, microchip lasers have attracted widespread attention due to their advantages of compact structure, high efficiency, and good beam quality. It has been demonstrated that by slightly tilting the two planar facets of a microchip laser to introduce asymmetry in the laser cavity, the light beam can be directed toward the crystal edge, allowing the generation of a transverse truncated Airy mode [117]. Additionally, on specially designed metal surfaces, non-diffracting plasmonic Airy beams can be produced from in-plane propagating surface plasmon polariton waves [116, 118]. The method for the nonlinear generation of Airy beams can be achieved by designing and fabricating an asymmetric quadratic nonlinear photonic structure, which allows the generated three-wave mixing output to possess a cubic phase front, which results in an accelerating Airy beam after performing a Fourier transform [118].

Unlike previously mentioned non-diffracting beams that propagate in straight lines, in the field of optical micromanipulation, the most captivating feature of the Airy beam is its ability to bend along a parabolic trajectory and bypass obstacles. These unique characteristics make Airy beams a powerful and novel tool in the field of optical trapping and manipulation. Figure 3.12 illustrates the Airy beam functioning as a micro-scale "snowblower", precisely transferring or clearing colloidal particles along parabolic trajectories from a given quadrant in the sample chamber [119]. This capability is termed "optically mediated particle clearing.“ As shown in Fig. 3.12a, the white overlay indicates the typical intensity distribution of the Airy beam. To highlight the "snowblowing" effect, the micrograph is divided into four colored sections. After two minutes of exposure, the Airy beam almost clears the green section, pushing particles along a parabolic trajectory toward the purple section. Figure 3.12b shows that after the Airy beam was rotated by 180° and exposed for one minute following the experiment in Fig. 3.12a. The direction of the beam is reversed, emptying the purple section and pushing particles back to the green section, demonstrating the impact of changing the direction of the Airy beam on particle distribution. This technique was applied to optically redistribute particles and mammalian cells between microwells [120]. Subsequently, numerical simulations investigated the ability of one-dimensional Airy beams to guide particles vertically along parabolic trajectories and the optical trapping capability of tightly focused 2D Airy beams [121, 122]. Optical experiments demonstrated that particles were stably trapped at the main intensity lobes of the focused beam [122]. By studying the optical force distribution exerted on microparticles by Airy beam array, which is composed of several Airy beams with different ballistic trajectories, it was found that increasing the number of beams and controlling the initial launch angle of the beams can significantly enhance the stability of optical trapping [123, 124].

3.3.4 Noval Self-Accelerating Beams

The propagation of the Airy beam follows a fixed parabolic path, resulting in its limited flexibility in practical applications. In recent years, there are other classes of structured beams with self-accelerating properties have been proposed, offering more possibilities for particle guidance. Figure 3.13a demonstrates two types of non-paraxial snake-like beams constructed by combining multiple Airy beams using the axial-plane Gerchberg–Saxton algorithm [85]. In Fig. 3.13a(I), the beam consists of two end-to-end linked Airy beams, while Fig. 3.13a(II) shows a configuration of four linked Airy beams. Unlike Airy beams, snake-like beams exhibit multiple curved turns during propagation. Using these two snake-like beams, particle guiding experiments were conducted, with an axial transport distance exceeding 100 μm. The particle trajectories were visually observed with the axial-plane imaging technique. In addition, in 2012, I. D. Chremmos et al. introduced Bessel-like optical beams with arbitrary trajectories [125, 126]. These beams combine the advantages of both non-accelerating and accelerating waves, maintaining the circular profile of standard Bessel beams while accelerating along arbitrary paths. The generation principle of the Bessel-like beam relies on designing a continuous focal line, which is formed by the apexes of conical ray bundles emitted from a series of expanding and shifting circles on the input plane (Fig. 3.13bI). Figure Fig. 3.13b(II, III) shows the trajectory of a Bessel-like optical beam and transverse profiles at different distances. Subsequently, they expanded their research and designed a diffraction-resistant curved singular beam that can propagate along arbitrary trajectories in space [127]. These beams carry OAM and feature a central "hole" where the light intensity is zero, similar to the characteristics of an optical vortex. When applied to optical manipulation, it was observed that microparticles are gradually captured onto the central main annulus with maximum light intensity, and exhibited

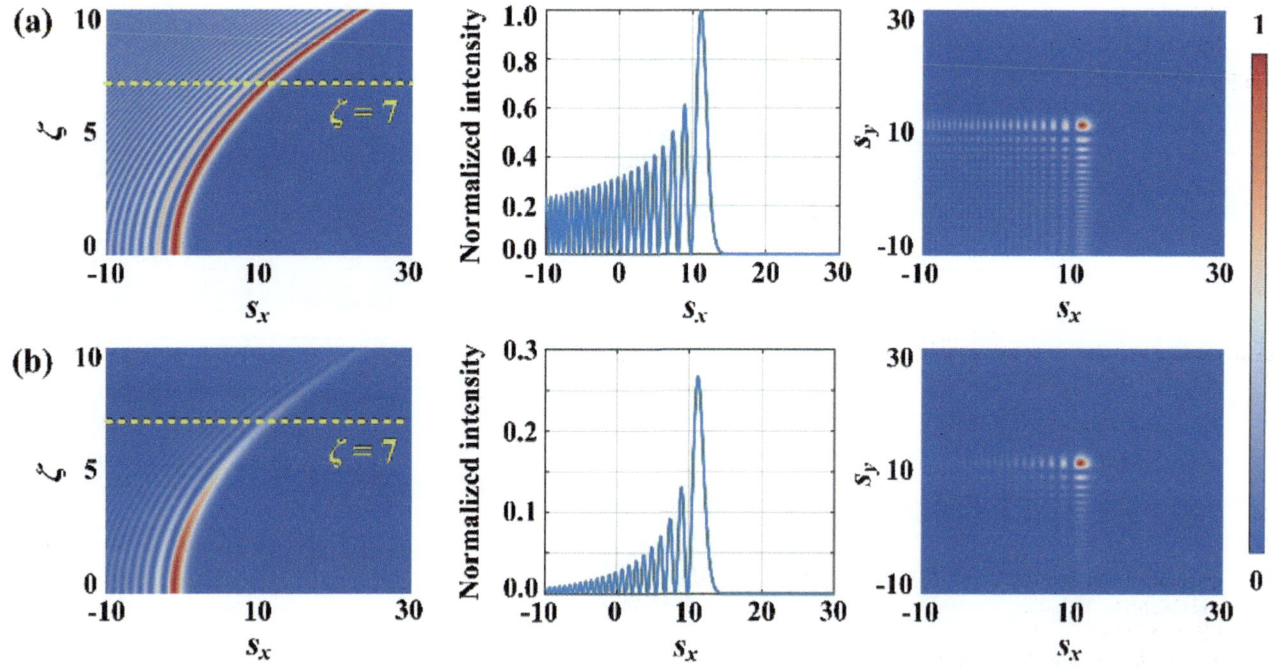

Fig. 3.11 Propagation dynamics of Airy beams. **a** Ideal Airy beam. **b** Finite-energy Airy beam with decay factor $a = 0.05$. The first column is the parabolic trajectory of the beams, and the corresponding transverse intensity profiles ($\zeta = 7$) and intensity distribution of 2D Airy beams are shown in the second and third columns, respectively

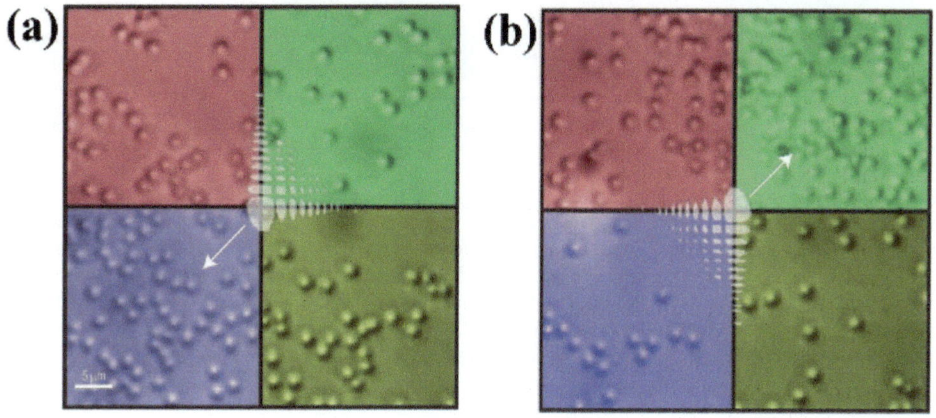

Fig. 3.12 Optically mediated particle clearing with Airy beam. **a** Colloidal particles are transported from the green region to the purple region. **b** After orienting the Airy beam in opposite direction, particles are back toward the green area (Adapted with permission from Springer Nature: Nat. Photonics [119])

three-dimensional spiral motion due to the transfer of OAM (Fig. 3.13c).

Abruptly autofocusing beams are a class of self-accelerating waves introduced in recent years [128–130], featured by the remarkable property that their intensity peak remains almost constant during propagation, while suddenly autofocuses and increases by several orders of magnitude as they approach the focal point. The fascinating optical characteristic holds significant potential in applications such as medical laser treatments and various nonlinear settings. Abruptly autofocusing beams were first proposed using circular Airy beams (CABs) [128]. These intriguing wave packets were subsequently observed experimentally [131] and their ability to guide and transport dielectric microparticles along the primary ring was explored [132]. As shown in Fig. 3.13d(I), the 3D intensity distribution of the autofocusing radially symmetric Airy beam is presented, with Fig. 3.13d(II–VI) corresponding to snapshots of the trapped microparticles in different transverse planes. It can be observed that the particles are captured by the primary ring of the beam, and the number of particles trapped within the ring decreases as the radius of the ring reduces (Fig. 3.13D(II–IV)). When the beam suddenly focuses into a bright spot, only one particle remains trapped at that location (Fig. 3.13d(V)). In the far field, the focused Airy beam turns into a Bessel beam [133], in which more particles are

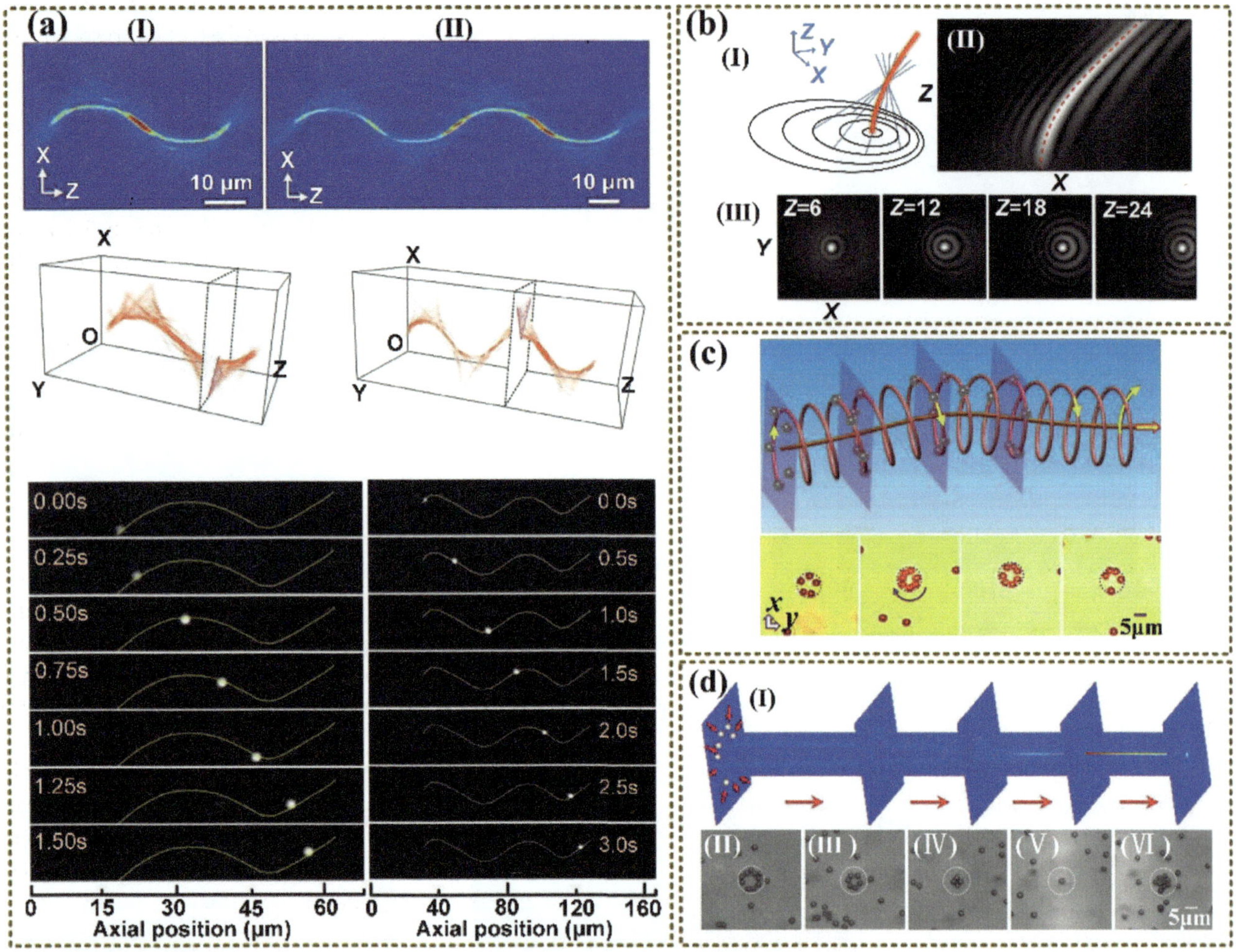

Fig. 3.13 Optical manipulation of self-accelerating beams. **a** Optical guiding with single-S (I) and double-S (II) snake-like beams (Used with permission of Reports on Progress in Physics, from Liang et al. [85], permission conveyed through Copyright Clearance Center, Inc). **b** Bessel-like optical beams. (I) Schematic of the principle. (II) The propagating trajectory of Bessel-like optical beams in X–Z plane. (III) transverse profile at different distances (Adapted with permission from [125] © Optical Society of America). **c** 3D microparticle spiraling driven with diffraction-resistant curved singular beam (Adapted with permission from [127] © Optical Society of America). **d** Particle guidance with an autofocusing radially symmetric Airy beam. (I) 3D intensity distribution of the beam. (II–VI) Snapshots of the trapped microparticles (Adapted with permission from [132] © Optical Society of America)

attracted (Fig. 3.13d(VI)). Beyond, by Fourier-transforming an appropriately apodized Bessel beam, circular abruptly autofocusing Airy beam or elegant paraboloid optical bottle can be created [134]. The fully closed optical bottle beam is capable of stably transporting absorbing particles from one container to another without being affected by ambient fluctuations [135]. Recently, Chen et al. introduced circular Pearcey beams (CPB) with abruptly autofocusing character, which have a stronger autofocused peak intensity and smaller focus distance compared to CAB [136]. Liang et al. demonstrated a circular Pearcey Airy beam (CPAB) by hybridizing CAB and CPB. The CPAB not only exhibits significantly enhanced trapping force but also possesses oscillating autofocus characteristic, i.e., multiple focusing peaks are formed along the direction of propagation, all of which are positioned to trap particles [137]. Moradi et al. performed the experiment of 3D optical trapping of submicron particles using a tightly focused CAB [138]. They pointed out that, compared to equivalent Gaussian beams with the same envelope or spot size, the tightly focused CAB offers higher trapping stiffness.

3.4 Optical Transport with Complex-Shaped Beams

Complex-shaped beams, characterized by tailored phase, amplitude, and polarization, such as generalized perfect optical vortices (GPOVs) [15], freestyle 3D laser traps [14,

139], knotted beams [140], solenoid beams [139, 141], and so on, have revolutionized optical manipulation. Using recent developments and advanced techniques as examples, we will explore the ideas, equations, and creation techniques of complicated structured light beams and the corresponding particle manipulations in this part.

3.4.1 Perfect Optical Vortices (POVs)

As introduced before, the LG beams with $p = 0$ are structured light fields with a helical (azimuthal) phase of the form $\exp(il\varphi)$ that carry $l\hbar$ OAM per photon, also called optical vortices (OVs) [23]. However, the OVs have "the strong dependence of the size of the central dark hollow of the generated vortex on its topological charge", which limits the order ($|l| \approx 100$) [142] and restricts the use of OVs in many applications. To break this limitation, A. S. Ostrovsky et al. introduced the concept of perfect optical vortices (POVs) [143]. The defining feature of POVs is that their ring radius remains constant, regardless of the topological charge, indicating a much higher intensity gradient [144]. This property makes them highly desirable for applications where a fixed beam size is necessary, such as in optical manipulation [13], optical encryption [145], and optical communication [146, 147].

The definition of the POV as an optical beam with the transverse distribution of complex amplitude given by the ideal model

$$g_v(\rho, \varphi) \equiv \delta(\rho - \rho_0) \exp(il\varphi), \tag{3.13}$$

where l represents the topological charge, (ρ, φ) are the polar coordinates, $\delta(\rho)$ is the Dirac function, ρ_0 denotes the radius of the vortex. According to Ref. [148], the POVs are seen as the Fourier transform property of a Bessel beam, which is generated by the combination of an axicon and a spiral phase plate. Noted that the ideal Bessel beam is impossible, Bessel-Gauss beam can be experimentally generated instead:

$$E(\rho, \phi) = J_l(k_r \rho) \exp(il\phi) \exp\left(-\frac{\rho^2}{\omega_g^2}\right), \tag{3.14}$$

where J_l is an lth order Bessel function of the first kind. k_r is the radial wave factor, ω_g is the beam waist of the Gaussian beam. Figure 3.14 shows the intensity profile (the first column), phase distribution (the second column), and the intensity plot (the third column) of POVs with topological charge $l = 5$ and 50.

Many efforts have been made to generate POVs, including SLMs, axicons, metasurface elements, and DMD. One common way to generate POVs is by the combination of an axicon and a vortex mask on the SLM [13, 149]. Additionally, a straightforward technique for generating POVs was proposed based on the width-pulse approximation of Bessel function [144]. In Ref. [150], POVs with the topological charge up to 90 were generated by employing a DMD. Detailed descriptions of POVs as Fourier transforms of Bessel beams and their experimental generation can be found in Refs. [148, 151]. Metasurface elements also provide efficient resolutions in the generation of POVs [152–154]. Besides, methods such as complex amplitude encoding [155] and free lens modulation [15] methods have been proposed to generate controllable, high-perfection POVs.

POVs have been experimentally observed and demonstrate significant potential in applications like optical trapping, optical communication, super-resolution imaging, and cryptography. Chen et al. first analyzed the dynamics of microparticles trapped within POVs and discovered the linear relationship between rotation rates and the topological charge (Fig. 3.15a), which has provided valuable insights into the behavior of particles under the influence of POVs [13] (Fig. 3.15b, c). In addition to the manipulation of high-refractive-index microparticles, Liang successfully employed POVs for the controllable manipulation of low-refractive-index microparticles by introducing "an extra point trap at the center of the vortex" [156], as shown in Fig. 3.15d–f.

Besides the integer topological charge, the topological charge value can also be nonintegral (when the phase step is not an integer multiple of 2π), referred to as a fractional vortex beam This type of beam induces complex amplitude and phase structures, thereby increasing the degrees of freedom for regulation [34, 157–159]. However, the phase appears discontinuous along the phase step, and the annular intensity ring is disrupted, resulting in a radial dark opening or low-intensity gap [160]. In 2017, TKACHENKO distributed the fractional phase discontinuity onto a number of smaller dislocations positioned periodically around the vortex [161], termed as fractional perfect optical vortices (FPOVs). The phase dislocations are distributed into N smaller dislocations. The phase mask for the SLM is defined as

$$\phi(l, \theta) = l_{\text{int}}\theta + (l - l_{\text{int}})[\theta \bmod (2\pi/N)] \tag{3.15}$$

where l_{int} represents the closest integer number to l, mod () is the remainder of the division. The intensity profile of experimental FPOVs is shown in Fig. 3.16.

The FPOV features a smooth intensity profile and allow for arbitrary topological charges extremely enriches the modes of POVs [162]. For instance, FPOVs can increase the optical communication capacity and overcome aperture size limitations [163–165]. Moreover, compared to the integer POVs, FPOVs enable more precise control of rotation speed through the interaction between light and matter [13, 166].

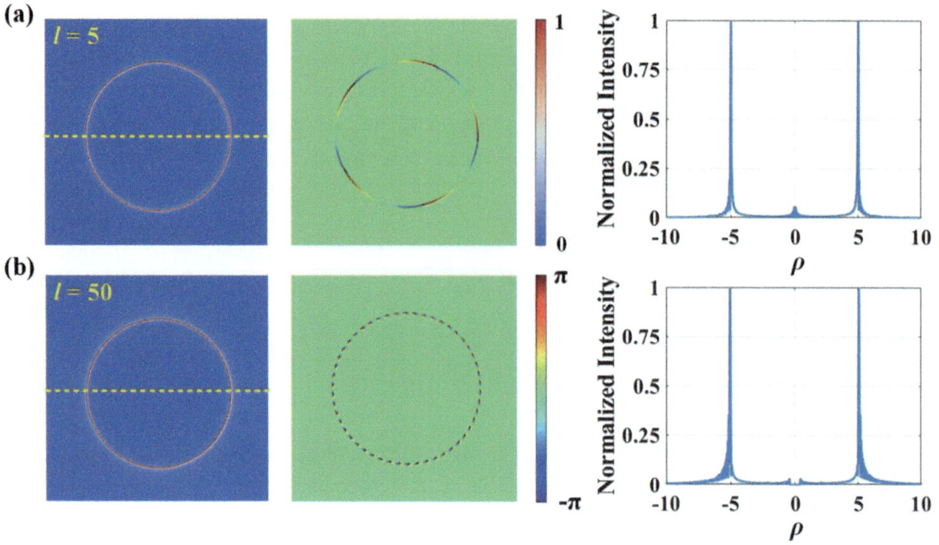

Fig. 3.14 Transverse intensity profiles of perfect optical vortices with different topological charges. **a** Topological charge $l = 5$. **b** Topological charge $l = 50$. The first column is the intensity profile of the beams, and the corresponding phase distribution and intensity plots are shown in the second and third columns, respectively

3.4.2 Generalized Perfect Optical Vortices (GPOVs)

Arbitrarily shaped POVs, also referred to as generalized POVs, share similar characteristics, such as a phase gradient along arbitrary trajectories and an intensity profile independent of the topological charge [15]. These light fields are extended into different intensity patterns, such as oval or triangular, no longer restricted to the circular shape. These alternately shaped POVs can provide additional possibilities in advanced applications, such as optical manipulation [12, 15], single-shot lithography [167], and optical communication [147, 168].

Taking the free lens modulation method as an example, the generalized expression of the transmission function of the digital lens can be [15]

$$t_l^{G}(r,\varphi) = P(r,\varphi)e^{-ik(r-\rho_0(\varphi))^2/2f(\varphi)}e^{il\varphi}, \tag{3.16}$$

where (r, φ) are the polar coordinates, $P(r, \varphi)$ represents the aperture function, $k = 2\pi/\lambda$ refers to the illumination wavenumber, and $\rho_0(\varphi)$ and $f(\varphi)$ denote the shape and focal distance of the designed digital lens. GPOVs with 2D polygon curves by setting the parameter $\rho_0(\varphi)$ to satisfy the relation

$$\rho_0(\varphi) = 1 - \frac{1}{p}\cos(q\varphi), \tag{3.17}$$

where p controls the smoothness of the polygon and q controls the shape of the curve. And the focal distance $f(\varphi)$ in Eq. (3.16) can satisfy the relation

$$f(\varphi) = f_0\frac{a - \sin(b\varphi)}{a}, \tag{3.18}$$

where a controls the slope and b controls the tilting times. Therefore, the intensity curve of 3D GPOVs takes the form

$$u_0(\varphi)|_{3D\ GPOV} = \left(1 - \frac{1}{p}\cos(q\varphi), \varphi, f_0\frac{a-\sin(b\varphi)}{a}\right). \tag{3.19}$$

Loading the phase maps of the free lenses on the SLM, various 3D GPOVs can be generated (Fig. 3.17a). Besides the arbitrarily shaped GPOVs (Fig. 3.17a, b), various complex structure beams have been reported, including double-ring POVs [169] (Fig. 3.17c), GPOV array [170] (Fig. 3.17d, e), and others [171].

GPOVs have emerged as a significant advancement in optical manipulation, offering unprecedented control over the motion trajectories of particles through the design of optical fields. Furthermore, the superposition of multiple generalized POVs results in "customized dark traps", which are highly quality and efficiently assembled structured beams [171]. The LRI particles can be confined/driven in arbitrary trajectories using the light fields with complex dark regions surrounded by high-intensity barriers, including rotation, translation, and sorting by size, as illustrated in Fig. 3.18.

3.4.3 Freestyle 3D Laser Traps

In the field of optical manipulation, the concept of "Freestyle 3D laser traps" plays a pivotal role in the precise control of light-driven particle dynamics. High intensity gradient (HIG) beams with a well-defined OAM distribution and a specified intensity profile are produced using the technique proposed by Rodrigo et al. [14, 139]. The complex function of the 3D curve satisfies

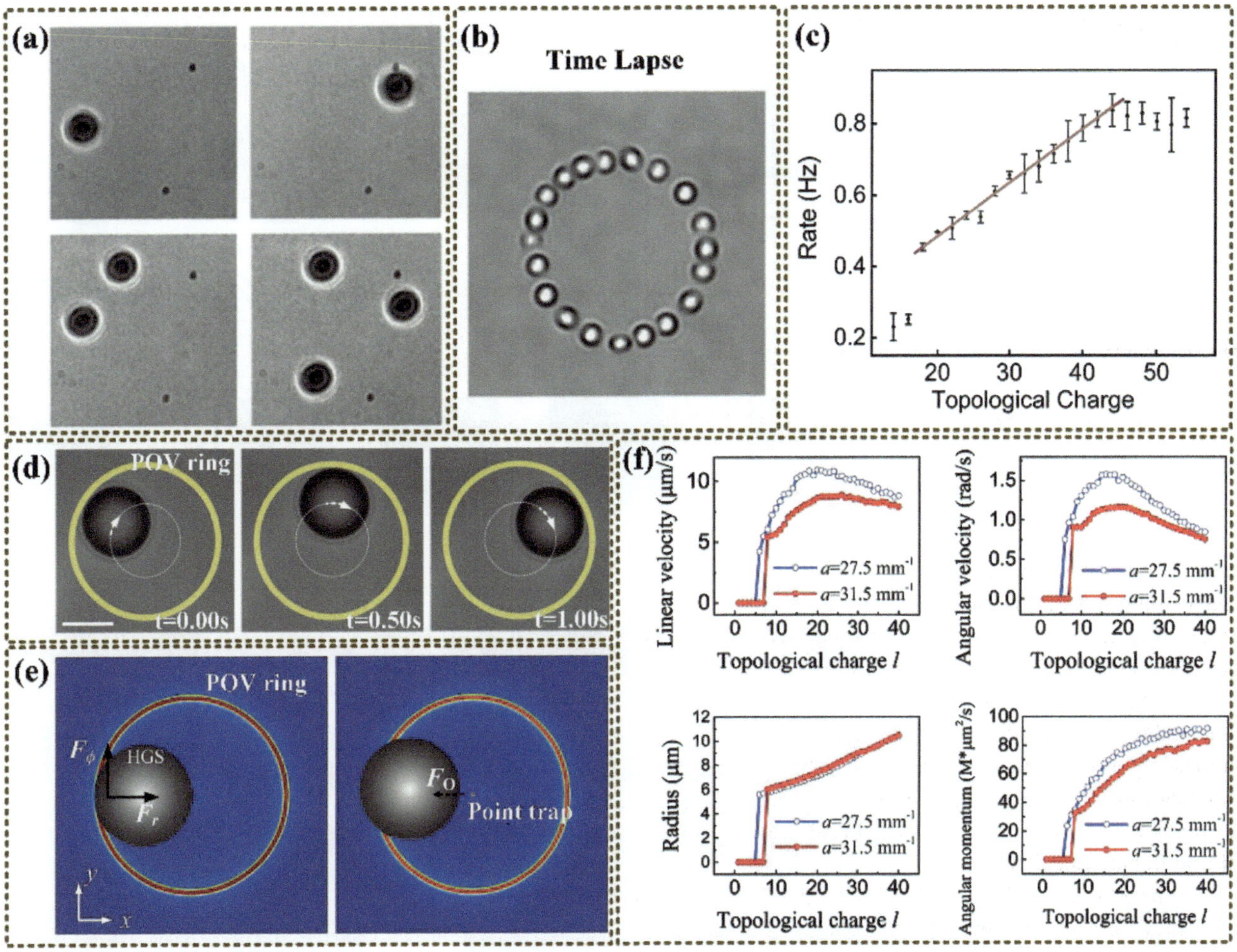

Fig. 3.15 Manipulation of particles with POVs. **a** show a single particle trapped by a POV (top) and multiple particles trapped by a POV after local OAM density correction (bottom) (Reprinted with permission from [13] © Optical Society of America). **b** Screenshot of the rotated particle in an annular POV. **c** Rotation rate against the topological charge (Reprinted with permission from[15] © Optical Society of America). **d** Rotating of a hollow glass sphere with a diameter of 14 μm. **e** Schematic analysis of rotation of a hollow glass sphere in two different light fields. **f** Rotating dynamics of a 12-μm-diameter hollow glass sphere in the POVs (Reprinted with permission from [156] © Optical Society of America)

$$H(\boldsymbol{r}) = \frac{1}{L}\int_0^T \phi(\boldsymbol{r}, t)\varphi(\boldsymbol{r}, t)\left|c_3'(t)\right|dt. \quad (3.20)$$

where $\mathbf{r} = (x, y)$, $\mathbf{c}_3(t) = (x_0(t), y_0(t), z_0(t))$ is the curve being the Cartesian coordinates. $L = \int_0^T \left|\boldsymbol{c}_3'(t)\right|dt$ denotes the length of the curve. The term $\phi(\mathbf{r}, t)$ shapes the trapping beam, which satisfies

$$\phi(\boldsymbol{r}, t) = \exp\left(\frac{i}{\rho^2}[yx_0(t) - xy_0(t)]\right)\exp\left(\frac{i2\pi m}{S(T)}S(t)\right), \quad (3.21)$$

where $t \in [0, T] S(t) = \int_0^t \left[x_0(\tau)y_0'(\tau) - y_0(\tau)x_0'(\tau)\right]d\tau$, ρ is a constant and m is the phase gradient parameter. And $\varphi(\mathbf{r}, t)$ is given by

$$\varphi(\boldsymbol{r}, t) = \exp\left(i\pi\frac{[x - x_0(t)]^2 + [y - y_0(t)]^2}{\lambda f^2}z_0(t)\right) \quad (3.22)$$

where f is the focal length, λ is the wavelength, and $z_0(t)$ is a defocusing distance defined along $\mathbf{c}_3(t)$. This beam shaping technique works as a "pencil tip" moves along the described curve and the generated HIG beams as shown in Fig. 3.19.

Figure 3.20 illustrates the experimental manipulation particles with HIG beams, called "freestyle 3D laser traps" [14]. This manipulation method enables the creation of freestyle traps that facilitate true 3D manipulation of particles. The trajectories of freestyle 3D laser traps including ring, triangle, solenoid, knotted, and toroidal-spiral laser traps, offering

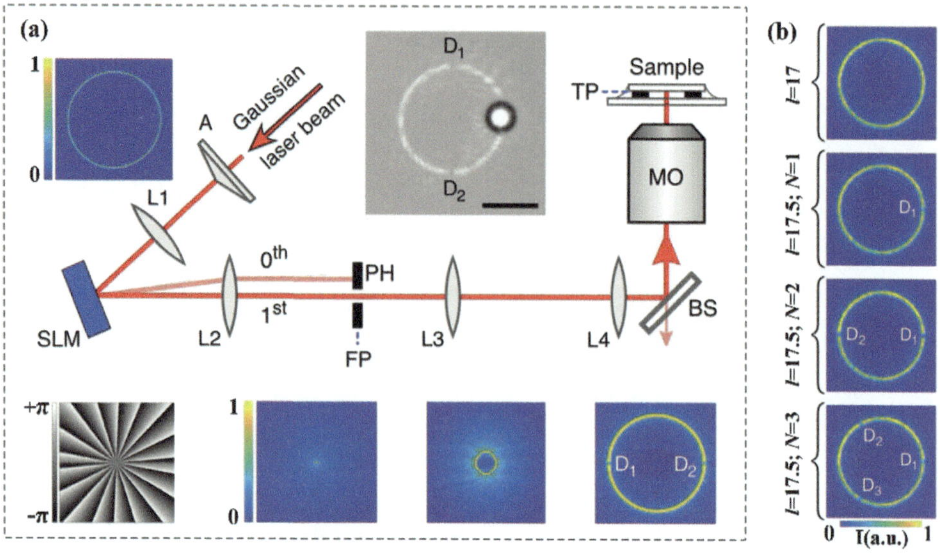

Fig. 3.16 Experimental Generation of fractional perfect optical vortices. **a** Sketch of the experimental setup and intensity patterns of FPOV generation by axicon. **b** Experimental PV rings with dislocations $N = 1, 2, 3$ (Adapted with permission from [161] © Optical Society of America)

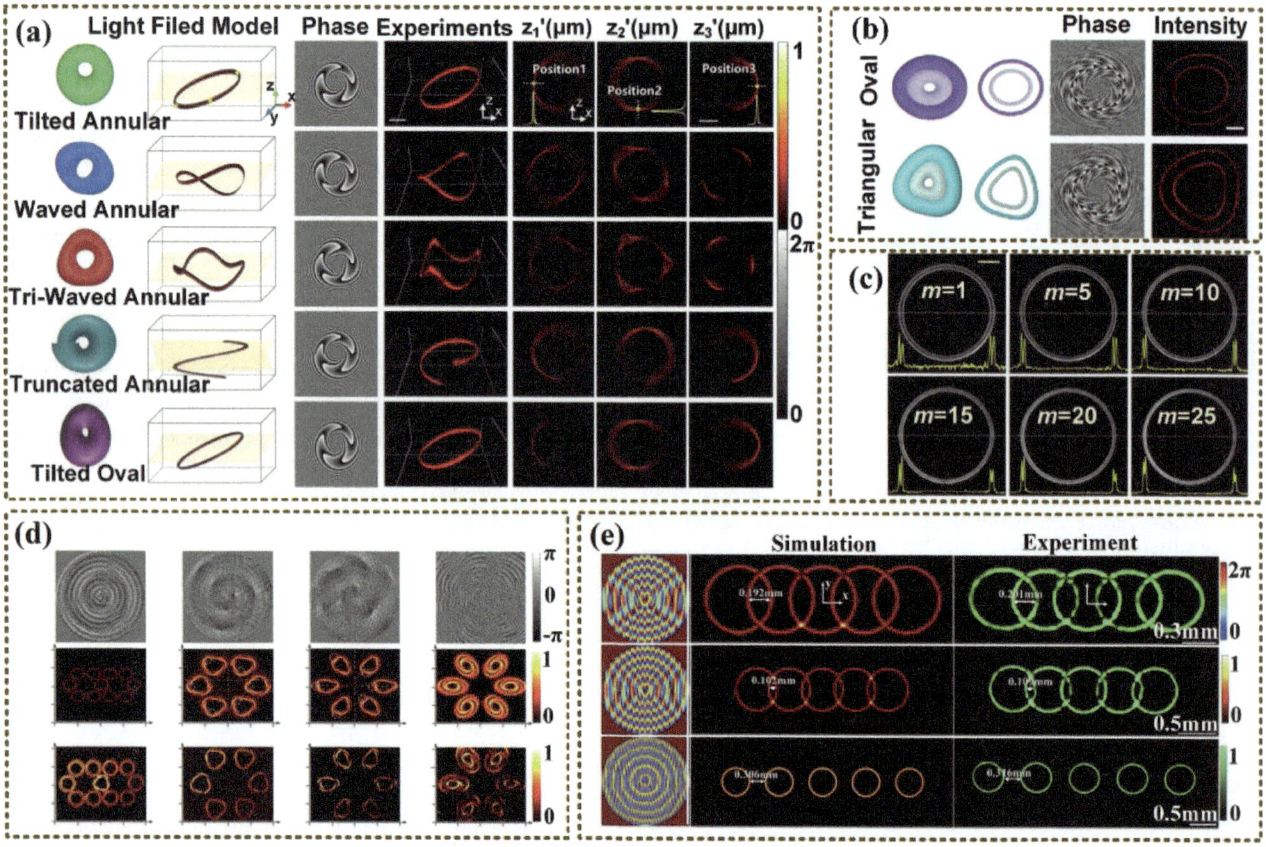

Fig. 3.17 Experimental Generation of generalized perfect optical vortices. **a** Generation of 3D GPOVs (Adapted with permission from [15] © Optical Society of America). **b** Free lens models, simulated light field models, phase maps, and experimentally generated light fields with $l = 10$ for oval and triangular dark traps, respectively. Scale bar: 10 μm (Adapted with permission from [171] © Optical Society of America). **c** Experimentally generated double-ring POVs for various topological charges (Reprinted with permission from [169] © Optical Society of America). **d** Vortex array composed of superposed double-ring vortex array (Image adapted with permission from Yang, Kaibo et al. [172], Copyright (2023), with permission from Elsevier). **e** Diagram of position control of 1D POV array (Adapted with permission from [170] © Optical Society of America)

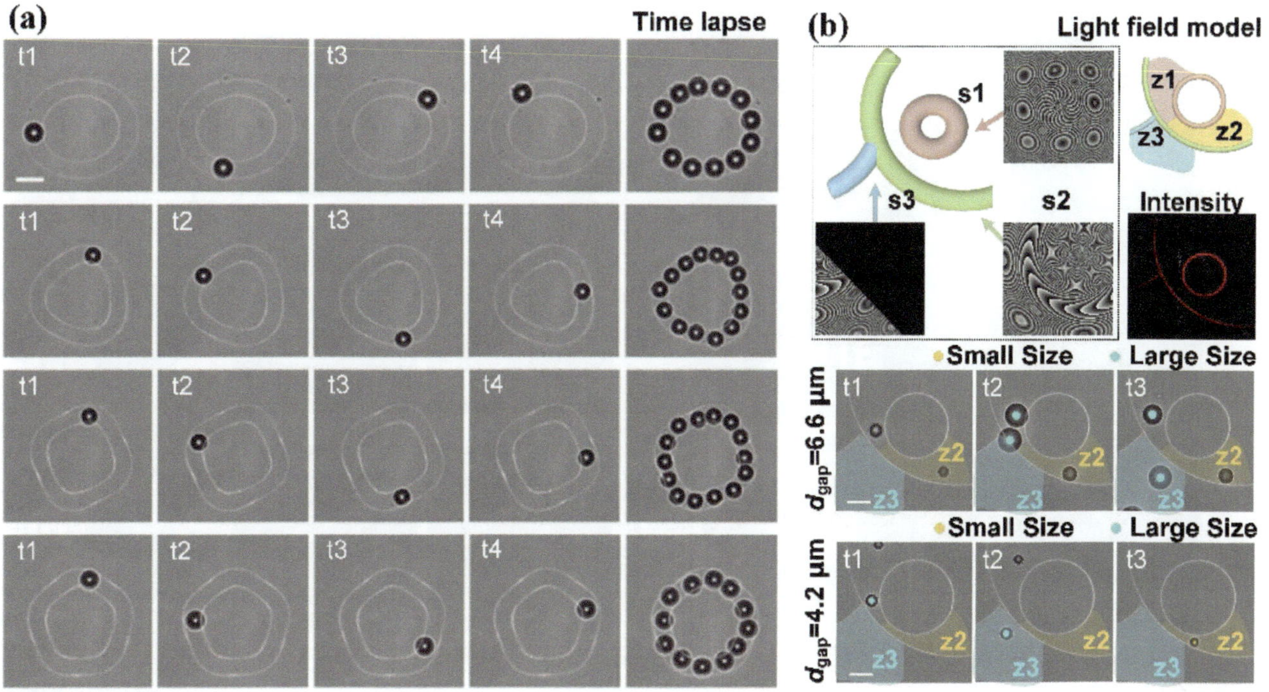

Fig. 3.18 Manipulation of microparticles with generalized perfect optical vortices [171]. **a** Screenshots (left) and time-lapse images (right) of an LRI particle rotating in customized dark traps. **b** Generation of dark traps for sorting and sorting experiments by size of LRI particles (Adapted with permission from [171] © Optical Society of America)

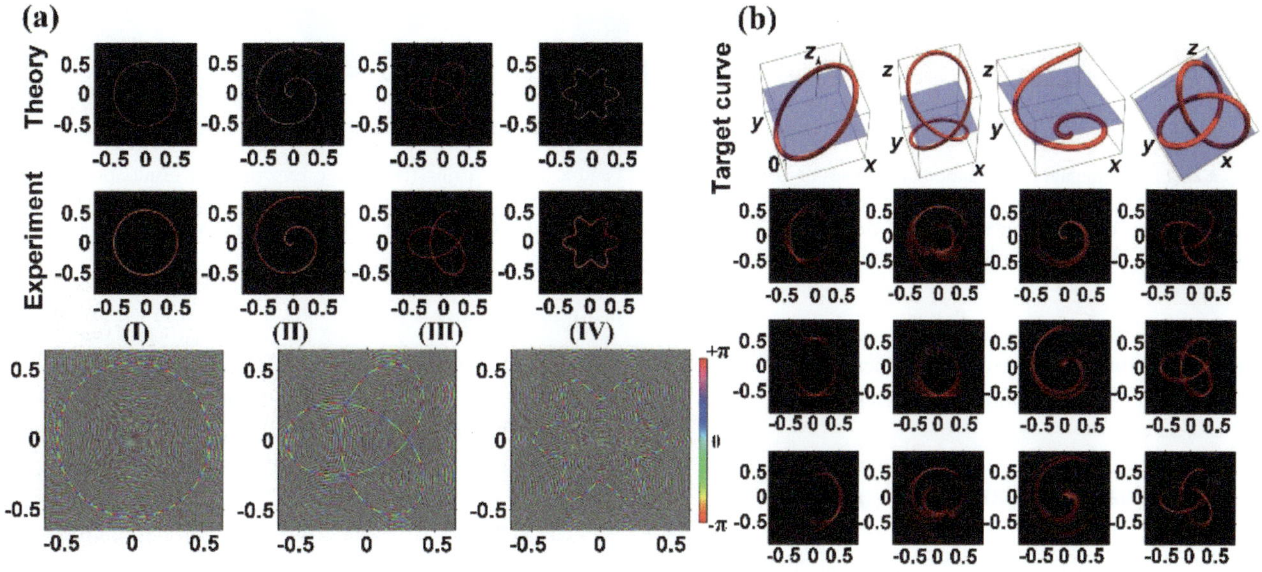

Fig. 3.19 Experimental Generation of high intensity gradient beams [139]. **a** Intensity distribution of the beam projected in the focal plane ($z = 0$) according to several closed and open 2D curves: (I) ring, (II) solenoid beam, (III) trefoil-knotted, and (IV) star-like curve. **b** Intensity distribution of the beams designed along 3D curves: tilted ring, Viviani's curve, Archimedean spiral, and trefoil-knotted curve (Adapted with permission from [139] © Optical Society of America)

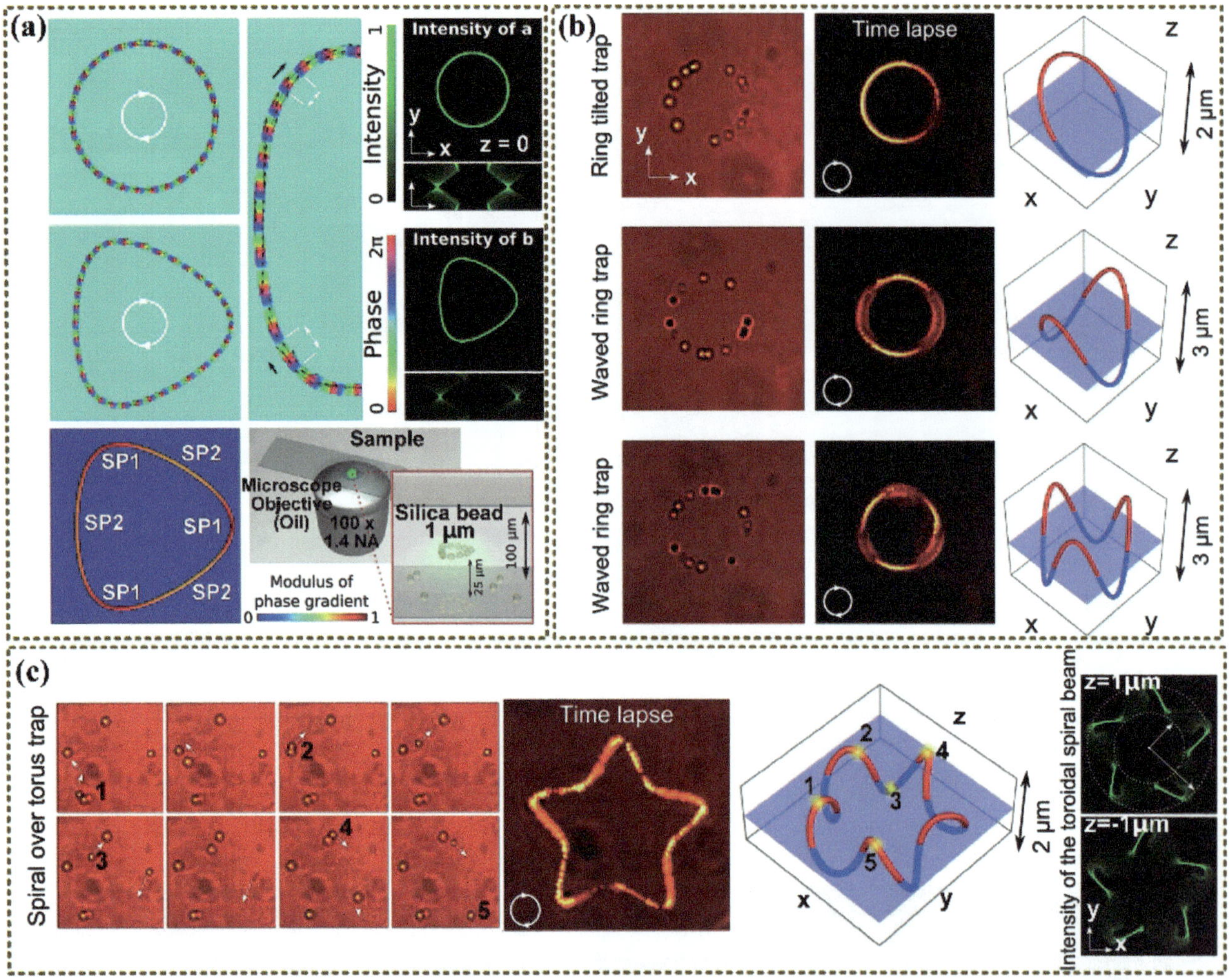

Fig. 3.20 Experimental manipulation of particles in freestyle 3D laser traps [14]. **a** Phase and intensity profile of the beam shaped in a ring and a triangle. **b** Experimental results of particles in 3D laser traps with radius $R = 5$ μm: a tilted ring (top), a waved ring (middle), and a tri-waved ring (bottom). **c** Experimental results of particles of the optically induced particle motion in a 3D toroidal-spiral laser trap (Adapted with permission from [14] © Optical Society of America)

greater flexibility in designing optical forces capable of confining and driving multiple particles.

3.4.4 Advanced Beam Shaping Techniques

In the rapidly advancing field of optical manipulation, a variety of techniques for the control of light fields have emerged, enabling the versatile manipulation of particles. Figure 3.21 shows some advanced beam shaping techniques to generate complex shaped beams. 3D solenoid beams, designed to create helical intensity profiles, can be employed to manipulate particles along helical trajectories [141]. Theoretical and experimental intensity profiles are shown in Fig. 3.21a. This technique is significant for its ability to guide particles along controlled paths, both along and against the optical axis. Additionally, knotted optical traps allow the manipulation of particles in a more intricate manner, which demonstrates an additional level of complexity in trapping potential. In Ref. [140], a pair of tilted ring traps was generated for the 3D optical trapping of multiple colloidal silica spheres (Fig. 3.21b). The real-time adaptation of the trapping beam's trajectory, enabling the manipulation of particles along arbitrary 3D paths is shown in Fig. 3.21c. The dynamic morphing technique is particularly important for its potential in real-time reconfigurable 3D optical transport [16]. In the field of nanoscale optical manipulation, Nan et al. combined modulation of intensity, phase, and polarization in structured light allows for versatile control over the manipulation of nanostructures [173] (Fig. 3.21d). This is particularly evident when a metal nanoparticle is confined within a linearly polarized laser field, where the optical force's direction is influenced

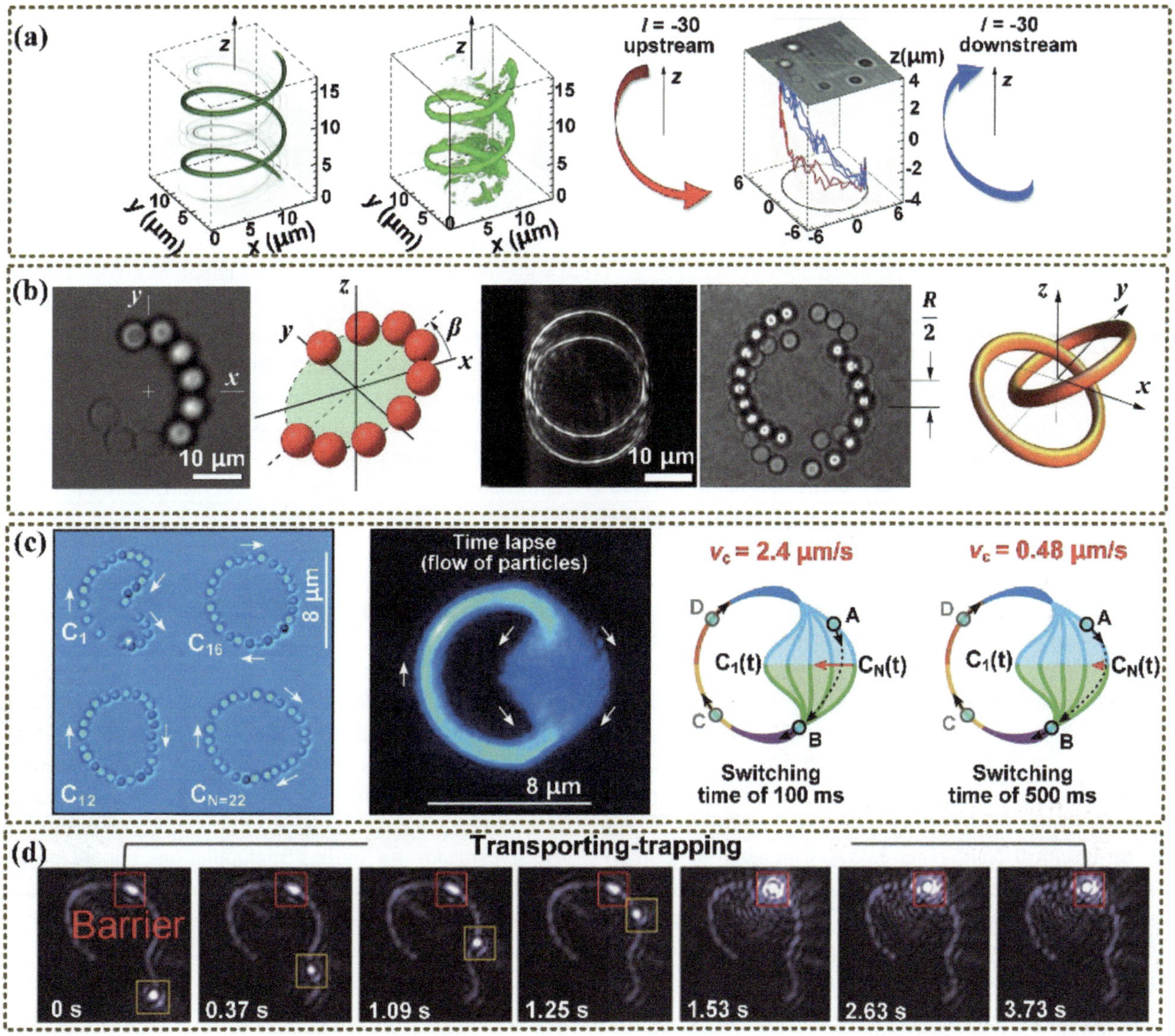

Fig. 3.21 Advanced beam shaping techniques and versatile manipulation of particles. **a** Generation of 3D solenoid beams and experimental optical transporting of microparticles results (Adapted with permission from [141] © Optical Society of America). **b** Generation of knotted beams and optical trapping and transporting of microparticles (Adapted with permission from [140] © Optical Society of America). **c** Experimental manipulation with dynamic morphing of a 2D curved laser trap (Adapted with permission from [16] © Optical Society of America). **d** Precisely manipulation of the particles along arbitrary trajectories (Reprinted with permission from Nan et al. [173], Copyright 2020 American Chemical Society)

by the particle's shape and the laser's intensity, phase, and polarization profiles.

3.5 Optical Manipulation with Cylindrical Vector Beams

As a transverse wave, the electric and magnetic fields of light are perpendicular to the propagation direction. Polarization refers to the vibration of the electric field, which is one of the most notable distinctions between transverse and longitudinal waves. Most traditional studies focus on spatially homogeneous polarization states, such as linear, circular, and elliptical polarization. In these states, the directions of the electric and magnetic fields of beams are homogeneously distributed in the beam cross section. Light beams exhibiting these polarization states are also defined as scalar beams. However, when the vibration direction of light varies non-uniformly in the plane perpendicular to the propagation direction, more complex vector light fields will emerge that significantly expand the boundaries of traditional scalar optics. Among these, the most notable are laser beams with cylindrical symmetry in polarization, known as cylindrical vector (CV) beams particularly those with radial and azimuthal polarization. This chapter will focus on several important optical micromanipulation applications of CV beams.

3.5.1 Optical Trapping with Cylindrical Vector Beam

CV beams are the solutions to the vector Helmholtz equation under cylindrical boundary conditions, with their polarization having cylindrical symmetry relative to the beam axis. The first row of Fig. 3.22 shows the far-field intensity distributions of the typical CV beams, where yellow arrows indicate the polarization direction. The polarization angle α represents the angle between the electric vector in the field and the radial direction. $\alpha = 0$ corresponds to radially polarization states (Fig. 3.22a), while $\alpha = \pi/2$ signifies azimuthal polarization (Fig. 3.22b). Generalized CV beams are the linear superposition of radial and azimuthal polarizations (Fig. 3.22c).

Although in the far-field the CV beams all exhibit a doughnut-shaped transverse intensity profile, their field distribution varies significantly when tightly focused by a high numerical aperture objective. Tightly focused radially polarized beam forms a smaller focal spot beyond the diffraction limit at the focal plane (Fig. 3.22d). Compared to spatially homogeneous polarization, radially polarized beams that feature large p polarization component enhance the axial trapping efficiency needed for stable trapping while reducing the scattering force, leading to the radially polarized beams being considered an ideal tool for trapping metallic particles [17, 18, 174]. Li et al. introduced a method for the selective trapping of particles with different chirality using a tightly focused vector beam with radially varying polarization [175]. The azimuthally polarized beam retains the doughnut mode under tightly focusing conditions (Fig. 3.22e), providing a distinct optical trapping scheme. High-refractive-index dielectric particles can be trapped in the ring structure, while low-refractive-index particles are trapped at the focal center [176, 177]. Based on these findings, Donato et al. demonstrated the optical trapping of nanotubes with radial and azimuthal polarization beams and proved the ability of CV beams to shape the trapping potential of one-dimensional nanostructures [178]. Zhong et al. revealed that using azimuthally polarized beams to trap core–shell magnetic microparticles can enhance the trapping efficiency in the propagation direction [19]. Recent studies indicated that metallic nanoparticles irradiated by a focused azimuthally polarized beam experience optical forces directing them toward a circular trapping region, without tangential force at equilibrium positions, allowing freely move around ring-shaped trajectory [179]. The intensity distribution of a tightly focused generalized CV beam with polarization angle α of 45° is shown in Fig. 3.22f. To expand the manipulation degrees of freedom beyond traditional scalar HOTs, N. Bhebhe et al. developed a vector HOTs system that utilizes digitally controlled higher-order Poincaré sphere beam arrays (Fig. 3.23) [180]. The innovation of this system lies in its ability to flexibly switch each beam in the array from scalar states to CV polarization states, enhancing the precision of manipulation and adapting a wide range of experimental requirements. In addition, advanced customization of the amplitude or phase of the incident vector beam can create a variety of special trapping potentials [181–183]. For instance, Otte et al. combined vector beams with phase vortices to customize tightly focused flower-shaped structured beams [184]. These beams can serve as advanced tools for capturing and orienting elongated zeolite-L nano-containers.

3.5.2 Long-Range Pulling Manipulation with Photophoretic Force

As mentioned in Sect. 2.1, the optical pulling force provided by Bessel beams allows objects to be drawn toward the light source. This phenomenon contrasts with traditional optical manipulations that push or trap objects, presenting a counterintuitive physical effect. Another light-induced force, known as the photophoretic force, also offers an effective means for pulling manipulation. For an absorptive particle irradiated by a tightly focused laser beam, in addition to experiencing radiation pressure, they are also subjected to laser-induced photophoretic force [185–188]. Due to differing radiation intensities on the two sides of the particle, the temperature on the illuminated side of the particle's surface is higher than that on the shadowed side. This temperature difference causes gas molecules on the illuminated side to impart greater momentum to the particle when they collide with and rebound from its surface, compared to gas molecules on the shadowed side. As a result, the photophoretic force is generated to drive the particle.

Figure 3.24 demonstrates the long-range (over tens of centimeters) pulling manipulation of gold-coated thin-walled spherical glass shells using photophoretic forces [91]. For such isotropic spherical particles, the direction of motion is determined by an asymmetry parameter (the J-factor), which is related to the distribution of the heat source over the particle volume. When $J < 0$, the particle moves along the direction of light propagation; conversely, if $J > 0$, the particle moves against the light beam. The sign of the J-factor can be controlled simply by adjusting the polarization state of the beam (Fig. 3.24a). Figure 3.24b shows a schematic of the experimental setup. The vector beam passes through two half-wave plates that switch the

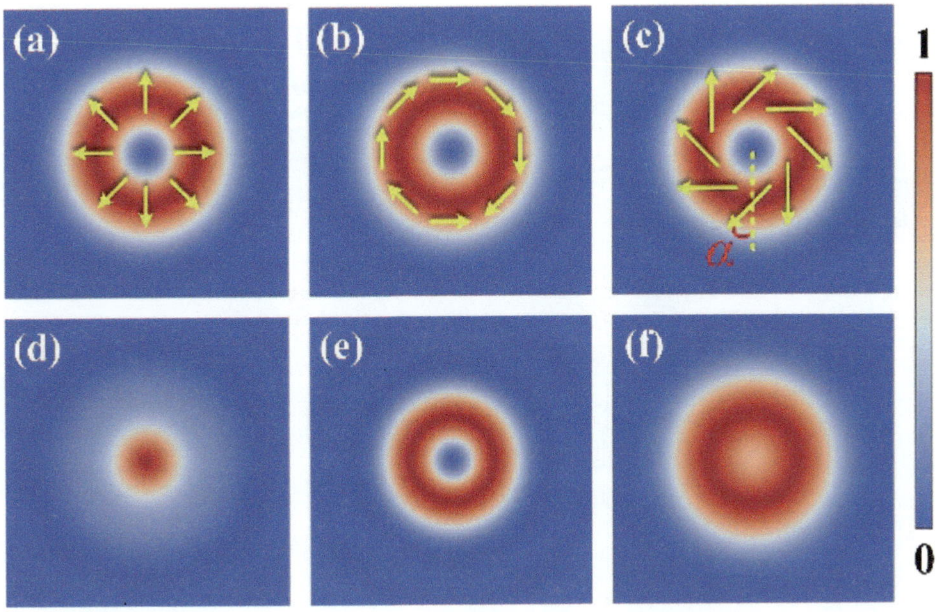

Fig. 3.22 Intensity distribution of CV beams in far-field (first row) and tightly focused (second row) cases. **a**, **d** Radially polarized mode. **b**, **e** Azimuthally polarized mode. **c**, **f** Generalized CV beams. α is the polarization angle

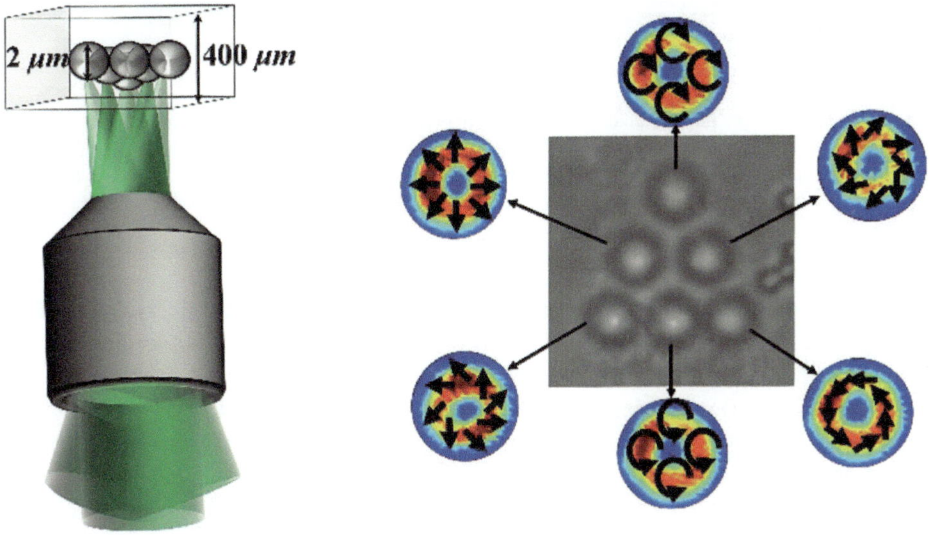

Fig. 3.23 Using vector beam arrays (left) with different polarization states to trap 2 μm silica beads (right) (Reprinted by permission from Springer Nature: Bhebhe et al. [180])

polarization state of the beam between azimuthal and radial polarization. Figure 3.24c, d show the motion of glass shells under different polarization states.

3.5.3 Imaginary Poynting Momentum Induced Particle Rotation

The Poynting vector is an important physical quantity used to describe the energy flux density in an electromagnetic field. Its magnitude represents the electromagnetic energy passing through a unit area per unit time, while its direction indicates the energy flow's direction. In optical tweezers, the gradient force depends on the spatial distribution of the light field and guides particles to the region where the light intensity is strongest. Meanwhile, the momentum transfer between photons and particles results in a force along the direction of light propagation, known as the radiation pressure [2]. This force is directly related to the real part of the Poynting vector, which is crucial for understanding the mechanism of optical trapping.

In recent years, the mechanical effects induced by the imaginary part of the Poynting vector (Imaginary Poynting momentum, IPM) have attracted widespread attention. IPM force is a lateral optical force that typically arises in special structured light fields such as evanescent waves [189] and interference fields [190]. It's worth noting that the CV beam carries IPM vortex with its direction completely aligned

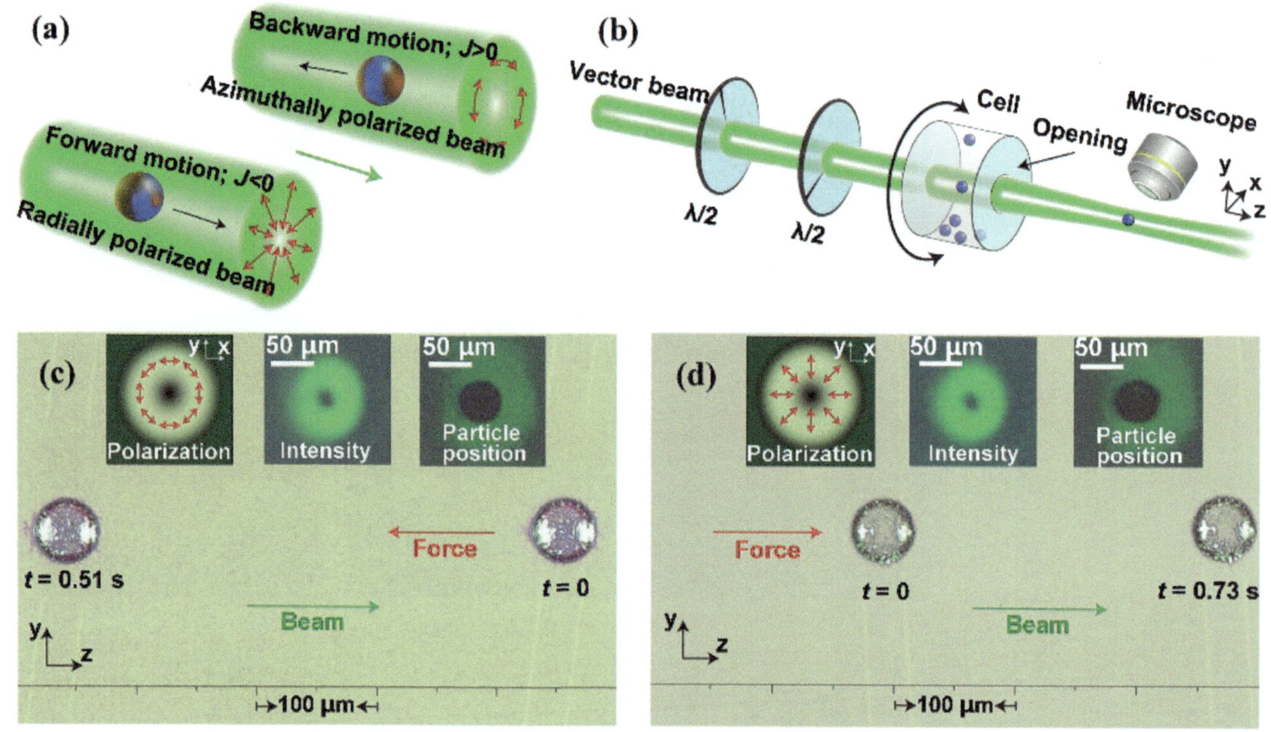

Fig. 3.24 Long-range pulling manipulation of gold-coated glass shells using cylindrical vector beams. **a** illustrates the schematic of polarization-controlled optical transport. The direction of a glass shell' s motion can be reversed by switching the polarization state of the beam. **b** Experimental setup. **c**, **d** show the snapshots of glass shell' s motion. The shell moves in opposite directions under azimuthally and radially polarized beams (Reprinted figure with permission from Springer Nature: Nature Photonics [91])

in the azimuthal direction, providing a new way to independently detect IPM forces (Fig. 3.25a, b) [191, 192]. In such a light field, the orbital angular momentum direction is distributed axially (Fig. 3.25c), while the Belinfante's spin momentum associated with the spin angular momentum is almost negligible, indicating that the field lacks orbital or spin angular momentum.

The IPM force was successfully observed in experiments using gold nanoparticles as probes [193]. The experimental setup is shown in Fig. 3.25d. An SLM was loaded with holograms to produce a Bessel-Gaussian beam. After passing through a half-wave plate and an S-waveplate, the linearly polarized light is modulated into cylindrical polarization. The light field focused by an objective lens exhibits a double-ring intensity profile. Figure 3.25e shows successive images where two gold particles are trapped on the inner and outer sides of the ring-shaped beam and rotate in opposite directions due to the IPM force. The IPM provides a novel mechanism to rotate particles without relying on orbital or spin angular momentum, expanding the degrees of freedom of optical micromanipulation.

3.5.4 Zero-Order-Free Beam Shaping Method and HOTs Apparatus

The SLM is one of the most powerful tools for beam shaping currently available, capable of modulating and generating almost all of the beams mentioned above. By precisely controlling the phase, intensity, and even polarization state of the light wave, the SLM enables the creation of complex light field distributions in space, offering significant flexibility for applications such as optical trapping, optical manipulation, and microstructural fabrication. Specifically, an SLM typically consists of an array of pixels, where each individual pixel can be controlled by electrical signal. By applying different voltages to each pixel, the SLM can adjust the phase or intensity of each part of the beam, thereby altering the beam's spatial profile. The operational modes of SLM include normal incidence (Fig. 3.26a) and oblique incidence (Fig. 3.26b). In HOTs, oblique incidence mode is usually used to minimize power loss.

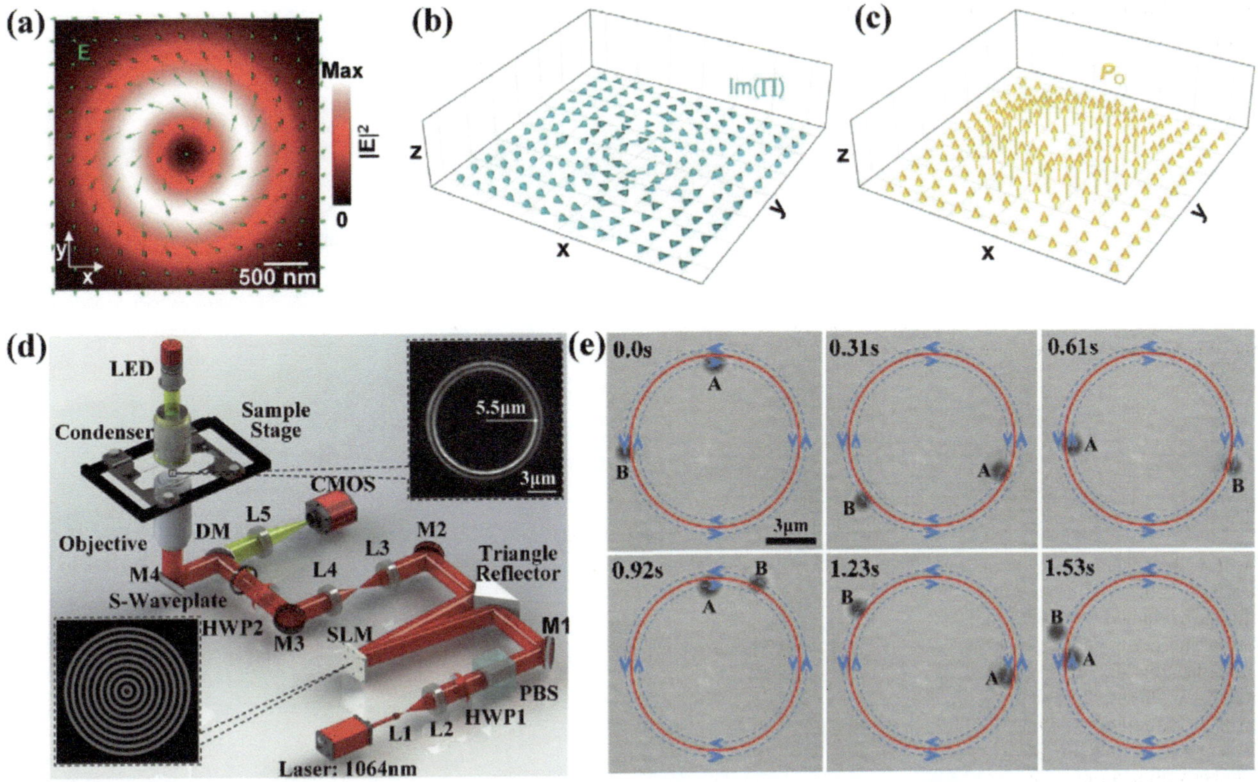

Fig. 3.25 The IPM in CV beams. **a** Intensity distribution and polarization direction of CV beam. **b**, **c** respectively represent the IPM density and orbital momentum density (Image adapted with permission from Xu et al. [191]. Copyright (2019) by the American Physical Society). **d** Experimental setup of IPM induced particle rotation. **e** Screenshot of two captured gold particles in reverse rotation around a circular orbit (Adapted from [193] under a Creative Commons CCBY 4.0 License)

However, a problem with SLM is that their pixelated structure causes parts of the beam to remain unmodulated. This unmodulated portion is known as "zero-order diffracted beams". The zero-order beam occupies part of the incident light's power, and due to its bright, highly localized nature, may affect the precise manipulation of HOTs. Especially when performing complex beam shaping, the power of the zero-order beam may even exceed that of the desired first-order beam, which is harmful to OTs applications.

To address this issue, several solutions have been proposed. The most common method involves using a spatial filter to block the unwanted zero-order beam in the Fourier plane (Fig. 3.26c) [194, 195]. Nevertheless, this approach often reduces the final reconstructed data region. Recently, we have proposed an innovative beam-shaping method that can completely suppress the zero-order beam [196, 197]. By using a homemade asymmetric triangle reflector and loading both grating and lens phases onto the SLM, this method ensures that only the first-order beam enters the relay system, while the zero-order beam is reflected backward, thus achieving the beam shaping without the burden of zero-order beam (Fig. 3.26d).

Furthermore, we have developed a self-designed zero-order free HOTs apparatus. Figure 3.26e, f show the light path diagram and the appearance of the apparatus. Compared to conventional HOTs apparatus, this device features a more compact form factor and higher manipulation precision, making it suitable for a wider range of applications, such as cell manipulation, force measurement, and microfluidics. Additionally, the design is simplified, easy to maintain, and highly practical with strong scalability, providing flexible solutions for various scientific experiments and industrial applications.

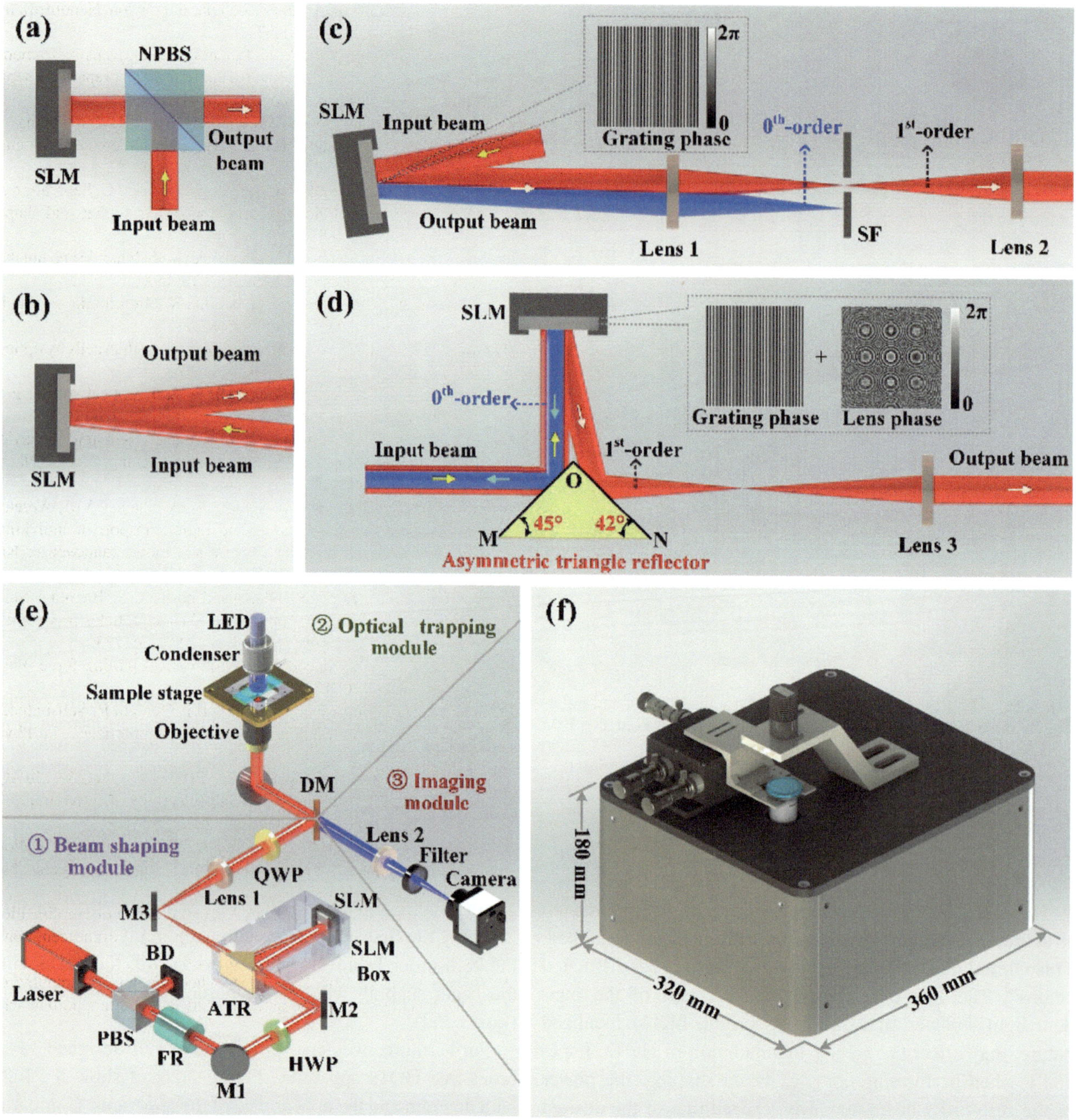

Fig. 3.26 Zero-order-free beam shaping method and HOTs apparatus. **a**, **b** are the two operational modes of SLM. **a** is normal incidence mode and **b** is oblique incidence mode. **c** Using spatial filter to suppress the zero-order beams. **d** Zero-order-free beam shaping method. **e** Light path diagram of the zero-order-free HOTs. BD: beam dump; BD: beam dump; BD: beam dump; DM: dichroic mirror. **f** The picture of the assembled HOTs apparatus (Adapted with permission from [197] © Optical Society of America)

3.6 Conclusion

In this chapter, we review the remarkable advances in the field of optical tweezers, focusing on the unprecedented manipulation capabilities brought about by the combination of structured beams, such as a variety of scalar beams, and vector beams, with optical tweezers technology. The introduction of structured light has enabled optical tweezers to not only make breakthroughs in manipulation accuracy, functionality, and dimensionality, but also greatly expand the application boundaries of optical tweezers across various fields. With the continuous advancement of beam shaping technologies and the ongoing evolution of optical tweezers, we believe that these developments will play an increasingly crucial role in advancing the understanding of light-matter interactions and fostering innovations in areas such as nanomanufacturing, biological and medical applications, and single-molecule spectroscopy.

Acknowledgements This work was supported in part by the National Key Research and Development Program of China (Nos. 2023YFF0722600, 2022YFF0712500); Natural Science Foundation of China (NSFC) (Nos. 62435007, 62135003, 62205267, 62205265); Natural Science Basic Research Program of Shaanxi (2024JC-YBMS494).

References

1. Ashkin A, Dziedzic JM, Bjorkholm JE, Chu S. Observation of a single-beam gradient force optical trap for dielectric particles. Opt Lett. 1986;11:288–90.
2. Ashkin A. Acceleration and trapping of particles by radiation pressure. Phys Rev Lett. 1970;24:156.
3. Dufresne ER, Grier DG. Optical tweezer arrays and optical substrates created with diffractive optics. Rev Sci Instrum. 1998;69:1974–7.
4. Liesener J, Reicherter M, Haist T, Tiziani HJ. Multi-functional optical tweezers using computer-generated holograms. Opt Commun. 2000;185:77–82.
5. Diekmann R, Wolfson DL, Spahn C, Heilemann M, Schüttpelz M, Huser T. Nanoscopy of bacterial cells immobilized by holographic optical tweezers. Nat Commun. 2016;7:13711.
6. Li C-Y, Cao D, Qi C-B, Kang Y-F, Song C-Y, Xu D-D, Zheng B, Pang D-W, Tang H-W. Combining holographic optical tweezers with upconversion luminescence encoding: imaging-based stable suspension array for sensitive responding of dual cancer biomarkers. Anal Chem. 2018;90:2639–47.
7. Sinjab F, Awuah D, Gibson G, Padgett M, Ghaemmaghami AM, Notingher I. Holographic optical trapping Raman micro-spectroscopy for non-invasive measurement and manipulation of live cells. Opt Express. 2018;26:25211–25.
8. Lee W, Kim H, Ahn J. Three-dimensional rearrangement of single atoms using actively controlled optical microtraps. Opt Express. 2016;24:9816–25.
9. Kim H, Lee W, Lee H-G, Jo H, Song Y, Ahn J. In situ single-atom array synthesis using dynamic holographic optical tweezers. Nat Commun. 2016;7:13317.
10. Gould OE, Box SJ, Boott CE, Ward AD, Winnik MA, Miles MJ, Manners I. Manipulation and deposition of complex, functional block copolymer nanostructures using optical tweezers. ACS Nano. 2019;13:3858–66.
11. Bergman J, Osunbayo O, Vershinin M. Constructing 3D microtubule networks using holographic optical trapping. Sci Rep. 2015;5:18085.
12. Curtis JE, Grier DG. Structure of optical vortices. Phys Rev Lett. 2003;90:133901.
13. Chen M, Mazilu M, Arita Y, Wright EM, Dholakia K. Dynamics of microparticles trapped in a perfect vortex beam. Opt Lett. 2013;38:4919–22.
14. Rodrigo JA, Alieva T. Freestyle 3D laser traps: tools for studying light-driven particle dynamics and beyond. Optica. 2015;2:812–5.
15. He M, Liang Y, Yun X, Wang Z, Zhao T, Wang S, Bianco PR, Lei M. Generalized perfect optical vortices with free lens modulation. Photon Res. 2023;11:27–34.
16. Rodrigo JA, Angulo M, Alieva T. Dynamic morphing of 3D curved laser traps for all-optical manipulation of particles. Opt Express. 2018;26:18608–20.
17. Kawauchi H, Yonezawa K, Kozawa Y, Sato S. Calculation of optical trapping forces on a dielectric sphere in the ray optics regime produced by a radially polarized laser beam. Opt Lett. 2007;32:1839–41.
18. Michihata M, Hayashi T, Takaya Y. Measurement of axial and transverse trapping stiffness of optical tweezers in air using a radially polarized beam. Appl Opt. 2009;48:6143–51.
19. Zhong M-C, Gong L, Li D, Zhou J-H, Wang Z-Q, Li YM. Optical trapping of core-shell magnetic microparticles by cylindrical vector beams. Appl Phys Lett. 2014;105.
20. Laabs H, Ozygus B. Excitation of Hermite Gaussian modes in end-pumped solid-state lasers via off-axis pumping. Opt Laser Technol. 1996;28:213–4.
21. Sato S, Ishigure M, Inaba H. Optical trapping and rotational manipulation of microscopic particles and biological cells using higher-order mode Nd: YAG laser beams. Electron Lett. 1991;20:1831–2.
22. Meyrath T, Schreck F, Hanssen J, Chuu C-S, Raizen M. A high frequency optical trap for atoms using Hermite-Gaussian beams. Opt Express. 2005;13:2843–51.
23. Allen L, Beijersbergen MW, Spreeuw R, Woerdman J. Orbital angular momentum of light and the transformation of Laguerre-Gaussian laser modes. Phys Rev A. 1992;45:8185.
24. Okida M, Omatsu T, Itoh M, Yatagai T. Direct generation of high power Laguerre-Gaussian output from a diode-pumped Nd: YVO4 1.3-μm bounce laser. Opt Express. 2007;15:7616–22.
25. Thirugnanasambandam M, Senatsky Y, Ueda K. Generation of very-high order Laguerre-Gaussian modes in Yb: YAG ceramic laser. Laser Phys Lett. 2010;7:637–43.
26. Qiao Z, Xie G, Wu Y, Yuan P, Ma J, Qian L, Fan D. Generating high-charge optical vortices directly from laser up to 288th order. Laser Photon Rev. 2018;12:1800019.
27. Uesugi Y, Kozawa Y, Sato S. Direct generation of the lowest-order vortex beam using a spot defect mirror in the ultraviolet region. Opt Lett. 2020;45:2115–8.
28. Zhao Y, Liu Q, Zhou W, Shen D. ~ 1 mJ pulsed vortex laser at 1645 nm with well-defined helicity. Opt Express. 2016;24:15596–602.
29. Li G, Xia K, Wang Z, Shen H, Shirakawa A, Ueda K-I, Li J. Conical refraction, for annular pumping of an efficient vortex Nd: YAG laser. Laser Phys Lett. 2017;14:075001.
30. He H-S, Chen Z, Dong J. Direct generation of vector vortex beams with switchable radial and azimuthal polarizations in a monolithic Nd: YAG microchip laser. Appl Phys Express. 2017;10:052701.
31. Li N, Xu B, Cui S, Qiu X, Luo Z, Xu H, Cai Z, Chen L, Moncorgé R. High-order vortex generation from CW and passively Q-switched Pr: YLF visible lasers. IEEE Photon Technol Lett. 2019;31:1457–60.
32. Dennis MR, O'holleran K, Padgett MJ. Singular optics: optical vortices and polarization singularities. In Progress in optics. Elsevier; 2009. p. 293–363.
33. Kim G-H, Jeon J-H, Ko K-H, Moon H-J, Lee J-H, Chang J-S. Optical vortices produced with a nonspiral phase plate. Appl Opt. 1997;36:8614–21.
34. Beijersbergen M, Coerwinkel R, Kristensen M, Woerdman J. Helical-wavefront laser beams produced with a spiral phaseplate. Opt Commun. 1994;112:321–7.
35. Ohtake Y, Ando T, Fukuchi N, Matsumoto N, Ito H, Hara T. Universal generation of higher-order multiringed Laguerre-Gaussian beams by using a spatial light modulator. Opt Lett. 2007;32:1411–3.
36. Ghai DP. Generation of optical vortices with an adaptive helical mirror. Appl Opt. 2011;50:1374–81.

37. Mao D, Zheng Y, Zeng C, Lu H, Wang C, Zhang H, Zhang W, Mei T, Zhao J. Generation of polarization and phase singular beams in fibers and fiber lasers. Adv Photon. 2021;3:014002–014002.
38. Simpson N, McGloin D, Dholakia K, Allen L, Padgett M. Optical tweezers with increased axial trapping efficiency. J Mod Opt. 1998;45:1943–9.
39. He H, Heckenberg N, Rubinsztein-Dunlop H. Optical particle trapping with higher-order doughnut beams produced using high efficiency computer generated holograms. J Mod Opt. 1995;42:217–23.
40. He H, Friese M, Heckenberg N, Rubinsztein-Dunlop H. Direct observation of transfer of angular momentum to absorptive particles from a laser beam with a phase singularity. Phys Rev Lett. 1995;75:826.
41. Padgett M, Allen L. Optical tweezers and spanners. Phys World. 1997;10:35.
42. Simpson N, Dholakia K, Allen L, Padgett M. Mechanical equivalence of spin and orbital angular momentum of light: an optical spanner. Opt Lett. 1997;22:52–4.
43. Simpson N, Allen L, Padgett M. Optical tweezers and optical spanners with Laguerre-Gaussian modes. J Mod Opt. 1996;43:2485–91.
44. Bishop AI, Nieminen TA, Heckenberg NR, Rubinsztein-Dunlop H. Optical microrheology using rotating laser-trapped particles. Phys Rev Lett. 2004;92:198104.
45. Leach J, Mushfique H, Keen S, Di Leonardo R, Ruocco G, Cooper J, Padgett M. Comparison of Faxén's correction for a microsphere translating or rotating near a surface. Phys Rev E Stat Nonlin Soft Matter Phys. 2009;79:026301.
46. Parkin SJ, Vogel R, Persson M, Funk M, Loke VL, Nieminen TA, Heckenberg NR, Rubinsztein-Dunlop H. Highly birefringent vaterite microspheres: production, characterization and applications for optical micromanipulation. Opt Express. 2009;17:21944–55.
47. Mohanty SK, Uppal A, Gupta PK. Self-rotation of red blood cells in optical tweezers: prospects for high throughput malaria diagnosis. Biotechnol Lett. 2004;26:971–4.
48. Elbez R, McNaughton BH, Patel L, Pienta KJ, Kopelman R. Nanoparticle induced cell magneto-rotation: monitoring morphology, stress and drug sensitivity of a suspended single cancer cell. PLoS ONE. 2011;6:e28475.
49. Oroszi L, Galajda P, Kirei H, Bottka S, Ormos P. Direct measurement of torque in an optical trap and its application to double-strand DNA. Phys Rev Lett. 2006;97:058301.
50. Forth S, Sheinin MY, Inman J, Wang MD. Torque measurement at the single-molecule level. Annu Rev Biophys. 2013;42:583–604.
51. Asavei T, Nieminen TA, Heckenberg NR, Rubinsztein-Dunlop H. Fabrication of microstructures for optically driven micromachines using two-photon photopolymerization of UV curing resins. J Opt A: Pure Appl Opt. 2009;11:034001.
52. Friese M, Rubinsztein-Dunlop H, Gold J, Hagberg P, Hanstorp D. Optically driven micromachine elements. Appl Phys Lett. 2001;78:547–9.
53. Paiè P, Zandrini T, Vázquez RM, Osellame R, Bragheri F. Particle manipulation by optical forces in microfluidic devices. Micromachines. 2018;9:200.
54. Zhang S, Elsayed M, Peng R, Chen Y, Zhang Y, Peng J, Li W, Chamberlain MD, Nikitina A, Yu S. Reconfigurable multi-component micromachines driven by optoelectronic tweezers. Nat Commun. 2021;12:5349.
55. Grier DG. A revolution in optical manipulation. Nature. 2003;424:810–6.
56. Shvedov VG, Desyatnikov AS, Rode AV, Krolikowski W, Kivshar YS. Optical guiding of absorbing nanoclusters in air. Opt Express. 2009;17:5743–57.
57. Shvedov V, Desyatnikov AS, Rode AV, Izdebskaya Y, Krolikowski W, Kivshar YS. Optical vortex beams for trapping and transport of particles in air. Appl Phys A. 2010;100:327–31.
58. Paterson L, MacDonald MP, Arlt J, Sibbett W, Bryant PE, Dholakia K. Controlled rotation of optically trapped microscopic particles. Science. 2001;292:912–4.
59. MacDonald MP, Paterson L, Volke-Sepulveda K, Arlt J, Sibbett W, Dholakia K. Creation and manipulation of three-dimensional optically trapped structures. Science. 2002;296:1101–3.
60. Bandres MA, Gutiérrez-Vega JC. Ince–gaussian beams. Opt Lett. 2004;29:144–6.
61. Schwarz UT, Bandres MA, Gutiérrez-Vega JC. Observation of Ince-Gaussian modes in stable resonators. Opt Lett. 2004;29:1870–2.
62. Schwarz UT, Bandres MA, Gutierrez-Vega JC. Formation of Ince-Gaussian modes in a stable laser oscillator. In: Laser resonators and beam control VIII. SPIE; 2005. p. 124–31.
63. Ohtomo T, Kamikariya K, Otsuka K, Chu S-C. Single-frequency Ince-Gaussian mode operations of laser-diode-pumped microchip solid-state lasers. Opt Express. 2007;15:10705–17.
64. Chu S-C, Otsuka K. Numerical study for selective excitation of Ince-Gaussian modes in end-pumped solid-state lasers. Opt Express. 2007;15:16506–19.
65. Lei J, Hu A, Wang Y, Chen P. A method for selective excitation of Ince-Gaussian modes in an end-pumped solid-state laser. Appl Phys B. 2014;117:1129–34.
66. Woerdemann M, Alpmann C, Denz C. Optical assembly of microparticles into highly ordered structures using Ince–Gaussian beams. Appl Phys Lett. 2011;98.
67. Ngcobo S, Litvin I, Burger L, Forbes A. A digital laser for on-demand laser modes. Nat Commun. 2013;4:2289.
68. Ren Y-X, Fang Z-X, Gong L, Huang K, Chen Y, Lu R-D. Dynamic generation of Ince-Gaussian modes with a digital micromirror device. J Appl Phys. 2015;117.
69. Ahmed H, Ansari MA, Paterson L, Li J, Chen X. Metasurface for engineering superimposed Ince-Gaussian beams. Adv Mater. 2024;2312853.
70. Durnin J, Miceli J Jr, Eberly JH. Diffraction-free beams. Phys Rev Lett. 1987;58:1499.
71. McGloin D, Dholakia K. Bessel beams: diffraction in a new light. Contemp Phys. 2005;46:15–28.
72. Durnin J. Exact solutions for nondiffracting beams. I. The scalar theory. JOSA A. 1987;4:651–4.
73. Scott G, McArdle N. Efficient generation of nearly diffraction-free beams using an axicon. Opt Eng. 1992;31:2640–3.
74. Sun Q, Zhou K, Fang G, Liu Z, Liu S. Generation of spiraling high-order Bessel beams. Appl Phys B. 2011;104:215–21.
75. Vetter C, Steinkopf R, Bergner K, Ornigotti M, Nolte S, Gross H, Szameit A. Realization of free-space long-distance self-healing Bessel beams. Laser Photonics Rev. 2019;13:1900103.
76. Yun X, Liang Y, He M, Guo L, Zhang X, Wang S, Zhao T, Kuang S, Lei M. High-efficiency generation of long-distance, tunable, high-order nondiffracting beams. Photon Res. 2024;12:2390–400.
77. Birch P, Ituen I, Young R, Chatwin C. Long-distance Bessel beam propagation through Kolmogorov turbulence. JOSA A. 2015;32:2066–73.
78. Chen WT, Khorasaninejad M, Zhu AY, Oh J, Devlin RC, Zaidi A, Capasso F. Generation of wavelength-independent subwavelength Bessel beams using metasurfaces. Light Sci Appl. 2017;6:e16259–e16259.
79. Fan Y, Cluzel B, Petit M, Le Roux X, Lupu A, De Lustrac A. 2D waveguided Bessel beam generated using integrated metasurface-based plasmonic axicon. ACS Appl Mater Interf. 2020;12:21114–9.
80. Zhi Z, Na Q, Xie Q, Chen B, Li Y, Liu X, Li X, Wang L, Lo G, Song J. On-chip generation of Bessel-Gaussian beam via concentrically

distributed grating arrays for long-range sensing. Light Sci Appl. 2023;12:92.
81. Sabatyan A, Meshginqalam B. Generation of annular beam by a novel class of Fresnel zone plate. Appl Opt. 2014;53:5995–6000.
82. Arlt J, Garcés-Chávez V, Sibbett W, Dholakia K. Optical micromanipulation using a Bessel light beam. Opt Commun. 2001;197:239–45.
83. Garcés-Chávez V, McGloin D, Melville H, Sibbett W, Dholakia K. Simultaneous micromanipulation in multiple planes using a self-reconstructing light beam. Nature. 2002;419:145–7.
84. Liang Y, Yan S, Yao B, Lei M. Direct observation and characterization of optical guiding of microparticles by tightly focused non-diffracting beams. Opt Express. 2019;27:37975–85.
85. Liang Y, Yan S, Wang Z, Li R, Cai Y, He M, Yao B, Lei M. Simultaneous optical trapping and imaging in the axial plane: a review of current progress. Rep Prog Phys. 2020;83:032401.
86. Volke-Sepulveda K, Garcés-Chávez V, Chávez-Cerda S, Arlt J, Dholakia K. Orbital angular momentum of a high-order Bessel light beam. J Opt B: Quantum Semiclas Opt. 2002;4:S82.
87. Khonina SN, Kotlyar V, Skidanov R, Soifer V, Jefimovs K, Simonen J, Turunen J. Rotation of microparticles with Bessel beams generated by diffractive elements. J Mod Opt. 2004;51:2167–84.
88. Brzobohatý O, Karásek V, Šiler M, Chvátal L, Čižmár T, Zemánek P. Experimental demonstration of optical transport, sorting and self-arrangement using a 'tractor beam.' Nat Photon. 2013;7:123–7.
89. Mitri F. Optical Bessel tractor beam on active dielectric Rayleigh prolate and oblate spheroids. JOSA B. 2017;34:899–908.
90. Mitri F, Li R, Guo L, Ding C. Optical tractor Bessel polarized beams. J Quant Spectrosc Radiat Transfer. 2017;187:97–115.
91. Shvedov V, Davoyan AR, Hnatovsky C, Engheta N, Krolikowski W. A long-range polarization-controlled optical tractor beam. Nat Photon. 2014;8:846–50.
92. Novitsky A, Qiu C-W, Wang H. Single gradientless light beam drags particles as tractor beams. Phys Rev Lett. 2011;107:203601.
93. Zamboni-Rached M. Stationary optical wave fields with arbitrary longitudinal shape by superposing equal frequency Bessel beams: Frozen Waves. Opt Express. 2004;12:4001–6.
94. Zamboni-Rached M, Recami E, Hernández-Figueroa HE. Theory of "frozen waves": modeling the shape of stationary wave fields. JOSA A. 2005;22:2465–75.
95. Vieira TA, Gesualdi MR, Zamboni-Rached M. Frozen waves: experimental generation. Opt Lett. 2012;37:2034–6.
96. Vieira TA, Gesualdi MR, Zamboni-Rached M, Recami E. Production of dynamic frozen waves: controlling shape, location (and speed) of diffraction-resistant beams. Opt Lett. 2015;40:5834–7.
97. Chen W, Ge X, Zhang Q, Natan RG, Fan JL, Scanziani M, Ji N. High-throughput volumetric mapping of synaptic transmission. Nat Methods. 2024;1–8.
98. Suarez RA, Ambrosio LA, Neves AA, Zamboni-Rached M, Gesualdi MR. Experimental optical trapping with frozen waves. Opt Lett. 2020;45:2514–7.
99. Gutiérrez-Vega JC, Iturbe-Castillo M, Chávez-Cerda S. Alternative formulation for invariant optical fields: Mathieu beams. Opt Lett. 2000;25:1493–5.
100. Alpmann C, Boguslawski M, Rose P, Wördemann M, Denz C. Tailored light fields: nondiffracting and self-similar beams for optical structuring and organization. In: Complex light and optical forces VI. SPIE; 2012. p. 123–34.
101. Gutiérrez-Vega J, Iturbe-Castillo M, Ramırez G, Tepichın E, Rodrıguez-Dagnino R, Chávez-Cerda S, New G. Experimental demonstration of optical Mathieu beams. Opt Commun. 2001;195:35–40.
102. Chávez-Cerda S, Padgett M, Allison I, New G, Gutiérrez-Vega JC, O'Neil A, MacVicar I, Courtial J. Holographic generation and orbital angular momentum of high-order Mathieu beams. J Opt B: Quant Semiclas Opt. 2002;4:S52.
103. Alvarez-Elizondo MB, Rodriguez-Masegosa R, Gutiérrez-Vega JC. Generation of Mathieu-Gauss modes with an axicon-based laser resonator. Opt Express. 2008;16:18770–5.
104. Ren Z, Hu H, Peng B. Generation of Mathieu beams using the method of 'combined axicon and amplitude modulation.' Opt Commun. 2018;426:226–30.
105. Hernández-Hernández R, Terborg R, Ricardez-Vargas I, Volke-Sepúlveda K. Experimental generation of Mathieu-Gauss beams with a phase-only spatial light modulator. Appl Opt. 2010;49:6903–9.
106. Alpmann C, Bowman R, Woerdemann M, Padgett M, Denz C. Mathieu beams as versatile light moulds for 3D micro particle assemblies. Opt Express. 2010;18:26084–91.
107. Yuan D, Yu S, Kou N. Generation of Mathieu beam using a single layer transmissive metasurface. Opt Commun. 2024;557:130346.
108. Lóxpez-Mariscal C, Gutiérrez-Vega JC, Milne G, Dholakia K. Orbital angular momentum transfer in helical Mathieu beams. Opt Express. 2006;14:4182–7.
109. Li X, Ren Z, Xu F, Song L, Lv X, Qian Y, Yu P. Generation of perfect helical Mathieu vortex beams. Opt Express. 2021;29:32439–52.
110. Lopez-Mariscal C, Gutierrez-Vega J, Garces-Chavez V, Dholakia K. Observation of the angular momentum transfer in the Mie regime using Mathieu beams. In: Optical trapping and optical micromanipulation II. SPIE; 2005. p. 468–77.
111. Siviloglou GA, Christodoulides DN. Accelerating finite energy Airy beams. Opt Lett. 2007;32:979–81.
112. Siviloglou GA, Broky J, Dogariu A, Christodoulides D. Observation of accelerating Airy beams. Phys Rev Lett. 2007;99:213901.
113. Siviloglou G, Broky J, Dogariu A, Christodoulides D. Ballistic dynamics of Airy beams. Opt Lett. 2008;33:207–9.
114. Kaminer I, Segev M, Christodoulides DN. Self-accelerating self-trapped optical beams. Phys Rev Lett. 2011;106:213903.
115. Dolev I, Kaminer I, Shapira A, Segev M, Arie A. Experimental observation of self-accelerating beams in quadratic nonlinear media. Phys Rev Lett. 2012;108:113903.
116. Porat G, Dolev I, Barlev O, Arie A. Airy beam laser. Opt Lett. 2011;36:4119–21.
117. Longhi S. Airy beams from a microchip laser. Opt Lett. 2011;36:716–8.
118. Li L, Li T, Wang S, Zhang C, Zhu S. Plasmonic Airy beam generated by in-plane diffraction. Phys Rev Lett. 2011;107:126804.
119. Baumgartl J, Mazilu M, Dholakia K. Optically mediated particle clearing using Airy wavepackets. Nat Photon. 2008;2:675–8.
120. Baumgartl J, Hannappel GM, Stevenson DJ, Day D, Gu M, Dholakia K. Optical redistribution of microparticles and cells between microwells. Lab Chip. 2009;9:1334–6.
121. Cheng H, Zang W, Zhou W, Tian J. Analysis of optical trapping and propulsion of Rayleigh particles using Airy beam. Opt Express. 2010;18:20384–94.
122. Zheng Z, Zhang B-F, Chen H, Ding J, Wang H-T. Optical trapping with focused Airy beams. Appl Opt. 2010;50:43–9.
123. Suarez RA, Neves AA, Gesualdi MR. Optimizing optical trap stiffness for Rayleigh particles with an Airy array beam. JOSA B. 2020;37:264–70.
124. Suarez RA, Neves AA, Gesualdi MR. Optical trapping with non-diffracting Airy beams array using a holographic optical tweezers. Opt Laser Technol. 2021;135:106678.
125. Chremmos ID, Chen Z, Christodoulides DN, Efremidis NK. Bessel-like optical beams with arbitrary trajectories. Opt Lett. 2012;37:5003–5.
126. Zhao J, Zhang P, Deng D, Liu J, Gao Y, Chremmos I, Efremidis NK, Christodoulides DN, Chen Z. Observation of self-accelerating

Bessel-like optical beams along arbitrary trajectories. Opt Lett. 2013;38:498–500.
127. Zhao J, Chremmos ID, Song D, Christodoulides DN, Efremidis NK, Chen Z. Curved singular beams for three-dimensional particle manipulation. Sci Rep. 2015;5:12086.
128. Efremidis NK, Christodoulides DN. Abruptly autofocusing waves. Opt Lett. 2010;35:4045–7.
129. Chremmos I, Efremidis NK, Christodoulides DN. Pre-engineered abruptly autofocusing beams. Opt Lett. 2011;36:1890–2.
130. Goutsoulas M, Efremidis NK. Precise amplitude, trajectory, and beam-width control of accelerating and abruptly autofocusing beams. Phys Rev A. 2018;97:063831.
131. Papazoglou DG, Efremidis NK, Christodoulides DN, Tzortzakis S. Observation of abruptly autofocusing waves. Opt Lett. 2011;36:1842–4.
132. Zhang P, Prakash J, Zhang Z, Mills MS, Efremidis NK, Christodoulides DN, Chen Z. Trapping and guiding microparticles with morphing autofocusing Airy beams. Opt Lett. 2011;36:2883–5.
133. Hwang C-Y, Kim K-Y, Lee B. Bessel-like beam generation by superposing multiple Airy beams. Opt Express. 2011;19:7356–64.
134. Chremmos I, Zhang P, Prakash J, Efremidis NK, Christodoulides DN, Chen Z. Fourier-space generation of abruptly autofocusing beams and optical bottle beams. Opt Lett. 2011;36:3675–7.
135. Zhang Z, Cannan D, Liu J, Zhang P, Christodoulides DN, Chen Z. Observation of trapping and transporting air-borne absorbing particles with a single optical beam. Opt Express. 2012;20:16212–7.
136. Chen X, Deng D, Zhuang J, Peng X, Li D, Zhang L, Zhao F, Yang X, Liu H, Wang G. Focusing properties of circle Pearcey beams. Opt Lett. 2018;43:3626–9.
137. Liang Y, Tan L, Liu N, Chen K, Liang H, Wu H, Luo B, Lu F, Chen H, Zou B. Tunable autofocusing and enhanced trapping forces with circular pearcey airy beams. Phys Rev Appl. 2023;19:014016.
138. Moradi H, Jabbarpour M, Abdollahpour D, Hajizadeh F. 3D optical trapping by a tightly focused circular airy beam. Opt Lett. 2022;47:4115–8.
139. Rodrigo JA, Alieva T, Abramochkin E, Castro I. Shaping of light beams along curves in three dimensions. Opt Express. 2013;21:20544–55.
140. Shanblatt ER, Grier DG. Extended and knotted optical traps in three dimensions. Opt Express. 2011;19:5833–8.
141. Lee S-H, Roichman Y, Grier DG. Optical solenoid beams. Opt Express. 2010;18:6988–93.
142. Roux FS. Optical vortex density limitation. Opt Commun. 2003;223:31–7.
143. Ostrovsky AS, Rickenstorff-Parrao C, Arrizón V. Generation of the "perfect" optical vortex using a liquid-crystal spatial light modulator. Opt Lett. 2013;38:534–6.
144. García-García J, Rickenstorff-Parrao C, Ramos-García R, Arrizón V, Ostrovsky AS. Simple technique for generating the perfect optical vortex. Opt Lett. 2014;39:5305–8.
145. Yang Q, Xie Z, Zhang M, Ouyang X, Xu Y, Cao Y, Wang S, Zhu L, Li X. Ultra-secure optical encryption based on tightly focused perfect optical vortex beams. Nanophotonics. 2022;11:1063–70.
146. Wang J. Advances in communications using optical vortices. Photon Res. 2016;4:B14–28.
147. Wan Z, Shen Y, Wang Z, Shi Z, Liu Q, Fu X. Divergence-degenerate spatial multiplexing towards future ultrahigh capacity, low error-rate optical communications. Light: Sci Appl. 2022;11:144.
148. Vaity P, Rusch L. Perfect vortex beam: Fourier transformation of a Bessel beam. Opt Lett. 2015;40:597–600.
149. Jabir M, Apurv Chaitanya N, Aadhi A, Samanta GK. Generation of "perfect" vortex of variable size and its effect in angular spectrum of the down-converted photons. Sci Rep. 2016;6:21877.
150. Chen Y, Fang Z-X, Ren Y-X, Gong L, Lu R-D. Generation and characterization of a perfect vortex beam with a large topological charge through a digital micromirror device. Appl Opt. 2015;54:8030–5.
151. Arrizón V, Ruiz U, Sánchez-de-la-Llave D, Mellado-Villaseñor G, Ostrovsky AS. Optimum generation of annular vortices using phase diffractive optical elements. Opt Lett. 2015;40:1173–6.
152. Pu M, Li X, Ma X, Wang Y, Zhao Z, Wang C, Hu C, Gao P, Huang C, Ren H. Catenary optics for achromatic generation of perfect optical angular momentum. Sci Adv. 2015;1:e1500396.
153. Liu Y, Ke Y, Zhou J, Liu Y, Luo H, Wen S, Fan D. Generation of perfect vortex and vector beams based on Pancharatnam-Berry phase elements. Sci Rep. 2017;7:44096.
154. Zhang Y, Liu W, Gao J, Yang X. Generating focused 3D perfect vortex beams by plasmonic metasurfaces. Adv Opt Mater. 2018;6:1701228.
155. Yang R, Jiang X, Yu J, Han J, Li Z, Zhang D, Shi Q, Zhu L. Controllable perfect optical vortex generated by complex amplitude encoding. Opt Lett. 2022;47:2101–4.
156. Liang Y, Lei M, Yan S, Li M, Cai Y, Wang Z, Yu X, Yao B. Rotating of low-refractive-index microparticles with a quasi-perfect optical vortex. Appl Opt. 2017;57:79–84.
157. Zhang H, Zeng J, Lu X, Wang Z, Zhao C, Cai Y. Review on fractional vortex beam. Nanophotonics. 2022;11:241–73.
158. Basistiy I, Soskin M, Vasnetsov M. Optical wavefront dislocations and their properties. Opt Commun. 1995;119:604–12.
159. Vasnetsov MV, Basistiy I, Soskin MS. Free-space evolution of monochromatic mixed screw-edge wavefront dislocations. In: International conference on singular optics. International Society for Optics and Photonics; 1998. p. 29–33.
160. Zeng R, Zhao Q, Shen Y, Liu Y, Yang Y. Structural stability of open vortex beams. Appl Phys Lett. 2021;119.
161. Tkachenko G, Chen M, Dholakia K, Mazilu M. Is it possible to create a perfect fractional vortex beam? Optica. 2017;4:330–3.
162. Bianchetti A, Etchepareborda P, Federico A. Determining the fractional topological charge shifting in perfect vortices from laser speckle. Opt Commun. 2019;441:74–9.
163. Willner AE, Huang H, Yan Y, Ren Y, Ahmed N, Xie G, Bao C, Li L, Cao Y, Zhao Z. Optical communications using orbital angular momentum beams. Adv Opt Photon. 2015;7:66–106.
164. Leach J, Yao E, Padgett MJ. Observation of the vortex structure of a non-integer vortex beam. New J Phys. 2004;6:71.
165. Xu Z, Gui C, Li S, Zhou J, Wang J. Fractional orbital angular momentum (OAM) free-space optical communications with atmospheric turbulence assisted by MIMO equalization. In: Photonic networks and devices. Optica Publishing Group; 2014. p. JT3A.
166. Tao S, Yuan X, Lin J, Peng X, Niu H. Fractional optical vortex beam induced rotation of particles. Opt Express. 2005;13:7726–31.
167. Jenness NJ, Hill RT, Hucknall A, Chilkoti A, Clark RL. A versatile diffractive maskless lithography for single-shot and serial microfabrication. Opt Express. 2010;18:11754–62.
168. Gibson G, Courtial J, Padgett MJ, Vasnetsov M, Pas'ko V, Barnett SM, Franke-Arnold S. Free-space information transfer using light beams carrying orbital angular momentum. Opt Express. 2004;12:5448–56.
169. Liang Y, Yan S, He M, Li M, Cai Y, Wang Z, Lei M, Yao B. Generation of a double-ring perfect optical vortex by the Fourier transform of azimuthally polarized Bessel beams. Opt Lett. 2019;44:1504–7.
170. Wang G, Kang X, Sun X, Li Z, Li Y, Chen K, Zhang N, Gao X, Zhuang S. Generation of perfect optical vortex arrays by an optical pen. Opt Express. 2022;30:31959–70.
171. He M, Liang Y, Yun X, Wang S, Zhao T, Guo L, Zhang X, Kuang S, Chen J, Lei M. Manipulation of low-refractive-index particles using customized dark traps. Photon Res. 2024;12:1334–43.

172. Yang K, Luo H, Li P, Wen F, Gu Y, Wu Z. Controlling spacing of double-ring perfect optical vortex using the Fourier transform of Bessel beam with axicon phase. Opt Laser Technol. 2023;158:108881.
173. Nan F, Yan Z. Synergy of intensity, phase, and polarization enables versatile optical nanomanipulation. Nano Lett. 2020;20:2778–83.
174. Kozawa Y, Sato S. Optical trapping of micrometer-sized dielectric particles by cylindrical vector beams. Opt Express. 2010;18:10828–33.
175. Li M, Yan S, Zhang Y, Zhang P, Yao B. Enantioselective optical trapping of chiral nanoparticles by tightly focused vector beams. JOSA B. 2019;36:2099–105.
176. Zhan Q. Trapping nanoparticles with cylindrical polarization. In: Optical trapping and optical micromanipulation. SPIE; 2004. p. 275–82.
177. Peng F, Yao B, Yan S, Zhao W, Lei M. Trapping of low-refractive-index particles with azimuthally polarized beam. JOSA B. 2009;26:2242–7.
178. Donato M, Vasi S, Sayed R, Jones P, Bonaccorso F, Ferrari A, Gucciardi P, Maragò O. Optical trapping of nanotubes with cylindrical vector beams. Opt Lett. 2012;37:3381–3.
179. Guo L, Liang Y, Yun X, He M, Zhao T, Wang S, Lei M. Annular optical trapping of metallic nanoparticles using the azimuthally polarized beam. Opt Lett. 2024;49:4206–9.
180. Bhebhe N, Williams PA, Rosales-Guzmán C, Rodriguez-Fajardo V, Forbes A. A vector holographic optical trap. Sci Rep. 2018;8:17387.
181. Wang H, Shi L, Lukyanchuk B, Sheppard C, Chong CT. Creation of a needle of longitudinally polarized light in vacuum using binary optics. Nat Photon. 2008;2:501–5.
182. Qin F, Huang K, Wu J, Jiao J, Luo X, Qiu C, Hong M. Shaping a subwavelength needle with ultra-long focal length by focusing azimuthally polarized light. Sci Rep. 2015;5:9977.
183. Otte E, Tekce K, Lamping S, Ravoo BJ, Denz C. Polarization nano-tomography of tightly focused light landscapes by self-assembled monolayers. Nat Commun. 2019;10:4308.
184. Otte E, Tekce K, Denz C. Customized focal light landscapes by complex vectorial fields for advanced optical trapping. In: Complex light and optical forces XII. SPIE; 2018. p. 115–22.
185. Kerker M, Cooke DD. Photophoretic force on aerosol particles in the free-molecule regime. JOSA. 1982;72:1267–72.
186. Rohatschek H. Direction, magnitude and causes of photophoretic forces. J Aerosol Sci. 1985;16:29–42.
187. Phuoc TX. A comparative study of the photon pressure force, the photophoretic force, and the adhesion van der Waals force. Opt Commun. 2005;245:27–35.
188. Lu J, Yang H, Zhou L, Yang Y, Luo S, Li Q, Qiu M. Light-induced pulling and pushing by the synergic effect of optical force and photophoretic force. Phys Rev Lett. 2017;118:043601.
189. Bliokh KY, Bekshaev AY, Nori F. Extraordinary momentum and spin in evanescent waves. Nat Commun. 2014;5:3300.
190. Bekshaev AY, Bliokh KY, Nori F. Transverse spin and momentum in two-wave interference. Phys Rev X. 2015;5:011039.
191. Xu X, Nieto-Vesperinas M. Azimuthal imaginary Poynting momentum density. Phys Rev Lett. 2019;123:233902.
192. Yun X, Liang Y, Guo L, He M, Zhao T, Wang S, Lei M. Imaginary Poynting momentum driven particle rotation by cylindrically polarized Gaussian beams; 2024. arXiv preprint arXiv:2408.07301.
193. Zhou Y, Xu X, Zhang Y, Li M, Yan S, Nieto-Vesperinas M, Li B, Qiu C-W, Yao B. Observation of high-order imaginary Poynting momentum optomechanics in structured light. Proc Natl Acad Sci. 2022;119:e2209721119.
194. Polin M, Ladavac K, Lee S-H, Roichman Y, Grier DG. Optimized holographic optical traps. Opt Express. 2005;13:5831–45.
195. Davis JA, Cottrell DM, Campos J, Yzuel MJ, Moreno I. Encoding amplitude information onto phase-only filters. Appl Opt. 1999;38:5004–13.
196. Liang Y, Yun X, He M, Wang Z, Wang S, Lei M. Zero-order-free complex beam shaping. Opt Lasers Eng. 2022;155:107048.
197. Yun X, Liang Y, He M, Guo L, Zhang X, Zhao T, Bianco PR, Lei M. Zero-order free holographic optical tweezers. Opt Express. 2023;31:19613–21.

4 Opto-Hydrodynamic Manipulation

Yang Shi and Hongbao Xin

Abstract

This chapter reviews the advanced technology of Opto-Hydrodynamic Manipulation, which elegantly combines optical and fluid dynamic principles to achieve precise control over microscale objects. The chapter begins by introducing the definition, significance, and historical context of opto-hydrodynamic manipulation, emphasizing its wide-ranging applications in fields such as biomedicine, nanotechnology, and microengineering. It then delves into the principles of optical forces arising from light-matter interaction, microscale fluid dynamics, and the theoretical frameworks of combined opto-hydrodynamic effects. The technological foundations, including optical tweezers and microfluidic systems, are discussed in detail, along with the latest advancements in light-induced hydrodynamic manipulation. The chapter highlights advanced applications in biomedicine and nanotechnology, such as light-induced cold Marangoni flow, 3D dynamic microswarm petals, and the opto-hydrodynamic diatombot. Finally, the chapter discusses the challenges and future directions of opto-hydrodynamic manipulation, including technical, biological, and medical challenges, as well as potential future research directions. This chapter aims to provide a comprehensive overview of the field of opto-hydrodynamic manipulation, showcasing its potential in scientific research and technological innovation.

Y. Shi · H. Xin (✉)
Guangdong Provincial Key Laboratory of Nanophotonic Manipulation, Institute of Nanophotonics, College of Physics and Optoelectronic Engineering, Jinan University, Guangzhou, P. R. China
e-mail: hongbaoxin@jnu.edu.cn

Keywords

Opto-hydrodynamic manipulation · Optical trapping · Hydrodynamic forces · Microfluidic systems · Non-invasive trapping

4.1 Introduction

Opto-hydrodynamic manipulation represents a sophisticated convergence of optical and fluid dynamic principles to exert precise control over microscale objects. Opto-hydrodynamic manipulation can be defined as the utilization of light to generate or synergize with hydrodynamic forces acting on microscale entities within a fluid medium, thereby enabling precise and flexible control. This technique combines the high-precision nature of optical forces and non-invasive nature of fluid forces with the long-distance transport capabilities of fluid systems. The significance of opto-hydrodynamic manipulation lies in its potential to revolutionize the way we handle and process microscale objects, allowing for a wide range of applications from fundamental research to practical applications in biomedicals, microengineering and materials science.

4.1.1 Historical Context and Evolution

The field of opto-hydrodynamic manipulation has evolved from pioneering work on optical trapping by Arthur Ashkin in the 1970s [1]. Ashkin's discovery that a focused laser beam could trap and manipulate microscopic particles opened new possibilities for the study of biological systems and the manipulation of matter at the microscale [2]. His groundbreaking work laid the foundation for subsequent advancements in optical tweezers, which have been further developed to include multi-beam systems capable of trapping multiple

B. Li et al. (eds.), *Biophotonic Manipulation*, Advances in Optics and Optoelectronics,
https://doi.org/10.1007/978-981-96-4982-2_4

particles simultaneously [3]. As technology progressed, the integration of hydrodynamic effects with optical forces expanded the capabilities of opto-hydrodynamic manipulation. Researchers began to explore how the movement of fluid could enhance the trapping and manipulation of particles in complex fluid environments, leading to novel applications in various scientific fields [4].

4.1.2 Applications in Biomedicine, Nanotechnology, and Microengineering

Advantages of opto-hydrodynamic manipulation include noninvasive control, which minimizes damage to sensitive biological samples, and the ability to manipulate objects in a label-free manner. The technique also offers high spatial resolution and the potential for real-time control, making it suitable for dynamic manipulation. Furthermore, the use of light as a manipulation tool allows for the integration with other optical techniques, such as fluorescence imaging, which can provide simultaneous monitoring of the manipulation process.

The applications of opto-hydrodynamic manipulation are vast and varied, spanning across different disciplines. In biomedicine, it is used for targeted drug delivery, where drugs are precisely delivered to diseased cells, reducing side effects and improving therapeutic outcomes [5]. Additionally, it facilitates the study of cellular mechanics, allowing researchers to measure the forces generated by cells and the mechanical properties of cellular components, which is crucial for understanding cell behavior in health and disease [6]. In nanotechnology, opto-hydrodynamic manipulation is employed in the assembly of nanostructures, such as the construction of photonic crystals and the integration of nanoelectronic components. It enables the precise placement of nanoparticles and the manipulation of nanowires, which is crucial for the development of next-generation electronic and optical devices [7, 8]. In microengineering, opto-hydrodynamic manipulation techniques are used to create microfluidic devices with integrated optical components. These devices allow for the control of fluid flow and particle movement within microchannels, with applications in lab-on-a-chip systems where complex chemical and biological assays are performed in a compact and automated manner [9, 10].

4.1.3 The Interdisciplinary Approach

Opto-hydrodynamic manipulation is inherently interdisciplinary, requiring expertise in physics, engineering, biology, and materials science. The development of this field relies on advancements in laser technology, microfluidic device fabrication, and the understanding of fluid dynamics at the microscale. Collaboration among researchers from different disciplines has led to the discovery of innovative techniques and applications that continually push the boundaries of this technology. This interdisciplinary approach is what drives the field forward, making opto-hydrodynamic manipulation a dynamic and rapidly evolving area of research. As new materials and methods are developed, the potential applications expand, promising to impact fields such as diagnostics, therapeutics, and materials engineering.

4.1.4 Chapter Overview

In this chapter, we will explore the principles that underpin opto-hydrodynamic manipulation, the technological foundations that support it, and the various applications that demonstrate its power and versatility. We will also discuss the challenges researchers face in pushing the boundaries of this technology and future perspectives that promise to further integrate opto-hydrodynamic manipulation into advanced systems with broader implications for science and technology.

4.2 Principles of Opto-hydrodynamic Manipulation

The principles of opto-hydrodynamic manipulation are rooted in the interaction of light with matter and the subsequent generation of forces and torques, as well as the dynamics of fluid flow at the microscale. This section explores the fundamental concepts governing these interactions and how they are harnessed for the precise manipulation of microscale objects.

4.2.1 Interaction of Light with Matter

Optical forces arise from the interaction of light with matter and can be categorized into two main types: gradient forces and scattering forces. The gradient force results from the intensity gradient of the light field, pulling particles toward the region of highest light intensity. This force is responsible for the initial trapping of particles within the focus of a laser beam. Conversely, the scattering force arises from the change in momentum of photons as they are scattered by the particle, resulting in a force that pushes the particle in the direction of light propagation. The capability to exert both radial and axial forces allows for the manipulation of particles in three dimensions. The balance and interplay between these forces are crucial for stable trapping and precise control of particle motion. Mathematically, these forces can be described using the Lorentz force law and related equations that account for the optical properties of the particle and the surrounding medium [11]. The total optical force exerted on the particle can be determined by integrating the time-averaged Maxwell stress tensor over the entire external surface of the particle. Understanding these forces is essential for designing effective optical manipulation systems.

4.2.2 Hydrodynamic Forces and Fluid Dynamics at the Microscale

Hydrodynamic forces are generated by the flow of fluids and the interaction of objects within these fluids. At the microscale, fluid dynamics are primarily governed by the Navier–Stokes equations, which describe the motion of fluid substances:

$$\rho\left(\frac{\partial \boldsymbol{u}}{\partial t} + \mathbf{u} \cdot \nabla \mathbf{u}\right) = -\nabla \mathrm{p} + \mu \nabla^2 \mathbf{u} + \mathbf{f} \tag{4.1}$$

where ρ represents the density of the fluid, $\boldsymbol{u}$ denotes the velocity vector of the fluid, p stands for the pressure of the fluid, μ indicates the dynamic viscosity of the fluid, and $\mathbf{f}$ represents the volume force acting on the fluid, such as gravity. However, due to the small length scales involved, inertial forces become less significant compared to viscous forces, leading to the dominance of Stokesian flow behaviour [12]. This regime is characterized by low Reynolds numbers, where the flow is smooth and laminar. In opto-hydrodynamic manipulation, hydrodynamic forces are often used in conjunction with optical forces. For instance, a focused laser beam can induce local heating in a fluid, creating a temperature gradient that leads to fluid flow through natural convection. This flow can be harnessed to transport particles or create localized flow fields for the manipulation of microscale objects [13]. The interaction between optical and hydrodynamic forces thus opens new avenues for controlling particle dynamics in complex environments.

4.2.3 Combined Opto-hydrodynamic Effects: Theoretical Frameworks

The combined effects of optical and hydrodynamic forces provide a rich and complex framework for manipulation at the microscale. The theoretical understanding of these effects involves integrating electromagnetic theory, fluid dynamics, and thermodynamics. One of the key challenges in this field is developing models that accurately predict the behavior of particles under the influence of both optical and hydrodynamic forces. This requires a multi-physics approach, where the equations governing light propagation (**Maxwell's equations**) are coupled with those describing fluid flow (**Navier–Stokes equations**) and heat transfer (**Thermal diffusion equation**). Such coupled models are essential for designing opto-hydrodynamic systems capable of performing complex manipulation tasks with high precision. Furthermore, these models aid in understanding the limitations and potential enhancements of current techniques and explore new methods for microscale manipulation. The theoretical principles of opto-hydrodynamic manipulation are validated through various experimental techniques. High-speed imaging and particle tracking allow researchers to observe the motion of particles under the influence of optical and hydrodynamic forces, providing valuable data for verifying theoretical models. Advanced imaging techniques, such as fluorescence microscopy and digital holography, enhance our ability to capture dynamic processes at the microscale.

Conclusion of Sect. 4.2

The principles of opto-hydrodynamic manipulation provide a robust foundation for the flexible control of microscale objects. Understanding the interplay between optical and hydrodynamic forces is crucial for developing new techniques and applications in this field. As our theoretical and experimental capabilities continue to advance, opto-hydrodynamic manipulation is poised to play an increasingly important role in science and technology.

4.3 Technological Foundations

The technological foundations of opto-hydrodynamic manipulation are built upon the principles of optical trapping and hydrodynamic forces. This section delves into the core technologies that enable the precise control of microscale objects through the synergistic application of light and fluid dynamics.

4.3.1 Optical Trapping and Tweezers

Optical trapping, the process of stably holding small dielectric particles in a focused laser beam using radiation pressure, forms the basis of opto-hydrodynamic manipulation. An optical tweezers system typically consists of a high-power laser, an optical microscope with a high numerical aperture objective lens, and a means of controlling the position of the trap, often an acousto-optic deflector (AOD) or a galvanometer mirror [14, 15]. The system allows for real-time visualization and manipulation of trapped particles, with the ability to create multiple traps for simultaneous manipulation of multiple particles. The high-power laser, usually a Nd:YAG or diode laser, provides the necessary intensity for trapping particles in the microscale range. The optical microscope, equipped with a high numerical aperture objective lens, focuses the laser beam to a diffraction-limited spot, creating the optical trap. The AOD or galvanometer mirror is used to scan the laser beam across the sample, allowing for the manipulation of particles in three dimensions.

4.3.2 Microfluidic Systems for Hydrodynamic Manipulation

Microfluidic systems provide a controlled environment for generating and manipulating hydrodynamic forces at the microscale. These systems, which can be fabricated using various materials and techniques, are designed to handle small volumes of fluid and can be integrated with optical components for combined opto-hydrodynamic manipulation. Microfluidic devices are typically fabricated using soft lithography or micromilling techniques. Soft lithography, derived from photolithography, involves using an elastomeric stamp to transfer a pattern onto a substrate, which is then used to create a mold for the microfluidic channels [16, 17]. In contrast, micromilling involves direct milling of a substrate, such as a silicon wafer, to create the channels and chambers of the microfluidic device [18]. These devices can be designed to create specific flow patterns, such as channels and chambers, which guide particles to desired locations or create localized flow fields for hydrodynamic trapping and manipulation. The design of the microfluidic device is crucial for the efficient generation of hydrodynamic forces and the precise control of particle movement. The integration of microfluidic systems with optical tweezers allows for precise control of particle movement in both space and time. By adjusting the flow rates within the microfluidic channels, particles can be transported to specific locations where they can be trapped and manipulated using optical forces. This combination of hydrodynamic and optical forces provides a powerful tool for manipulating particles in complex environments, enhancing the flexibility and functionality of the system.

4.3.3 Light-Induced Hydrodynamic Manipulation

Opto-hydrodynamic manipulation can be effectively achieved by utilizing light-induced hydrodynamic processes to control target objects. This technique capitalizes on the interaction between light and matter, generating heat and fluid dynamics that enable precise movement and positioning of particles. The photothermal effects and optical forces involved in this process are increasingly acknowledged for their potential in manipulating micro- and nanoscale objects through light-induced hydrodynamic forces [8, 19–22].

The principle of photothermal-induced hydrodynamic manipulation is founded on the absorption of light by specific materials, which generates heat. This heat creates temperature gradients in the surrounding medium, leading to fluid motion through thermophoresis, thermo-osmotic flows, and Marangoni flow. Thermophoresis involves the movement of particles along a temperature gradient, while thermo-osmotic flows are driven by pressure gradients resulting from temperature-induced changes in fluid density. Marangoni flow arises from variations in surface tension due to temperature differences. Together, these mechanisms enable precise control of fluid dynamics and particle movement at micro- and nanoscale levels. In contrast, the principle of optical force-induced hydrodynamic manipulation relies on the transfer of momentum to the liquid through optical forces acting on the target. This interaction generates a localized flow field and associated hydrodynamic effects, enhancing the precision of particle and fluid manipulation. This system involves a multi-directional coupling of optical forces, the target, and the fluid, making it inherently more complex. However, because the photothermal effect is not utilized, this approach minimizes damage to the sample. Additionally, the dominance of the localized flow field makes it less sensitive to the boundaries of the liquid environment, resulting in greater universality and improved potential for in vivo manipulation.

Conclusion of Sect. 4.3

The technological foundations of opto-hydrodynamic manipulation are diverse and rapidly evolving. As our understanding of the principles of optical and hydrodynamic forces deepens, new techniques and devices are being developed to push the boundaries of what is possible in the manipulation of microscale objects. The integration of these technologies with microfluidics and other forms of manipulation opens up a world of possibilities for applications in research and industry.

4.4 Light-Induced Hydrodynamic Manipulation

4.4.1 Overview

The realm of opto-hydrodynamic manipulation has expanded beyond traditional microscale control, leading to a myriad of advanced applications that harness the synergistic effects of light and fluid dynamics. These applications represent the forefront of a technological revolution, where the micro and nanoscale worlds are manipulated with unprecedented precision and sophistication. This section focuses specifically on light-induced hydrodynamic manipulation, delineating an overview of these avant-garde applications, thereby laying the groundwork for a more profound examination of their ramifications and prospective applications.

One of the most compelling applications is the light-induced cold Marangoni flow (CMF) for microswarm actuation, which enables control of microswarms with high universality and intelligent biomimetic behaviors. This approach is particularly advantageous for biomedical tasks that require photothermal damage-free actuation, such as collective gene delivery and targeted single-cell chemotherapy. The CMF's ability to isolate the microswarm from photothermal interference makes it ideal for sensitive biological applications.

Another groundbreaking application is the formation of 3D dynamic microswarm petals through opto-hydrodynamic strategies, which represents a significant advancement in the field. These petals, formed by the reorganization of microparticles into firework-like structures, exhibit unparalleled controllability and reconfigurability. This technology opens up new possibilities for executing complex spatiotemporal tasks, particularly in biomedical and microengineering applications.

Lastly, the wake-riding effect-inspired opto-hydrodynamic diatombot (OHD) offers an innovative approach to the non-invasive trapping and removal of nano-biothreats. This biocompatible system utilizes the rotational motion of a diatom to generate a localized flow field, effectively trapping and removing contaminants from biological environments. The OHD's non-invasive nature and ability to operate within bio-microenvironments make it a promising tool for applications in cell culturing and biomanufacturing.

4.4.2 Light-Induced Cold Marangoni Flow for Microswarm Actuation

Light-induced cold Marangoni flow (CMF) is an advanced technique for actuating microswarms, particularly in sensitive biological environments [21]. This method is notable for its capacity to manipulate microswarms without causing photothermal damage, making it particularly advantageous for biomedical applications such as drug delivery and targeted cell chemotherapy. This section explores the principles underlying CMF, its generation and control mechanisms, and its significant implications for collective drug delivery and targeted therapies.

Principle of Light-Induced CMF

The generation of cold Marangoni flow (CMF) involves focusing a laser beam, typically at a wavelength of 1064 nm, onto an oil region surrounding a water droplet (Fig. 4.1a). The oil efficiently absorbs the laser energy, resulting in localized heating. As the temperature rises at the water–oil interface, a temperature gradient develops, creating a corresponding

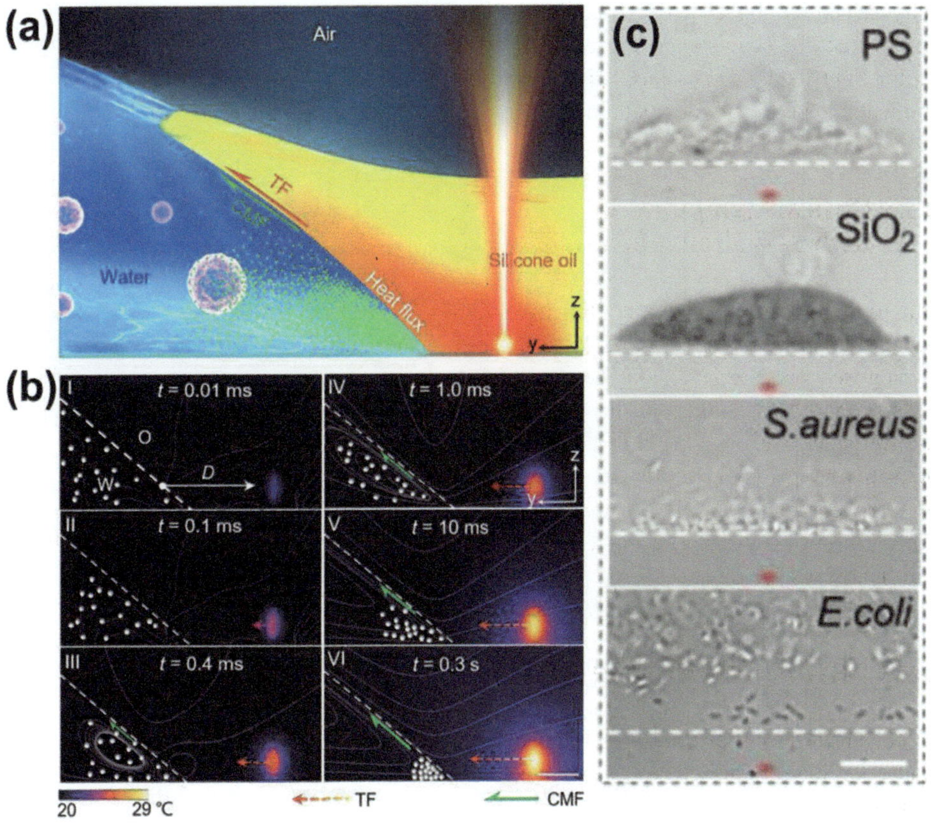

Fig. 4.1 **a** Schematic illustration of swarm formation at the W/O interface by light-induced CMF, and further for targeted drug delivery and single-cell chemotherapy. **b** Simulated evolution of thermal dynamics (temperature and flow fields) during light irradiation on oil near W/O interface. Scale bar: 10 μm. **c** Swarm formation for different particle individuals. Scale bars: 5 μm (Shi et al. [21], Copyright (2022), with permission from John Wiley and Sons)

surface tension gradient. Notably, the surface tension at the water/silicone oil (W/O) interface increases abnormally with the rise in temperature, a phenomenon known as thermal shrinkage. This results in a reversed Marangoni flow, where individual entities migrate from the cooler surrounding area toward the higher temperature region, ultimately forming a microswarm (Fig. 4.1b). This mechanism enables precise control over the movement and organization of both living and non-living entities in a fluid environment (Fig. 4.1c).

Controlling CMF involves adjusting several parameters, particularly laser power and the distance between the laser focus and the water–oil interface. Increasing laser power amplifies the temperature gradient, enhancing both flow velocity and actuation strength of the microswarms. Conversely, the distance to the interface affects flow characteristics: shorter distances produce stronger, more localized flows, while greater distances may weaken actuation. By optimizing these parameters, researchers can achieve a high degree of precision in manipulating microswarms. This capability for fine-tuning is crucial for tailoring actuation to specific applications, allowing effective customization of swarm behaviors based on varying experimental conditions.

Intelligent Biomimetic Swarming Behaviors

Microswarms actuated by cold Marangoni flow (CMF) exhibit a range of intelligent biomimetic behaviors that are crucial for their biomedical applications (Fig. 4.2). These behaviors include collective migration, size-based self-organization, and group rejection, all driven by the interaction between the cold flow and the individual agents within the swarm. The intelligent behaviors of CMF-actuated microswarms closely resemble those found in natural swarming phenomena. For instance, these microswarms can collectively migrate toward a light source or self-organize based on particle size. Such capabilities are vital for targeted drug delivery, where precise control over swarm movement and organization is essential. By mimicking these natural behaviors, researchers can develop advanced drug delivery systems that navigate complex biological environments with greater intelligence. The ability to effectively steer microswarms within living systems promises significant advancements in therapeutic delivery methods.

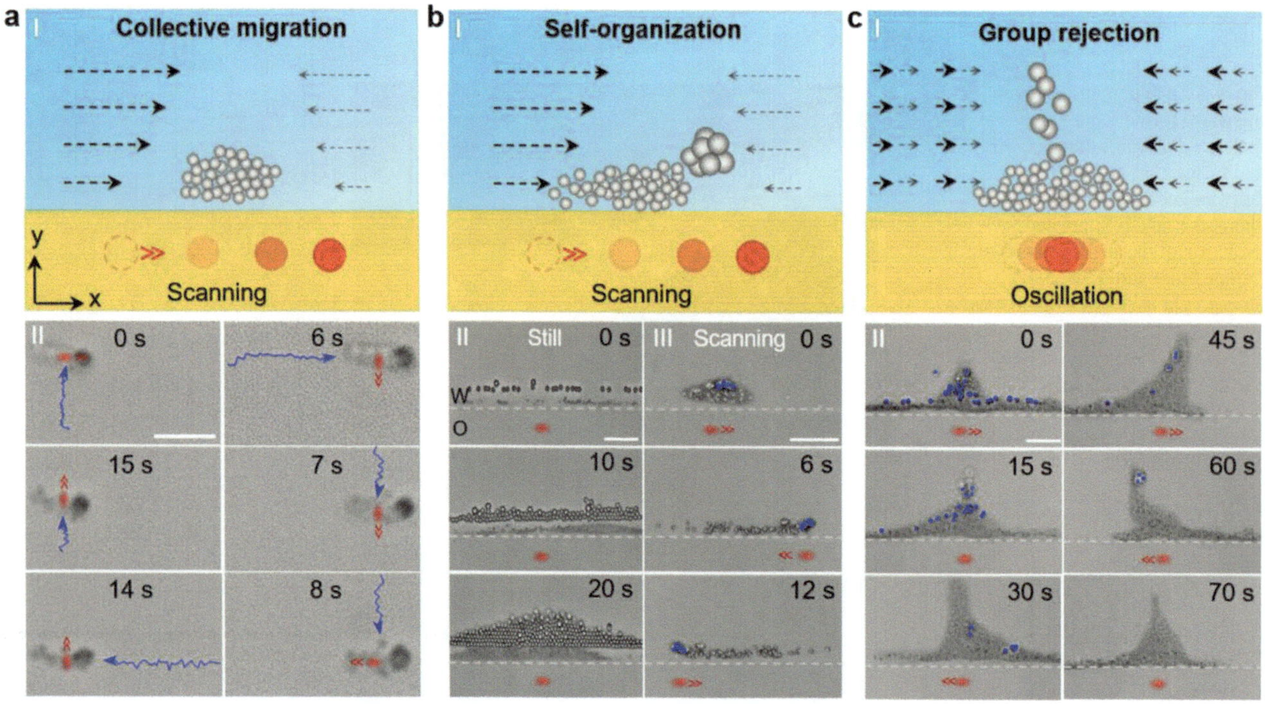

Fig. 4.2 **Intelligent biomimetic swarming behaviors.** **a** Collective migration. I: Schematic illustration of the collective migration along with the movement of the light irradiation (red spots indicated), black arrows indicate the flow direction and strength. II: Experimental images of swarm migration (SiO_2, $d = 200$ nm). **b** Swarm self-organization. I: Schematic illustration of the dynamic self-organization of swarms. Experimental images showing self-organization of swarms (PS, $d_1 = 500$ nm, $d_2 = 1.0$ μm) with the light spot II) keeping still and III) scanning. Blue dots indicate the 1.0 μm particles, white dashed lines indicate the W/O interface. **c** Swarm group rejection. I: Schematic illustration. II: Experimental images showing expelling large particles (PS, $d = 1.0$ μm) from small particle (PS, $d = 500$ nm) swarm. Blue dots indicate the 1.0 μm particles. Scale bar: 10 μm (Shi et al. [21], Copyright (2022), with permission from John Wiley and Sons)

Applications in Collective Drug Delivery and Targeted Chemotherapy

A significant advantage of cold Marangoni flow (CMF) is its ability to isolate the heat generated by a laser from the water phase, resulting in a "cold" flow. This characteristic is particularly beneficial for light- or temperature-sensitive materials, including biological cells. The cold flow preserves the integrity of microswarm components and ensures that biological systems remain unharmed during manipulation, making it crucial for applications in delicate environments where maintaining cell viability is paramount. Additionally, light-induced CMF enhances the practical applicability of microswarms in biomedical scenarios, such as collective drug delivery and targeted chemotherapy. CMF-actuated microswarms can effectively transport drug-loaded particles to target cells or tissues through directed movement. The cold flow generated by CMF ensures that the drugs remain active during transport, preserving their efficacy and preventing degradation that may occur with conventional delivery methods (Fig. 4.3a). This capability is particularly advantageous for sensitive therapeutics, such as RNA molecules or biologic drugs, which can be adversely affected by heat. The precision of CMF in directing microswarm movement enhances targeting accuracy, improving the overall effectiveness of drug delivery systems. This precise control also enables targeted chemotherapy, selectively treating cancer cells while sparing healthy ones (Fig. 4.3b). By directing drug-loaded microswarms to tumor sites, researchers can significantly enhance cancer treatment efficacy, reducing the side effects commonly associated with traditional chemotherapy.

Conclusion

Light-induced cold Marangoni flow presents a novel and effective method for microswarm actuation that is both gentle and controllable. Its capability to function without causing photothermal damage positions it as a promising tool for advanced biomedical applications, especially in collective drug delivery and targeted cell chemotherapy. As research in this area progresses, we can expect further advancements in the control and application of CMF-actuated microswarms. This innovative technology not only has the potential to enhance therapeutic strategies but also paves the way for new solutions in healthcare and beyond. By continuing to explore

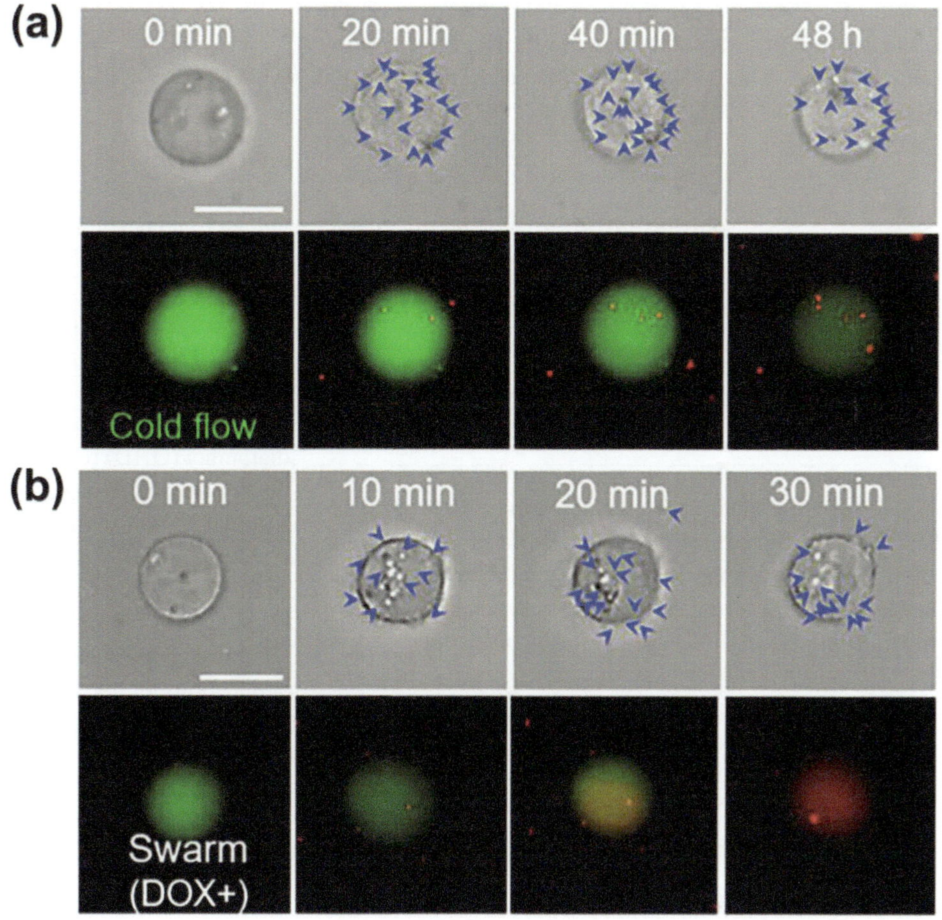

Fig. 4.3 **a** Photothermal damage-free gene delivery by CMF-induced microswarm. **b** Collective drug delivery and targeted cell chemotherapy. Scale bar: 10 μm (Shi et al. [21], Copyright (2022), with permission from John Wiley and Sons)

the capabilities of cold Marangoni flow, researchers can unlock new possibilities for manipulating microscale systems, ultimately improving patient outcomes and advancing the field of biomedical engineering.

4.4.3 Opto-hydrodynamic Driven 3D Dynamic Microswarm Petals

The emergence of opto-hydrodynamically driven 3D dynamic microswarm petals represents a significant advancement in microrobotics, particularly for applications requiring precise spatiotemporal control of microscale objects [19]. This section explores the intricate methodologies, distinctive features, and promising applications of these dynamic microswarms, emphasizing their potential to revolutionize various scientific and engineering fields.

Formation and Actuation of 3D Dynamic Microswarm Petals

The formation of 3D dynamic microswarm petals is achieved through an innovative opto-hydrodynamic strategy that leverages the photothermal gradient-induced Marangoni effect. This process begins by irradiating the water–air interface with a tapered optical fiber (TOF) emitting light at a wavelength of 1.55 μm. When the laser light interacts with the water surface, it is absorbed, creating a localized temperature gradient. This temperature differential generates the Marangoni effect, resulting in fluid flow that traps particles and assembles them into a 3D dynamic structure (Fig. 4.4a). The symmetry and strength of this flow field can be modulated by varying the laser power and adjusting the position of the TOF.

This approach not only organizes randomly distributed microparticles into intricate, firework-like 3D structures but also demonstrates a remarkable degree of controllability and reconfigurability. By adjusting the angle between the TOF and the water–air interface (θ), the shape and movement of the microswarm can be finely manipulated (Fig. 4.4b). This capability enables a variety of complex behaviors and tasks, including altering flow patterns to produce different structural configurations of the microswarm petals. Moreover, the microswarms exhibit dynamic behaviors such as stable migration and deformation, achievable simply by moving the TOF. This ability to manipulate microswarms in three dimensions opens avenues for complex spatiotemporal tasks, including coordinated movements and adaptive responses to environmental changes. The dynamic nature of these microswarms marks a significant advancement over traditional methods, where limited mobility and adaptability often constrained the applications of microscale systems.

Applications in Biomedical and Microengineering

The opto-hydrodynamic strategy enables the creation of biomicroswarms composed of living microorganisms, effectively regulating biological activities. For example, the rapid spinning of these living microorganism microswarms can control bacterial division processes, leading to breakthroughs in our understanding of cell behavior and growth dynamics (Fig. 4.5). This capability is particularly crucial in microbiology and biotechnology, where precise control over microbial populations can have profound implications for both research and industrial applications.

Conclusion

In summary, the development of opto-hydrodynamically driven 3D dynamic microswarm petals marks a significant advancement in microrobotics. Their ability to perform complex spatiotemporal tasks with precision and adaptability opens up a wide range of potential applications in both biomedical and microengineering fields. As research in this area progresses, these microswarms may become integral to advancements in targeted drug delivery, microsurgery, and environmental monitoring at the microscale. The future of microrobotics is promising with the introduction of these dynamic microswarm petals. Their unique capabilities not only enhance our understanding of microscale phenomena but also offer innovative solutions to longstanding challenges across various scientific and engineering disciplines. Continued exploration and development of these technologies could lead to groundbreaking discoveries and applications that transform our approach to complex problems in health, industry, and beyond.

4.4.4 Wake-Riding Effect-Inspired Opto-hydrodynamic Diatombot

The wake-riding effect-inspired opto-hydrodynamic diatombot (OHD) represents a groundbreaking innovation in the non-invasive trapping and removal of nano-biothreats [22]. Inspired by the remarkable strategies employed by marine organisms, this technology integrates advanced principles of fluid dynamics and optical manipulation to tackle critical challenges in bio-microenvironments. This chapter explores the design philosophy, operational principles, and potential applications of the OHD, emphasizing its significance in biomedicine, biotechnology, and environmental monitoring.

Design Philosophy and Inspiration

The core inspiration for the opto-hydrodynamic diatombot (OHD) comes from the wake-riding effect, a natural

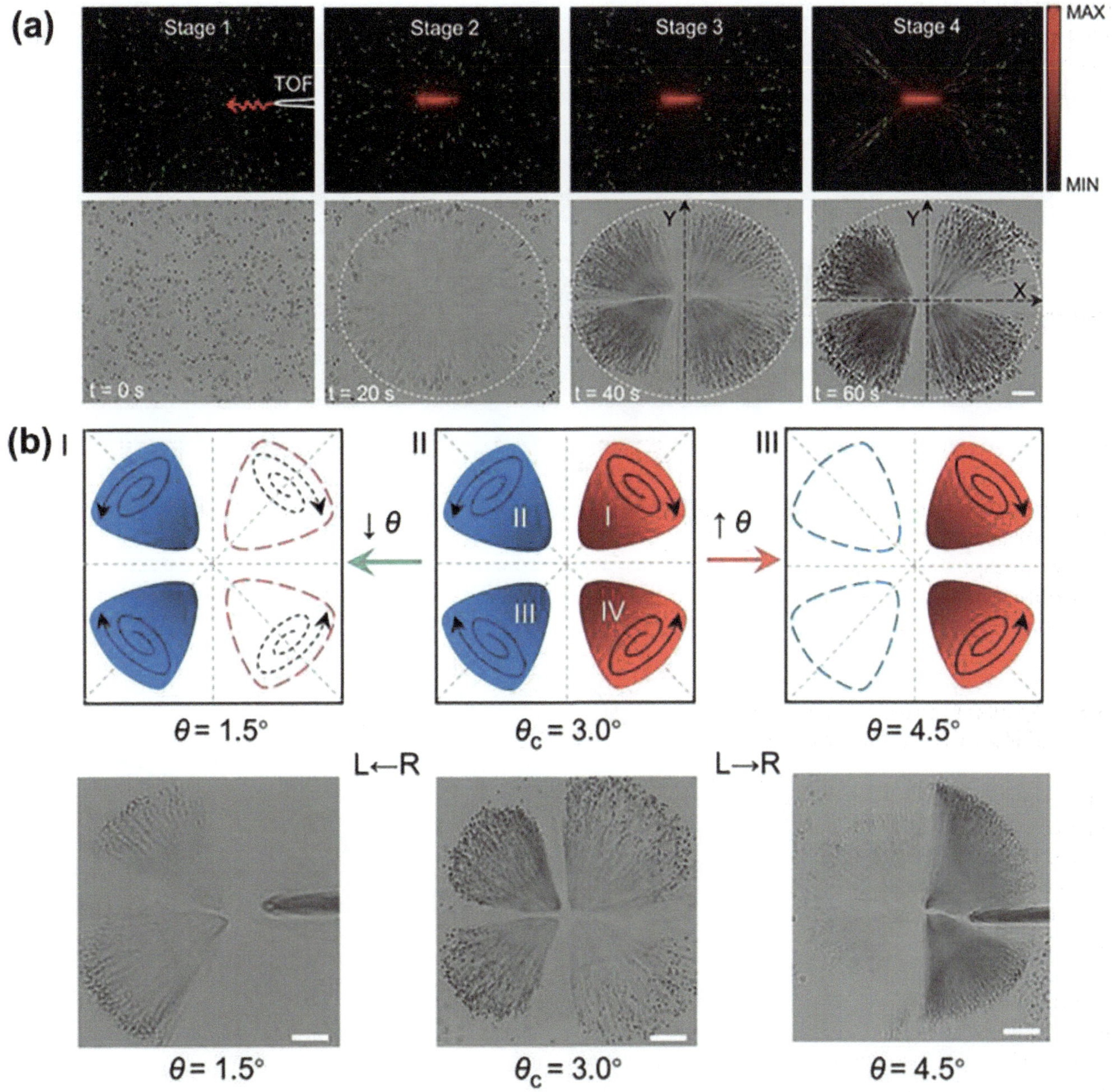

Fig. 4.4 **a** Simulation (upper panel) and experiments (lower panel) showing microswarm formation. Stage 1: Before the laser lunched, the 3-μm PS were randomly distributed in the water owing to Brownian motion. Stages 2–4: The single vortex gradually splits into four smaller vortices symmetrically distributed in four quadrants. Color bar: temperature, scale bar: 50 μm. **b** Reconfigurable microswarm petals by tuning the TOF angle. Scale bars: 50 μm (Li et al. [19], Copyright (2023), with permission from John Wiley and Sons)

phenomenon where marine animals, such as dolphins, use the fluid dynamics generated by moving objects to enhance their locomotion while conserving energy (Fig. 4.6a). This biological strategy serves as a model for the OHD, which mimics these natural behaviors to navigate and manipulate microscale threats in complex fluid environments (Fig. 4.6b). By leveraging biomimetic principles, the OHD aims to replicate the efficiency and functionality of these natural systems, providing innovative solutions to modern challenges.

The design of the OHD is fundamentally anchored in the morphology of Phaeodactylum tricornutum (PTB), a common diatom characterized by its triradiate shape (Fig. 4.6c). This unique form offers a structural basis for the OHD, enabling it to generate localized flow fields through rotational motion (Fig. 4.6d). Such biomimetic design not only enhances the device's operational efficiency but also allows it to perform intricate fluid dynamic maneuvers, similar to the strategies observed in dolphins and other marine organisms.

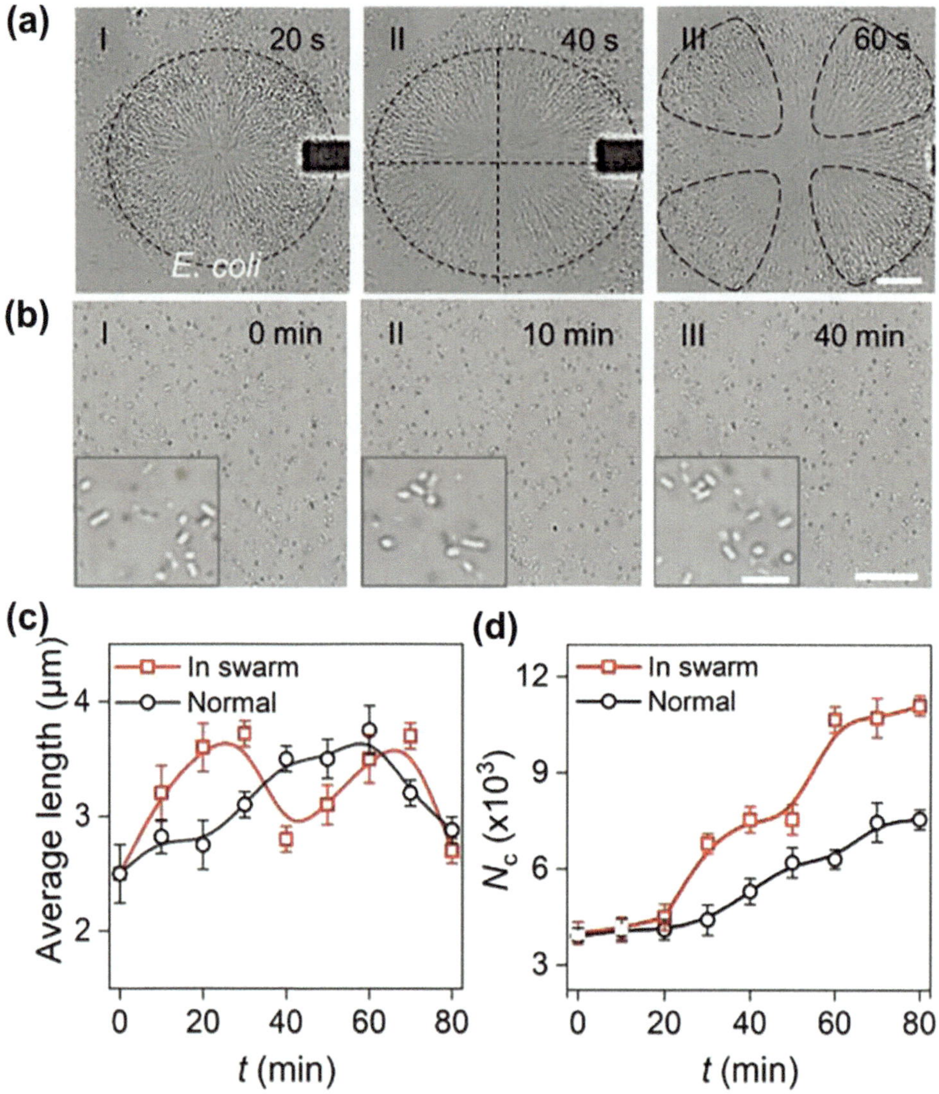

Fig. 4.5 **a** Formation of *E. coli* bio-microswarm. Scale bar: 50 μm. **b** Representative images of *E. coli* growth over time, scale bar: 50 μm. Insets show the enlarged view, scale bar: 10 μm. **c** Changes in the average length of *E. coli* over time. **d** Changes in the individual number of *E. coli* microswarms over time (Li et al. [19], Copyright (2023), with permission from John Wiley and Sons)

Operational Principles

The operational principles of the opto-hydrodynamic diatombot (OHD) rely on the synergistic application of optical forces and hydrodynamic interactions. Central to the OHD's functionality is optical trapping, which employs a focused laser beam to generate forces that facilitate the rotation and navigation of the diatombot within a fluidic environment. The rotational motion of the OHD creates a localized flow field or "wake," which plays a crucial role in trapping and transporting nano-biothreats like viruses and bacteria. This hydrodynamic force is generated by the diatombot's movement through the fluid, allowing it to effectively "surf" the wake it creates (Fig. 4.7a–c). This innovative approach not only enhances the capture of contaminants but also significantly improves the overall efficiency of the trapping process (Fig. 4.7d). The interplay between optical and hydrodynamic forces establishes a robust mechanism for addressing the complexities of nano-biothreat management.

Non-invasive Trapping and Removal of Nano-biothreats

One of the most compelling advantages of the opto-hydrodynamic diatombot (OHD) is its non-invasive strategy for managing nano-biothreats in sensitive bio-microenvironments, such as cell cultures. The OHD operates without compromising the integrity of cultured cells or other biological samples, making it an invaluable tool in various biomedical applications. Its design enables precise navigation within microfluidic channels or biological samples, allowing for targeted trapping and removal of nano-biothreats through the application of opto-hydrodynamic forces (Fig. 4.8).

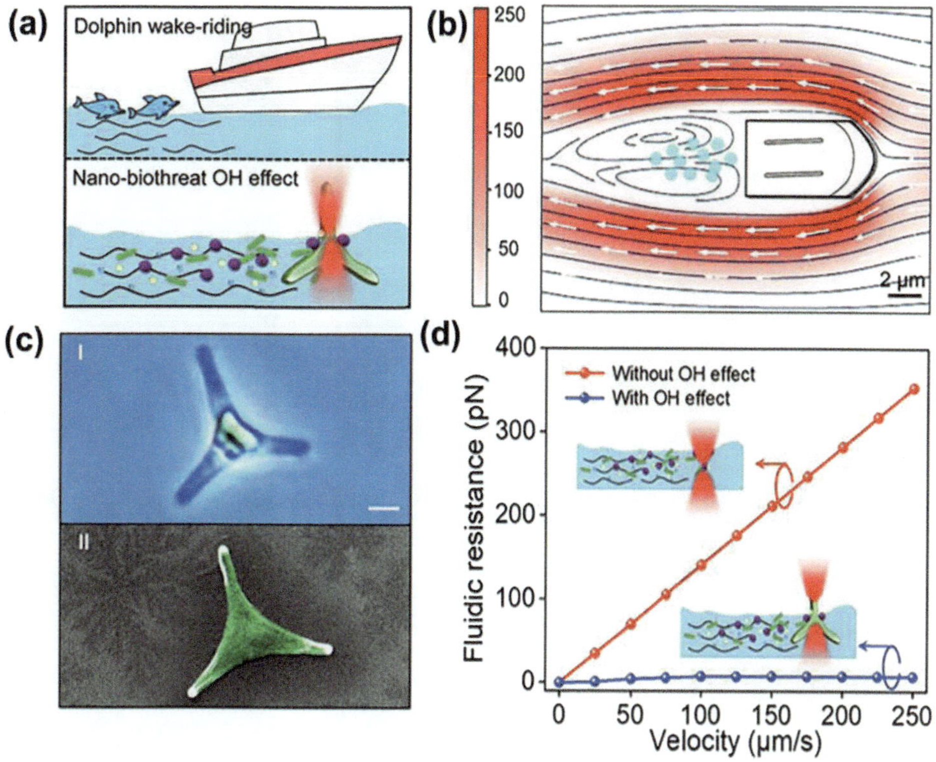

Fig. 4.6 **a** Schematic illustration of the OH effect for nano-biothreat removal via OHD. **b** Simulated flow field of moving microboat with nano-biothreat behind. **c** Phase contrast microscope image (I) and scanning electron microscope image (II) of PTB. **d** Fluidic resistance exerted on the nano-biothreat as a function of forward velocity. (Xiong et al. [22], Copyright (2023), with permission from John Wiley and Sons.)

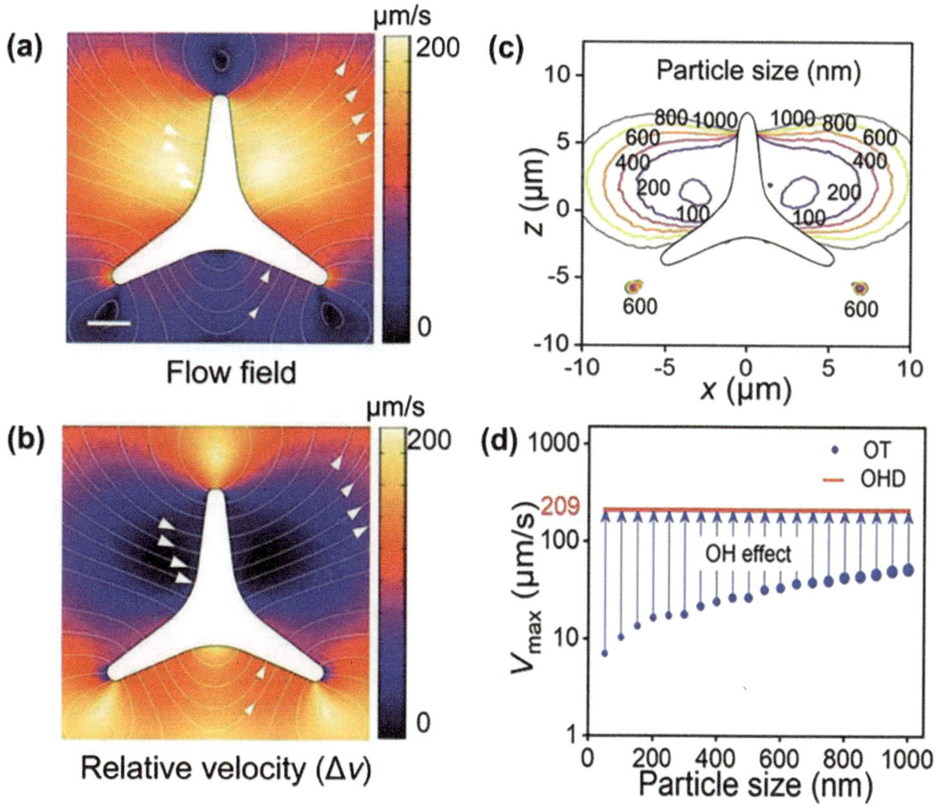

Fig. 4.7 Numerical simulation and calculation of nanoscale trapping via OH effect of OHD. **a** 2D flow velocity generated by moving OHD. **b** 2D relative flow velocity to OHD. **c** Calculated effective trapping range for particles with different sizes in the flow field generated by the OHD. **d** Maximum velocity of nanoparticles that can be effectively trapped by OT and OHD. The arrows indicate the velocity transition of different-sized particles via the OH effect (Xiong et al. [22], Copyright (2023), with permission from John Wiley and Sons)

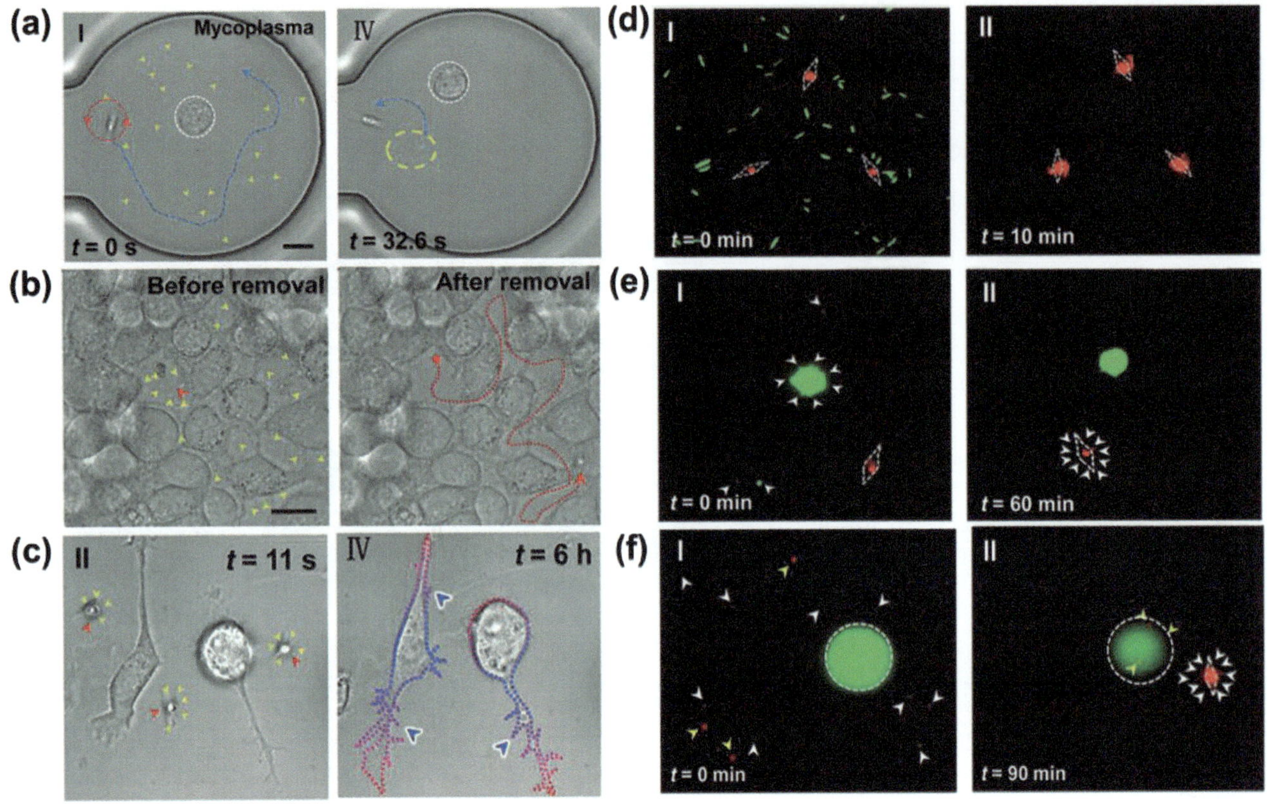

Fig. 4.8 **a** Trapping and removal of nano-biothreats in a microfluidic channel via OHD, yellow arrows indicate released nano-biothreats. **b** Removal of invaded *E. coli* bacteria in real dense cell cultures with cell confluency of about 90%. **c** Microscopic images showing the living states of *E. coli* contaminated primary hippocampal neurons with the treatment of OHD array treatment. Yellow arrows indicate *E. coli*, blue arrows indicate the boundaries of neurons after culturing of 6 h, and red arrows indicate OHD. **d** Fluorescent image showing the killing of *E. coli* with OHD array, live *E. coli* in green and dead *E. coli* in red. White dashed rhombuses indicate the position of the OHD. **e** Fluorescent image showing the viability of HL-60 cells contaminated with *E. coli* with treatment of chi-OHD. White arrows indicate *E. coli*. **f** Fluorescent image showing gene delivery for contaminated HL-60 cells treated with chi-OHD. White dashed circle indicates the boundary of the HL-60 cell, and yellow arrows indicate the siRNA-loaded MSNs. (Xiong et al. [22], Copyright (2023), with permission from John Wiley and Sons.)

This targeted approach significantly minimizes disruption to the surrounding environment, enhancing the efficacy of the removal process. By selectively engaging with contaminants, the OHD helps maintain the delicate balance required for successful cell culture and other biological experiments. Its non-invasive nature ensures that biological samples remain unharmed during trapping and removal, which is particularly crucial in applications where the viability and functionality of cells are paramount, such as in drug testing and biological research. The ability to operate without damaging the cells involved opens new avenues for research and development, leading to more reliable experimental outcomes.

Conclusion

The wake-riding effect-inspired opto-hydrodynamic diatombot (OHD) represents a significant advancement in microscale manipulation and control. Its non-invasive approach to managing nano-biothreats, combined with its potential for targeted applications in biomedicine and environmental science, positions the OHD as a powerful tool for addressing various critical challenges. As research in this area continues to evolve, the OHD may become an indispensable component in the toolkit for mitigating microscale threats across a wide array of sensitive environments.

The ongoing exploration of the OHD's capabilities promises to yield further innovations in managing nano-biothreats and enhancing our understanding of fluid dynamics in biological contexts. As the integration of technology and biology progresses, the OHD stands at the forefront, poised to revolutionize our strategies for maintaining the integrity of biological systems while tackling environmental challenges.

4.5 Challenges and Future Perspectives

As opto-hydrodynamic manipulation continues to evolve, it faces a series of challenges that must be addressed to fully realize its potential. This section discusses these challenges and explores the future perspectives that could shape the development and application of this technology.

4.5.1 Technical Challenges

The field of opto-hydrodynamic manipulation is fraught with technical hurdles that require innovative solutions.

Precision and Control

Achieving the level of precision and control necessary for complex manipulation tasks remains a primary challenge. While current technologies allow for the manipulation of particles with nanometer precision, there is an ongoing need for even greater control, especially in three-dimensional spaces. Fine-tuning the manipulation parameters in real-time to account for dynamic changes in the environment is essential for applications requiring high precision.

Integration of Modalities

Integrating opto-hydrodynamic manipulation with other modalities, such as magnetic and electric fields, presents compatibility and synchronization challenges. Developing systems that can seamlessly integrate these different forces is crucial for expanding the capabilities of opto-hydrodynamic manipulation. This integration can enhance the versatility and effectiveness of manipulation strategies across various applications.

Scaling up

Scaling up the technology for use in larger systems or for manipulating larger objects poses significant challenges. Current systems are well-suited for microscale manipulation, but extending this to the macroscale requires overcoming technical barriers related to force generation, control, and fluid dynamics at larger scales.

4.5.2 Biological and Medical Challenges

The application of opto-hydrodynamic manipulation in biological and medical fields presents unique challenges.

Biocompatibility

Ensuring that manipulation processes are biocompatible and do not harm living cells or tissues is critical. This necessitates a deep understanding of the interactions between light, fluid dynamics, and biological matter. Research must focus on optimizing laser parameters and fluid conditions to minimize potential damage to biological samples.

Real-Time Monitoring

Real-time monitoring of the manipulation process is essential for medical applications, such as targeted drug delivery. Developing imaging and sensing technologies that provide immediate feedback is a significant challenge. These technologies must be capable of integrating seamlessly with manipulation systems to enhance precision and effectiveness.

4.5.3 Future Directions

Despite these challenges, the future of opto-hydrodynamic manipulation is promising, with several potential directions for research and development. The development of new materials that respond to light and fluid forces in novel ways could expand the applications for opto-hydrodynamic manipulation. This includes materials capable of changing their properties in response to external stimuli, such as temperature or light intensity, allowing for enhanced functionality and versatility in manipulation tasks. As nanotechnology advances, opto-hydrodynamic manipulation could play a crucial role in the assembly of nanostructures and the manipulation of individual atoms and molecules. This capability would significantly enhance the precision of nanofabrication processes and the development of nanoscale devices. Incorporating automation and artificial intelligence into opto-hydrodynamic systems could enhance their capabilities, enabling more complex and adaptive manipulation tasks. Machine learning algorithms could optimize the manipulation process in real-time, allowing for greater efficiency and responsiveness to environmental changes.

Conclusion of Sect. 4.5

The field of opto-hydrodynamic manipulation is at the forefront of microscale control, offering vast potential for scientific and technological advancement. While challenges remain, the future perspectives are promising, with the potential to revolutionize fields from medicine to nanotechnology. As research progresses, it is crucial to address these challenges and consider the ethical and societal implications of this powerful technology.

Acknowledgements This work was supported by the National Key Research and Development Program of China (2023YFF0613700), the National Natural Science Foundation of China (No. 12204196, 61975065).

References

1. Ashkin A. Phys Rev Lett. 1970;24:156.
2. Omori R, Kobayashi T, Suzuki A. Opt Lett. 1997;22:816.
3. Lynge Eriksen R, Ricardo Daria V, Glückstad J. Opt Exp. 2002;10:597.
4. Vasantham S, Kotnala A, Promovych Y, Garstecki P, Derzsi L. Lab Chip. 2023;24:517.
5. Martin JC, Chang C, Boschetti G, Ungaro R, Giri M, Grout JA, Gettler K, Chuang L, Nayar S, Greenstein AJ, Dubinsky M, Walker L, Leader A, Fine JS, Whitehurst CE, Mbow Ml, Kugathasan S, Denson LA, Hyams JS, Friedman JR, Desai PT, Ko HM, Laface I, Akturk G, Schadt EE, Salmon H, Gnjatic S, Rahman AH, Merad M, Cho JH, Kenigsberg E. Cell. 2019;178:1493.
6. Ehrlicher A, Betz T, Stuhrmann B, Koch D, Milner V, Raizen MG, Käs J. Proc Natl Acad Sci. 2002;99:16024.
7. Liu S, Wang J, Shao J, Ouyang D, Zhang W, Liu S, Li Y, Zhai T. Adv Mater. 2022;34:202200734.
8. Shi Y, Liu L, Huang J, Xiong J, Zhong S, Zhu G, Li X, He Z, Pan T, Xin H, Li B. Adv Mater. 2024. https://doi.org/10.1002/adma.202412895.
9. Landenberger B, Höfemann H, Wadle S, Rohrbach A. Lab Chip. 2012;12:3177.
10. Enger J, Goksör M, Ramser K, Hagberg P, Hanstorp D. Lab Chip. 2004;4:196.
11. Zhao X, Zhao N, Shi Y, Xin H, Li B. Micromachines. 2020;11:114.
12. John BF, Bossis G. Annu Rev Fluid Mech. 1988;20:111.
13. Xin H, Li X, Li B. Opt Express. 2011;19:17065.
14. Chen X, Wu T, Gong Z, Guo J, Liu X, Zhang Y, Li Y, Ferraro P, Li B. Light: Sci Appl. 2021;10:242.
15. Shan X, Ding L, Wang D, Wen S, Shi J, Chen C, Wang Y, Zhu H, Huang Z, Wang SSJ, Zhong X, Liu B, Reece P, Ren W, Hao W, Lu X, Lu J, Su Q, Chang L, Sun L, Jin D, Jiang L, Wang F. Nat Photon. 2024. https://doi.org/10.1038/s41566-024-01462-7.
16. Shi Y, Liang L, Zuo Y, Zhu X, Yang Y, Xin H, Li B. J Phys Chem Lett. 2020;11:4747.
17. Shi Y, Liu HL, Zhu XQ, Zhu JM, Zuo YF, Yang Y, Jiang FH, Sun CJ, Zhao WH, Han XT. Lab Chip. 2018;18:2994.
18. Guckenberger DJ, de Groot TE, Wan AMD, Beebe DJ, Young EWK. Lab Chip. 2015;15:2364.
19. Li X, Shi Y, Pan T, Lu D, Zhu G, Xiong J, Wang D, He Z, Huang J, Xin H. Laser Photon Rev. 2023;18:2300480.
20. Pan T, Shi Y, Zhao N, Xiong J, Xiao Y, Xin H, Li B. Adv Func Mater. 2022;32:2111038.
21. Shi Y, Wang D, Xiao Y, Pan T, Lu D, Zhu G, Xiong J, Li B, Xin H. Laser Photon Rev. 2022;16:202200533.
22. Xiong J, Shi Y, Pan T, Lu D, He Z, Wang D, Li X, Zhu G, Xin H, Li B. Adv Sci. 2022;10:2301365.

5 Optical Biomanipulation Using Optofluidic Techniques

Yuzhi Shi, Zhanshan Wang, and Xinbin Cheng

Abstract

Optical manipulation in a static environment promises numerous biological applications. This mostly deployed environment endows an idea testbed with various advantages, such as stability, highly controllable, reduced motion artifacts, high precision, easy configuration, integration, etc. However, some intrinsic deficiencies must be noticed, such as low throughput, lack of dynamic control, heat accumulation, etc. A fluidic environment provides alternative strategies for biomedical studies by leveraging its merits of high throughput, multi-phase control, efficient interaction, etc. In this chapter, we center our attention on advances in biomedical applications using optical tweezers in optofluidic environments. We begin by elaborating on some fundamental physics of forces in optofluidics systems, describing how those forces interplay and play important roles in optical manipulation. We then outline some typical examples of optical manipulation in optofluidics, including bioparticle transporting, sorting, detection, interaction, measurement, etc. This chapter concludes with our vision of the perspectives of ongoing research directions and how this field can evolve to facilitate more prosperous applications in the future.

Keywords

Optofluidics · Microchannel · Optical sorting · Biointeraction · Microfluidic mixing · Particle transporting · Antibodies · Biodetection

5.1 Introduction

Optofluidics, which fuse two classic while intriguing research fields of optics and microfluidics [1, 2], have garnered significant attention since it came to public notice in the early 2010s. Consequently, this technique possesses the advantages of both fields, and achieves capabilities that are unattainable by bio-optical manipulation in still environments. For instance, flow streams are proficient in transporting and delivering biosamples to target positions, facilitating further manipulation detection. Fast flow guarantees fast analysis and high throughput. An excellent example is the fluorescence-activated cell sorter [3], which combines various techniques, such as microfluidics, fluorescence, optical or electric forces, etc. The technique and relevant machines can handle a great diversity of samples and achieve a high flow rate, in μL/s and up to 100,000 cells per second. In optofluidics, there are multiple forces, including forces from optics, fluidics, gravity, and thermal dynamics. In the first section, we would like to view those forces and discuss how those forces interact and play significant roles in optical tweezers.

5.1.1 Forces in Optofluidics

Analysing forces on particles is critical to comprehending the dynamics of particles in optofluidics. Particularly, the behaviour of particles in force fields can be characterized

Y. Shi (✉) · Z. Wang · X. Cheng (✉)
Institute of Precision Optical Engineering, School of Physics Science and Engineering, Tongji University, Shanghai, China
e-mail: yzshi@tongji.edu.cn

X. Cheng
e-mail: chengxb@tongji.edu.cn

Technology Innovation Center of Mass Spectrometry for State Market Regulation, Center for Advanced Measurement Science, National Institute of Metrology, Beijing, China

MOE Key Laboratory of Advanced Micro-Structured Materials, Shanghai, China

Shanghai Institute of Intelligent Science and Technology, Tongji University, Shanghai, China

B. Li et al. (eds.), *Biophotonic Manipulation*, Advances in Optics and Optoelectronics,
https://doi.org/10.1007/978-981-96-4982-2_5

by the renowned Langevin equation [4–6], which can be expressed as

$$m\frac{d\mathbf{v}}{dt} + \mathbf{F}_{\text{drag}} + \mathbf{F}_{\text{optical}} + \mathbf{F}_{\text{Brownian}} + \mathbf{F}_{\text{external}} = 0, \quad (5.1)$$

where η is the viscosity of the medium; m, a and $\mathbf{v}$ denote the mass, radius and velocity of the particle, respectively; $\mathbf{F}_{\text{drag}}$, $\mathbf{F}_{\text{optical}}$, $\mathbf{F}_{\text{Brownian}}$ and $\mathbf{F}_{\text{external}}$ represent the fluidic drag force, optical force, Brownian force, and external force, respectively. The first term of Eq. (5.1) signifies the inertial term of the nanoparticle; the second term is the fluidic drag force [7–9], which can be expressed as $\mathbf{F}_{\text{optical}} = 6\pi\eta a(\mathbf{v}-\mathbf{u})$, where $\mathbf{u}$ is the velocity of the liquid. $\mathbf{F}_{\text{Brownian}}$ is the random force with zero mean, which can be written as $\mathbf{F}_{\text{Brownian}} = \sqrt{2S}\mathbf{W}(t)$, where $\mathbf{W}(t)$ is a white noise with $\langle\mathbf{W}(t)\rangle = 0$, and $2S$ is the intensity of the noise.

In microfluidics, the flow stream is normally a laminar flow, which refers to a flow regime where fluid flows in smooth, orderly layers or streamlines, with minimal mixing between layers. The laminar flow has a low Reynolds Number (Re) and minimal turbulence, which is feasible for the precise control of the mixing process and steady transport of samples at the microscopic scale. Indeed, the Reynold number is a dimensionless parameter determined by the ratio of inertia force to viscous force, which can be used to characterize the fluid flow regime in microfluidic systems. Its expression can be given as [10]

$$\text{Re} = \frac{F_{\text{inertia}}}{F_{\text{viscos}}} = \frac{\rho_f V_{\text{avg}} D_h}{\eta}, \quad (5.2)$$

where ρ_f is the density of the fluid, V_{avg} is the averaged flow velocity, D_h is the hydraulic diameter of the microchannel, and η is the dynamic viscosity of the fluid. The most common laminar flow in the microchannel exhibits a Re $\ll$ 1. As Re increases, flow can transition from laminar to a regime where vortices may form, depending on the channel design and flow conditions.

In a microfluidic channel, the fluid profile is typically parabolic under laminar flow conditions. The parabolic profile is uniform across the microchannel at any given cross-section with the highest at the centre of the channel and decreases towards the walls. Consequently, the shear rate is highest at the channel walls and lowest at the centre. This can influence how cells or particles experience forces within the channel.

One issue worth noting is that a fluidic lift force may arise due to the interactions between the particle and the laminar flow field in the channel. The velocity gradient near the channel walls creates a pressure difference that can lift particles away from the walls. The lift force has various types depending on different conditions of the flow and particles. For instance, the well-known Magnus lift force is a phenomenon that occurs when a rotating object moves through a fluid, which results from the interaction between the object's rotation and the fluid flow, creating a lift force perpendicular to both the direction of motion and the axis of rotation. The Magnus lift force can be expressed as [11]

$$F_{ML} = \frac{1}{2}\rho_f V_r^2 A C_L, \quad (5.3)$$

where V_r is the fluid velocity relative to the particle, A is the reference area (which is typically the cross-section of the particle), and C_L is the lift coefficient depending on the rotation rate, flow conditions, and the shape of the object. For a sphere with a radius a and an angular velocity $\mathbf{\Omega}$ and moving through a fluid with velocity $\mathbf{V}_f$, the Magnus lift force can be approximately written as

$$\mathbf{F}_{ML} = \pi\rho_f a^3(\mathbf{V}_f - \mathbf{V}_p) \times \mathbf{\Omega}, \quad (5.4)$$

where V_p is the velocity of the particle. From Eq. (5.4), we can comprehend that the Magnus lift force can be significantly affected by the relative velocity of the fluid, the size and rotation velocity of the particle. Another type of lift force is called the Saffman lift force, which emerges when a spherical particle moves through a viscous fluid with a velocity gradient, which exerts a difference in pressure on two sides of the particle. Its expression can be given as

$$\mathbf{F}_{SL} = 4ka^3(\mathbf{V}_f - \mathbf{V}_p)\sqrt{\eta\rho_f\gamma}, \quad (5.5)$$

where γ is the shear rate, and k is a constant ~ 1.615. It can be seen from Eqs. (5.4) and (5.5) that the Magnus lift force and Saffman lift force are both proportional to the velocity gradient of the flow.

In a laminar flow, the profile of flow velocity is parabolic, which may generate a typical type of Saffman lift force, called the "shear gradient force" [12]. This force directs the particle towards the side with a higher relative velocity. The presence of a wall can also induce a special type of lift force, which arises from two reasons. One is that the flow is frozen near the wall and lagged behind the flow in the microchannel, generating a velocity gradient, and thus the Saffman lift force. Another issue is that the streamlines around the sphere are altered when the sphere is close to the wall. The presence of the nearby wall causes asymmetry in the wake vorticity distribution generated at the sphere's surface. This asymmetric wake vorticity results in higher pressure on the wall side, creating a lift force that pushes the sphere away from the wall. The balance of the wall-induced lift force and Saffman lift force make the particle move steady at a distance away from the wall, as shown in Fig. 5.1.

In microfluidics, there are other ways to utilize the fluidic forces. Good examples include the Dean flow, which arises due to the centrifugal forces acting on the fluid in a curved or

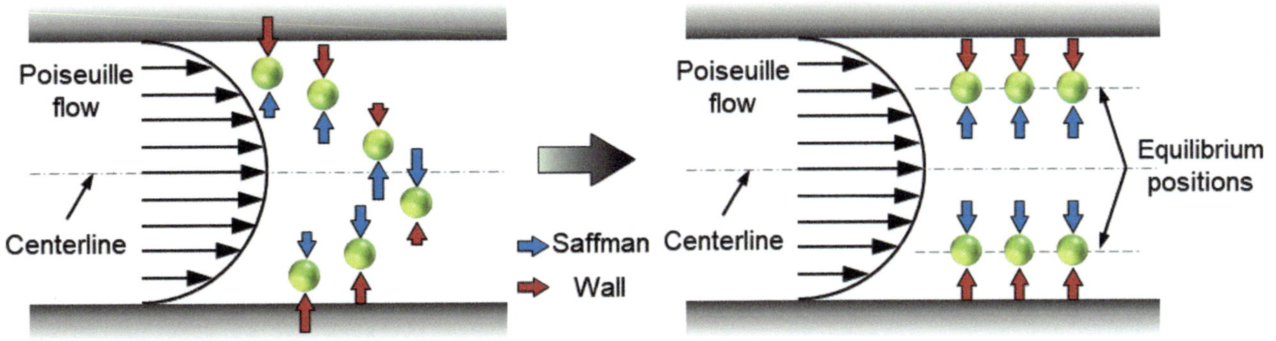

Fig. 5.1 Schematic illustration of lift force in the microchannel [12]. The wall-induced lift force pushes the particle away from the wall. The Saffman lift force induced by the shear gradient makes the particle shift away from the centre of the microchannel. The two types of forces balance with each other and make particles move in equilibrium positions in the microchannel. (Reprinted from Geislinger et al. [12], Copyright (2014), with permission from ELsevier)

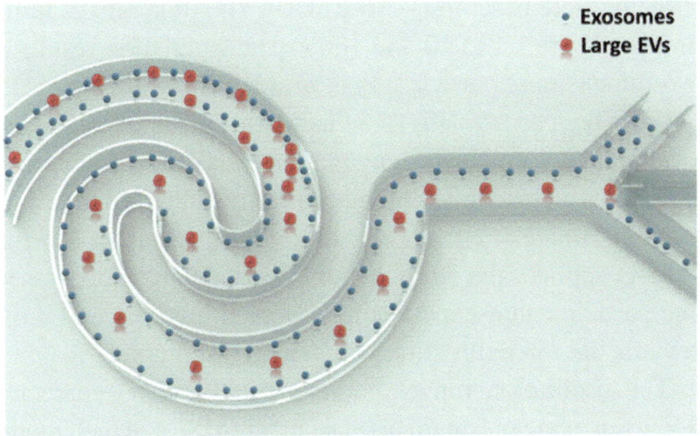

Fig. 5.2 Microfluidic sorting using the dean flow [13]. (Reprinted from Zhou et al. [13], Copyright (2019), with permission from American Chemical Society)

rotating cylindrical pipe or channel. In the Dean flow, at least three kinds of forces are employed, e.g., the centrifuge force, lift force and drag force. Since different particles experience varying fluidic forces, they can move along distinct trajectories and be separated through careful design, as shown in Fig. 5.2.

The optical forces on a dipole particle that satisfies $ka \ll 1$ (k is the wavenumber in the medium) in an electromagnetic field can be given as

$$\mathbf{F} = \mathbf{F}_e + \mathbf{F}_m + \mathbf{F}_{em} = \frac{1}{2}\mathrm{Re}\left\{ \underbrace{\mathbf{p}\cdot(\nabla)\mathbf{E}^*}_{\sim(ka)^3} + \underbrace{\mathbf{m}\cdot(\nabla)\mathbf{H}^*}_{\sim(ka)^5 \text{ for } \mu_p=\mu} - \underbrace{\frac{ck^4}{6\pi\sqrt{\varepsilon\mu}}(\mathbf{p}\times\mathbf{m}^*)}_{\sim(ka)^8 \text{ for } \mu_p=\mu} \right\}, \tag{5.6}$$

where ε and μ are the permittivity and permeability of the medium, respectively; ε_p and μ_p are the permittivity and permeability of the particle, respectively; $\mathbf{p}$ and $\mathbf{m}$ are the electric and magnetic dipole moments of the particle, respectively, which can be expressed as

$$\mathbf{p} = \alpha_e \mathbf{E}, \mathbf{m} = \alpha_m \mathbf{H}, \tag{5.7}$$

where α_e and α_m are the electric and magnetic polarizabilities for a spherical dipolar particle, respectively, which can be calculated using the electric and magnetic first Mie coefficients $a_1{}^{(1)}$ and $b_1{}^{(1)}$, respectively, as

$$\alpha_e = \frac{i6\pi\varepsilon}{k^3} a_1^{(1)} = \frac{a_e^0}{1 - \frac{ik^3 a_e^0}{6\pi\varepsilon}}, \quad \alpha_m = \frac{i6\pi}{\mu k^3} b_1^{(1)} = \frac{a_m^0}{1 - \frac{i\mu k^3 a_e^0}{6\pi}}, \tag{5.8}$$

where $a_e^0 = 4\pi\varepsilon a^3 \frac{\varepsilon_p - \varepsilon}{\varepsilon_p + 2\varepsilon}$ and $a_m^0 = 4\pi\mu^{-1} a^3 \frac{\mu_p - \mu}{\mu_p + 2\mu} + O\left([ka]^5\right)$ are quasistatic limits.

In Eq. (5.5), $\mathbf{F}_e$, $\mathbf{F}_m$ and $\mathbf{F}_{em}$ denote the optical force on the electric dipole, magnetic dipole and electric–magnetic dipole interaction, respectively. Normally, for a non-magnetic particle, $\mathbf{F}_m$ and $\mathbf{F}_{em}$ are much smaller than $\mathbf{F}_e$, thus, $\mathbf{F}_e$ has been widely utilized to calculate optical forces on dielectric particles such as polystyrene particles or cells.

In this section, we outline some typical forces in optofluidic systems, while some other forces such as the gravity and electrophoresis forces can also be deployed. In the next section, we will discuss how these forces affect particle dynamics, especially in optical trapping and sorting.

5.1.2 Particle Dynamics in Optofluidics

Analysing forces on particles in an optofluidics environment is crucial to manoeuvring the particle dynamics and achieving the multifunctional manipulation of particles. Apart from the

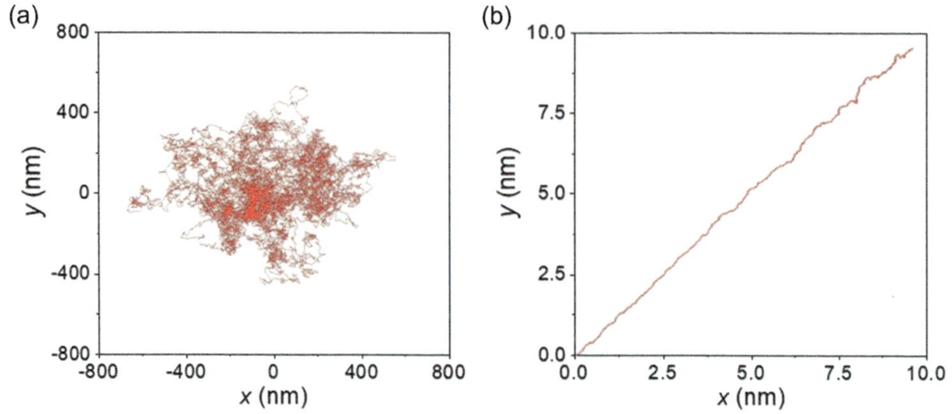

Fig. 5.3 Trajectories of particles in an optofluidic trapping system. **a** The 10-μm particle only experiences an optical trapping force with an optical stiffness of 1×10^{-7} N/m. The particle has a strong Brownian motion. The temperature is set to 300 K. **b** Two additional forces 1 pN are applied to the x and y directions, which become a dominant factor to drive the movement of the particle. The simulated time ranges in (**a**) and (**b**) are set to 10 s and 1 s, respectively

fluidic forces and optical forces mentioned in the previous section, the particle may experience various such as the thermal Brownian force [14], and the dielectrophoresis force. In this section, we shall discuss how the particle behaves in a force field. For a particle in a force field, the kinetics of the particle can be analysed using the Langevin equation, which can be expressed as

$$m\frac{d^2}{dt^2}r(t) = -\gamma\frac{d}{dt}r(t) + F(r) + \chi(t), \tag{5.9}$$

where m is the mass of the particle, r is the particle position, $F(r)$ is the external force field, γ is the particle friction coefficient, which for a spherical particle of radius a moving in a fluid with a viscosity η, can be determined by Stokes' law as $\gamma = 6\pi\eta a$. And $\chi(t)$ is the random force with zero mean, which can be written as $\chi(t) = \sqrt{2S}W(t)$, where $W(t)$ is a white noise with $\langle W(t)\rangle = 0$, and $2S$ is the intensity of the noise. The term on the left side of Eq. (5.8) is the inertial term, which can be omitted in an overdamped system; the first term on the right side of Eq. (5.8) is the fluidic drag force, which is associated with the viscosity of the liquid and particle velocity. In the low Reynolds number regime, the overdamped Langevin equation can be obtained as

$$\gamma\frac{d}{dt}r(t) = F(r) + \xi(t), \tag{5.10}$$

where $\xi(t) = \sqrt{2D}W(t)$ is a white noise with intensity 2D, where $D = k_B T/\gamma$ is the diffusion coefficient. Here, k_B is the Boltzmann constant, and T is the temperature.

Assuming that in an optical force field where the particle experiences only the optical force $\mathbf{F}_{\text{opt}} = k_1\mathbf{x} + k_2\mathbf{y}$, we write the two-dimensional Langevin equation as

$$\begin{cases} \dfrac{dx(t)}{dt} = \dfrac{k_x x(t)}{\gamma} + \sqrt{2D}W_x(t), \\ \dfrac{dy(t)}{dt} = \dfrac{k_y x(t)}{\gamma} + \sqrt{2D}W_y(t). \end{cases} \tag{5.11}$$

We then simulate the particle trajectories in this force field. k_x and k_y are both set to 1×10^{-7} N/m. The initial position of the particle is set to (− 10 μm, 0). The time ranges in the simulation are set to 10 and 100 s. The simulation results of two-dimensional particle trajectories are shown in Fig. 5.3.

It can be seen from Fig. 5.3 that when a particle is subjected to an optical trapping system without external forces, it tends to oscillate around the trapping centre ($x = 0$; $y = 0$) in a range of approximately 1 μm [15]. In contrast, when an external force is applied, the Brownian motion is hugely suppressed. The particle can escape from the optical trapping and move towards the force direction with a mere oscillation.

The oscillation range of the particle can be characterized using a mean square displacement (MSD), which unanimously unfolds a complete map for the joint realm of sorting and trapping. A system can be firstly classified into two possible categories: overdamped $\left(\omega_0 \le 1/2\tau_p\right)$ and underdamped $\left(\omega_0 > 1/2\tau_p\right)$, where $\omega_0 = \sqrt{k/m}$ is the resonant frequency with trapping stiffness k and $\tau_p = m/\gamma_f$ is the momentum relaxation time of a nanoparticle with mass m and Stokes viscous drag coefficient. The overdamped and underdamped systems can be divided by the red dashed line expressed as $\lg(\eta) = \frac{1}{2}\lg(k) + \lg\left(\frac{\sqrt{m}}{3\pi R}\right)$ in the diagram. Optical trapping on the right side of the roadmap relies on the tight focus ($k \sim 10^{-6} - 10^{-4}$ N/m) to attract nanoparticles to the focal point either in a liquid or gas. When a nanoparticle (e.g., $d = 80$ nm) is trapped in the gas with a viscosity η of 10^{-5} Pa s and drag coefficient of 7.54×10^{-12} N s/m, the system is underdamped. Experiments involving optical trapping of nanoparticles are often conducted in liquids with viscosities of approximately 10^{-3} Pa s, in which $1/2\tau_p$ could easily turn to 7.3×10^7 Hz and the system becomes overdamped. Conventional optical chromatography builds potential wells with low stiffness ($10^{-8} - 10^{-6}$ N/m), and has been widely used to sort microparticles and spores (Fig. 5.4).

In addition to the highly overdamped region, we discover a nearly unexplored region with stiffness from 10^{-10} to 10^{-8} N/m on the leftmost side of the diagram, which represents a

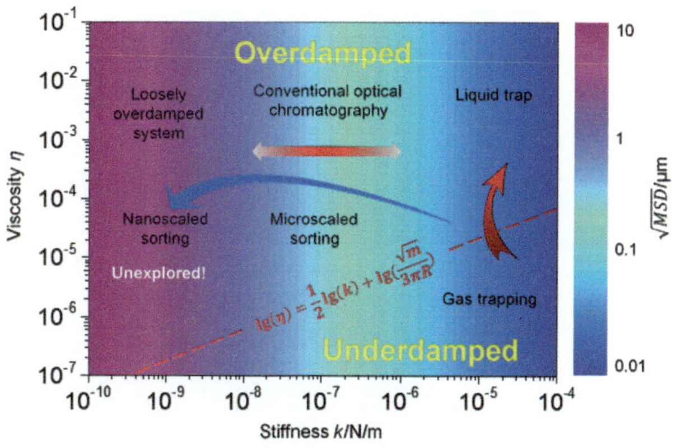

Fig. 5.4 The damping diagram unfolds the complete map of trapping and sorting particles in optofluidic systems: classic trapping, microscale sorting, and nanoscale sorting [15]. (Reprinted from Shi et al. [15], Copyright (2018), with permission from The American Association for the Advancement of Science)

"loosely overdamped" system $\omega_0 \ll 1/2\tau_p$. Surprisingly, this long-ignored region actually empowers single-nm precision linear sorting with marked displacement position offset. This clearly marks the valuable significance of our physical model and damping diagram established.

5.1.3 Optical Sorting in Optofluidic Systems

One profound example of optical sorting is optical chromatography, which utilizes the balance of the optical extinction and fluidic drag force, serving as a paradise for the real-time measurement of the size of single particles in a chip [16]. Using this method, sizes of 1 and 3 μm polystyrene particles were measured [17]. However, there is no demonstration of measuring the size of a single gold nanoparticle using this method because the separate trapping of nanoparticles along the light beam is a precondition for the measurement. In order to separate gold nanoparticles, a small divergence of the light beam is required (NA < 0.1), which has been accomplished using a photonic crystal fibre (NA ~ 0.05) [18]. Moreover, the tight focus of the beam (usually < 2 μm) is also necessary for a stable trap of nanoparticle [19]. The two conditions are usually opposite against each other in most of the light beams, especially for Gaussian beams [20]. That is why optical chromatography is normally only used for the separation of microparticles and spores. In 2018, Shi et al. proposed a nontrivial scheme for optical chromatography using the quasi-Bessel beam, which is configured by passing a Gaussian beam through a microlens. As shown in Fig. 5.5, the microlens have two open angles θ_1 and θ_2, Quasi-Bessel beam and discrete interference pattern will generate after the lens due to the interference effect, which creates a "loosely overdamped" system. These characters serve as a paradigm for the trapping and sorting of nanoparticles. The tight focus ensures the trapping stabilities of nanoparticles, preventing them from escaping the beam by the Brownian motion. Therefore, the low NA facilitates the high-resolution separate trapping of nanoparticles along the beam.

5.1.4 Single-Particle Sorting via Far-Field Optical Manipulation

As shown in Fig. 5.6, this work marks a significant advancement in the precision of single-particle sorting [15]. This research addresses a critical challenge in the field: the sorting of sub-50-nm nanoparticles with single-nanometre precision. Traditional optical trapping regimes, which deal with trapping stiffness larger than 10^{-8} N/m, have struggled with this task. By establishing a damping roadmap and synchronizing optical force with the flow drag force, the authors ventured into such a "loosely overdamped" realm (stiffness, 10^{-10} to 10^{-8} N/m), achieving remarkable results.

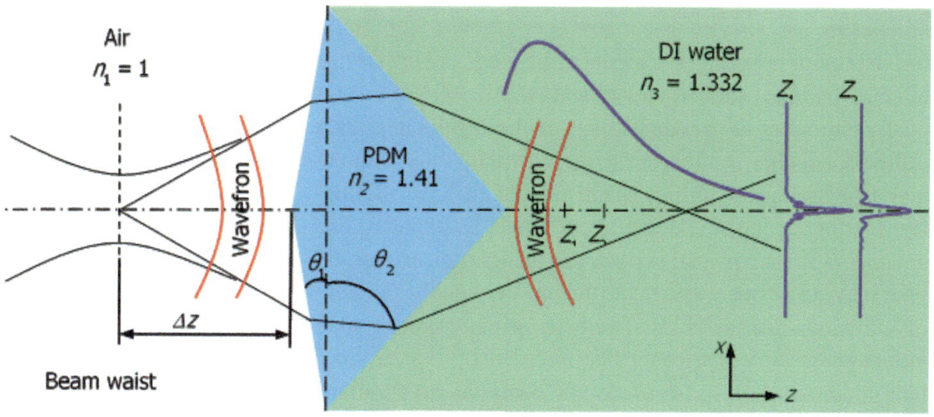

Fig. 5.5 The micro-quadrangular lens has two open angles (θ_1 and θ_2) for the generation of the quasi-Bessel beam and other intriguing interference patterns

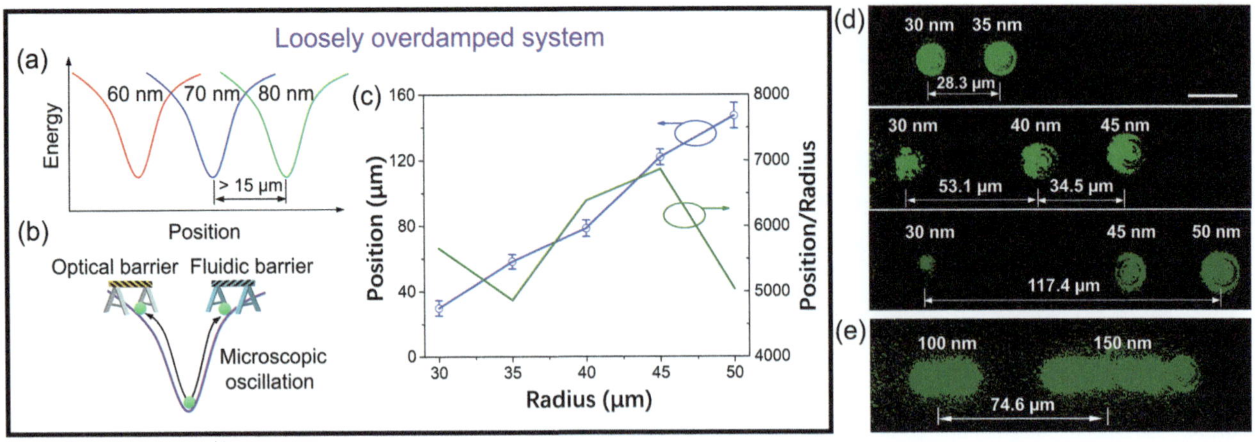

Fig. 5.6 Sorting of nanoparticles and microorganisms with nanometer-level precision within a loosely overdamped system [15]. **a** A loosely overdamped system for nanoparticle sorting; **b** Nanoparticles oscillate in microscale under the control of synchronized optical and fluidic barriers; **c** The sorting distance increases in direct proportion to the radius as observed in experiments; **d** Nanoparticles of varying sizes trapped within the microchannel; **e** The process of isolating polystyrene nanoparticles that have radii of 100 and 150 nm. (Reprinted from Shi et al. [15], Copyright (2018), with permission from The American Association for the Advancement of Science)

The physics of the synchronized dual barriers associated with the loosely overdamped system is described, where the optical barrier is created by the dominant extinction force, and the fluidic barrier is actuated via the drag force. This system allows for the confinement of nanoparticles even in potential wells with trapping stiffness as low as 10^{-10} N/m, which is a significant departure from traditional optical trapping methods that rely on higher stiffness.

The authors show that gold nanoparticles of different sizes can be separated with large distances (> 15 μm) between them, and the trapping distance increases linearly with the size of the nanoparticles. They successfully sorted single gold nanoparticles with radii from 30 to 50 nm, as well as 100- and 150-nm polystyrene nanoparticles with unprecedented precision. The use of a quasi-Bessel optical profile and the creation of loosely overdamped potential wells in the microchannel allowed for the separation, positioning, and microscopic oscillation of these nanoparticles. This work not only advances our fundamental understanding of particle kinetics in optical systems but also opens up new opportunities in tumour targeting, intracellular imaging, and the sorting of small particles such as viruses and DNA.

Transitioning to the application of these principles in a different context, the following study explores the manipulation of biological entities at the single-cell level. Biological binding agents, including antibodies [21], peptides, and aptamers [22], are pivotal in pathogen recognition and inhibition, playing a critical role in the diagnostics and therapeutics of infectious diseases [23]. The heterogeneity among individual bacteria, which cannot be unveiled by conventional bulk methods [24], has made single bacterial isolation and detection an essential technique. There is a strong demand for screening biological binding agents at the single-cell level to uncover new insights into complex biological interactions.

Traditional binding assays for screening pathogen antibody libraries encompass precipitation assays [25], agglutination assays [26], enzyme-linked immunosorbent assays (ELISA) [27], SPR [28], Western blot [29], and fluorescence-activated cell sorting (FACS) [30]. These technologies are time-consuming, involving multiple sequential testing steps and requiring large sample volumes. Microfluidic systems that employ bead or surface capture have been reported for antibody-based screening of specific bacteria [31]. However, these methods struggle with controlling one-on-one particle-cell interactions, and thus, no conventional method can effectively screen antibodies at the single-bacteria level.

Building upon the precision achieved in single-particle manipulation, Shi et al. [7] introduced an optofluidic lattice that enables the control of particle interactions, including aggregation and multi-hopping, as shown in Fig. 5.7. This study significantly reduced the mean residence time of a single particle from 7 s, as predicted by Kramer's theory, to 0:6 s due to mechanical interactions among aggregated particles.

One of the most intriguing aspects of this work is the optofluidic lattice's capability to perform single-bacteria-level screening of biological binding agents, such as antibodies. Individual bacterial cells were first conjugated with biotin-labelled antibodies and, subsequently, trapped in the optical potential wells. Then, a streptavidin-conjugated microparticle passed by and bound to the bacteria. The microparticle–bacterium complex would hop away from the potential well. With high-affinity antibodies, a large percentage of bacterial cells would bind to the microparticles and hop away and vice versa. For antibodies with high specificity, only the target bacteria would hop away with the

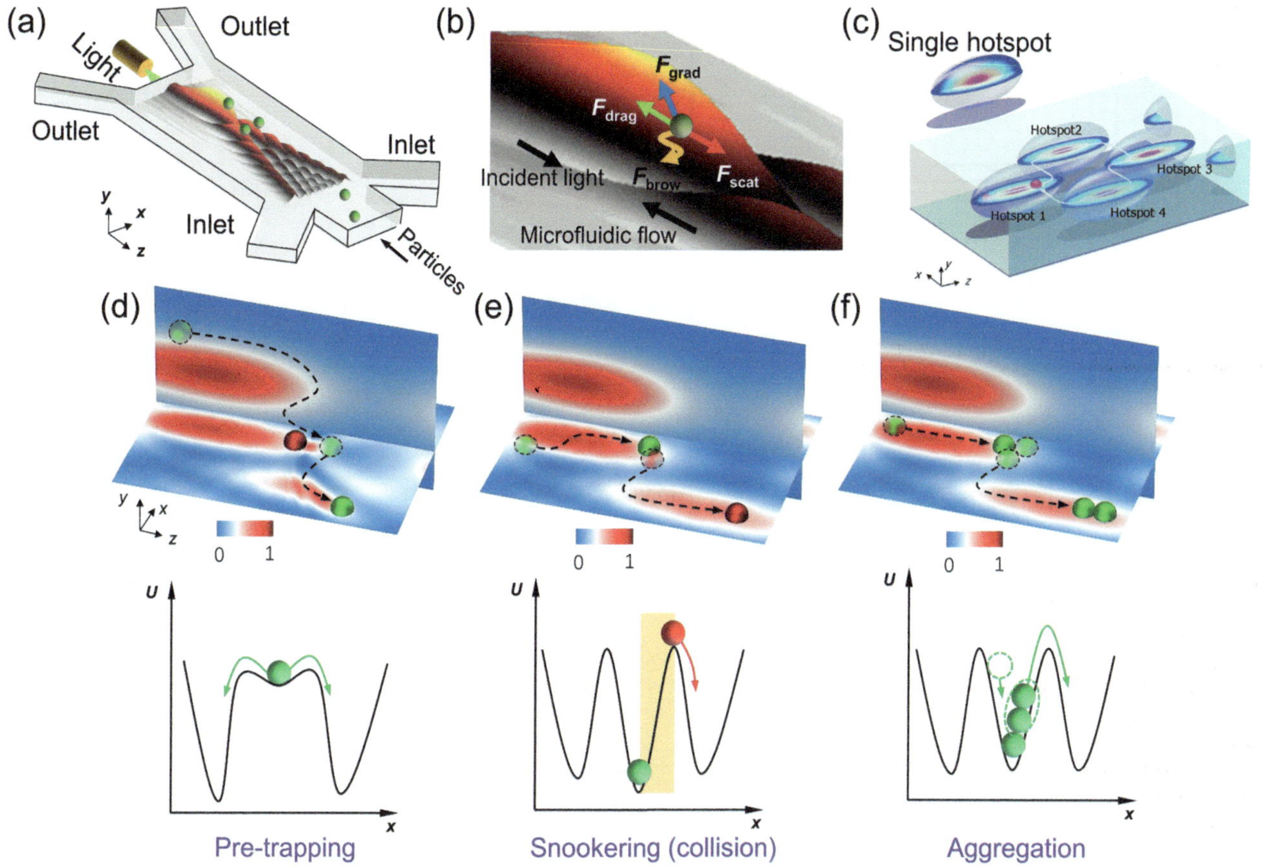

Fig. 5.7 Single-particle sorting in an optofluidic lattice [7]. **A** Creating a two-dimensional lattice within an optofluidic chip. **b** The forces exerted on a particle within the hotspot region. **c** Illustration of the realization of controllable particle hopping loop around the ellipsoid hotspots; **d** particle bypassing; **e** snookering-like particle collision; **f** particle aggregation. (Reprinted from Shi et al. [7], Copyright (2018), with permission from Springer Nature)

microparticle whereas nontarget bacteria remained trapped. In contrast, antibodies with low specificity would bind to non-target cells, causing the non-target cells to hop away from the potential well. The binding efficiency of antibodies could be determined directly, selectively, quantitatively, and efficiently. This is particularly important for diagnostics and therapeutics of infectious diseases, where the heterogeneity between individual bacteria cannot be revealed by conventional bulk approaches.

The optofluidic lattice established by Shi et al. harnesses the tunable control of optical and hydrodynamic forces to drive particles between hotspots through different hopping mechanisms, including particle bypassing, collision, and aggregation. This holistic framework for steering the kinetics and synergies between multiple particles represents a new technique for harnessing Brownian force to create microscopic motors.

The practical application of this technology is showcased through the screening of antibodies for *Escherichia coli* (*E. coli*) and *Shigella flexneri* (*S. flexneri*) at the single-bacteria level. The platform allows for the direct measurement of binding affinity and specificity, providing a semi-quantitative assessment based on the sophisticated but controllable multi-hopping phenomenon.

While optical sorting techniques may be constrained by relatively lower speeds compared to microfluidics and acoustics [32–34], they possess a distinct advantage in their ability to manipulate nanoscale particles with precision. The integration of optical sorting with complementary techniques could unlock new avenues for the rapid sorting and sensing of bioparticles [35].

Optoelectronic tweezers exemplify the synergy between optical fields and dielectrophoretic forces [36, 37], leveraging projected optical images to stimulate electric fields and exert forces. This approach offers significant benefits in manipulating nanoscale objects with minimal light and electric power requirements.

In a recent advancement, Gan et al. [38] have engineered an optoelectronic tweezers micro-well system capable of sorting cells at an impressive flow rate of 20 μL/min, as depicted in

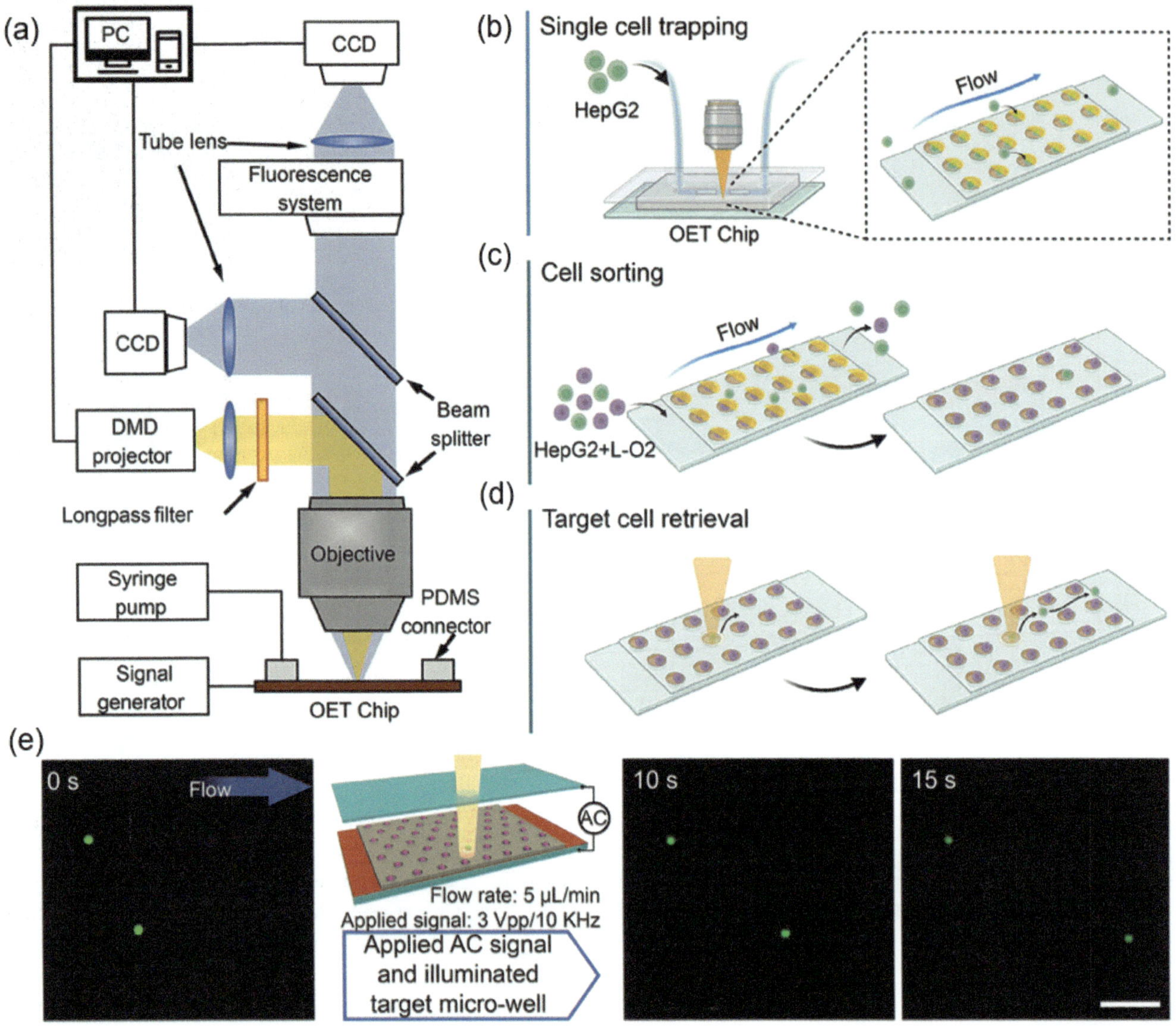

Fig. 5.8 Optoelectronic tweezers micro-well system for highly efficient single-cell trapping, dynamic sorting, and retrieval [38]. **a** Schematic of the experimental setup; **b** Single-cell trapping and cell array formation; **c** Cell sorting from a HepG2 and L-O2 cell mixture; **d** Cell retrieval facilitated by light-induced negative DEP force; **e** The fluorescence image sequences of the selective retrieval of a HepG2 cell. (Reprinted from Gan et al. [38], Copyright (2018), with permission from Wiley-VCH GmbH)

Fig. 5.8. This system enables the entrapment and confinement of cells within discrete microwells, boasting a trapping efficiency that exceeds 91.9%. By modulating the frequency and power of the applied signal, HepG2 cells can be selectively trapped within the microwells while L-O2 cells are released. Conversely, HepG2 cells can be released from the microwells upon altering the signal parameters. This dual capability underscores the system's versatility in the manipulation and analysis of single cells.

In summary, these works above stand as a testament to the potential of optofluidic systems in revolutionizing the way we manipulate and detect bioparticles. Their findings not only enrich our understanding of particle kinetics but also offer new possibilities for probing and utilizing biomolecule interactions at unprecedented levels of precision.

5.1.5 Multi-particle Sorting via Near-Field Optical Manipulation

In practical applications such as biomedical and nanoscience, high-throughput multi-particle sorting is often the only way to meet the demand. Methods for nanoparticle sorting are

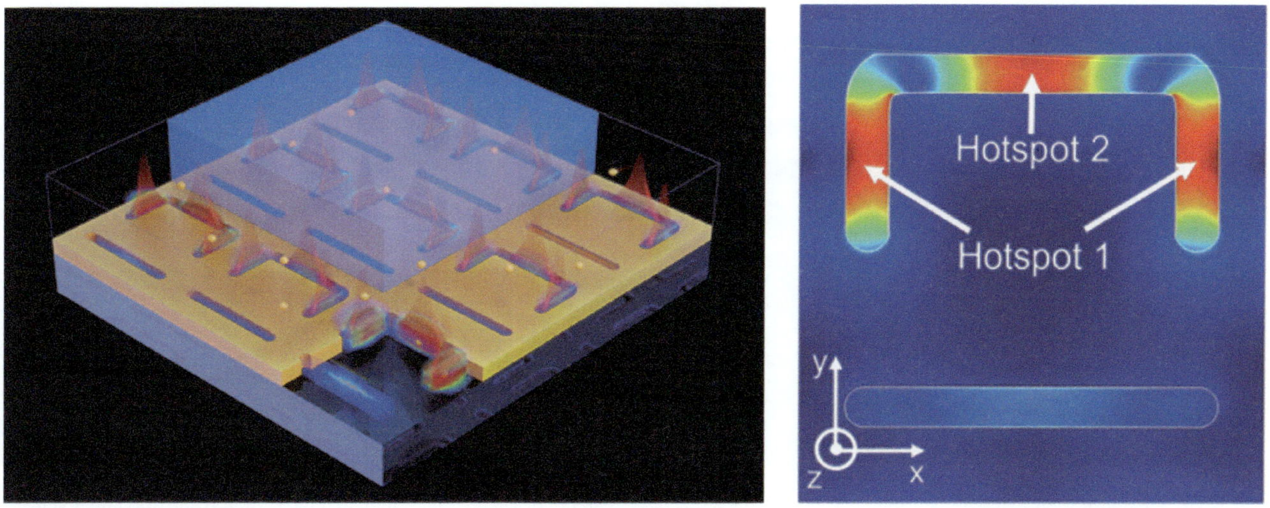

Fig. 5.9 Sorting of nanoparticles with a plasmonic optical tweezer based on the self-induced back-action effect and Fano resonance [49]. (Reprinted from Bouloumis et al. [49], under Copyright license CC BY)

essential to ensure a homogeneous size distribution for reliable reactions in size-sensitive applications [39]. Significant efforts from various fields have been dedicated to achieving nanoscale particle sorting. Examples include microfluidic centrifugal forces [40], acoustic sorting [41], electrophoresis [42], and optical forces [43]. Most optical methods, such as holographic optical tweezers [44], flow cytometry [45], interference patterns [46], and near-field photonics [47], sort nanoparticles without trapping, which limits their manipulation capabilities for real-time analysis.

Optical forces can trap [47] and sort metallic nanoparticles through surface plasmon resonances [48]; however, they are only effective for sorting dielectric nanoparticles larger than 200 nm. Additionally, optical sorting requires tightly focused beams, which generate large intensity gradients capable of trapping or deviating nanoparticles. The number of nanoparticles that can be effectively handled by these beams is limited, typically to a few 200 and 300 nm polystyrene nanoparticles, risking the overlap of particle clusters, which is not conducive to the realization of multi-particle sorting.

Employing optical potential well arrays represents an effective strategy for enhancing the capabilities of plasmonic optical tweezers, particularly in augmenting the trapping stiffness of nanoparticles. Bouloumis et al. [49] have recently introduced a metamaterial plasmonic tweezers system that demonstrates an impressive trapping stiffness of up to 4.48 ± 0.2 (fN/nm)(mW/μm^2) for 20-nm gold nanoparticles, as illustrated in Fig. 5.9. This innovative device is not only capable of achieving high stiffness but also selective trapping of various bioparticles, as reported by the authors. Moreover, the study delves into the fascinating self-induced back-action effect, where the presence of nanoparticles can substantially alter the trapping light field. This interaction can modify the trapping stiffness and, in some cases, even reverse the direction of the optical force [50, 51]. It is noteworthy that the use of plasmonic optical tweezers for trapping can result in a heating effect, which may pose a potential risk to bioparticles.

To solve this issue, dielectric materials could be considered. The difficulties arise from the controversy on the requirement of a tightly focused light spot in order to create strong optical forces while a large area is kept for the sorting. To overcome this problem, Shi et al. [52] developed an optical potential-well array, which enables large-scale, high-selectivity trapping and sorting of nanoparticles, as shown in Fig. 5.10. This method stands out for its ability to handle a broad size range of nanoparticles, from 100 to 500 nm, by creating a near-field array of potential wells with interconnected tiny hotspots. The energy depths of these wells are differentiated, allowing for the selective and massive trapping of nanoparticles within a microchannel. By precisely tuning the laser power, this technology can sort nanoparticles based on size, with larger particles being trapped in deeper wells. This capability for large-scale sorting is particularly remarkable, as it overcomes previous limitations in optical tweezers that could only handle a few particles at a time without efficient sorting.

The methodology involves directing laser light into a silicon nano-waveguide pair array, which splits the light into multiple beams that generate coupling hotspots, creating the optical potential wells. The optical gradient force, which is essential for trapping nanoparticles, increases exponentially with the size of the nanoparticle. This allows for the selective trapping of larger nanoparticles in deeper potential wells

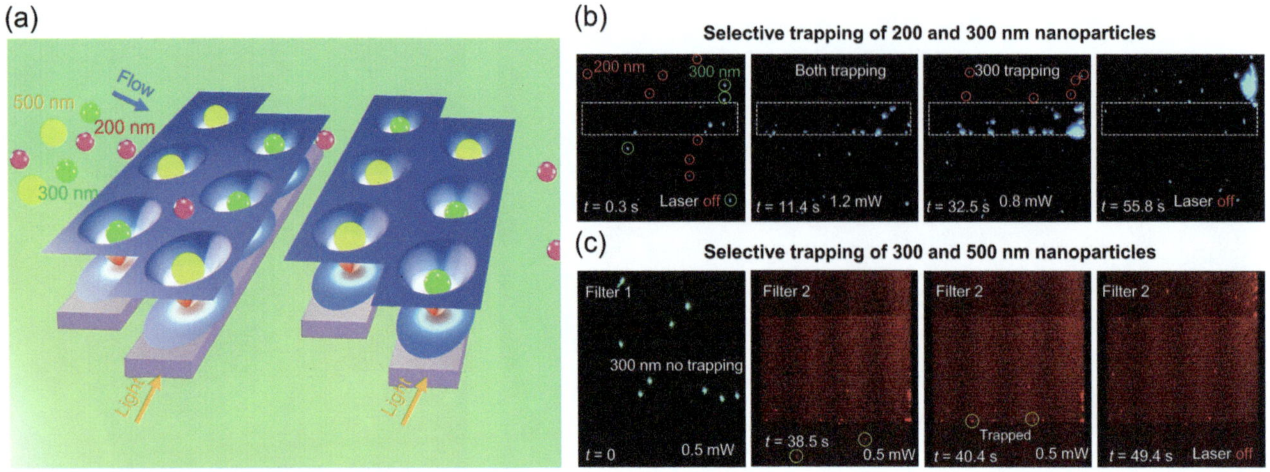

Fig. 5.10 Demonstration of separately trapping nanoparticles on the surface of the nanowaveguide pairs [52]. **a** Diagrams illustrating the capture of 300 and 500 nm nanoparticles, as well as the release of 200 nm nanoparticles, within an array of potential wells; **b** Selective capture of nanoparticles with diameters of 200 and 300 nm; **c** Selective capture of nanoparticles with sizes of 300 and 500 nm. (Reprinted from Shi et al. [52], Copyright (2020), with permission from American Chemical Society)

and the release of smaller ones. Experimental results demonstrate the successful large-scale trapping of 300 and 500 nm nanoparticles and the release of 200 nm nanoparticles by adjusting the laser power. This large-scale sorting capability is a significant breakthrough, offering a powerful tool for real-time analysis and sorting of multiple nanoparticles and biomolecules, such as bacteria and particle-labelled exosomes and viruses, which has broad implications for physical and biomedical sciences.

Despite the fact that optical forces are known to vary with the morphology of particles, no method has yet been developed to leverage these forces for the shape-based separation of bioparticles [53]. This oversight stems from the minimal force variation attributed to shape in traditional single-beam optical trapping systems, especially for particles with diameters < 1 μm, as dictated by Rayleigh and Mie scattering theories. Furthermore, the majority of optical force-based studies have relied on single laser beams or discrete laser spots, which limits their ability to manipulate particles. This approach has regrettably side-lined the complex dynamics of rod-shaped particles within dual hotspots, which offer a rich array of freedom in movement. The diverse shapes of microbes, predominantly spherical or rod-like, have been largely ignored in sorting techniques. Coccus (spherical bacteria) and Bacillus (rod-shaped bacteria), two widespread types of bacteria found in bodily fluids and aquatic environments, are significant in the context of human disease. The ability to distinguish between these microbial shapes could greatly enhance our understanding of disease processes. However, there remains a significant gap in the availability of efficient sorting techniques in the fields of microfluidics, optics, and acoustics for the shape-specific separation of bacteria without the need for labelling.

As shown in Fig. 5.11, Shi et al. [47] introduced an innovative approach to particle sorting with the development of an optofluidic nanophotonic sawtooth array (ONSA) that enables shape-selective sieving of submicron particles. This technology generates sawtooth-like light fields through light coupling, imposing different optical torques on particles based on their shapes. Using unstained *S. aureus* and *E. coli* as a model system, the study demonstrated efficient shape-selective sorting, with more than 95% of *S. aureus* retained within ONSA and more than 97% of *E. coli* removed. This nanophotonic chip represents a significant advancement in shape-selective sorting of submicron particles and expands the capabilities of optofluidics-based particle manipulation.

The application of waveguides for sorting biological particles marked a significant milestone in leveraging near-field optical forces for precise particle manipulation. By exploiting the strong optical gradient forces generated near the waveguide surface, researchers were able to separate particles based on size, refractive index, and other properties, demonstrating the potential of waveguide-based systems for high-precision sorting in microfluidic environments. Building on this foundation, the focus naturally shifted toward more complex applications, such as manipulating nanoscale biological entities like viruses. However, due to the small size and low refractive index of viruses, direct manipulation using waveguides proved challenging [15, 47, 54–56]. To overcome this, researchers developed indirect manipulation strategies, coupling viruses to larger nanoparticles through immunolabelling protocols. This approach effectively enhanced the interaction between the virus-particle complex and the optical

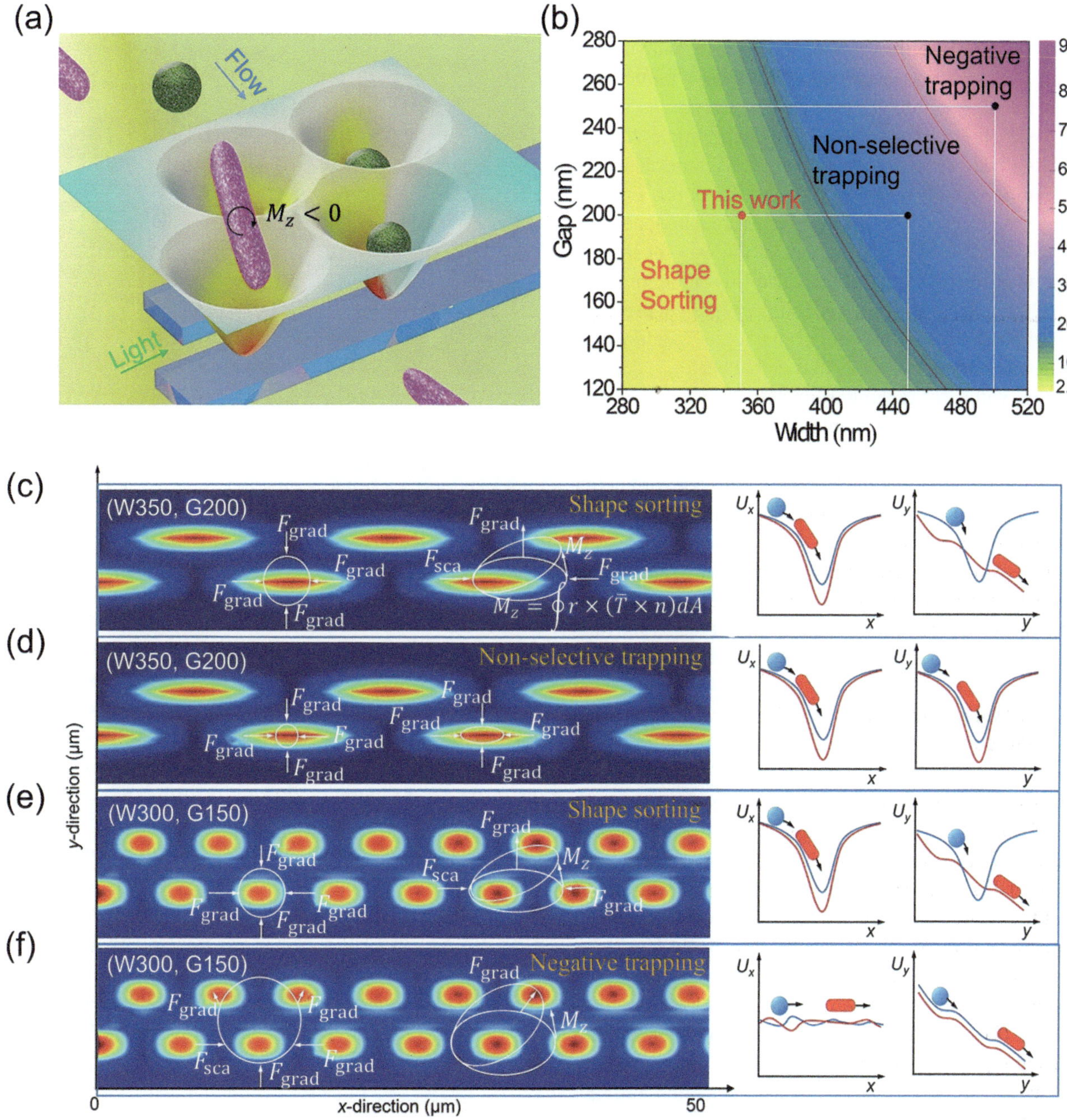

Fig. 5.11 Design of the ONSA for the manipulation of spherical (*S. aureus*) and rod-shaped (*E. coli*) bacteria [47]. **a** Depiction of the sorting process and the dynamics of bacteria within the ONSA. **b** Schematic representation of the versatile optical manipulation capabilities, which can be achieved by altering the coupling lengths through the adjustment of the width and gap between two waveguides; **c** Shape sorting of normal bioparticles (width 350 nm, gap 200 nm); **d** Nonselective trapping of small bioparticles (width 350 nm, gap 200 nm); **e** Shape sorting of normal particles (width 300 nm, gap 150 nm); **f** Negative trapping of large bioparticles (width 300 nm, gap 15 nm). (Reprinted from Shi et al. [47], Copyright (2019), with permission from American Chemical Society)

forces generated by the waveguide, enabling efficient trapping, transportation, and analysis of viruses. This transition from sorting biological particles to indirectly manipulating viruses showcased the adaptability of waveguide-based systems and set the stage for more advanced techniques in virus handling and virology research.

5.1.6 Virus Manipulation in Optofluidics

Precise control over virus behaviour, such as trapping, transporting, positioning, and sorting, enables researchers to study viral pathogenesis, host-virus interactions, and infection dynamics at an unprecedented level of detail [57–59]. Unlike traditional methods, optofluidic platforms offer non-invasive, label-free manipulation, preserving the integrity and viability of viruses during experiments [60–62]. This is particularly valuable for studying highly sensitive viruses, such as SARS-CoV-2, where precise control over individual particles or large populations can aid in understanding infection mechanisms, developing diagnostics, and testing antiviral drugs. Furthermore, optofluidic systems facilitate high-throughput analysis and single-virus-level precision, paving the way for innovations in vaccine development and real-time monitoring of viral evolution [63, 64]. These capabilities highlight the transformative potential of optofluidic technologies in addressing pressing challenges in virology and global health.

Viruses, with sizes ranging from 40 to 300 nm, are notoriously difficult to isolate and manipulate due to their nanoscale dimensions and low refractive index. Traditional optical trapping techniques, such as optical tweezers, struggle to generate sufficient optical forces for nanoscale particles. To address this limitation, in 2021, Shi et al. [65] developed an innovative optofluidic chip that uses nanowaveguide arrays to manipulate viruses indirectly by tethering them to larger nanoparticles. This method, combining immunolabelling protocols and fluorescence detection, provides a scalable and efficient platform for virus trapping and detection.

Nanowaveguides are subwavelength optical structures capable of confining light and generating strong evanescent fields along their edges [52]. These fields exert optical gradient forces that attract particles towards the waveguide surface. However, due to their small size (90–100 nm), viruses do not interact strongly enough with the optical field to be directly trapped. To overcome this, Shi et al. employ immunolabelling protocols to bind viruses to larger polystyrene nanoparticles (500 nm in diameter), effectively increasing their interaction with the optical forces.

Figure 5.12 illustrates the structural design of the optofluidic chip, which integrates an array of nanowaveguide pairs with a shallow microfluidic channel. The nanowaveguide pairs are arranged at regular intervals, creating multiple optical "hotspots" through light coupling. These hotspots generate high-intensity evanescent fields, which act as optical potential wells to trap virus-particle complexes. Viruses are conjugated with larger polystyrene nanoparticles using an immunolabelling protocol. This conjugation amplifies the optical forces exerted on the virus-particle complex, making it possible to trap and manipulate the complex in an optical field. The trapping mechanism relies on nanowaveguides embedded within an optofluidic chip. When light is coupled into adjacent nanowaveguides, localized optical "hotspots" are formed due to the coupling of evanescent waves. These hotspots create optical potential wells with high gradient forces, capable of pulling particles from the fluid flow and trapping them stably on the waveguide surface. The shallow microchannel design ensures that virus-particle complexes remain within the evanescent field, enhancing the effectiveness of the optical forces. The dimensions of the waveguides and the microchannel are optimized to maximize the interaction between light and the virus-particle complex, enabling efficient trapping even in dynamic flow conditions.

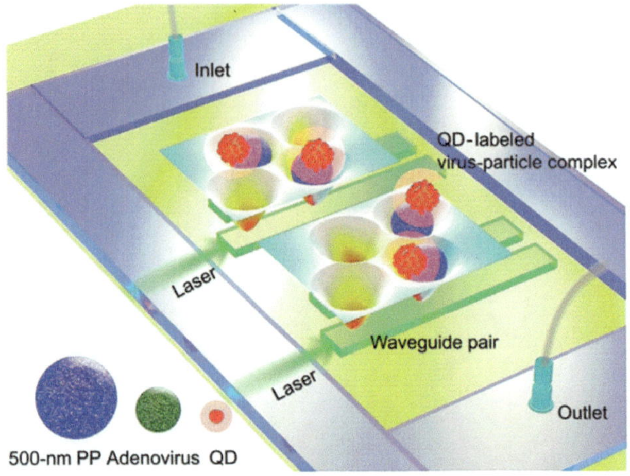

Fig. 5.12 Optical detection of single viruses in an optofluidic chip [65]. (Reprinted from Shi et al. [65], Copyright (2021), with permission from American Chemical Society)

To visualize the trapped viruses, fluorescence quantum dots are attached to the virus-particle complexes during the immunolabelling process. These quantum dots emit light under optical excitation, allowing real-time imaging of the trapped complexes using a charge-coupled device (CCD) camera. The fluorescence intensity is proportional to the number of viruses attached to each nanoparticle, enabling precise quantification of virus concentration in the sample. This fluorescence-based detection system, combined with the optical trapping capability of the waveguides, provides a comprehensive platform for virus manipulation and analysis.

This indirect trapping method not only overcomes the limitations of directly manipulating viruses but also enables versatile applications in virology. By leveraging the strong optical fields generated by nanowaveguides and the amplification provided by nanoparticles, this approach achieves high trapping efficiency and sensitivity. It is particularly well-suited for studying low-concentration virus samples, early disease diagnostics, and virology research requiring precise manipulation and quantification of single viruses. However, there are still some limitations in using waveguides [47, 52, 66–68] to indirectly manipulate viruses. These methods require additional sample preparation and rely on auxiliary particles, limiting their ability to interact directly with viruses. To overcome these challenges, Shi et al. have turned to photonic crystal-based approaches, which utilize precisely engineered nanostructures to generate intense optical forces capable of directly trapping and manipulating viruses without the need for binding to larger particles. This leap not only simplifies the experimental process but also significantly enhances the precision and versatility of virus manipulation, allowing researchers to study viruses at their native scale while exploring new functionalities such as sorting, patterning, and concentration. The shift from waveguides to photonic crystals thus signifies a natural progression toward more advanced, direct, and multifunctional virus manipulation techniques [55, 65, 69, 70].

Photonic crystals are engineered nanostructures with periodic dielectric patterns that can manipulate light at subwavelength scales. By leveraging the unique optical phenomena within photonic crystals, such as high-Q resonances, bound states in the continuum (BIC), and leaky modes, Shi et al. [71] have developed systems capable of generating intense optical gradient forces for the direct manipulation of viruses. This approach eliminates the need for auxiliary nanoparticles, simplifying the experimental process while significantly enhancing the precision and functionality of virus manipulation.

The conceptual design of virus manipulation using a nanocavity array is shown in Fig. 5.13a. The array consists of nanoholes, each with a diameter of 110 nm, illuminated by a broad Gaussian laser beam (wavelength 532 nm, diameter > 10 μm). When virus particles flow into the illuminated region, those smaller than the nanohole diameter are captured within the cavities due to the strong optical gradient forces generated by localized light modes. Larger particles, however, are unable to be trapped and are carried away by the fluid flow, demonstrating the size-selective trapping capability of this system. Figure 5.13b illustrates the layered structure of the multifunctional virus manipulation chip. A shallow microchannel with a depth of 1 μm ensures that particles are positioned close enough to interact with the near-field optical forces. The chip is composed of multiple material layers, including a thick PDMS layer for encapsulation, a thin quartz glass layer for optical imaging, and Si_3N_4 nanocavities for virus trapping. This integrated design allows for efficient virus manipulation and real-time observation under a dark-field microscope.

In this system, photonic crystal nanocavities are designed to confine light tightly within subwavelength regions, creating "Trapping Modes" that generate optical hotspots with high intensity and strong optical gradient forces. Unlike waveguides, which rely on evanescent fields along their surfaces, photonic crystal nanocavities achieve a balance between light confinement and controlled leakage, enabling efficient interaction with viruses even at their native nanoscale dimensions. The trapping force generated by these nanocavities is sufficient to directly capture viruses as small as 40 nm, such as SARS-CoV-2 particles, without the need for immunolabelling or particle conjugation.

The key innovation lies in the use of "leaky modes" within the photonic crystal band structure. These modes allow a controlled amount of light to escape from the nanocavity, ensuring that viruses in the microfluidic channel interact effectively with the optical field. This interaction creates a highly localized optical potential well that attracts viruses to the nanocavity centre. The nanocavities are fabricated using silicon nitride (Si_3N_4), a low-loss dielectric material that minimizes heat generation, preserving the structural integrity and viability of viruses during manipulation. Additionally, the photonic crystal design is integrated into a shallow microchannel (1 μm in depth), ensuring that viruses remain within the effective range of the optical forces.

The multifunctionality of photonic crystal-based virus manipulation is demonstrated through a variety of applications. Viruses can be trapped, transported, and positioned with high precision, enabling single-virus studies and patterning. By dynamically adjusting the position of the laser beam or the stage, viruses can be arranged into specific spatial configurations, such as linear or grid patterns, for studying their interactions or infection dynamics. The size-selective trapping capability of photonic crystals also allows for virus sorting, with only particles smaller than the nanocavity size being effectively captured. This feature is particularly useful for analysing heterogeneous virus populations or isolating specific subtypes of viruses for further study. One of the most significant advancements enabled by photonic crystals is the ability to concentrate large numbers of viruses simultaneously. By increasing the laser power, multiple nanocavities can be activated, creating an array of optical traps that capture hundreds or even thousands of viruses within a single operation. This capability is particularly beneficial for low-concentration virus samples, where traditional methods struggle to achieve sufficient sensitivity. Furthermore, the negligible heat generation of the photonic crystal system ensures that viruses remain viable for downstream applications, such as infection assays or vaccine development.

The use of photonic crystals for direct virus manipulation not only enhances trapping efficiency but also opens new avenues for virology research. The ability to manipulate viruses at their native scale without auxiliary particles provides an unprecedented level of control, enabling researchers to study virus behaviour, host–pathogen interactions, and the effects of viral load with unparalleled precision. The integration of photonic crystals with real-time imaging systems, such as fluorescence or dark-field microscopy, allows for simultaneous manipulation and visualization of viruses, further enhancing the analytical capabilities of this approach.

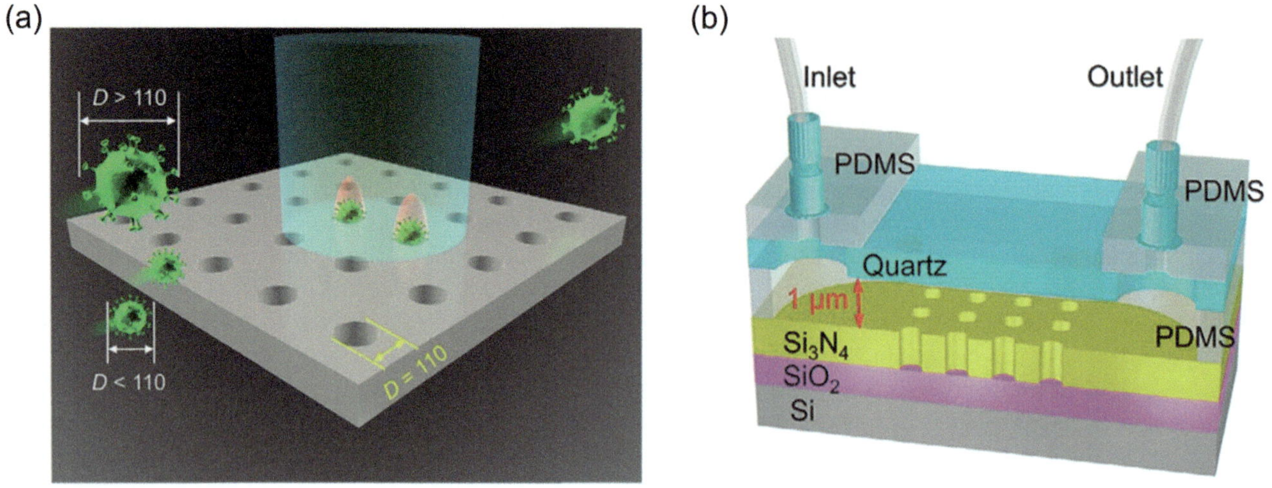

Fig. 5.13 Virus manipulation with resonator nanohole arrays [71]. **a** Schematic illustration of nanocavities. **b** Schematic illustration of a multifunctional virus manipulation chip. (Reprinted from Shi et al. [71], Copyright (2022), with permission from Wiley-VCH GmbH)

5.1.7 Conclusion and Outlook

The development of optofluidic manipulation and detection technologies has progressed significantly, transitioning from far-field single-particle manipulation to near-field multi-particle manipulation, followed by indirect virus manipulation using waveguides, and culminating in direct virus manipulation with photonic crystals. Far-field techniques, such as traditional optical tweezers, are initially effective for microscale particles but face limitations in manipulating nanoscale biological particles like viruses due to weak optical forces. Near-field approaches, utilizing nanostructures such as waveguides, improve optical force strength and enable the manipulation of multiple particles simultaneously. However, when applied to viruses, these methods required indirect manipulation, relying on immunolabelling protocols to bind viruses to larger nanoparticles for efficient optical trapping. While this waveguide-based approach addresses some challenges, it introduces added complexity and limited direct interaction with viruses. The advent of photonic crystal-based methods overcomes these hurdles by utilizing engineered nanocavities with "Trapping Modes" to directly manipulate viruses. This advancement eliminates the need for auxiliary particles, enhances precision, and enables multifunctional manipulation, including trapping, sorting, and concentration, paving the way for ground-breaking applications in virology and nanomedicine.

Looking ahead, the future of this field centres on enhancing precision, scalability, and functionality. Advances in nanophotonic design and fabrication will improve trapping efficiency and extend the capability to manipulate even smaller particles, such as sub-40 nm viruses. Integration with lab-on-a-chip platforms and real-time spectroscopic analysis will enable high-throughput, automated systems for virus detection, diagnostics, and drug screening. Additionally, applying these technologies to complex biological environments, such as live cells or tissues, will provide deeper insights into virus behaviour and infection mechanisms. Coupling artificial intelligence with optofluidic systems could further optimize particle manipulation and data analysis, enabling intelligent and adaptive control. With these advancements, optofluidic manipulation is poised to become a transformative tool in virology, diagnostics, and nanomedicine, offering unprecedented potential for tackling global health challenges.

Acknowledgements This work was supported by National Key Research and Development Program of China (2023YFF0613600), National Natural Science Foundation of China (61925504, 62205246, 62475192, 61621001, 62192770, 62192772, 12274296, 62020106009 and 62111530053), Shanghai Pilot Program for Basic Research, Science and Technology Commission of Shanghai Municipality (17JC1400800, 20JC1414600, 21JC1406100 and 22ZR1432400), the "Shu Guang" project supported by Shanghai Municipal Education Commission and Shanghai Education (17SG22), and Fundamental Research Funds for the Central Universities.

References

1. Psaltis D, Quake SR, Yang C. Developing optofluidic technology through the fusion of microfluidics and optics. Nature. 2006;442:381.
2. Erickson D, Sinton D, Psaltis D. Optofluidics for energy applications. Nat Photonics. 2011;5:583.
3. Sweetlove LJ, Obata T, Fernie AR. Systems analysis of metabolic phenotypes: what have we learnt? Trends Plant Sci. 2014;19:222.
4. Mainardi F, Pironi P. The fractional Langevin equation: Brownian motion Revisited; 2008;10:140.
5. Thomas PJ. On the influence of the Basset history force on the motion of a particle through a fluid. Phys Fluids A. 1992;4:2090.

6. Quinn RA, Goree J. Single-particle Langevin model of particle temperature in dusty plasmas. Phys Rev E. 2000;61:3033.
7. Shi YZ, et al. Sculpting nanoparticle dynamics for single-bacteria-level screening and direct binding-efficiency measurement. Nat Commun. 2018;9:815.
8. Minopoli A, Wagner S, Erben E, Liao W, Stoev ID, Lauga E, Kreysing M. ISO-FLUCS: symmetrization of optofluidic manipulations in quasi-isothermal micro-environments. eLight 2023;3:16.
9. Stoev ID, Seelbinder B, Erben E, Maghelli N, Kreysing M. Highly sensitive force measurements in an optically generated, harmonic hydrodynamic trap. eLight 2021;1:7.
10. Uruba V. Reynolds number in laminar flows and in turbulence. AIP Conf Proc. 2019;2118: 020003.
11. Traut M, Gilbert P, Walsh C, Bows A, Filippone A, Stansby P, Wood R. Propulsive power contribution of a kite and a Flettner rotor on selected shipping routes. Appl Energ. 2014;113:362.
12. Geislinger TM, Franke T. Hydrodynamic lift of vesicles and red blood cells in flow—from Fåhræus & Lindqvist to microfluidic cell sorting. Adv Colloid Interface Sci. 2014;208:161.
13. Zhou Y, Ma Z, Tayebi M, Ai Y. Submicron particle focusing and exosome sorting by wavy microchannel structures within viscoelastic fluids. Anal Chem. 2019;91:4577.
14. Hida T. Brownian Motion. In: Hida T, editors. Brownian motion, vol. 11. US: Springer;1980. p.44.
15. Shi Y, et al. Nanometer-precision linear sorting with synchronized optofluidic dual barriers. Sci Adv. 2018;4:eaao0773.
16. Kaneta T, Ishidzu Y, Mishima N, Imasaka T. Theory of optical chromatography. Anal Chem. 1997;69:2701.
17. Imasaka T, Kawabata Y, Kaneta T, Ishidzu Y. Optical chromatography. Anal Chem. 1995;67:1763.
18. Ashok P, Marchington R, Mthunzi P, Krauss T, Dholakia K. Optical chromatography using a photonic crystal fiber with on-chip fluorescence excitation. Opt Express. 2010;18:6396.
19. Yu-Xuan R, Jian-Guang W, Man C, Huang L, Yin-Mei L. Stability of novel time-sharing dual optical tweezers using a rotating tilt glass plate. Chin Phys Lett. 2010;27: 028703.
20. Paschotta, R. Field guide to lasers, vol. 12; 2008.
21. James LC, Tawfik DS. Structure and kinetics of a transient antibody binding intermediate reveal a kinetic discrimination mechanism in antigen recognition. Proc Natl Acad Sci. 2005;102:12730.
22. Chen L, et al. The isolation of an RNA aptamer targeting to p53 protein with single amino acid mutation. Proc Natl Acad Sci. 2015;112:10002.
23. Keefe AD, Pai S, Ellington A. Aptamers as therapeutics. Nat Rev Drug Discov. 2010;9:537.
24. Gawad C, Koh W, Quake SR. Single-cell genome sequencing: current state of the science. Nat Rev Genet. 2016;17:175.
25. Schmourlo G, Mendonça-Filho RR, Alviano CS, Costa SS. Screening of antifungal agents using ethanol precipitation and bioautography of medicinal and food plants. J Ethnopharmacol. 2005;96:563.
26. Schaut RG, Corbeil LB, Blake CN, Brewer MT. Development of a bead-agglutination assay for rapid detection of Tritrichomonas foetus. Vet Parasitol. 2017;243:188.
27. de Wildt RMT, Mundy CR, Gorick BD, Tomlinson IM. Antibody arrays for high-throughput screening of antibody–antigen interactions. Nat Biotechnol. 2000;18:989.
28. Brogioni B, Berti F. Surface plasmon resonance for the characterization of bacterial polysaccharide antigens: a review. MedChemComm. 2014;5:1058.
29. Liu ZQ, Mahmood T, Yang PC. Western blot: technique, theory and trouble shooting. N Am J Med Sci. 2014;6:160.
30. Hoogenboom HR. Selecting and screening recombinant antibody libraries. Nat Biotechnol. 2005;23:1105.
31. Bessette PH, Hu X, Soh HT, Daugherty PS. Microfluidic library screening for mapping antibody epitopes. Anal Chem. 2007;79:2174.
32. Ren L, et al. A high-throughput acoustic cell sorter. Lab Chip. 2015;15:3870.
33. Ozcelik A, Rufo J, Guo F, Gu Y, Li P, Lata J, Huang TJ. Acoustic tweezers for the life sciences. Nat Methods. 2018;15:1021.
34. Wu M, et al. Isolation of exosomes from whole blood by integrating acoustics and microfluidics. Proc Natl Acad Sci. 2017;114:10584.
35. Wang X, Chen S, Kong M, Wang Z, Costa KD, Li RA, Sun D. Enhanced cell sorting and manipulation with combined optical tweezer and microfluidic chip technologies. Lab Chip. 2011;11:3656.
36. Zhang S, et al. Optoelectronic tweezers: a versatile toolbox for nano-/micro-manipulation. Chem Soc Rev. 2022;51:9203.
37. Wu MC. Optoelectronic tweezers. Nat Photonics. 2011;5:322.
38. Gan C, Zhang J, Chen B, Wang A, Xiong H, Zhao J, Wang C, Liang S, Feng L. Optoelectronic tweezers micro-well system for highly efficient single-cell trapping, dynamic sorting, and retrieval. Small. 2024;20:2307329.
39. Hoshyar N, Gray S, Han H, Bao G. The effect of nanoparticle size on in vivo pharmacokinetics and cellular interaction. Nanomedicine (London, England) 2016;11:673
40. Arosio P, Müller T, Mahadevan L, Knowles TPJ. Density-gradient-free microfluidic centrifugation for analytical and preparative separation of nanoparticles. Nano Lett. 2014;14:2365.
41. Wu M, Ozcelik A, Rufo J, Wang Z, Fang R, Jun Huang T. Acoustofluidic separation of cells and particles. Microsyst Nanoeng. 2019;5:32
42. Hanauer M, Pierrat S, Zins I, Lotz A, Sönnichsen C. Separation of nanoparticles by gel electrophoresis according to size and shape. Nano Lett. 2007;7:2881.
43. Nan F, Yan Z. Sorting metal nanoparticles with dynamic and tunable optical driven forces. Nano Lett. 2018;18:4500.
44. MacDonald MP, Spalding GC, Dholakia K. Microfluidic sorting in an optical lattice. Nature. 2003;426:421.
45. Wang MM, et al. Microfluidic sorting of mammalian cells by optical force switching. Nat Biotechnol. 2005;23:83.
46. Jákl P, Čižmár T, Šerý M, Zemánek P. Static optical sorting in a laser interference field. Appl Phys Lett. 2008;92: 161110.
47. Shi Y, et al. Nanophotonic array-induced dynamic behavior for label-free shape-selective bacteria sieving. ACS Nano. 2019;13:12070.
48. Ndukaife JC, Kildishev AV, Nnanna AGA, Shalaev VM, Wereley ST, Boltasseva A. Long-range and rapid transport of individual nano-objects by a hybrid electrothermoplasmonic nanotweezer. Nat Nanotechnol. 2016;11:53.
49. Bouloumis TD, Kotsifaki DG, Nic Chormaic S. Enabling Self-Induced Back-Action Trapping of Gold Nanoparticles in Metamaterial Plasmonic Tweezers. Nano Lett. 2023;23:4723.
50. Zhu T, et al. Self-induced backaction optical pulling force. Phys Rev Lett. 2018;120: 123901.
51. Juan ML, Gordon R, Pang Y, Eftekhari F, Quidant R. Self-induced back-action optical trapping of dielectric nanoparticles. Nat Phys. 2009;5:915.
52. Shi Y, Zhao H, Chin LK, Zhang Y, Yap PH, Ser W, Qiu C-W, Liu AQ. Optical potential-well array for high-selectivity, massive trapping and sorting at nanoscale. Nano Lett. 2020;20:5193.
53. Neuman KC, Nagy A. Single-molecule force spectroscopy: optical tweezers, magnetic tweezers and atomic force microscopy. Nat Methods. 2008;5:491.
54. Yang Y, Liu J, Zhou X. A CRISPR-based and post-amplification coupled SARS-CoV-2 detection with a portable evanescent wave biosensor. Biosens Bioelectron. 2021;190: 113418.
55. Pang Y, Song H, Kim JH, Hou X, Cheng W. Optical trapping of individual human immunodeficiency viruses in culture fluid reveals

heterogeneity with single-molecule resolution. Nat Nanotechnol. 2014;9:624.
56. Ashkin A, Dziedzic JM. Optical trapping and manipulation of viruses and bacteria. Science. 1987;235:1517.
57. Gupta A, Akin D, Bashir R. Single virus particle mass detection using microresonators with nanoscale thickness. Appl Phys Lett. 2004;84:1976.
58. Ozcelik D, Parks JW, Wall TA, Stott MA, Cai H, Parks JW, Hawkins AR, Schmidt H. Optofluidic wavelength division multiplexing for single-virus detection. Proc Natl Acad Sci USA. 2015;112:12933.
59. Leong JA, Garapin AC, Jackson N, Fanshier L, Levinson W, Bishop JM. Virus-specific ribonucleic acid in cells producing rous sarcoma virus: detection and characterization. J Virol. 1972;9:891.
60. Black JA, Ganjalizadeh V, Parks JW, Schmidt H. Multi-channel velocity multiplexing of single virus detection on an optofluidic chip. Opt Lett. 2018;43:4425.
61. Mitra A, Deutsch B, Ignatovich F, Dykes C, Novotny L. Nano-optofluidic detection of single viruses and nanoparticles. ACS Nano. 2010;4:1305.
62. Shi Y, et al. On-chip optical detection of viruses: a review. Adv Photonics Res. 2021;2:2000150.
63. Soler M, Estevez MC, Cardenosa-Rubio M, Astua A, Lechuga LM. How nanophotonic label-free biosensors can contribute to rapid and massive diagnostics of respiratory virus infections: COVID-19 case. ACS Sens. 2020;5:2663.
64. Harvey WT, et al. SARS-CoV-2 variants, spike mutations and immune escape. Nat Rev Microbiol. 2021;19:409.
65. Shi Y, et al. Trapping and detection of single viruses in an optofluidic chip. ACS Sens. 2021;6:3445.
66. Yang X, Liu Y, Oulton RF, Yin X, Zhang X. Optical forces in hybrid plasmonic waveguides. Nano Lett. 2011;11:321.
67. Lin S, Schonbrun E, Crozier K. Optical manipulation with planar silicon microring resonators. Nano Lett. 2010;10:2408.
68. Erickson D, Serey X, Chen YF, Mandal S. Nanomanipulation using near field photonics. Lab Chip. 2011;11:995.
69. Mandal S, Serey X, Erickson D. Nanomanipulation using silicon photonic crystal resonators. Nano Lett. 2010;10:99.
70. Masuda T, Maruyama H, Honda A, Arai F. Virus enrichment for single virus infection by using 3D insulator based dielectrophoresis. PLoS ONE. 2014;9: e94083.
71. Shi Y, et al. Multifunctional virus manipulation with large-scale arrays of all-dielectric resonant nanocavities. Laser Photonics Rev. 2022;16:2100197.

Advances in Optoelectronic Tweezers: Principles, Applications, and Future Directions

6

Rongxin Fu, Bingrui Xu, Fan Yang, Gong Li, Zonghao Li, Hang Li, Yao Lu, and Shuailong Zhang

Abstract

The rapid progress in micromanipulation technologies has created exciting new possibilities for the actuation, selection, and assembly of both biological and non-biological nano/micro-objects, with applications in microfabrication, cell analysis, tissue engineering, biochemical sensing, and nano/micro-machinery. A wide range of precise, flexible, and high-throughput manipulation techniques have been developed to meet these diverse needs. Among these, optoelectronic tweezers (OET) stand out as a cutting-edge technology that combines light stimuli with electric fields, utilizing the photoconductive effect of semiconductor materials. OET uses optical modulation of electric fields to directly control micro-object behavior in a flexible and non-invasive manner. This light-induced electrokinetic effect provides key advantages, including programmability, flexibility, versatility, high throughput, and easy integration with other characterization systems, distinguishing OET from many other manipulation techniques. In recent years, OET has gained significant attention, with rapid advancements across a range of scientific and engineering fields. This chapter aims to provide a comprehensive review of OET technology, exploring its underlying mechanisms, experimental setups, applications in both biological and non-biological contexts, commercialization efforts, and future prospects.

Keywords

Optoelectronic tweezers · Micromanipulation · Optoelectronic manipulation technology · Dielectrophoresis

R. Fu (✉) · H. Li
School of Medical Technology, Beijing Institute of Technology, Beijing, China
e-mail: furongxin@bit.edu.cn

R. Fu · H. Li · S. Zhang (✉)
Zhengzhou Research Institute, Beijing Institute of Technology, Henan, China
e-mail: shuailong.zhang@bit.edu.cn

R. Fu · H. Li · Y. Lu · S. Zhang
Chongqing Institute of Microelectronics and Microsystems, Beijing Institute of Technology, Chongqing, China

B. Xu · F. Yang · S. Zhang
Beijing Advanced Innovation Center for Intelligent Robots and Systems, School of Mechatronical Engineering, Beijing Institute of Technology, Beijing, China

G. Li · Z. Li · Y. Lu · S. Zhang
School of Integrated Circuits and Electronics, Beijing Institute of Technology, Beijing, China

6.1 Introduction

Optoelectronic tweezers (OET) is an optical manipulation technique developed by Ming C. Wu and his team in 2005 [1]. Since then, it has gained widespread use in manipulating a variety of micro/nano-scale objects made from different materials, including carbon nanotubes [2], semiconductor nanowires [3–5], nanoparticles [3, 6, 7], DNA molecules [8], proteins [9], polystyrene and metallic microbeads [10–16], cells, as well as electronic and photonic components [17–19]. The ability to generate strong electrokinetic forces in the nN range and manipulate multiple targets simultaneously has made OET particularly valuable in recent micromanipulation research. OET manipulates nano/micro-scale objects by locally generating electrokinetic forces through light patterns projected onto semiconductor photoconductive substrates. Over the past decade, OET has been developed with a variety of materials, structures, and functions, enhancing its versatility for a wide range of applications and scientific research. This section introduces OET technology, exploring its device structures, various experimental setups, working mechanisms and diverse applications.

B. Li et al. (eds.), *Biophotonic Manipulation*, Advances in Optics and Optoelectronics,
https://doi.org/10.1007/978-981-96-4982-2_6

6.2 Principles and Configuration of Optoelectronic Tweezers (OET)

6.2.1 Standard OET Device

The OET device and system, as shown in Fig. 6.1a, b, consists of a photoconductive layer and conductive plates [20, 21]. The top and bottom plates are coated with indium tin oxide (ITO), serving as electrically conductive layers, while an additional hydrogenated amorphous silicon (a-Si:H) layer on the bottom plate acts as the photoconductive layer. The functionality of OET relies on the unique properties of the photoconductive material, which can absorb light and convert photonic energy into electrical energy through the photoelectric effect. This process generates electron–hole pairs in the illuminated region, leading to an increase in conductivity in that region, while the dark region remains an insulator. The conductivity difference between these regions generates the manipulation forces in the OET system. By projecting a light pattern onto the photoconductive layer, force is generated to precisely manipulate particles within the liquid medium. This method allows for the parallel manipulation of multiple particles using programmable light patterns, as demonstrated in Fig. 6.1b. The light wavelengths typically range from violet to red, matching the absorption spectra of the photoconductive materials, such as a-Si:H [6, 20, 22–24], titanyl phthalocyanine (TiOPc) [25, 26], bulk heterojunction (BHJ) polymers [27, 28], and silicon-based phototransistors [29, 30]. Upon light projection, the photoconductive layer experiences photoexcitation and generates electron–hole carriers via the photoelectric effect. This increases the conductivity of the illuminated region, causing a significant impedance difference between the illuminated and dark regions in the photoconductive layer. This impedance difference induces a voltage drop in the liquid medium above the illuminated region, generating a non-uniform electric field that produces the electrokinetic manipulation force observed in the experiment. To assemble the device, a double-sided adhesive tape is used as a spacer, which not only facilitates assembly but also creates an enclosed chamber to hold the liquid medium and enable manipulation of micro/nano-objects [21].

6.2.2 OET Experimental Setup

Various experimental setups have been developed for OET research. A standard OET system, as shown in Fig. 6.1c, d [11, 20], consists of several key components: a DMD (Digital Micromirror Device) projector, an upright microscope equipped with a camera, filters, and a motorized positioning stage, along with a function generator, an amplifier, and a computer. The DMD projector, controlled via a computer interface, generates light patterns which are projected through the microscope objective onto the photoconductive layer of the OET device. Long-pass and short-pass filters are used to ensure sufficient light power reaches the OET device, while allowing clear visualization of manipulated objects through the microscope. The function generator provides an AC signal, which is amplified and used to drive the OET device.

Depending on the experimental setup, light patterns can be projected onto the OET device [12, 15, 22, 25, 26] (Fig. 6.1c) either from above or below the device [31, 32] (Fig. 6.1d). The light source is also a crucial component of the OET system, which operates effectively with low optical density. Commercial DMD and LCD projectors are commonly used [4, 11, 15, 22, 23, 33], along with self-built optical systems combining a DMD chip and light sources like LEDs, lasers, or Hg lamps. These light sources act as optical displays that project dynamic light patterns onto the OET device, generating "light-activated virtual electrodes" for programmable manipulation. Shown in Fig. 6.1e, recent advancements have demonstrated that an LCD display can also serve as a direct light source for OET.

6.3 Physical Mechanism of OET

The manipulation forces in OET arise from the interaction between micro- and nano-objects and the non-uniform electric field, driven by various electrokinetic effects, including dielectrophoresis (DEP), electrophoresis (EP), AC electroosmosis (ACEO), and AC electrothermal (ACET) [20, 21, 34], as illustrated in Fig. 6.2. A thorough understanding and effective utilization of these mechanisms are crucial for optimizing OET parameters and enhancing its performance across a range of applications.

6.3.1 Dielectrophoresis (DEP)

DEP is an electrokinetic phenomenon in which polarizable particles move under the influence of a non-uniform electric field, driven by interactions between their induced dipole moment and the field gradient (Fig. 6.2a). In OET systems, DEP forces are generated dynamically by projecting spatially patterned light onto a photoconductive substrate, creating localized non-uniform electric fields in the medium. This phenomenon, termed optically induced DEP (ODEP), is the principal and dominant mechanism enabling precise and flexible particle manipulation in OET devices.

For the quantitative analysis of DEP forces, two primary methods are applied based on particle geometry. For spherical particles, the classical dipole approximation is the most commonly employed approach. The time-averaged DEP force

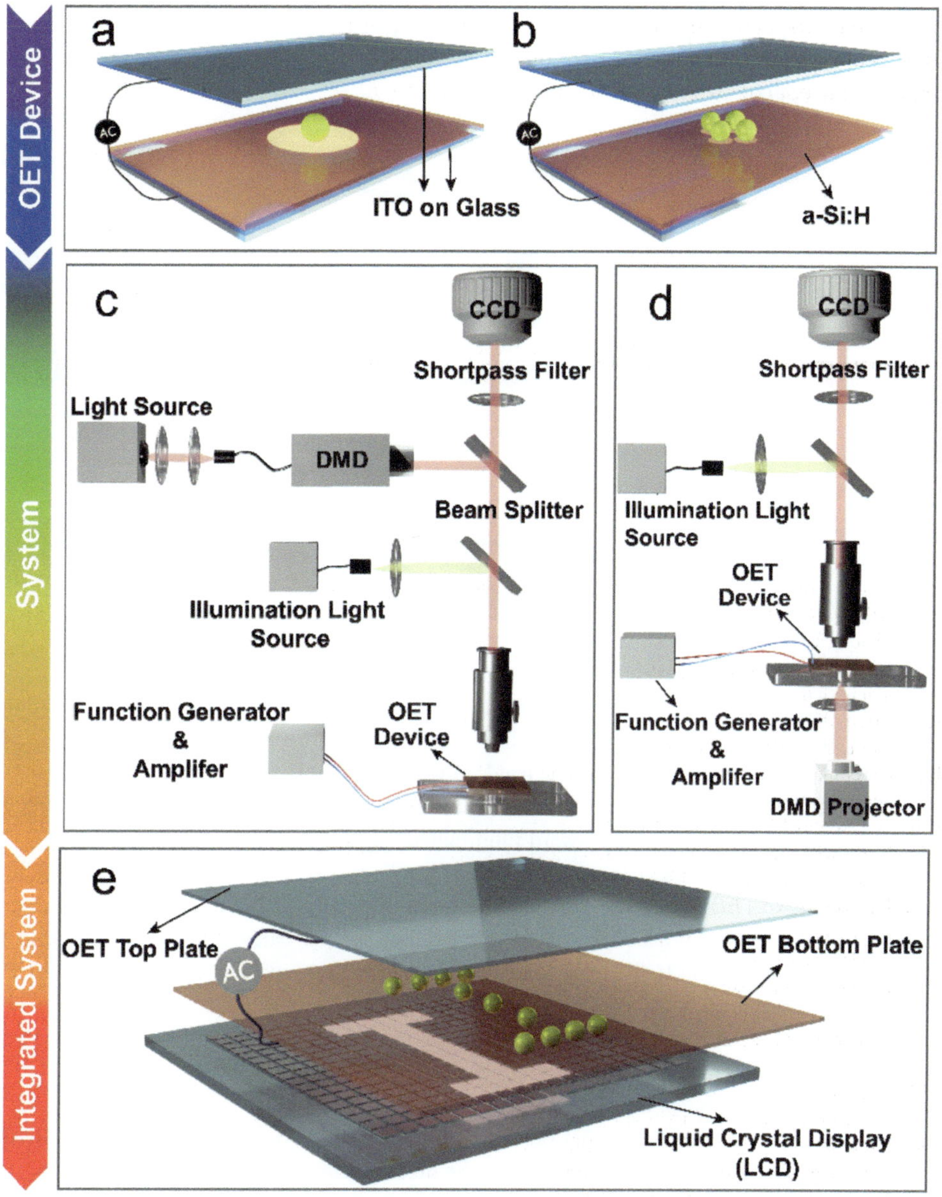

Fig. 6.1 **OET device structure and experimental setup.** **a** Diagram illustrating the use of a standard OET device to manipulate a single object. **b** Diagram depicting the use of a standard OET device for the parallel manipulation of multiple objects. **c** Diagram of an OET setup where the light pattern is projected from the top of the device. **d** Diagram of an OET setup where the light pattern is projected from the bottom of the device. **e** Diagram of an integrated OET setup, where an LCD display is attached to the bottom plate of the OET device to project light patterns. Reproduced from Ref. [21] with permission from John Wiley and Sons, copyright 2024

acting on a spherical particle in a dielectric medium is described as [35, 36]:

$$F_{DEP} = 2\pi r^3 \varepsilon_m Re[K(\omega)] \nabla |E_{rms}|^2 \tag{6.1}$$

where r is the radius of spherical particle; ε_m is the permittivity of liquid medium; ω is the angular frequency of applied AC voltage; E_{rms} is the root-mean-square value of the non-uniform electric field; $Re[K(\omega)]$ is the real part of Clausius–Mossotti (CM) factor. It provides a foundational framework for optimizing DEP-based particle manipulation strategies and is defined as follows:

$$K(\omega) = \left[\frac{\varepsilon_p^* - \varepsilon_m^*}{\varepsilon_p^* + 2\varepsilon_m^*} \right] \tag{6.2}$$

where ε_p^* is the complex permittivity of the spherical particle; ε_m^* is the complex permittivity of the liquid medium. The expressions of ε_p^* and ε_m^* are described as follows:

$$\varepsilon_p^* = \varepsilon_p - j\frac{\sigma_p}{\omega} \tag{6.3}$$

$$\varepsilon_m^* = \varepsilon_m - j\frac{\sigma_m}{\omega} \tag{6.4}$$

where j is the imaginary unit $\left(j = \sqrt{-1}\right)$; where the ε_p and ε_m are the permittivity of the spherical particle and liquid medium, respectively; σ_p and σ_m are the conductivity of the spherical particle and liquid medium, respectively.

The permittivity (ε_m and ε_p) and conductivity (σ_p and σ_m) of the particle and surrounding medium, along with

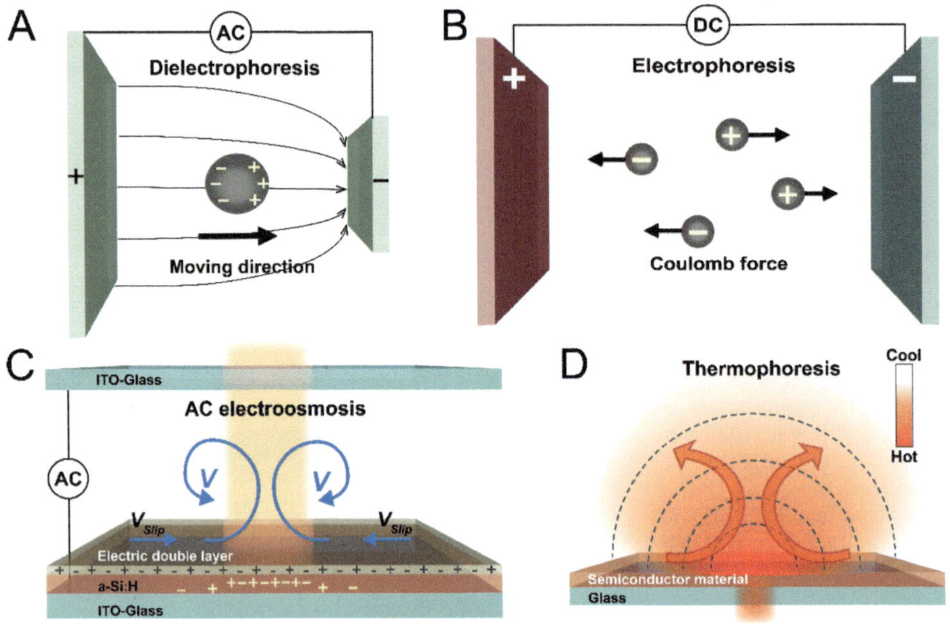

Fig. 6.2 Different physical mechanisms existing in OET. **a** Dielectrophoresis, **b** Electrophoresis, **c** AC electroosmosis, **d** Thermophoresis. Reproduced from Ref. [21] with permission from John Wiley and Sons, copyright 2024

the angular frequency (ω), collectively define the Clausius–Mossotti (CM) factor, a measure of polarizability that dictates the direction of the DEP force. When $Re[K(\omega)] > 0$, the particle is more polarizable than the medium, resulting in positive DEP, which attracts particles toward illuminated regions with stronger electric fields. Conversely, $Re[K(\omega)] < 0$ signifies negative DEP, where less polarizable particles are repelled toward dark regions with weaker electric fields. In cases where $Re[K(\omega)] = 0$, the DEP force vanishes, leaving the particles stationary.

For spherical particles, the DEP force is typically calculated using the classical dipole approximation (Eq. (6.1)). For non-spherical particles, the DEP force can also be determined using an alternative approach rooted in Lorentz law and Maxwell stress tensor analysis. In such cases, the momentum change of a dielectric non-spherical object under the influence of DEP forces in an electromagnetic field can be described as follows [19, 37, 38]:

$$F_{DEP} = \frac{d}{dt}(\boldsymbol{P_R} + \boldsymbol{P_F}) = \oint_S (\boldsymbol{T} \cdot \boldsymbol{n}) dS \tag{6.5}$$

where $\boldsymbol{P_R}$ is the momentum of the object, $\boldsymbol{P_F}$ is the total momentum of electromagnetic field, S is the enclosing surface, $\boldsymbol{n}$ is the unit tensor, and $\boldsymbol{T}$ is the Maxwell stress tensor.

6.3.2 Electrophoresis (EP)

EP describes the motion and separation of charged micro-objects relative to a fluid under the influence of a direct current (DC) or low-frequency electric field (below 10 Hz), shown in Fig. 6.2b. This phenomenon arises from the Coulomb force, expressed as [36]:

$$F_{Coulomb} = Eq \tag{6.6}$$

where E is the electric field strength, and q is the net charge of the micro-object. Biological molecules and cells typically carry a negative charge, a property exploited in techniques such as capillary or gel EP to separate DNA fragments. In OET systems, EP forces are relevant only when the device operates with DC or low-frequency AC signals.

6.3.3 AC Electroosmosis (ACEO)

ACEO plays a crucial role in OET, especially for long-distance particle manipulation. ACEO is a frequency-dependent fluid flow phenomenon driven by electroosmotic stress, which results from the interaction between the electric field and the electrical double layer (EDL) surrounding the electrodes [34, 39] (Fig. 6.2c). When an electrical potential is applied to an ionic fluid, it generates an EDL. If a tangential electric field component is present within the EDL, ions in the layer move in response to this field. The resulting ion velocity, known as the slip velocity, is defined by the Helmholtz–Smoluchowski equation [39]:

$$v_{SLIP} = -\frac{\varepsilon_l \zeta E_t}{\eta} \tag{6.7}$$

where v_{SLIP} is the slip velocity, ε_l is the permittivity of the liquid medium, ζ is the zeta potential of the EDL, E_t is the tangential component of electric field, η is the fluid viscosity. LACE has been used in OET devices for nanoparticle trapping, where a virtual electrode created by patterned light eliminates the need for traditional metal electrodes [6, 40–42]. In addition, ACEO is more effective for the manipulation of smaller objects such as nanoparticles and biomolecules, and over long distance [39].

6.3.4 AC Electrothermal (ACET)

Optically induced temperature gradients, referred to as thermophoresis, also play a significant role in OET manipulation, complementing the forces generated by non-uniform electric fields (Fig. 6.2d). Upon illumination, the a-Si:H coating on the bottom plate absorbs optical energy and dissipates it through mechanisms such as electron–hole pair generation and phonon interactions. This energy dissipation creates a localized heat source within the a-Si:H layer, resulting in spatial differences in fluid permittivity and conductivity between illuminated and non-illuminated regions. When exposed to an AC electric field, such fluid, exhibiting permittivity and conductivity gradients, experiences a net body force that drives motion. This phenomenon, known as the AC electrothermal effect (ACET) or thermophoresis, is mathematically expressed as follows [43–45]:

$$F_{ACET} = \frac{1}{2}Re\left[\left[\frac{\sigma_m \varepsilon_m}{\sigma_m + i\omega\varepsilon_m}\right](\kappa_\varepsilon - \kappa_\sigma)(\nabla T \cdot E)E^*\right] - \frac{1}{2}|E|^2\kappa_\varepsilon\kappa_\sigma\nabla T \quad (6.8)$$

where σ_m and ε_m represent the medium conductivity and permittivity respectively, T is the temperature, E is the electric field, E^* is the complex conjugate of the electric field, ω represents the angular frequency of external AC electric field, κ_ε and κ_σ represent change of the conductivity and permittivity per-unit temperature respectively. ACET has limited influence on the performance of OET because the optical power density of OET is normally below 1 W/cm^2.

6.4 OET for Biological Applications

OET has found extensive use in a wide array of biomedical applications, including cell sorting [1, 46–50], cell analysis [31, 51, 52], DNA transfection [53], and bio-fabrication [54, 55]. Unlike traditional optical tweezers, which rely on high-intensity light beams to manipulate biological samples, OET leverages low-power light-induced electrokinetic forces. This fundamental difference minimizes optically-induced heating and reduces physiological damage to cells, offering a safer and more effective solution for manipulating sensitive biological systems. This section delves into the transformative roles of OET in biomedical research, providing a comprehensive overview of its diverse applications and the underlying mechanisms driving these innovations.

OET technique enables rapid micropatterning of poly(ethylene glycol) diacrylate (PEGDA) hydrogels on modified a-Si:H substrates [56, 57], as shown in Fig. 6.3a, b [58]. This approach uses projection light patterns and AC voltage to create precise hydrogel patterns within seconds, confining MCF-7 breast cancer cell growth to bare a-Si:H regions [58].

Figure 6.3c, d illustrate the concentration and movement of RBCs using an LCD pattern that forms the word "KAIST", with positive DEP directing the RBCs into the bright areas [59]. Figure 6.3e–h illustrate the selection and storage of MCF-7 breast cancer cells and ARPE-19 human retinal pigment epithelial cells in p-OET traps [11]. As shown in Fig. 6.3i–j, OET can manipulate swimming cells, such as *Chlamydomonas reinhardtii* (*C. reinhardtii*) [60]. Under the influence of positive DEP force and using programmable light patterns, an array of live *C. reinhardtii* cells was successfully arranged, forming a micro-motor array with self-rotating capabilities. Additionally, other cell types, can also self-rotate in the OET system, a feature that has been leveraged for fluidic flow control, electrical property measurement, and distinguishing cells based on species and viability [52, 61].

OET can be applied for cell fusion [62, 63], where different cell types are merged to form a hybrid cell. This process is vital for applications such as monoclonal antibody production, cell reprogramming, cancer immunotherapy, and tissue generation [63–65]. Figure 6.3k shows the process of cell fusion in OET [62]. Another key application of OET is cell lysis, a vital technique for extracting proteins and nucleic acids from within cells for various research purposes [66–69]. When the electric field strength exceeds a certain threshold, the pores induced by electroporation on the cell membrane do not reseal, leading to cell lysis. Leveraging this mechanism, OET has been used for light-induced cell electrolysis. Figure 6.3l illustrates selective cell lysis based on cell shape [66].

OET has also been used to manipulate various biochemical substances, including polysaccharides, proteins, and fluorophores. Figure 6.3n, m show microscope images that illustrate the temporal change in the chemical concentration of fluorescein isothiocyanate (FITC)-labeled dextran in the illuminated region of the OET device [9]. AC voltage draws FITC-dextran from the dark to the illuminated region, causing a fluorescence increase due to higher concentration. The local concentration of biochemical substances can be dynamically modulated in both temporal and spatial manners by controlling the applied AC signal and light pattern [9, 70].

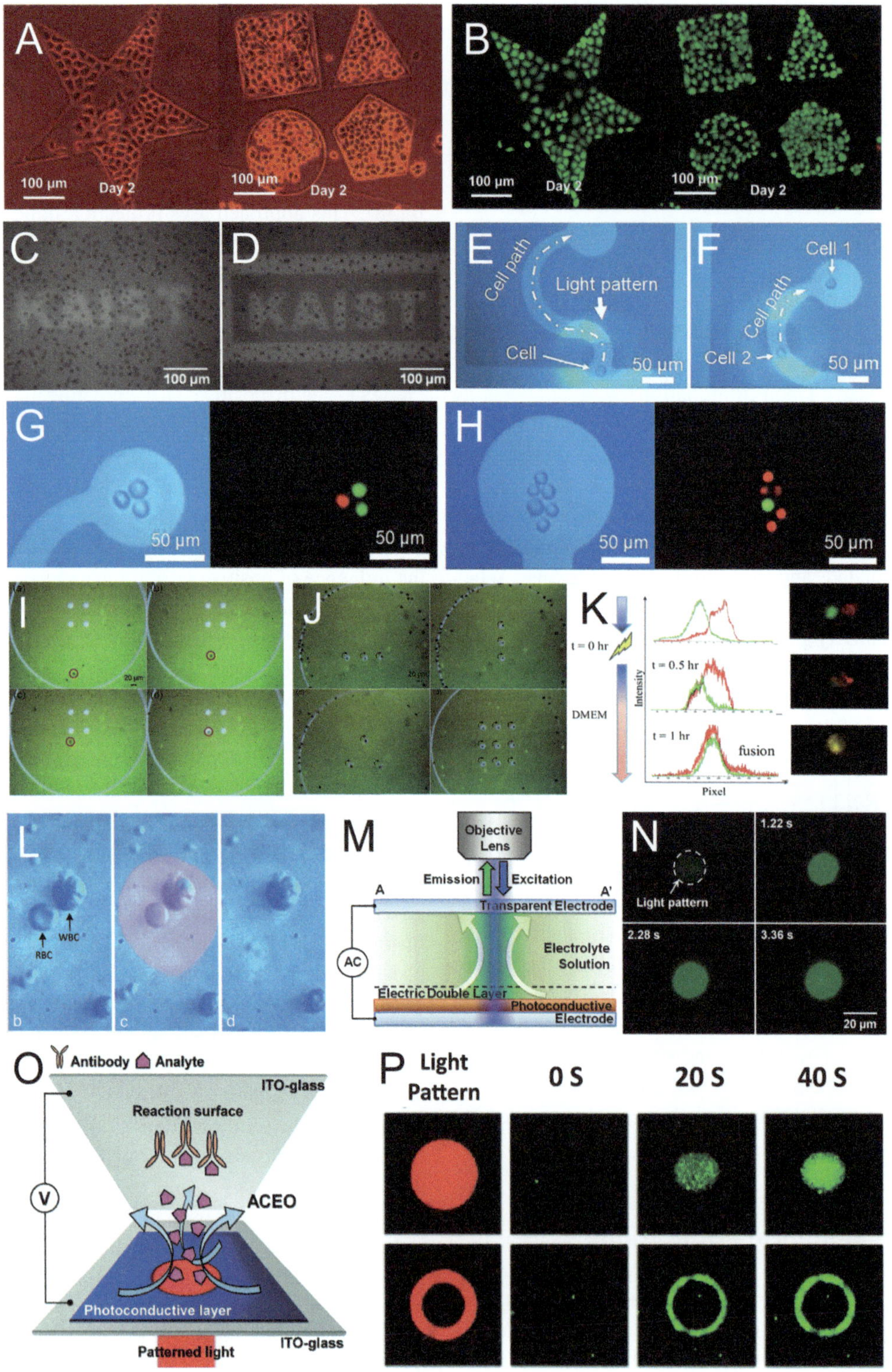
A
100 μm
Day 2
100 μm
Day 2
B
100 μm
Day 2
100 μm
Day 2
C
100 μm
D
100 μm
E
Cell path
Light pattern
Cell
50 μm
F
Cell 1
Cell path
Cell 2
50 μm
G
50 μm
50 μm
H
50 μm
50 μm
I
J
K
t = 0 hr
t = 0.5 hr
t = 1 hr
DMEM
Intensity
fusion
Pixel
L
WBC
RBC
b
c
d
M
Objective Lens
Emission
Excitation
A
A'
Transparent Electrode
AC
Electrolyte Solution
Electric Double Layer
Photoconductive
Electrode
N
Light pattern
1.22 s
2.28 s
3.36 s
20 μm
O
Antibody
Analyte
ITO-glass
Reaction surface
V
ACEO
Photoconductive layer
Patterned light
ITO-glass
P
Light Pattern
0 S
20 S
40 S

◄**Fig. 6.3** **OET-based manipulation and assembly of biological objects.** **a** Optical microscope image obtained using phase contrast imaging, showing cell staining with Calcein-AM / PI. **b** Corresponding fluorescent images, with viable cells stained green and dead cells stained red. Reproduced from Ref. [58] with permission from the Royal Society of Chemistry, copyright 2014. **c** Concentration of RBCs using an LCD image forming the word "KAIST." **d** After 60 s, the RBCs are moved to the bright area of the LCD image by positive DEP. Reproduced from Ref. [59] with permission from John Wiley and Sons, copyright 2008. **e–h** Selection and storage of MCF-7 breast cancer cells (prelabeled green by Cell Tracker dye) and ARPE-19 human retinal pigment epithelial cells (prelabeled red by Cell Tracker dye) in p-OET traps. Reproduced from ref. [11] with permission from John Wiley and Sons, copyright 2018. **i** Microscope images showing the assembly of *C. reinhardtii* cells into array format using light patterns. **j** Microscope images showing different micromotor arrays consisting of OET-assembled self-rotating *C. reinhardtii* cells. Reproduced from Ref. [60] with permission from Royal Society of Chemistry, copyright 2017. **k** Fluorescence intensity profile after generating the virtual electrode. Reproduced from Ref. [62] with permission from Springer Nature, copyright 2016. **l** Bright-field microscope images showing the use of light to selectively lyse cell. Reproduced from Ref. [66] with permission from John Wiley and Sons, copyright 2014. **m** Schematic diagram illustrating the cross section of the optoelectrofluidic device. **n** Microscopic images of FITC-dextran concentration changes in the illuminated region. Reproduced from Ref. [9] with permission from American Chemistry Society, copyright 2009. **o** Schematic illustration of an OET-based immunoreaction system. **p** Experimental results for the confirmation of optically induced ACEO flow using nanoparticles. Reproduced from Ref. [71] with permission from the Royal Society of Chemistry, copyright 2016

In addition to controlling chemical concentrations, OET has been utilized as an immunoreaction platform for biological assays [71–74]. Figure 6.3o illustrates the schematic of the OET-based immunoreaction system [71]. The OET top plate is functionalized with capture antibodies, serving as an antibody-immobilized assay spot. To investigate the formation of optically-induced ACEO within this system, experiments using fluorescent nanoparticles were conducted shown in Fig. 6.3p. The presence of ACEO was confirmed by observing the accumulation of nanoparticles at the center of the optically-induced virtual electrode, driven by the counter-rotating convective flows. Experimental results revealed temporal changes in nanoparticle concentration at the irradiated region, confirming the successful generation of ACEO flow [71].

6.5 OET for Non-biological Applications

In this section, we will overview the use of OET to manipulate a variety of non-biological nano/micro-objects for applications in micro-assembly and microfabrication.

6.5.1 Manipulation and Assembly of Micro- and Nano-scale Objects

An important application of OET is the assembly of microscale electronic and photonic components into functional devices [4, 7, 11, 14, 15, 18, 75–78]. Figure 6.4a shows an "OET" pattern formed by $Sn_{62}Pb_{36}Ag_2$ solder microbeads, which are drawn to the illuminated region [15]. Figure 6.4b illustrates the process of assembling solder beads into a straight line using OET to connect two isolated metal electrodes [79]. After assembly, freeze-drying was employed to remove the liquid medium and stabilize the solder beads in place. The assembled beads were then heated and melted to form connections. Figure 6.3c, d show microscope images of multiple and single 15 μm-diameter polystyrene microbeads being manipulated by OET, respectively [10]. Shown in Fig. 6.4e, two polystyrene particles, measuring 10 and 24 μm, are fractionated along the z-axis as they pass through the sorter path, with their trajectories redirected by an optical wedge at the end of the path due to the asymmetrical optical patterns [1]. Polystyrene microbeads can also be assembled into a variety of topographical micropatterns (TMPs) in OET [3]. As shown in Fig. 6.4f, polystyrene microbeads were assembled to form a TMP featuring stylized caricature of Isaac Newton with an apple [16]. After the assembly, a method based on freeze drying was developed to safely remove the liquid medium while keep the assembled TMP in place. As shown in Fig. 6.4g, the OET device chamber contains both the nano/micro-objects to be manipulated and a photopolymerizable PEGDA hydrogel solution, which solidifies upon light exposure [18]. As shown in Fig. 6.4h, OET can dynamically pattern nanoparticles within seconds [6]. This capability enables the creation of various arbitrary nanoparticle patterns and arrays. Additionally, by patterning 90 nm gold nanoparticles, surface-enhanced Raman spectroscopy (SERS) hotspots are generated, achieving enhancement factors exceeding 10^7 and sensitivities down to picomolar concentrations [6]. OET enable the manipulation of nanowires (Fig. 6.4i), including semiconducting and metallic types with diameters below 20 nm, using significantly lower optical power than traditional optical tweezers, allowing faster transport and real-time assembly, as well as the separation of different nanowire types for heterogeneous integration in nanoscale devices [3].

6.5.2 OET-Based Manipulation of Artificial Micromachines

OET distinguishes from other optical micromanipulation techniques by generating stronger manipulation forces at comparable light intensities. This unique capability makes OET ideal for the parallel manipulation of multiple objects, including micromachines with complex structures or those composed of various components. The versatility of OET

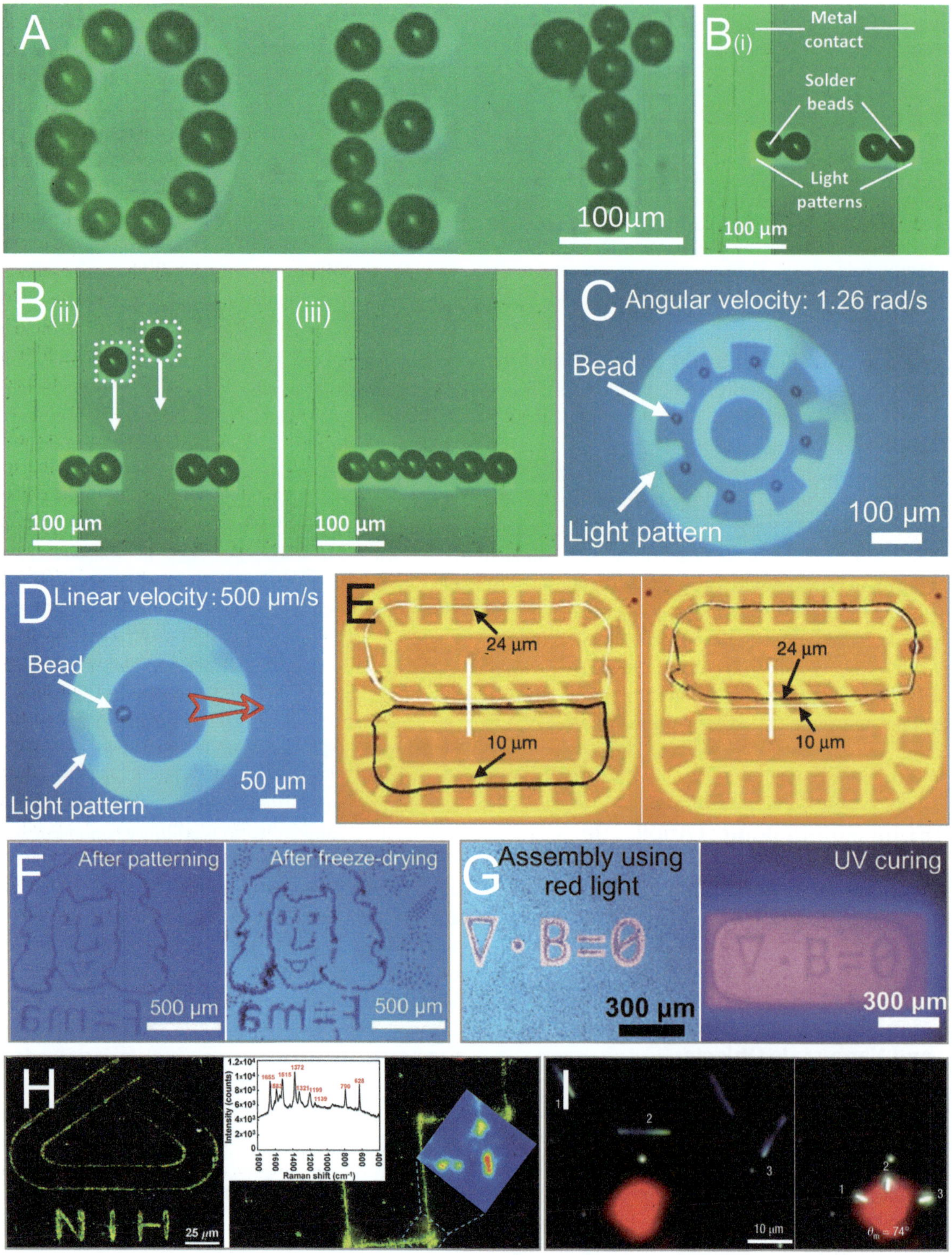
A
100µm
B(i)
Metal contact
Solder beads
Light patterns
100 µm
B(ii)
100 µm
(iii)
100 µm
C Angular velocity: 1.26 rad/s
Bead
Light pattern
100 µm
D Linear velocity: 500 µm/s
Bead
Light pattern
50 µm
E
24 µm
10 µm
24 µm
10 µm
F
After patterning
500 µm
After freeze-drying
500 µm
G
Assembly using red light
300 µm
UV curing
300 µm
H
25 µm
Raman shift (cm⁻¹)
Intensity (counts)
I
10 µm

◂**Fig. 6.4** **OET-based manipulation and assembly of non-biological objects**. **a** "OET" pattern formed by solder beads attracted to the illuminated region. Reproduced from Ref. [15] with permission from AIP Publishing, copyright 2016. **b** Assembly of solder beads to form a straight line to connect two isolated metal electrodes. Reproduced from Ref. [79] with permission from Optica Publishing Group, copyright 2017. **c** The use of a "Roulette"-shaped light pattern to rotate 15 μm-diameter polystyrene microbeads. **d** Moving a 15 μm-diameter polystyrene microbead at a linear velocity of 500 μm/s using a "doughnut"-shaped light pattern. Reproduced from Ref. [10] with permission from Optica Publishing Group, copyright 2019. **e** polystyrene particles pass through the sorter path and are fractionated in the z-direction. Reproduced from Ref. [1] with permission from Springer Nature, copyright 2005. **f** Bright-field microscope images showing the illumination of the microbead suspension with a light pattern depicting a negative, mirror image of a stylized caricature of Isaac Newton with an apple, after assembly and freeze-drying. Reproduced from Ref. [16] with permission from John Wiley and Sons, copyright 2019. **g** Bright-field microscopy images of carbon nanoparticle TMPs **d** after OET assembly and **f** after UV curing. Reproduced from Ref. [18] with permission from John Wiley and Sons, copyright 2021. **h** Large area patterning of nanoparticles using NanoPen. Measurement of Rhodamine 6G (R6G) Raman spectrum using the NanoPen patterned gold nanoparticle. Reproduced from Ref. [6] with permission from American Chemical Society, copyright 2009. **i** OET manipulation of silver nanowires. Reproduced from Ref. [3] with permission from Springer Nature, copyright 2008

has made it a powerful tool for advancing micromachinery research, enabling the development of sophisticated, optically controlled micromachines for diverse applications [80–83]. Figure 6.5a shows optical microscope and corresponding SEM images of nano-kirigami microrotors with 3D deformed blades [80]. The "Nano-kirigami" approach enables the transformation of flat 2D sheets or membranes into micromotors that can move and rotate within the microscopic domain. Metallic nano-kirigami microrotors are easily controlled by OET, offering precise adjustments to both their speed and direction. Figure 6.5b displays diverse movement patterns [80].

Beyond passive microrotors, OET also enables the guidance of self-propelling micromotors. Figure 6.5c illustrates a schematic of SiO_2-Pt Janus micromotors controlled by OET [81]. These micromotors achieve self-propulsion in an AC electric field through induced charge electrophoresis. Within an OET system, the strong electric field at the edge of a light pattern generates a positive DEP force that traps the micromotor, directing it to self-propel along the light pattern's boundary, as shown in Fig. 6.5d, e. This OET-driven steering method offers high versatility and compatibility with diverse micromotors, presenting a practical approach for precise and programmable micromotor control in complex environments [81].

In addition to self-propelling Janus micromotors, OET proves to be an effective tool for guiding microrobots in performing a diverse array of complex tasks [20, 79, 82, 83]. Figure 6.5f illustrates a gear-shaped microrobot featuring a semi-enclosed chamber with eight symmetrically arranged teeth along its edge [82]. A valuable application of this microrobot is its ability to load, transport, and deliver payloads, making it ideal for handling microparticles and cells (Fig. 6.5g) [82]. A notable advantage of this OET-driven microrobotic system is its ability to control multiple microrobots simultaneously for parallel operations, as illustrated in Fig. 6.5h. This superior performance has facilitated applications such as single-cell selection, clonal expansion, single-cell analysis, studies of cell–cell interactions, and the extraction of valuable microtissues from complex biological samples [82]. These findings highlight the potential of OET-driven microrobotic technology as a powerful tool for advancing research in cell biology and life sciences. OET-driven micro-gears can also form a micro-gear train, where an active gear drives passive gears [70], as illustrated in Fig 6.5i. This micro-gear train is highly reconfigurable, allowing for various combinations of active and passive gears.

Figure 6.5j shows a schematic of a touchless micro-feed-roller based on two OET-driven micro-gears rotating in opposite directions (based on negative DEP force) [83]. In addition to micro-feed-roller, OET-driven micro-rack-and-pinion was also built. As shown in Fig. 6.5k, the OET-based micro-rack-and-pinion structure effectively converts rotary motion into linear motion [83]. The reconfigurable nature of the light pattern makes it easy to assemble these sophisticated, multi-component micromechanical systems, highlighting their significant potential in fields such as micro-electro-mechanical systems (MEMS), microfluidics, micromanipulation, and beyond.

6.6 Integration of OET with Microfluidics and Micromanipulation Technologies

6.6.1 Integrating Conventional DEP with OET

OET has been combined with micromanipulation technologies for various applications. For instance, conventional DEP technology was integrated with OET for nanowire assembly [84]. Nanowires were first moved near the metal electrodes using OET and then attracted and arranged in the gap between the two metal electrodes through DEP forces generated by applying an AC bias to the metal electrodes.

Patterned optoelectronic tweezers (p-OET) offer an alternative for integrating conventional DEP with OET [11]. As shown in Fig. 6.6a, p-OET modifies the conventional OET design by patterning the photoconductive layer to expose

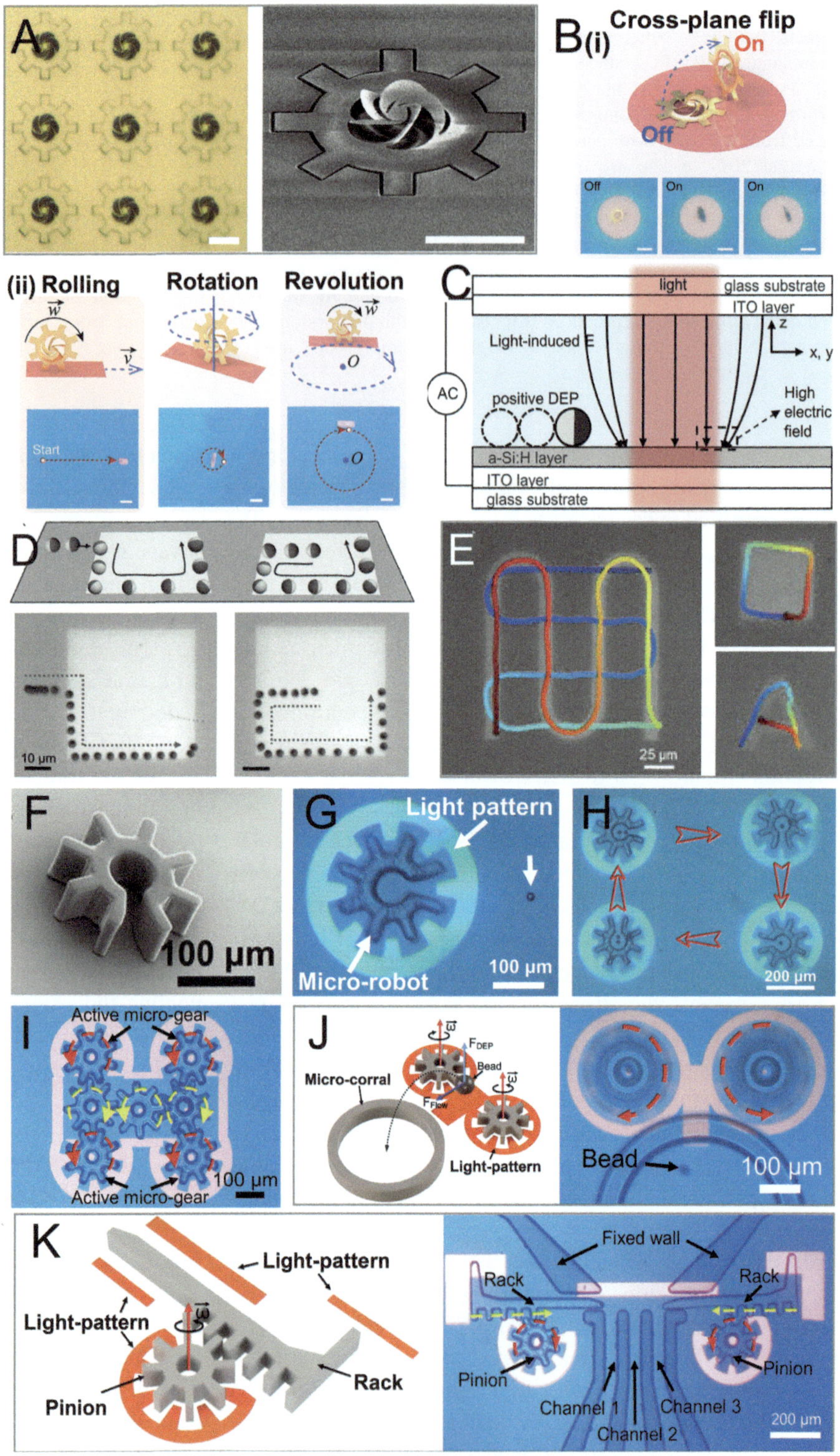

Fig. 6.5 Manipulation of artificial micromachines with OET. **a** Optical microscope image and SEM image of nano-kirigami microrotors. **b** The schematic and experimental images depict the cross-plane flip of a microrotor. Reproduced from Ref. [80] with permission from American Association for the Advancement of Science, copyright 2024. **c** Schematic of using OET to manipulate SiO_2-Pt Janus micromotor via positive DEP. **d** Schematic and microscope images showing the Janus micromotors moving along the edge of light pattern. **e** Microscope images showing the use of different light pattens to steer Janus micromotor along different paths. Reproduced from Ref. [81] with permission from ACS Publications, copyright 2023. **f** SEM image of acogwheel-shaped microrobot. **g** Microscope image showing the use of microrobot to load a microparticle. **h** Microscope image showing the parallel manipulation of four microrobots with OET. Reproduced from Ref. [82] with permission from Proceedings of the National Academy of Sciences of the United States of America, copyright 2019. **i** Microscope image of an OET-driven micro-gear train made up by an active and a passive micro-gear. **j** Schematic and microscope images of a touchless micro-feed-roller based on two OET-driven micro-gears. **k** Schematic of an OET-driven micro-rack-and-pinion system and a microscope image showing two such structures functioning together as a microfluidic valve. Reproduced from Ref. [83] with permission from from Springer Nature, copyright 2021

regions of the ITO electrode beneath. Simulations (Fig. 6.6b) reveal strong non-uniform electric fields at the edges of the exposed ITO, inducing DEP forces that push unwanted particles away from areas of interest, solving the "particle-crowding" issue in conventional OET. Additionally, particles can be trapped in regions where photoconductor projections are surrounded by exposed electrodes (Fig. 6.6c). The p-OET format thus enables precise light-pattern-driven particle manipulation while incorporating DEP-based trapping, even after the light source is turned off, thereby minimizing light-induced heating and improved flexibility.

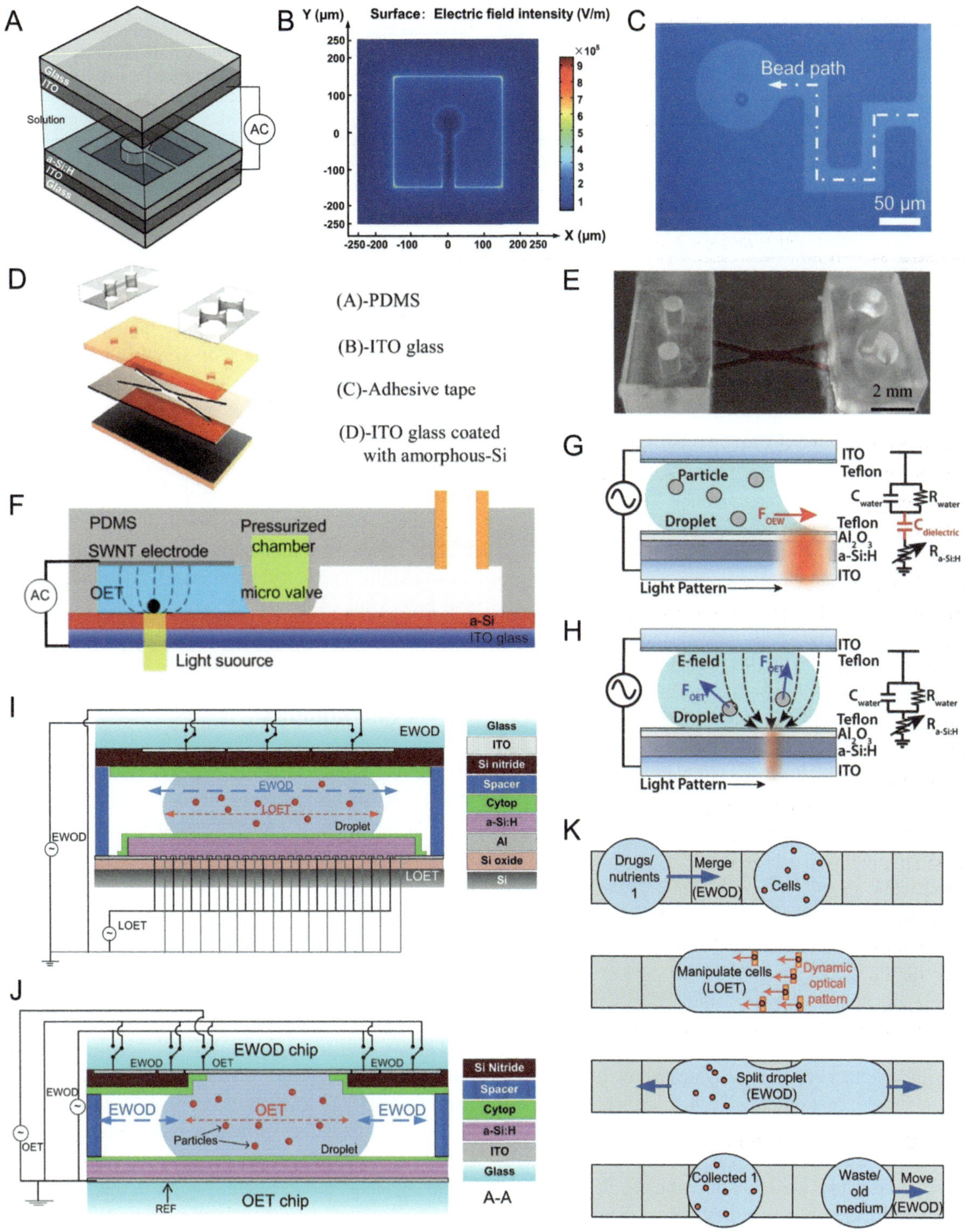
A
Glass
ITO
Solution
AC
a-Si:H
ITO
Glass
B
Y (μm)
Surface: Electric field intensity (V/m)
X (μm)
C
Bead path
50 μm
D
(A)-PDMS
(B)-ITO glass
(C)-Adhesive tape
(D)-ITO glass coated with amorphous-Si
E
2 mm
F
PDMS
SWNT electrode
Pressurized chamber
AC
OET
micro valve
a-Si
ITO glass
Light suource
G
Particle
Droplet
ITO
Teflon
Teflon
a-Si:H
ITO
Light Pattern
H
E-field
Droplet
ITO
Teflon
Teflon
a-Si:H
ITO
Light Pattern
I
EWOD
Droplet
LOET
Glass
ITO
Si nitride
Spacer
Cytop
a-Si:H
Al
Si oxide
Si
J
EWOD chip
OET
Particles
Droplet
REF
OET chip
Si Nitride
Spacer
Cytop
a-Si:H
ITO
Glass
A-A
K
Drugs/ nutrients 1
Merge (EWOD)
Cells
Manipulate cells (LOET)
Dynamic optical pattern
Split droplet (EWOD)
Collected 1
Waste/ old medium
Move (EWOD)

◂**Fig. 6.6 OET integration with microfluidics and micromanipulation technologies.** **a** Schematic of a p-OET device. **b** Simulations on the electric field and its distribution in a p-OET device (AC voltage: 7 Vpp, 25 kHz). **c** Microscope images showing the use of a p-OET device to manipulate and trap a microbead. **a–c** Reproduced from Ref. [11] with permission from John Wiley and Sons, copyright 2018. **d** Schematic design of an OET-microfluidic device based on double-sided adhesive tape with microchannels. **e** Image of an OET-microfluidic device. **d** and **e** Reproduced from Ref. [46] with permission from Royal Society of Chemistry, copyright 2013. **f** Schematic design of an OET-microfluidic device with PDMS top plate and microchannels. Reproduced from Ref. [86] with permission from Royal Society of Chemistry, copyright 2013. **g** Cross section of a device integrating OET and DMF operation. **h** Using lateral OET as the bottom plate, the integrated device allows OET and DMF operation simultaneously. **i** An application of the integrated device in cellular biology. **g–i** Reproduced from Ref. [90] with permission from Royal Society of Chemistry, copyright 2009. **j** Schematic of the device working in the OEW mode (f < 100 kHz), causing aqueous droplets in the vicinity to move towards the light pattern. **k** Diagram of the same device operating in the OET mode (f > 100 kHz). Particles within the droplet experience a DEP force when in the vicinity of light illuminated region. **j** and **k** Reproduced from Ref. [94] with permission from Royal Society of Chemistry, copyright 2011

Another OET system utilizing a non-uniform background electric field is achieved by patterning the geometry of the physical electrode layer beneath the photoconductive layer to create field gradients along a specific axis [12]. The resulting background DEP forces trap particles along the rim of the optical spot, leading to improved resolution over conventional OET technology.

6.6.2 Integration OET with Channel Microfluidic

The integration of OET with channel microfluidics was realized shortly after its invention. Channel microfluidics enable precise fluid control in sub-millimeter channels, allowing continuous delivery and extraction of samples to the OET region for high-throughput sorting, separation, and processing. Integration involves fabricating a spacer with microchannel features to connect the top and bottom OET plates, forming an enclosed channel.

One method fabricates channels on double-sided adhesive tape, sandwiched between the OET plates (Fig. 6.6d) [46]. Figure 6.6e shows a device based on this design, with laser cutting used to create hollow structures in the tape. However, issues like burnt or melted edges arise with laser cutting. To address this, a "cookie-cutter" approach uses a custom metal mold for manual punching. Recently, Xurography, a blade-based cutting technique, has enabled higher-resolution features with smoother edges, improving on previous methods [82]. These simple, cost-effective techniques make adhesive tape-based microfluidic integration widely used for OET.

Poly(dimethylsiloxane) (PDMS), a silicon-based organic polymer, is commonly used for microfluidic device fabrication. By incorporating gold mesh or carbon nanotubes into PDMS [85], an OET device with PDMS microfluidic channels was developed. Unlike other OET-microfluidic devices, where the spacer is fabricated separately, this device uses PDMS structures for both the OET top plate and the spacer [86], as shown in Fig. 6.6f. The integrated pneumatic-controlled valve allows for single-cell isolation and extraction for mRNA analysis.

Although this device offers superior performance, its fabrication is more complex than methods using double-sided tape or SU8 materials. To simplify fabrication, hybrid microfluidic systems have been developed by interfacing PDMS microfluidic devices with double-sided tape-based OET devices [69, 87, 88]. In this setup, PDMS devices handle fluidic control, while OET devices manage target sorting/processing. The two modules are stacked, aligning their inlet and outlet holes, combining the advantages of PDMS device for fluid control and OET for micromanipulation, creating a powerful platform for continuous cell sorting and processing [49].

6.6.3 OET Integration with Digital Microfluidics

Digital microfluidics (DMF) is a liquid-handling technology that enables precise control over discrete micro-droplets on an array of electrode [89], offering an alternative to the conventional channel microfluidics paradigm, where fluids are transported through enclosed micro-channels. The structure of DMF shares similarities with that of OET. The DMF bottom plate comprises an array of electrodes, a dielectric layer, and a hydrophobic coating, while the DMF top plate is an ITO-coated glass slide with a hydrophobic coating, resembling the OET device configuration.

The working principle of DMF relies on the electrowetting-on-dielectric (EWOD) effect [89], which controls the hydrophobicity of a substrate using an electric field. Without an electric field, the surface remains hydrophobic, and a liquid droplet retains a spherical shape with a high contact angle. When an electric field is applied, the surface becomes hydrophilic, flattening the droplet and reducing its contact angle. By localizing this polarization, an interfacial tension gradient is created, enabling precise droplet movement across the DMF surface. Essentially, activating a DMF electrode changes the region above it from hydrophobic to hydrophilic, attracting nearby micro-droplets.

While DMF excels at manipulating micro-droplets (e.g., moving, splitting, merging, and dispensing), it cannot manipulate micro-objects within the droplets. This limitation drives the need to integrate DMF with OET, combining both technologies for versatile fluidic control and effective micromanipulation on a single chip, opening new possibilities for various applications.

The dielectric layer in DMF devices presents challenges for OET operation. One solution is to partially remove the dielectric layer to integrate DMF with OET, as shown in Fig. 6.6g [90]. In this design, the bottom plate is identical to that of a standard OET device, while the top plate features patterned ITO electrodes for DMF. The dielectric layer is selectively removed from one electrode, creating a cross-section typical of OET at that spot, and inverted for the other electrodes. This configuration allows droplet manipulation via EWOD or particle manipulation via OET in distinct areas.

Further enhancement involves using a lateral OET device as the bottom plate, enabling simultaneous DMF control and OET manipulation, as shown in Fig. 6.6h. This integrated DMF-OET system is ideal for both droplet control and micro-object manipulation. In cellular biology applications (Fig. 6.6i), it can create isolated microenvironments for individual cells, where (1) drugs, nutrients, and salts are introduced independently to each droplet, (2) cells are manipulated via parallel LOET with real-time image feedback, and (3–4) waste or old medium is removed by splitting droplets using EWOD, offering a precise solution for cell culture and manipulation.

6.6.4 OET Integration with Optoelectrowetting

An alternative solution for simultaneous droplet control and micro-object manipulation is the optoelectrowetting (OEW) effect [91–93]. OEW uses light-induced "virtual electrodes" to selectively convert a hydrophobic surface into a hydrophilic one, attracting nearby micro-droplets. Unlike DMF, which uses physical electrodes, OEW uses virtual electrodes created by light. Figure 6.6j illustrates an OEW device with a top plate similar to that of a DMF device and a bottom plate resembling an OET device, with an added dielectric layer and hydrophobic coating [94]. The device can operate in OEW mode to move droplets or in OET mode to manipulate micro-particles within droplets (Fig. 6.6k), depending on the AC frequency and light pattern. Like OET, OEW devices rely on the photoelectric effect of photoconductor to generate non-uniform electric field [95]. Single-sided lateral OEW devices, which apply the electric field laterally, have also been developed for micro-droplet actuation [96, 97]. This method requires low light power density (less than 1 W/cm^2), making it an energy-efficient approach for controlling droplets and micro-particles.

6.7 The Commercialization of OET

Driven by the need to utilize OET for various applications, significant efforts have been made to commercialize the technology for both fundamental research and biopharmaceutical applications. The OptoBot® 500, developed by Optoseeker Biotechnology (Shenzhen) Co., Ltd. [98], is an advanced and automated desktop OET platform that uses Digital Micromirror Device (DMD) projection technology for precise manipulation of micro- and nanoscale targets such as metal particles, dielectric particles, nanomaterials, and cells (Fig. 6.7a). This machine is primarily designed for researchers conducting OET-related studies. Key features include fully functional OET device, high-precision DMD projection, stable LED light source, ultra-high-resolution imaging, and flexible graphical editing functions. This versatility makes the OptoBot® 500 suitable for applications in micro- and nanomanipulation, droplet control, and micro-assembly. Its software, OptoMind® 500, enhances user control over target movement and graphical editing, expanding its range of applications in fundamental research.

In the field of biological applications, Ming Wu, the inventor of OET, founded Berkeley Lights Inc. with Igor Khandros and William Davidow in 2011 [99]. This California-based company (acquired by Bruker in 2023) has successfully commercialized fully automated OET instruments to enable the functional screening and selection of individual cells for antibody discovery, cell line development [100–103], cell therapy development and synthetic biology [99]. The Berkeley Lights Beacon platform (now belongs to Bruker) uses OET technology and a microfluidic chip with up to 14,000 "NanoPen" chambers, connected by micro-channels for cell, bead, and reagent handling (Fig. 6.7b) [104]. NanoPen volumes range from 250 pL to 2 nL, enabling rapid phenotypic characterization of single cells.

A typical workflow for cell analysis and selection is illustrated in Fig. 6.7C using the Beacon platform [102, 105]. Cells are loaded onto an OptoSelect Chip (OET-microfluidic chip) and positioned in NanoPens via light-induced DEP using the Beacon system. After clonal expansion, reagents are introduced to characterize various phenotypic traits, such as secreted proteins and surface protein expression. Selected clonal colonies are then exported for off-chip processing, including genomic analysis and/or downstream expansion. This method links the phenotypic characteristics of individual cells to their genotypic profiles.

The OET manipulation system's flexibility supports a wide range of applications, including discovering antibody therapeutic candidates directly from B cells, and developing

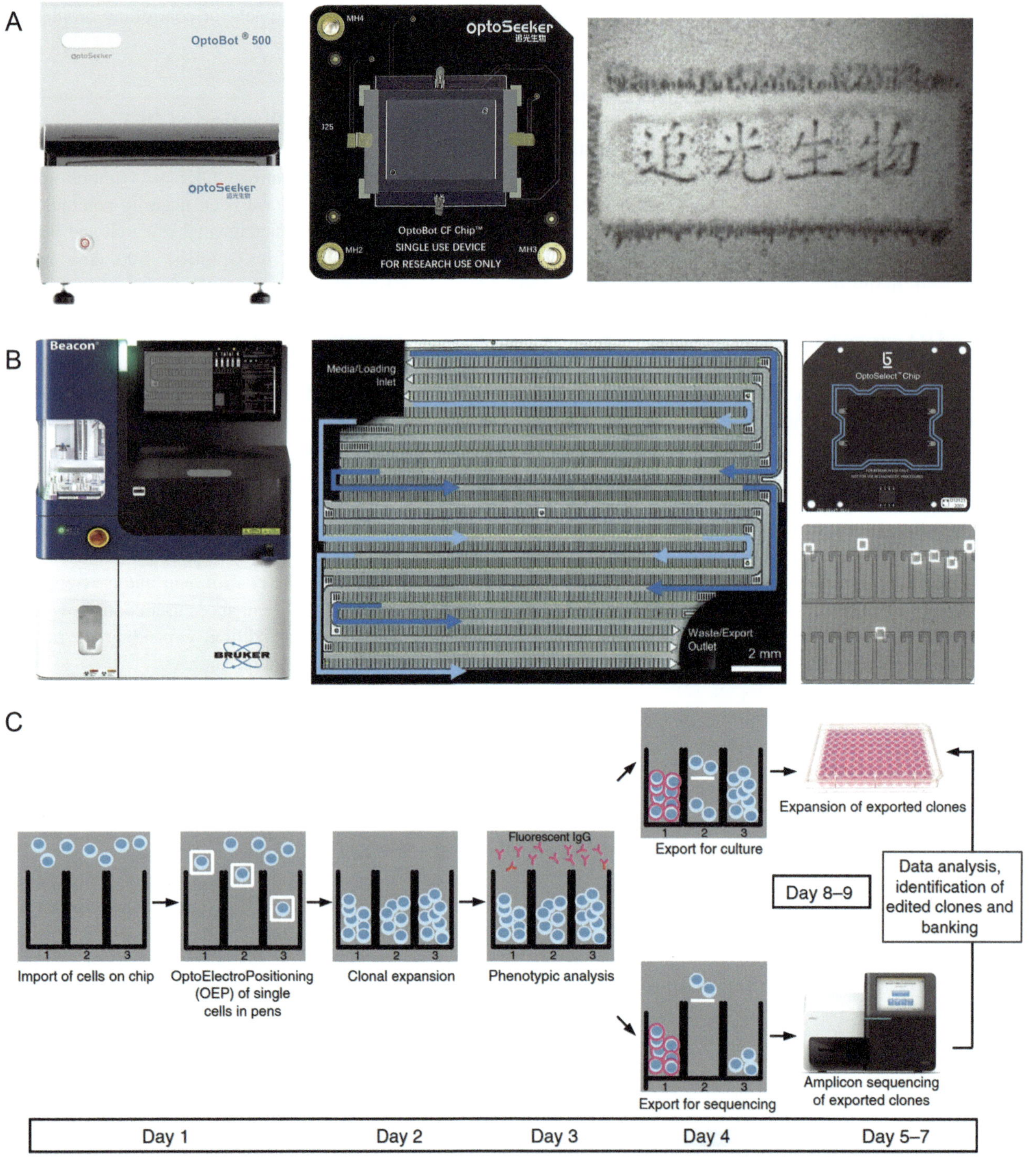

Fig. 6.7 **The commercialization of OET.** **a** OptoBot® 500 Optoelectronic Tweezer System featuring the device and the micro-patterns created from microparticles. Reproduced from Ref. [99] with permission from Optoseeker Biotechnology (Shenzhen) Co., Ltd., copyright 2024. **b** Beacon system for cell screening and selection. Photograph of the Beacon system. The samples are loaded on a proprietary OET-microfluidic chip shown on the right. Reproduced from Ref. [104] with permission from IEEE, copyright 2019 and Ref. [105] with permission from Royal Society of Chemistry, copyright 2021. **c** Workflow for cell manipulation, analysis and recovery using the Beacon system. **b**, **c** Reproduced from Ref. [103] with permission from Springer Nature, copyright 2018

potent cellular therapeutics by selecting functional T cells [99]. It significantly accelerates cell-based product discovery and development [100]. For instance, traditional methods like hybridoma workflow take 2–3 months to discover antibodies, while the OET manipulation system reduces screening time to just several days based on single B cell workflow. Additionally, it requires fewer cells and reagents, enabling faster and more cost-effective identification of the best cells. In summary, the OET technology plays a key role in advancing fundamental research, cellular therapeutics, and biopharmaceuticals.

6.8 Conclusions and Perspectives

This chapter provides a comprehensive review of OET technology, exploring its underlying mechanisms, experimental setups, applications in both biological and non-biological contexts and commercialization efforts. As a highly versatile tool, OET enables non-contact manipulation of micro- and nanoscale particles, cells, and biomolecules through light-induced electric fields, offering exceptional programmability and multifunctionality. Compared to traditional optical tweezers, OET operates efficiently at lower optical intensities, significantly reducing the risk of sample damage, making it particularly suitable for handling biological samples. Moreover, OET is capable of independently manipulating multiple targets simultaneously, thereby greatly enhancing operational efficiency, which is especially beneficial for high-throughput screening and the analysis of complex samples. By utilizing mainly the principles of light-induced DEP, OET generates dynamic electric fields through light patterns to facilitate precise movement, rotation, and aggregation of targets. Its seamless integration with microfluidic technologies further enables applications such as single-cell manipulation, rapid screening, and precision medicine.

To better utilize the advantages of OET, it is of great importance to integrate OET with other devices/sensor/tools to detect, analyze, and measure micro/nano-objects. This requires systematically combining OET devices with other predominant characterization techniques, such as atomic force microscopy, confocal microscopy, and Raman microscopy to achieve multiparameter simultaneous acquisition, detection, and analysis of targeted samples. Moreover, fully automated and cost-effective OET system is required. As the scale and complexity of experiments continue to grow, the demand for real-time data analysis and precise operation in OET systems has increased significantly [106–111]. In this context, the integration of artificial intelligence (AI) presents new opportunities for the advancement of OET technology. By combining AI algorithms with OET systems, it becomes possible to automate image processing, target recognition, and operation path optimization, thereby significantly enhancing system efficiency and accuracy. For instance, AI can rapidly analyze high-resolution microscopic images, accurately identify and locate targets at the micro- and nanoscale, and predict their dynamic behavior within complex electric fields. Moreover, machine learning-based optimization algorithms can adjust light field distributions in real time, enabling automatic parallel manipulation of multiple targets and expanding OET's potential applications in high-throughput screening and multi-scale assembly [106–111].

In summary, this chapter presents a comprehensive review of the OET technology, which offers tremendous research benefits by providing a programmable, flexible and versatile approach for parallel and multi-scaled micromanipulation. With its distinct advantages and expanding application scope, OET technology is becoming an indispensable tool in scientific research and industrial production, and it is expected to play a pivotal role in the fields of microfabrication, materials science, biomedicine and beyond.

Acknowledgements This work was supported by National Key R&D Program of China (Grant No. 2024YFC3406900, 2023YFE0112400, 2022YFA1207100 and 2023YFF0721500), National Natural Science Foundation of China (Grant No. 62103050, 62105177, 21904008), Young Elite Scientists Sponsorship Program by CAST (Grant No. 2023QNRC001), Beijing Municipal Natural Science Foundation (Grant No. L24030, 4242060, 2242018), Chongqing Municipal Natural Science Foundation (Grant No. 2024NSCQ-MSX3784), the BIT Research and Innovation Promoting Project (Grant No. 2023CX01002), Open Research Fund of State Key Laboratory of Optoelectronic Materials and Technologies (Sun Yat-sen University, Grant No. OEMT-2024-KF-03). R. F., B. X and F. Y. contributed equally to this work.

References

1. Chiou PY, Ohta AT, Wu MC. Massively parallel manipulation of single cells and microparticles using optical images. Nature. 2005;436(7049):370–2. https://doi.org/10.1038/nature03831.
2. Pauzauskie PJ, Jamshidi A, Valley JK, Satcher JH Jr, Wu MC. Parallel trapping of multiwalled carbon nanotubes with optoelectronic tweezers. Appl Phys Lett. 2009;95(11): 113104. https://doi.org/10.1063/1.3212725.
3. Jamshidi A, Pauzauskie PJ, Schuck PJ, Ohta AT, Chiou P-Y, Chou J, Yang P, Wu MC. Dynamic manipulation and separation of individual semiconducting and metallic nanowires. Nat Photonics. 2008;2(2):86–9. https://doi.org/10.1038/nphoton.2007.277.
4. Zhang S, Cooper JM, Neale SL. Assembling silver nanowires using optoelectronic tweezers. In: SPIE OPTO: advanced fabrication technologies for micro/nano optics and photonics IX 9759, 97590S; 2016. https://doi.org/10.1117/12.2211085.
5. Lim MB, Hanson JL, Vandsburger L, Roder PB, Zhou X, Smith BE, Ohuchi FS, Pauzauskie PJ. Copper- and chloride-mediated synthesis and optoelectronic trapping of ultra-high aspect ratio palladium nanowires. J Mater Chem A. 2018;6(4):5644–51. https://doi.org/10.1039/c7ta07324k.

6. Jamshidi A, Neale SL, Yu K, Pauzauskie PJ, Schuck PJ, Valley JK, Hsu H-Y, Ohta AT, Wu MC. NanoPen: dynamic, low-power, and light-actuated patterning of nanoparticles. Nano Lett. 2009;9(8):2921–5. https://doi.org/10.1021/nl901239a.
7. Ota S, Wang S, Wang Y, Yin X, Zhang X. Lipid bilayer-integrated optoelectronic tweezers for nanoparticle manipulations. Nano Lett. 2013;13(6):2766–70. https://doi.org/10.1021/nl400999f.
8. Lin Y-H, Chang C-M, Lee G-B. Manipulation of single DNA molecules by using optically projected images. Opt Express. 2009;17(17):15318–29. https://doi.org/10.1364/OE.17.015318.
9. Hwang H, Park J-K. Dynamic light-activated control of local chemical concentration in a fluid. Anal Chem. 2009;81(14):5865–70. https://doi.org/10.1021/ac901047v.
10. Zhang S, Li W, Elsayed M, Tian P, Clark AW, Wheeler AR, Neale SL. Size-scaling effects for microparticles and cells manipulated by optoelectronic tweezers. Opt Lett. 2019;44(17):4171–4. https://doi.org/10.1364/OL.44.004171.
11. Zhang S, Shakiba N, Chen Y, Zhang Y, Tian P, Singh J, Chamberlain MD, Satkauskas M, Flood AG, Kherani NP, Yu S, Zandstra PW, Wheeler AR. Patterned optoelectronic tweezers: a new scheme for selecting, moving, and storing dielectric particles and cells. Small. 2018;14(45):1803342. https://doi.org/10.1002/smll.201803342.
12. Zaman MA, Padhy P, Cheng Y-T, Galambos L, Hesselink L. Optoelectronic tweezers with a non-uniform background field. Appl Phys Lett. 2020;117(17): 171102. https://doi.org/10.1063/5.0020446.
13. Zaman MA, Wu M, Ren W, Jensen MA, Davis RW, Hesselink L. Resolution improvement of optoelectronic tweezers using patterned electrodes. Appl Phys Lett. 2023;123(4): 041104. https://doi.org/10.1063/5.0160939.
14. Zhang S, Juvert J, Cooper JM, Neale SL. Manipulating and assembling metallic beads with optoelectronic tweezers. Sci Rep. 2016;6:32840. https://doi.org/10.1038/srep32840.
15. Zhang S, Liu Y, Juvert J, Tian P, Navarro J-C, Cooper JM, Neale SL. Use of optoelectronic tweezers in manufacturing—accurate solder bead positioning. Appl Phys Lett. 2016;109(22): 221110. https://doi.org/10.1063/1.4971348.
16. Zhang S, Zhai Y, Peng R, Shayegannia M, Flood AG, Qu J, Liu X, Kherani NP, Wheeler AR. Assembly of topographical micropatterns with optoelectronic tweezers. Adv Opt Mater. 2019;7(20):1900669. https://doi.org/10.1002/adom.201900669.
17. Tien M-C, Ohta AT, Yu K, Neale SL, Wu MC. Heterogeneous integration of InGaAsP microdisk laser on a silicon platform using optofluidic assembly. Appl Phys A. 2009;95(4):967–72. https://doi.org/10.1007/s00339-009-5125-2.
18. Zhang S, Li W, Elsayed M, Peng J, Chen Y, Zhang Y, Zhang Y, Shayegannia M, Dou W, Wang T, Sun Y, Kherani NP, Neale SL, Wheeler AR. Integrated assembly and photopreservation of topographical micropatterns. Small. 2021;17(37):2103702. https://doi.org/10.1002/smll.202103702.
19. Juvert J, Zhang S, Eddie I, Mitchell CJ, Reed GT, Wilkinson JS, Kelly A, Neale SL. Micromanipulation of InP lasers with optoelectronic tweezers for integration on a photonic platform. Opt Express. 2016;24(16):18163–75. https://doi.org/10.1364/OE.24.018163.
20. Zhang S, Xu B, Elsayed M, Nan F, Liang W, Valley JK, Liu L, Huang Q, Wu MC, Wheeler AR. Optoelectronic tweezers: a versatile toolbox for nano-/micro-manipulation. Chem Soc Rev. 2022;51(22):9203–42. https://doi.org/10.1039/d2cs00359g.
21. Xu B, Zhao Y, Chen X, Fu R, Li H, Xie S, Liu H, Li Y, Zhang S, Li B. Power micromachines with light. Laser Photon Rev. 2024;19:240–0791. https://doi.org/10.1002/lpor.202400791.
22. Zhang S, Elsayed M, Peng R, Chen Y, Zhang Y, Neale SL, Wheeler AR. Influence of light pattern thickness on the manipulation of dielectric microparticles by optoelectronic tweezers. Photon Res. 2022;10(2):550–6. https://doi.org/10.1364/PRJ.437528.
23. Zhang S, Nikitina A, Chen Y, Zhang Y, Liu L, Flood AG, Juvert J, Chamberlain MD, Kherani NP, Neale SL, Wheeler AR. Escape from an optoelectronic tweezer trap: experimental results and simulations. Opt Express. 2018;26(5):5300–9. https://doi.org/10.1364/OE.26.005300.
24. Ren W, Zaman MA, Wu M, Jensen MA, Davis RW, Hesselink L. Microparticle electrical conductivity measurement using optoelectronic tweezers. J Appl Phys. 2023;134(11): 113104. https://doi.org/10.1063/5.0169565.
25. Yang S-M, Yu T-M, Huang H-P, Ku M-Y, Hsu L, Liu C-H. Dynamic manipulation and patterning of microparticles and cells by using TiOPc-based optoelectronic dielectrophoresis. Opt Lett. 2010;35(12):1959–61. https://doi.org/10.1364/OL.35.001959.
26. Yang S-M, Yu T-M, Huang H-P, Ku M-Y, Tseng S-Y, Tsai C-L, Chen H-P, Hsu L, Liu C-H. Light-driven manipulation of picobubbles on a titanium oxide phthalocyanine-based optoelectronic chip. Appl Phys Lett. 2011;98(15): 153512. https://doi.org/10.1063/1.3580760.
27. Wang W, Lin Y-H, Guan R-S, Wen T-C, Guo T-F, Lee G-B. Bulk-heterojunction polymers in optically-induced dielectrophoretic devices for the manipulation of microparticles. Opt Express. 2009;17(20):17603–13. https://doi.org/10.1364/OE.17.017603.
28. Wang W, Lin Y-H, Wen T-C, Guo T-F, Lee G-B. Selective manipulation of microparticles using polymer-based optically induced dielectrophoretic devices. Appl Phys Lett. 2010;96(11): 113302. https://doi.org/10.1063/1.3358193.
29. Hsu H-Y, Ohta AT, Chiou P-Y, Jamshidi A, Neale SL, Wu MC. Phototransistor-based optoelectronic tweezers for dynamic cell manipulation in cell culture media. Lab Chip. 2010;10(2):165–72. https://doi.org/10.1039/b906593h.
30. Yang Y, Mao Y, Shin K-S, Chui CO, Chiou P-Y. Self-locking optoelectronic tweezers for single-cell and microparticle manipulation across a large area in high conductivity media. Sci Rep. 2016;6:22630. https://doi.org/10.1038/srep22630.
31. Liu N, Lin Y, Peng Y, Xin L, Yue T, Liu Y, Ru C, Xie S, Dong L, Pu H, Chen H, Li WJ, Sun Y. Automated parallel electrical characterization of cells using optically-induced dielectrophoresis. IEEE Trans Automat Sci Eng. 2020;17(2):1084–92. https://doi.org/10.1109/TASE.2019.2963044.
32. Li Y, Lai SHS, Liu N, Zhang G, Liu L, Lee G-B, Li WJ. Fabrication of high-aspect-ratio 3D hydrogel microstructures using optically induced electrokinetics. Micromachines. 2016;7(4):65. https://doi.org/10.3390/mi7040065.
33. Hwang H, Park Y-H, Park J-K. Optoelectrofluidic control of colloidal assembly in an optically induced electric field. Langmuir. 2009;25(11):6010–4. https://doi.org/10.1021/la9005604.
34. Valley JK, Jamshidi A, Ohta AT, Hsu HY, Wu MC. Operational regimes and physics present in optoelectronic tweezers. J Microelectromech Syst. 2008;17(2):342–50. https://doi.org/10.1109/JMEMS.2008.916335.
35. Pethig R. Review article—dielectrophoresis: status of the theory, technology, and applications. Biomicrofluidics. 2010;4(2):022811. https://doi.org/10.1063/1.3456626
36. Hwang H, Park J-K. Optoelectrofluidic platforms for chemistry and biology. Lab Chip. 2011;11(1):33–47. https://doi.org/10.1039/c0lc00117a.
37. Kumar S, Hesketh PJ. Interpretation of ac dielectrophoretic behavior of tin oxide nanobelts using Maxwell stress tensor approach modeling. Sens Actuat B Chem. 2012;161(1):1198–208. https://doi.org/10.1016/j.snb.2011.12.010.
38. Wang X, Wang X-B, Gascoyne PRC. General expressions for dielectrophoretic force and electrorotational torque derived using the Maxwell stress tensor method. J Electrostat. 1997;39(4):277–95. https://doi.org/10.1016/S0304-3886(97)00126-5.

39. Chiou PY, Ohta AT, Jamshidi A, Hsu HY, Wu MC. Light-actuated AC electroosmosis for nanoparticle manipulation. J Microelectromech Syst. 2008;17(3):525–31. https://doi.org/10.1109/JMEMS.2008.916342.
40. Zarowna-Dabrowska A, Neale SL, Massoubre D, McKendry J, Rae BR, Yin H, Henderson RK, Rose MJ, Dawson MD, Cooper JM, Gu E. Miniaturised optoelectronic tweezers controlled by GaN micro light emitting diode arrays. In: 2010 23rd annual meeting of the IEEE photonics society; 2010; pp. 102–03 (2010). https://doi.org/10.1109/PHOTONICS.2010.5698778.
41. Liang W, Liu L, Lai SHS, Wang Y, Lee G-B, Li WJ. Rapid assembly of gold nanoparticle-based microstructures using optically-induced electrokinetics. Opt Mater Express. 2014;4(11):2368–80. https://doi.org/10.1364/OME.4.002368.
42. Hwang H, Park J-K. Rapid and selective concentration of microparticles in an optoelectrofluidic platform. Lab Chip. 2009;9(2):199–206. https://doi.org/10.1039/b811740c.
43. Kumar A, Kwon J-S, Williams SJ, Green NG, Yip NK, Wereley ST. Optically modulated electrokinetic manipulation and concentration of colloidal particles near an electrode surface. Langmuir. 2010;26(7):5262–72. https://doi.org/10.1021/la904661y.
44. Castellanos A, Ramos A, González A, Green NG, Morgan H. Electrohydrodynamics and dielectrophoresis in microsystems: scaling laws. J Phys D Appl Phys. 2003;36(20):2584. https://doi.org/10.1088/0022-3727/36/20/023.
45. Williams SJ, Kumar A, Wereley ST. Electrokinetic patterning of colloidal particles with optical landscapes. Lab Chip. 2008;8(11):1879–82. https://doi.org/10.1039/b810787d.
46. Huang S-B, Wu M-H, Lin Y-H, Hsieh C-H, Yang C-L, Lin H-C, Tseng C-P, Lee G-B. High-purity and label-free isolation of circulating tumor cells (CTCs) in a microfluidic platform by using optically-induced-dielectrophoretic (ODEP) force. Lab Chip. 2013;13(7):1371–83. https://doi.org/10.1039/c3lc41256c.
47. Zhang Y, Zhao J, Yu H, Li P, Liang W, Liu Z, Lee GB, Liu L, Li WJ, Wang Z. Detection and isolation of free cancer cells from ascites and peritoneal lavages using optically induced electrokinetics (OEK). Sci Adv. 2020;6(32):96-eaba28. https://doi.org/10.1126/sciadv.aba9628.
48. Ohta AT, Garcia M, Valley JK, Banie L, Hsu H-Y, Jamshidi A, Neale SL, Lue T, Wu MC. Motile and non-motile sperm diagnostic manipulation using optoelectronic tweezers. Lab Chip. 2010;10(23):3213–7. https://doi.org/10.1039/c0lc00072h.
49. Lee G-B, Wu H-C, Yang P-F, Mai JD. Optically induced dielectropheresis sorting with automated medium exchange in an integrated optofluidic device resulting in higher cell viability. Lab Chip. 2014;14(15):2837–43. https://doi.org/10.1039/c4lc00466c.
50. Chiu T-K, Chou W-P, Huang S-B, Wang H-M, Lin Y-C, Hsieh C-H, Wu M-H. Application of optically-induced-dielectrophoresis in microfluidic system for purification of circulating tumour cells for gene expression analysis—cancer cell line model. Sci Rep. 2016;6:32851. https://doi.org/10.1038/srep32851.
51. Zhao Y, Lai HSS, Zhang G, Lee G-B, Li WJ. Rapid determination of cell mass and density using digitally controlled electric field in a microfluidic chip. Lab Chip. 2014;14(22):4426–34. https://doi.org/10.1039/c4lc00795f.
52. Liang W, Zhao Y, Liu L, Wang Y, Li WJ, Lee G-B. Determination of cell membrane capacitance and conductance via optically induced electrokinetics. Biophys J. 2017;113(7):1531–9. https://doi.org/10.1016/j.bpj.2017.08.006.
53. Wang C-H, Lee Y-H, Kuo H-T, Liang W-F, Li W-J, Lee G-B. Dielectrophoretically-assisted electroporation using light-activated virtual microelectrodes for multiple DNA transfection. Lab Chip. 2014;14(3):592–601. https://doi.org/10.1039/c3lc51102b.
54. Yang W, Yu H, Li G, Wang Y, Liu L. High-throughput fabrication and modular assembly of 3D heterogeneous microscale tissues. Small. 2017;13(5):1602769. https://doi.org/10.1002/smll.201602769.
55. Gi HJ, Han D, Park J-K. Optoelectrofluidic printing system for fabricating hydrogel sheets with on-demand patterned cells and microparticles. Biofabrication. 2017;9(1): 015011. https://doi.org/10.1088/1758-5090/aa564c.
56. Liang X, Chen Z, Deng Y, Liu D, Liu X, Huang Q, Arai T. Field-controlled microrobots fabricated by photopolymerization. Cyborg Bionic Syst. 2023;4:0009. https://doi.org/10.34133/cbsystems.0009.
57. Zhou H, Dong G, Gao G, Du R, Tang X, Ma Y, Li J. Hydrogel-based stimuli-responsive micromotors for biomedicine. Cyborg Bionic Syst. 2022;9852853. https://doi.org/10.34133/2022/9852853.
58. Liu N, Liang W, Liu L, Wang Y, Mai JD, Lee G-B, Li WJ. Extracellular-controlled breast cancer cell formation and growth using non-UV patterned hydrogels via optically-induced electrokinetics. Lab Chip. 2014;14(7):1367–76. https://doi.org/10.1039/c3lc51247a.
59. Hwang H, Choi Y-J, Choi W, Kim S-H, Jang J, Park J-K. Interactive manipulation of blood cells using a lens-integrated liquid crystal display based optoelectronic tweezers system. Electrophoresis. 2008;29(6):1203–12. https://doi.org/10.1002/elps.200700415.
60. Xie S, Wang X, Jiao N, Tung S, Liu L. Programmable micrometer-sized motor array based on live cells. Lab Chip. 2017;17(12):2046–53. https://doi.org/10.1039/c7lc00017k.
61. Liang W, Wang Y, Zhang H, Liu L. Characterization of the self-rotational motion of stored red blood cells by using optically-induced electrokinetics. Opt Lett. 2016;41(12):2763–6. https://doi.org/10.1364/OL.41.002763.
62. Yang P-F, Wang C-H, Lee G-B. Optically-induced cell fusion on cell pairing microstructures. Sci Rep. 2016;6:22036. https://doi.org/10.1038/srep22036.
63. Hsiao YC, Wang CH, Lee WB, Lee GB. Automatic cell fusion using optically-induced dielectrophoresis and optically-induced localized electric field on a structure-free microfluidic chip. In: 2017 IEEE 12th international conference on nano/micro engineered and molecular systems (NEMS); 2017. pp. 7–10.
64. He W, Huang L, Feng Y, Liang F, Ding W, Wang W. Highly integrated microfluidic device for cell pairing, fusion and culture. Biomicrofluidics. 2019;13(5):054109. https://doi.org/10.1063/1.5124705.
65. Skelley AM, Kirak O, Suh H, Jaenisch R, Voldman J. Microfluidic control of cell pairing and fusion. Nat Methods. 2009;6(2):147–52. https://doi.org/10.1038/NMETH.1290.
66. Kremer C, Witte C, Neale SL, Reboud J, Barrett MP, Cooper JM. Shape-dependent optoelectronic cell lysis. Angew Chem Int Edit. 2014;53(3):842–6. https://doi.org/10.1002/anie.201307751.
67. Witte C, Kremer C, Chanasakulniyom M, Reboud J, Wilson R, Cooper JM, Neale SL. Spatially selecting a single cell for lysis using light-induced electric fields. Small. 2014;10(15):3026–31. https://doi.org/10.1002/smll.201400247.
68. Lin Y-H, Lee G-B. An integrated cell counting and continuous cell lysis device using an optically induced electric field. Sens Actuators B Chem. 2010;145(2):854–60. https://doi.org/10.1016/j.snb.2010.01.019.
69. Huang S-H, Hung L-Y, Lee G-B. Continuous nucleus extraction by optically-induced cell lysis on a batch-type microfluidic platform. Lab Chip. 2016;16(8):1447–56. https://doi.org/10.1039/c5lc01284h.
70. Hwang H, Park J-K. Measurement of molecular diffusion based on optoelectrofluidic fluorescence microscopy. Anal Chem. 2009;81(21):9163–7. https://doi.org/10.1021/ac9021709.
71. Han D, Park J-K. Optoelectrofluidic enhanced immunoreaction based on optically-induced dynamic AC electroosmosis. Lab Chip. 2016;16(7):1189–96. https://doi.org/10.1039/c6lc00110f.

72. Han D, Park JK. Microarray-integrated optoelectrofluidic immunoassay system. Biomicrofluidics. 2016;10(3): 034106. https://doi.org/10.1063/1.4950787.
73. Han D, Hwang H, Park J-K. Optoelectrofluidic behavior of metal–polymer hybrid colloidal particles. Appl Phys Lett. 2013;102(5): 054105. https://doi.org/10.1063/1.4790622.
74. Hwang H, Chon H, Choo J, Park J-K. Optoelectrofluidic sandwich immunoassays for detection of human tumor marker using surface-enhanced Raman scattering. Anal Chem. 2010;82(18):7603–10. https://doi.org/10.1021/ac101325t.
75. Liu N, Li MI, Liu L-YD, Yang YD, Mai J, Pu H, Sun Y, Li WJ. Single-step fabrication of electrodes with controlled nanostructured surface roughness using optically-induced electrodeposition. J Micromech Microeng. 2018;28(2):025011. https://doi.org/10.1088/1361-6439/aaa1f8.
76. Liu N, Wang F, Liu L, Yu H, Xie S, Wang J, Wang Y, Lee G-B, Li WJ. Rapidly patterning micro/nano devices by directly assembling ions and nanomaterials. Sci Rep. 2016;6:32106. https://doi.org/10.1038/srep32106.
77. Liu N, Wei F, Liu L, Lai HSS, Yu H, Wang Y, Lee G-B, Li WJ. Optically-controlled digital electrodeposition of thin-film metals for fabrication of nano-devices. Opt Mater Express. 2015;5(4):838–48. https://doi.org/10.1364/OME.5.000838.
78. Li P, Liu N, Yu H, Wang F, Liu L, Lee G-B, Wang Y, Li WJ. Silver nanostructures synthesis via optically induced electrochemical deposition. Sci Rep. 2016;6:28035. https://doi.org/10.1038/srep28035.
79. Zhang S, Liu Y, Qian Y, Li W, Juvert J, Tian P, Navarro J-C, Clark AW, Gu E, Dawson MD, Cooper JM, Neale SL. Manufacturing with light—micro-assembly of opto-electronic microstructures. Opt Express. 2017;25(23):28838–50. https://doi.org/10.1364/OE.25.028838.
80. Hong X, Xu B, Li G, Nan F, Wang X, Liang Q, Dong W, Dong W, Sun H, Zhang Y, Li C, Fu R, Wang Z, Shen G, Wang Y, Yao Y, Zhang S, Li J. Optoelectronically navigated nano-kirigami microrotors. Sci Adv. 2024;10(17):eadn7582. https://doi.org/10.1126/sciadv.adn7582.
81. Chen X, Chen X, Elsayed M, Edwards H, Liu J, Peng Y, Zhang HP, Zhang S, Wang W, Wheeler AR. Steering micromotors via reprogrammable optoelectronic paths. ACS Nano. 2023;17(6):5894–904. https://doi.org/10.1021/acsnano.2c12811.
82. Zhang S, Scott EY, Singh J, Chen Y, Zhang Y, Elsayed M, Chamberlain MD, Shakiba N, Adams K, Yu S, Morshead CM, Zandstra PW, Wheeler AR. The optoelectronic microrobot: a versatile toolbox for micromanipulation. Proc Natl Acad Sci U S A. 2019;116(30):14823–8. https://doi.org/10.1073/pnas.1903406116.
83. Zhang S, Elsayed M, Peng R, Chen Y, Zhang Y, Peng J, Li W, Chamberlain MD, Nikitina A, Yu S, Liu X, Neale SL, Wheeler AR. Reconfigurable multi-component micromachines driven by optoelectronic tweezers. Nat Commun. 2021;12(1):5349. https://doi.org/10.1038/s41467-021-25582-8.
84. Lin YH, Ho KS, Yang CT, Wang JH, Lai CS. A highly flexible platform for nanowire sensor assembly using a combination of optically induced and conventional dielectrophoresis. Opt Express. 2014;22(11):13811–24. https://doi.org/10.1364/OE.22.013811.
85. Wang F, Yu H, Liu N, Mai JD, Liu L, Lee GB, Li WJ. Non-ultraviolet-based patterning of polymer structures by optically induced electrohydrodynamic instability. Appl Phys Lett. 2013;103(21): 214101. https://doi.org/10.1063/1.4830001.
86. Huang KW, Wu YC, Lee JA, Chiou PY. Microfluidic integrated optoelectronic tweezers for single-cell preparation and analysis. Lab Chip. 2013;13(18):3721–7. https://doi.org/10.1039/c3lc50607j.
87. Chen YS, Lai CP, Chen C, Lee GB. Isolation and recovery of extracellular vesicles using optically-induced dielectrophoresis on an integrated microfluidic platform. Lab Chip. 2021;21(8):1475–83. https://doi.org/10.1039/D1LC00093D.
88. Lee GB, Chang CJ, Wang CH, Lu MY, Luo YY. Continuous medium exchange and optically induced electroporation of cells in an integrated microfluidic system. Microsyst Nanoeng. 2015;1:15007. https://doi.org/10.1038/micronano.2015.7.
89. Ng AH, Li BB, Chamberlain MD, Wheeler AR. Digital microfluidic cell culture. Annu Rev Biomed Eng. 2015;17:91–112. https://doi.org/10.1146/annurev-bioeng-071114-040808.
90. Shah GJ, Ohta AT, Chiou EP, Wu MC. EWOD-driven droplet microfluidic device integrated with optoelectronic tweezers as an automated platform for cellular isolation and analysis. Lab Chip. 2009;9(12):1732–9. https://doi.org/10.1039/b821508a.
91. Baigl D. Photo-actuation of liquids for light-driven microfluidics: state of the art and perspectives. Lab Chip. 2012;12(19):3637–53. https://doi.org/10.1039/c2lc40596b.
92. Chiou PY, Park SY, Wu MC. Continuous optoelectrowetting for picoliter droplet manipulation. Appl Phys Lett. 2008;93(22): 221110. https://doi.org/10.1063/1.3039070.
93. Pei SN, Valley JK, Wang YL, Wu MC. Distributed circuit model for multi-color light-actuated opto-electrowetting microfluidic device. J Lightwave Technol. 2015;33(16):3486–93. https://doi.org/10.1109/JLT.2015.2405076.
94. Valley JK, Pei SN, Jamshidi A, Hsu HY, Wu MC. A unified platform for optoelectrowetting and optoelectronic tweezers. Lab Chip. 2011;11(7):1292. https://doi.org/10.1039/c0lc00568a.
95. Yu TM, Yang SM, Fu CY, Liu MH, Hsu L, Chang HY, Liu CH. Integration of organic opto-electrowetting and poly (ethylene) glycol diacrylate (PEGDA) microfluidics for droplets manipulation. Sens Actuators B Chem. 2013;180:35–42. https://doi.org/10.1016/j.snb.2011.12.059.
96. Chuang HS, Kumar A, Wereley ST. Open optoelectrowetting droplet actuation. Appl Phys Lett. 2008;93(6): 064104. https://doi.org/10.1063/1.2970047.
97. Park SY, Teitell MA, Chiou EP. Single-sided continuous optoelectrowetting (SCOEW) for droplet manipulation with light patterns. Lab Chip. 2010;10(13):1655–61. https://doi.org/10.1039/c001324b.
98. Optoseeker Biotechnology (Shenzhen) Co., Ltd. https://www.optoseeker.com/index.jsp
99. Bruker Cellular Analysis. https://www.berkeleylights.com/
100. Jorgolli M, Nevill T, Winters A, Chen I, Chong S, Lin FF, Mock M, Chen C, Le K, Tan C, Jess P. Nanoscale integration of single cell biologics discovery processes using optofluidic manipulation and monitoring. Biotechnol Bioeng. 2019;116(9):2393–411. https://doi.org/10.1002/bit.27024.
101. Le K, Tan C, Gupta S, Guhan T, Barkhordarian H, Lull J, Stevens J, Munro T. A novel mammalian cell line development platform utilizing nanofluidics and optoelectro positioning technology. Biotechnol Prog. 2018;34(6):1438–46. https://doi.org/10.1002/btpr.2690.
102. Le K, Tan C, Le H, Tat J, Zasadzinska E, Diep J, Zastrow R, Chen C, Stevens J. Assuring clonality on the beacon digital cell line development platform. Biotechnol J. 2020;15(1):1900247. https://doi.org/10.1002/biot.201900247.
103. Mocciaro A, Roth TL, Bennett HM, Soumillon M, Shah A, Hiatt J, Chapman K, Marson A, Lavieu G. Light-activated cell identification and sorting (LACIS) for selection of edited clones on a nanofluidic device. Commun Biol. 2018;1:41. https://doi.org/10.1038/s42003-018-0034-6.
104. Wu MC. Optoelectronic tweezers—An optofluidic platform for digital cell biology. In: 2019 international conference on optical MEMS and nanophotonics; 2019. pp. 94–5. https://doi.org/10.1109/OMN.2019.8925125.
105. Matthew R, Lin K-C, Mobilia KC, Sackmann EK, Kurz V, Navidi AH, King J, Onorato RM, Chao LK, Wu T,

Jiang H, Valley JK, Lionberger TA, Leavell MD. High-throughput optofluidic screening for improved microbial cell factories via real-time micron-scale productivity monitoring. Lab Chip. 2019;21(15):2901–12. https://doi.org/10.1039/D1LC00389E.

106. Mennillo L, Bendkowski C, Elsayed M, Edwards H, Zhang S, Pawar V, Wheeler AR, Stoyanov D, Shaw M. Adaptive autonomous navigation of multiple optoelectronic microrobots in dynamic environments. IEEE Robot Autom Lett. 2022;7(4):11102–9. https://doi.org/10.1109/LRA.2022.3194308.
107. Bendkowski C, Mennillo L, Xu T, Elsayed M, Stojic F, Edwards H, Zhang S, Morshead C, Pawar V, Wheeler AR, Stoyanov D, Shaw M. Autonomous object harvesting using synchronized optoelectronic microrobots. In: 2021 IEEE/RSJ international conference on intelligent robots and systems (IROS); 2021. pp 7498–504. https://doi.org/10.1109/IROS51168.2021.9636475.
108. Liu J, Wang H, Liu M, Zhao R, Zhao Y, Sun T, Shi Q. POMDP-based real-time path planning for manipulation of multiple microparticles via optoelectronic tweezers. Cyborg Bionic Syst. 2022; 9890607. https://doi.org/10.34133/2022/9890607.
109. Zhao J, Gan C, Zhang J, Liang S, Yang J, Feng L. Deep learning assisted automated separation platform of single cells and microparticles using optoelectronic tweezers. In: 2022 WRC symposium on advanced robotics and automation (WRC SARA); 2022. pp. 275–79. https://doi.org/10.1109/WRCSARA57040.2022.9903963.
110. Wang A, Gan C, Han H, Xiong H, Zhao J, Wang C, Feng L. Dynamic adaptive imaging system on optoelectronic tweezers platform. In: 2024 IEEE international conference on robotics and automation (ICRA); 2022. pp. 15622–27. https://doi.org/10.1109/ICRA57147.2024.10611608.
111. Du M, Liu F, Li G. Precision manipulation by an optically induced dielectrophoresis system based on an improved A-star algorithm. Appl Opt. 2024;63(20):5237–43. https://doi.org/10.1364/AO.516613.10.1109/WRCSARA57040.2022.9903963.

Part II

Biological Applications of Optical Manipulation Techniques

Microsphere-Assisted Technology for Optical Manipulation, Detection and Imaging

7

Xixi Chen, Heng Li, Tong He, Xingdi Luo, Yuchao Li, and Baojun Li

Abstract

With the advancement of nanoscience and technology, researchers have increasingly focused on the microscopic world at the nanoscale. To overcome the constraints of optical diffraction limit for better prepare and observe nanoscale samples, optical systems are evolving towards higher optical resolution, greater sensitivity, and increased integration. In recent years, researchers have employed artificial dielectric materials and natural biomaterials as microsphere lenses to facilitate conventional optical systems, enabling them to achieve nano-focusing of light beams and efficiently collect weak light signals. Additionally, the simple structure of the microsphere allows for rapid integration with various devices, fostering the development of more flexible and efficient optical systems. Consequently, the emerging microsphere-assisted technology demonstrates significant potential for applications in the fields of super-resolution optical imaging, signal enhancement, optical manipulation, and nanofabrication.

Keywords

Microsphere lenses · Microsphere-assisted technology · Manipulating microlens · Bio-microlenses · Optical applications

X. Chen · H. Li · T. He · X. Luo · Y. Li · B. Li (✉)
Guangdong Provincial Key Laboratory of Nanophotonic Manipulation, Institute of Nanophotonics, College of Physics and Optoelectronic Engineering, Jinan University, Guangzhou, China
e-mail: baojunli@jnu.edu.cn

7.1 Introduction

The effective utilization of the information and energy conveyed by light has been a significant area of inquiry for researchers throughout history [1]. Researchers have engineered, fabricated, and assembled transparent optical dielectric materials to develop advanced optical systems capable of capturing, modulating, and utilizing optical signals [2, 3]. Nonetheless, the optical diffraction limit has constrained the precision of conventional optical systems, impeding progress from the micron scale to the nanometer scale in various domains, including life sciences, materials manufacturing, and information technology [4, 5]. Consequently, overcoming the optical diffraction limit has emerged as a significant focus in optical research [6–12]. Researchers have discovered that micron-sized spherical structures can facilitate the surpassing of the optical diffraction limit in conventional optical systems [13, 14]. In contrast to other techniques that achieve sub-diffraction-limit resolution, microsphere-assisted technology does not necessitate alterations to the existing optical system. Instead, it simply requires the integration of appropriate microsphere lenses between the optical system and the sample. This approach enhances the optical signal modulation capabilities of the system while remaining cost-effective. Furthermore, with the convergence of optical technology and innovations in life sciences, numerous natural biomaterials exhibiting lensing effects have been identified and utilized to create bio-microlenses. These bio-microlenses are characterized by high biocompatibility and do not require nanomanufacturing processes [15]. Leveraging the exceptional optical properties of microspheres, microsphere-assisted technology can be integrated with a wide range of optical systems, thereby advancing research applications across diverse fields, including super-resolution optical imaging, signal enhancement, optical manipulation, and laser nano-fabrication [14, 16, 17].

B. Li et al. (eds.), *Biophotonic Manipulation*, Advances in Optics and Optoelectronics,
https://doi.org/10.1007/978-981-96-4982-2_7

7.2 Types of Microsphere-Assisted Technology

In the microsphere-assisted technology, microsphere's structural configuration and material characteristics significantly influence its optical performance. Additionally, the microscopic size of the microlens necessitates advanced manipulation techniques to control its distribution, marking this as a critical area of research. This section provides a concise overview of microsphere-assisted technology, emphasizing the materials and control methods employed in current studies.

7.2.1 Composition of the Microsphere Lenses

In the existing research of microlenses, common geometries include hemispherical, cylindrical and spherical shapes. Compared with other shapes of microlenses, the larger curvature of spherical microlenses can provide tighter light focusing and higher optical imaging magnification for research. Since 2001, when Leiderer's research team identified the near-field focusing capabilities of dielectric microparticles on incident light [18], the optical properties of microsphere lenses have garnered substantial research interest [13, 16, 19]. Benefiting from the advancements in the preparation process of microsphere materials, commonly utilized microsphere materials in research encompass silica (SiO_2, $n = 1.46$), plexiglass (PMMA, $n = 1.49$), polystyrene (PS, $n = 1.59$), barium titanate ($BaTiO_2$, BTG, $n = 1.92$), titanium dioxide (TiO_2, $n = 2.55$), and other commercially available solid microsphere materials. Additionally, glycerol and other organic liquid materials are employed for fabricating adjustable and variable-focus microsphere lenses [20, 21]. Recent studies indicate that research teams are actively developing novel microsphere materials to enhance microlens performance. Notably, in 2022, Tong's research team reported the preparation and investigation of microspheres using arsenic trisulfide (As_2S_3, $n = 2.31$), which addressed the challenge of constructing high refractive index materials into micron-sized spherical structures [22].

In recent years, advancements in biophotonics research have necessitated higher biocompatibility standards for microlens materials utilized in biological studies. The majority of microlens materials employed in current research are solid dielectric substances, which, upon entering biological environments, exhibit limited natural degradability within the timeframe of the research. This limitation poses significant challenges to the safety of biological samples. Consequently, the advancement of biomaterials for the fabrication of microlenses has increasingly become a necessity. In 2012, Sun and colleagues employed laser direct writing technology to fabricate protein-based hemispherical microlenses [23]. Subsequently, in 2015, Ferraro's research group successfully developed photofluidic microlenses using red blood cells (refractive index $n = 1.37 \sim 1.42$) [24]. In 2016, Wang's team utilized natural spider silk ($n = 1.55$) to create cylindrical biological microlenses with a diameter of 6.8 μm [25]. Li's research team had employed spherical bacteria, yeast, free stem cells, and lipid droplets of biological cell to construct biological magnifiers for optical imaging and manipulation [26–28]. Collectively, these studies lay the groundwork for the development of biocompatible spherical microlenses.

7.2.2 The Manipulation Methods of the Microsphere

The working range of microsphere-assisted technology is limited by the dimensions of the microsphere in practical applications. Consequently, precise control over the number and positioning of microlenses can enhance the efficiency of this technology. For instance, microlens arrays, composed of multiple microlenses with uniform specifications arranged in a regular pattern, can extend the effective working range [20, 29–31]. One prevalent method for improving the flexibility and positional accuracy in sample detection involves using external mechanical devices to manipulate the microsphere's position. Such devices include flexible components like the cantilever of an atomic force microscope (AFM), fiber probes, and objective lenses [32–35]. Concurrently, researchers have developed various non-contact manipulation techniques for controlling microspheres, including chemical actuation [36, 37] (Fig. 7.1a), optical manipulation [21, 27, 38–40] (Fig. 7.1b), and acoustic manipulation [41] (Fig. 7.1c). Notably, optical manipulation techniques, characterized by the high spatial and temporal precision of the focused beam, facilitate the accurate capture and manipulation of microlenses. Notably, the utilization of optical techniques for microlens manipulation, coupled with the high spatial and temporal precision of focused beams, facilitates the accurate capture and handling of individual microlenses. This capability is crucial for the flexible assembly of microlens arrays and the precise control of microlenses within biological environments.

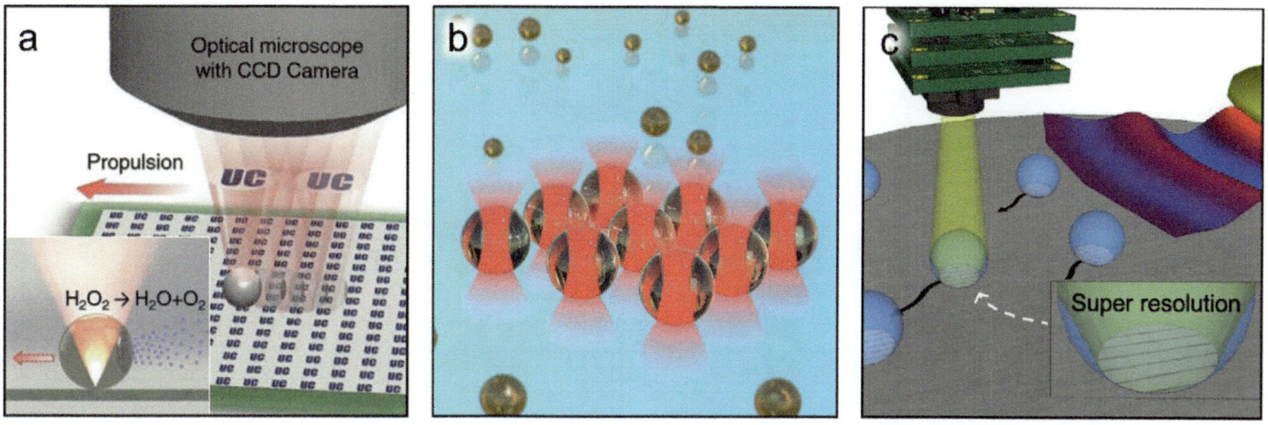

Fig. 7.1 **a** Schematic illustration of the swimming microrobot optical nanoscopy. (Image adapted with permission from Li et al. [36], Copyright (2016), with permission from American Chemical Society) **b** Optical trapping and manipulation of multiple droplet microlenses arranged in an array. (Image adapted from the Open Access paper Chen et al. [21] under Copyright (2008) Optical Society of America) **c** Flexural acoustic waves are transmitted by the piezoelectric transducer for microparticle manipulation during the scanning process. (Image adapted with permission from Jin et al. [41], Copyright (2020), with permission from American Chemical Society)

7.3 Applications of Microsphere-Assisted Technology

7.3.1 Super-Resolution Optical Imaging

The advent of the optical microscope marked a significant milestone in enabling humanity to explore the microscopic realm. Its ultrafast imaging capabilities and broad sample applicability render optical microscopy an indispensable auxiliary tool across diverse research domains, including biomedicine and materials chemistry. Nevertheless, as scientific inquiry advanced, it became apparent that the magnification and observational capacities of optical microscopy are constrained by the classical optical diffraction limit [4, 5]. Microsphere-assisted technology, when integrated with conventional optical microscopy systems, can significantly enhance imaging resolution while preserving the inherent advantages of optical imaging. The optical microscope obtains fine images of the sample by imaging the virtual or real image plane formed by the microsphere. While the elucidation of the physical mechanisms underlying super-resolution optical imaging facilitated by microspheres is still under development, researchers have empirically demonstrated the superior imaging capabilities of these microspheres [14]. By modulating the focusing characteristics of the microsphere, the researchers can make the microsphere work in contact and non-contact mode to suit different application requirements. In contact mode, the base of the microsphere is either in direct contact with the sample surface or positioned within the optical near-field range ($< \lambda$), that enabling efficient collection of evanescent wave signals from the sample surface. The integration of the microsphere with an optical imaging system can achieve an imaging resolution of 50 nm ($\lambda/8$) under white light illumination [42] (Fig. 7.2a). Furthermore, the combination of microsphere-assisted technology with a conventional scanning laser microscope has achieved an imaging resolution of 25 nm ($\lambda/17$) [43] (Fig. 7.2b).

As the study of microsphere lenses advanced, researchers identified several limitations associated with the direct placement of microspheres on sample surfaces for imaging purposes. The arbitrary positioning of microspheres often results in inaccurate imaging of the target object, and there is a potential risk of sample contamination, as the microspheres that come into direct contact with the sample during experiments are challenging to recover effectively. Consequently, various methods for controlling microspheres were developed to facilitate non-contact imaging. In this non-contact mode, precise regulation of the distance between the microsphere, the sample, and the objective lens is essential for investigating the imaging mechanism of microspheres, thereby optimizing their optical performance. In 2018, Hong's research team reported that the highest imaging resolution achieved with microsphere-assisted imaging technology in non-contact mode reached 23 nm [44]. The microsphere's super-resolution imaging capability has been utilized for imaging biological samples, facilitating the observation of subcellular structures such as adenovirus (Fig. 7.2c), cytoskeleton, chromosomes, lysosomes, and neutrophils [27, 45, 46].

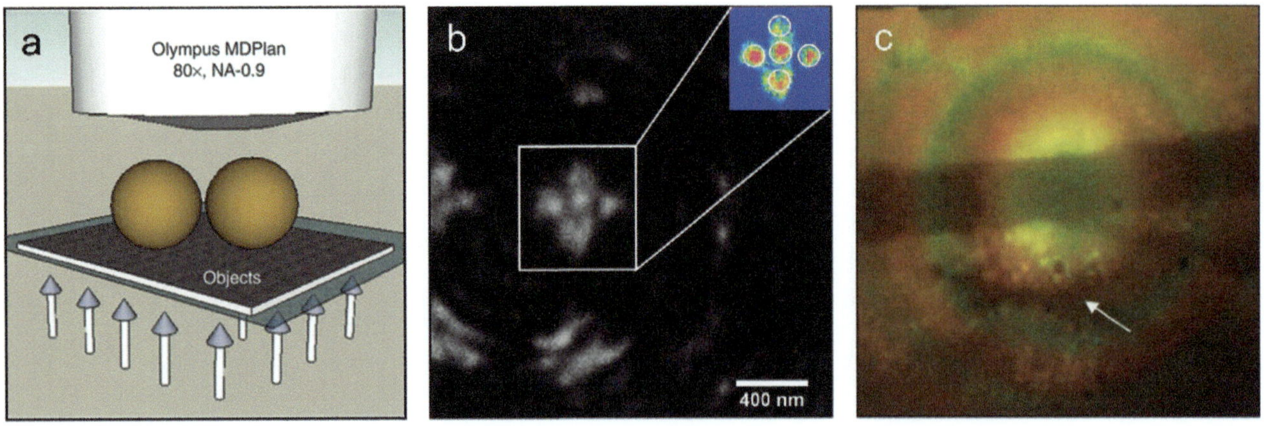

Fig. 7.2 **a** Schematic of the transmission mode microsphere integrated with a classical optical microscope. (Image adapted with permission from Wang et al. [42], Copyright (2011), with permission from Springer Nature) **b** Images of gold quintuplet nanodots (25 nm) on a glass substrate obtained using a conventional scanning laser confocal microscope with a microsphere. (Image adapted with permission from Yan et al. [43], Copyright (2014), with permission from American Chemical Society) **c** Image of adenovirus clusters captured by a conventional optical microscope with a microsphere. (Image adapted the Open Access paper Li et al. [46], under Copyright (2013) the Authors)

7.3.2 Signal Enhancement and Signal Detection

The optical properties of the microsphere, including photonic nanojets (PNJ) [16], directional antennas [47], and whispering-gallery modes [48], significantly enhance light-matter interactions and improve the detection efficiency of optical systems for weak signals. Furthermore, the microsphere's broad spectral response properties allow for the modulation of various optical signals. Coupled with its three-dimensional symmetric geometry, which influences both the excitation and detected light within the optical path, this enhancement mechanism has led to improved optical signal enhancement for Raman scattering, backscattering signal (Fig. 7.3a), and fluorescence signal (Fig. 7.3b) [24, 30, 31, 34, 49].

Simultaneously, microsphere-assisted technology has been employed in the investigation of biological samples. For instance, the optical field localization capability of microspheres has been utilized to enhance the precision and efficacy of nerve cell stimulation in optogenetic studies [50, 51]. To facilitate efficient and visualized tracking of cellular physiological states, researchers have transformed cellular physiological signals into optical signals for observation. The microsphere is internalized within the cells, enabling real-time monitoring of cellular states through the whispering gallery mode generated by the internalized microsphere. Current studies have successfully achieved cell localization during cell migration [52, 53] and real-time monitoring of the beating state of cardiomyocytes [54].

7.3.3 Optical Manipulation

Among the traditional optical manipulation methods, non-contact and non-invasive manipulation methods using focused beams have emerged as vital tools for investigating micro-nano structures [55, 56]. Nevertheless, the optical diffraction limit poses a significant challenge, as it restricts the ability of conventional optical tweezers to focus trapping light into a spot smaller than the diffraction limit. This limitation results in instability when manipulating samples smaller than the spot size within the optical trap. Consequently, traditional optical tweezers struggle to stably manipulate nanoscale samples with diameters ranging from 1 to 100 nm, such as Rayleigh nanoparticles (diameter $< \lambda$), biomolecules, and organelles. To address the limitations in optical manipulation accuracy, advanced nano-manipulation techniques, including plasmonic tweezers [57], photonic crystal optical tweezers [58], and slot waveguides [59], have been developed. However, these methods necessitate specialized structural designs and intricate fabrication processes, making it challenging to achieve precise control of nanoparticles in three-dimensional directions.

The identification of nanofocusing properties in microspheres has facilitated the optical manipulation of nanoscale samples by enabling the focusing of beams beyond the subdiffraction limit in free space. Li's research team proposed integrating a microsphere with an optical fiber probe to develop a high-precision and highly flexible optical manipulation tool. In 2016, the team demonstrated the ability to flexibly control the position of the microsphere using the fiber probe,

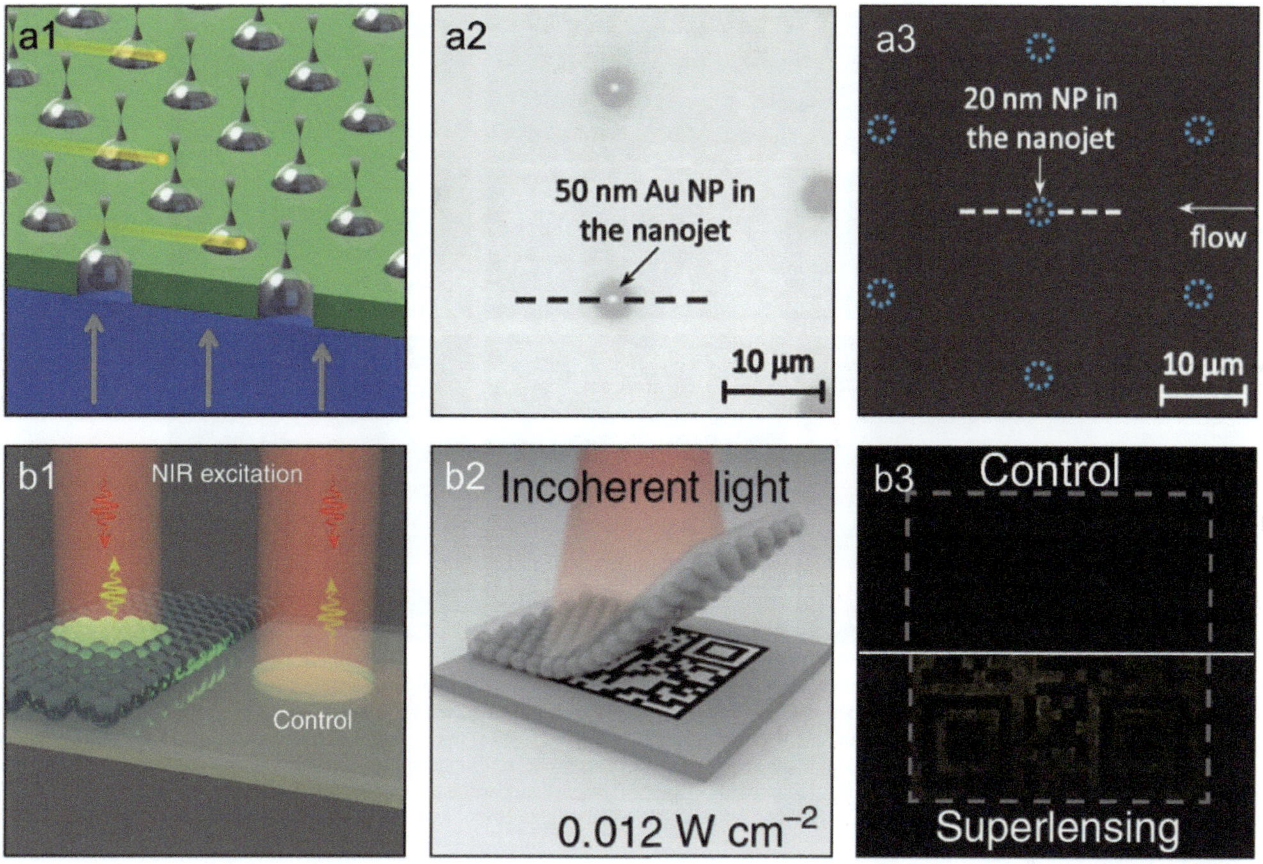

Fig. 7.3 **a** Dielectric microspheres are patterned in a microfabricated well template on a glass substrate, forming a microlens array. An optical microscope with a low-NA objective is placed beneath them and is thus used for the optical detection of nanoparticles. (Image adapted with permission from Yang et al. [31], Copyright (2015), with permission from American Chemical Society) **a1** Schematic illustration of the optical detection of nanoparticles. **a2** Detection of Au nanoparticles during their motion through a nanojet. **a3** Detection of fluorescent nanoparticles during their motion through a nanojet. **b** Upconversion luminescence amplification through dielectric superlensing. **b1** Schematic illustration of the microsphere array enhancement upconversion emission intensity. **b2** Schematic illustration of the experimental design for upconversion-based encrypted barcoding. **b3** Photographic images, taken under incoherent light illumination, of an encrypted quick-response code with and without the microsphere array. (Image adapted the Open Access paper Liang et al. [49], under Copyright (2019) the Authors)

while employing the microsphere to focus the light output from the probe, thereby creating a highly localized photonic nanojet (PNJ) for the stable capture of nanoparticles and DNA molecules measuring 85 nm in size [34] (Fig. 7.4a). In the same year, the team advanced the technique by fabricating a microsphere array on the flat end face of the fiber, enhancing the application efficiency of the microspheres. This configuration allows for the direct formation of a large-area PNJ array when the captured light traverses the fiber, thereby enabling the efficient capture of nanoparticles and cells [60] (Fig. 7.4b). Subsequently, the biocompatibility of the microspheres was optimized, enabling subwavelength beam focusing through the utilization of spherical biological cells as microsphere lenses [28]. The study further demonstrated that the focusing of the capture light in free space by biological microlenses can enhance the localization accuracy of captured nanoparticles to below 100 nm [26].

7.3.4 Laser Nano-fabrication

In the laser cleaning experiment, researchers observed that when the laser irradiated nearly spherical micro-particles, it resulted in the formation of holes on the substrate with pore sizes smaller than the optical diffraction limit [18, 61, 62]. This phenomenon presents a new challenge to be addressed in laser cleaning experiments; however, it constitutes a significant discovery in the field of laser nano-

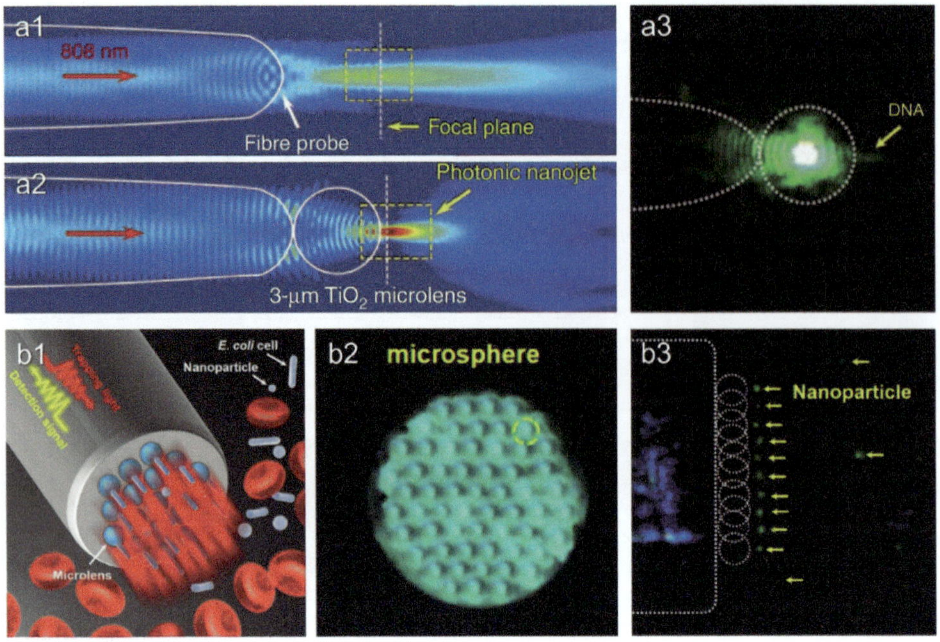

Fig. 7.4 **a1** The electric field intensity distributions of the fiber probes without microsphere. **a2** The electric field intensity distributions of fiber probes with a 3-μm TiO_2 microsphere. **a3** The dark-field light-scattering image of the DNA molecules trapped by a dielectric microsphere bound with an optical fiber probe. (Image adapted the Open Access paper Li et al. [34], under Copyright (2016) the Authors) **b1** Schematic illustration of selective trapping and detection of multiple nanoparticles and subwavelength cells by a parallel photonic nanojet array. **b2** Optical microscope images of microsphere array on a compact fiber probe. **b3** Optical microscope images showing trapped 190 nm fluorescent nanoparticles. (Image adapted with permission from Li et al. [60], Copyright (2016), with permission from American Chemical Society)

fabrication. Consequently, the nano-focusing properties of the microsphere were rapidly integrated into the optical systems used in laser nano-fabrication, enhancing the precision and efficiency of nano-processing through concentrated etching lasers. In the initial stages of the research, the microsphere was also directly placed on the substrate surface to be processed, with both direct and off-axis irradiation techniques employed to control the lithography pattern [63] (Fig. 7.5a).

To enhance the precision of lithography, researchers have introduced techniques such as mechanical cantilevering and optical manipulation to achieve precise control over the distribution of microsphere positions [37–39, 64]. These methods complement the optimization of the photonic nanojet (PNJ) characteristics of the microsphere lens, ensuring its operation at an optimal position (Fig. 7.5b). Furthermore, the precise positioning of the microsphere lens lays the technical foundation for creating complex lithographic patterns (Fig. 7.5c). Additionally, researchers have discovered that red blood cells in biomaterials can serve as photolithography masks, facilitating the control of laser-induced refractive index changes in lithium niobate crystals [62]. This finding suggests novel applications for biomaterials in the field of optics.

7.4 Summary and Outlook

The microsphere has emerged as a significant focus in research due to its ability to surpass the optical diffraction limit. Advances in microsphere processing and manipulation technologies have facilitated its versatile application in fields such as biomedicine and materials chemistry. In the realm of super-resolution optical imaging, traditional optical systems enhanced with microspheres can effectively overcome the constraints of the optical diffraction limit, achieving super-resolution imaging capabilities. Current research utilizing microsphere-assisted technology has achieved an imaging resolution of approximately 23 nm, significantly broadening the scope of optical imaging research. Additionally, microsphere-assisted technology has developed two distinct imaging modes contact and non-contact to accommodate various imaging requirements. The optical properties of microspheres, such as photonic nanojets, directional antennas, and whispering-gallery modes, significantly enhance light-matter interactions. These properties have garnered considerable interest in the field of signal detection and facilitate the probing of nonlinear optical properties of materials. Additionally, they contribute to the enhanced

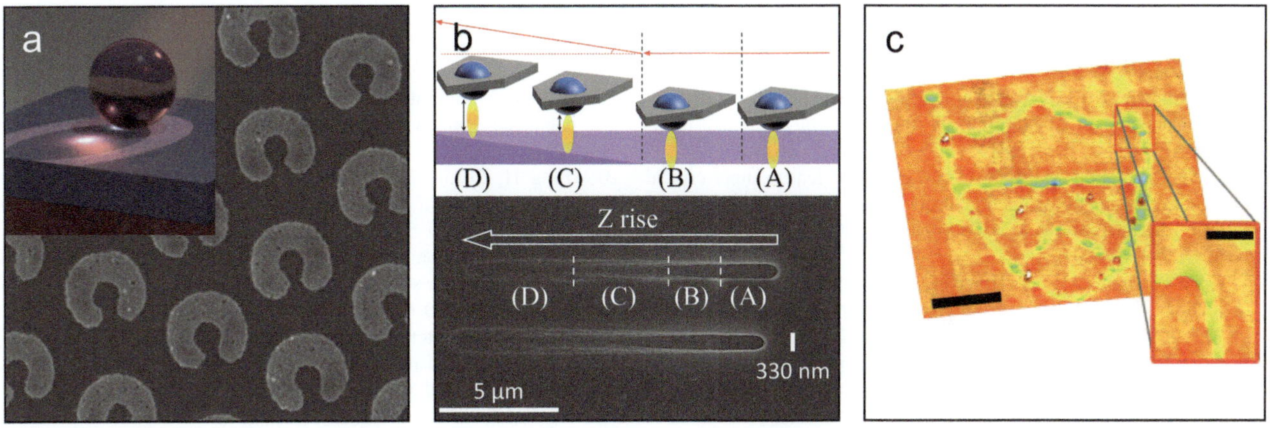

Fig. 7.5 **a** Schematic illustrations of the fabrication procedures using angled nanospherical-lens lithography. (Image adapted with permission from Chien et al. [63], Copyright (2017), with permission from American Chemical Society) **b** The top panel is a sketch of the cantilever during an out-of-plane scanning. The bottom panel represents the SEM image of two lines from an out-of-plane scan. (Image adapted the Open Access paper Jacassi et al. [64], under Copyright (2017) the Authors) **c** The Princeton University logo is nanopatterned on polyimide film using microsphere-assisted technology. (Image adapted with permission from McLeod et al. [38], Copyright (2008), with permission from Springer Nature)

detection of weak signals. The nano-focusing capabilities of microspheres offer a simple yet high-performance solution for fields requiring high optical precision. Microspheres have been successfully integrated into optical manipulation technologies, enabling the three-dimensional stable manipulation of single molecules. Furthermore, in optical etching applications, microspheres effectively reduce the focusing spot of processing lasers, achieving a lithography accuracy of nearly 50 nm.

Following extensive research and exploration over several decades, microlens-assisted technology has achieved significant advancements in both fundamental research and the practical application of scientific findings. In the domain of nanofabrication, the integration of microspheres with existing optical systems has facilitated the development of microsphere-based laser processing instruments for pre-processing and microsphere-based nano-optical microscopes for post-inspection. These innovations have enabled the advancement of high-performance optical instrumentation for the lithographic processing of nanostructures and high-resolution real-time imaging inspection, all at a reduced cost. Consequently, commercial products have been developed, and efforts are ongoing to enhance their operational performance.

Simultaneously, microlens-assisted technology has been integrated with research focused on the design of micro-optical devices. Owing to the diminutive size of microspheres and their facile integration with micro-optical components, microspheres are employed in the construction of high-performance, miniaturized optical devices. This is particularly advantageous in biological research, as the miniaturized optical devices constructed from microspheres can be easily introduced or implanted into the intricate environments within organisms. These devices are instrumental in advancing biophotonics technologies and devices aimed at real-time monitoring and precise therapeutic interventions.

Acknowledgements This work was supported by National Natural Science Foundation of China (Nos. 62135005 and 12304322). X.C. and H.L. contributed equally to this work.

References

1. Plantzos D. Crystals and lenses in the Graeco-Roman world. Am J Archaeol. 1997;101(3):451–64. https://doi.org/10.2307/507106.
2. Uluç K, Kujoth GC, Başkaya MK. Operating microscopes: past, present, and future. Neurosurg Focus. 2009;27(3):E4. https://doi.org/10.3171/2009.6.FOCUS09120.
3. Kingslake R. Optical system design. New York: Academic Press; (2021). https://doi.org/10.1016/B978-0-12-408660-9.
4. Hecht J. A short history of laser development. Appl Opt. 2010;49(9):99–122. https://doi.org/10.1364/AO.49.000F99.
5. Yen A. Rayleigh or Abbe? Origin and naming of the resolution formula of microlithography. J Micro/Nanolithogr MEMS, MOEMS. 2020;19(4):040501. https://doi.org/10.1117/1.jmm.19.4.040501.
6. Rayleigh L. Investigations in optics, with special reference to the spectroscope. London Edinb Dublin Philos Mag J Sci. 1879;8(49):261–74. https://doi.org/10.1080/14786447908639684.
7. Huang B, Babcock H, Zhuang X. Breaking the diffraction barrier: super-resolution imaging of cells. Cell. 2010;143(7):1047–58. https://doi.org/10.1016/j.cell.2010.12.002.
8. Zhang X, Liu Z. Superlenses to overcome the diffraction limit. Nat Mater. 2008;7:435–41. https://doi.org/10.1038/nmat2141.
9. Zheludev NI. What diffraction limit? Nat Mater. 2008;7:420–2. https://doi.org/10.1038/nnano.2009.115.

10. Gramotnev DK, Bozhevolnyi SI. Plasmonics beyond the diffraction limit. Nat Photonics. 2010;4:83–91. https://doi.org/10.1038/nphoton.2009.282.
11. Miroshnichenko A. Deep learning beats the optical diffraction limit. Nat Nanotechnol. 2019;14:198–9. https://doi.org/10.1038/s41565-018-0357-y.
12. Qin F, Li X, Hong M. From super-osciallatory lens to super-critical lens: surpassing the diffraction limit via light field modulation. Opto Elec Eng. 2017;44(8):757–71. https://doi.org/10.3969/j.issn.1003-501X.2017.08.001.
13. Chen L, Zhou Y, Li Y, Hong M. Microsphere enhanced optical imaging and patterning: From physics to applications. Appl Phys Rev. 2019;6:021304. https://doi.org/10.1063/1.5082215.
14. Chen L, Zhou Y, Zhou R, Hong M. Microsphere-toward future of optical microscopes. iScience. 2020;23(6):101211. https://doi.org/10.1016/j.isci.2020.101211.
15. Miccio L, Memmolo P, Merola F, Mugnano M, Ferraro P. Optobiology: live cells in optics and photonics. J Phys Photonics. 2020;3:012003. https://doi.org/10.1088/2515-7647/abac19.
16. Li H, Song W, Zhao Y, Cao Q, Wen A. Optical trapping, sensing, and imaging by photonic nanojets. Photonics. 2021;8(10):434. https://doi.org/10.3390/photonics8100434.
17. Surdo S, Duocastella M, Diaspro A. Nanopatterning with photonic nanojets: review and perspectives in biomedical research. Micromachines. 2021;12(3):256. https://doi.org/10.3390/mi12030256.
18. Mosbacher M, Münzer H-J, Zimmermann J, Solis J, Boneberg J, Leiderer P. Optical field enhancement effects in laser-assisted particle removal. Appl Phys A. 2001;72:41–4. https://doi.org/10.1007/s003390000715.
19. Luk'yanchuk BS, Domínguez RP, Minin I, Minin O, Wang Z. Refractive index less than two: photonic nanojets yesterday, today and tomorrow. Opt Mater Express. 2017;7(6):1820–47. https://doi.org/10.1364/ome.7.001820
20. Jia B, Wang F, Chan H, Zhang G, Li WJ. In situ printing of liquid superlenses for subdiffraction-limited color imaging of nanobiostructures in nature. Microsyst Nanoeng. 2019;5:1. https://doi.org/10.1038/s41378-018-0040-3.
21. Chen X, Wu T, Gong Z, Li Y, Zhang Y, Li B. Subwavelength imaging and detection using adjustable and movable droplet microlenses. Photonics Res. 2020;8(3):225–34. https://doi.org/10.1364/prj.377795.
22. Xie Y, Cai D, Pan J, Zhou N, Guo X, Wang P, Tong L. Chalcogenide microsphere-assisted optical super-resolution imaging. Adv Opt Mater. 2022;10(6):2102269. https://doi.org/10.1002/adom.202102269.
23. Sun Y-L, Dong WF, Yang R-Z, Zhang L, Chen Q-D, Sun H-B. Dynamically tunable protein microlenses. Angew Chem Int Ed. 2012;51:1558–62. https://doi.org/10.1002/anie.201105925.
24. Miccio L, Memmolo P, Merola F, Netti PA, Ferraro P. Red blood cell as an adaptive optofluidic microlens. Nat Commun. 2015;6:6502. https://doi.org/10.1038/ncomms7502.
25. Monks JN, Yan B, Hawkins N, Vollrath F, Wang Z. Spider silk: mother nature's bio-superlens. Nano Lett. 2016;16(9):5842–5. https://doi.org/10.1021/acs.nanolett.6b02641.
26. Li Y, Liu X, Li B. Single-cell biomagnifier for optical nanoscopes and nanotweezers. Light Sci Appl 2019;8:61. https://doi.org/10.1038/s41377-019-0168-4.
27. Chen X, Wu T, Gong Z, Guo J, Liu X, Zhang Y, Li Y, Ferraro P, Li B. Lipid droplets as endogenous intracellular microlenses. Light Sci Appl 2021;10:242. https://doi.org/10.1038/s41377-021-00687-3.
28. Chen X, Li H, Wu T, Gong Z, Guo J, Li Y, Li B, Ferraro P, Zhang Y. Optical-force-controlled red-blood-cell microlenses for subwavelength trapping and imaging. Biomed Opt Express. 2022;13(5):2995–3004. https://doi.org/10.1364/boe.457700.
29. Darafsheh A, Guardiola C, Palovcak A, Finlay JC, Cárabe A. Optical super-resolution imaging by high-index microspheres embedded in elastomers. Opt Lett. 2015;40(1):5–8. https://doi.org/10.1364/ol.40.000005.
30. Xing C, Yan Y, Feng C, Xu J, Dong P, Guan W, Zeng Y, Zhao Y, Jiang Y. Flexible microsphere-embedded film for microsphere-enhanced Raman spectroscopy. ACS Appl Mater Interfaces. 2017;9(38):32896–906. https://doi.org/10.1021/acsami.7b09884.
31. Yang H, Cornaglia M, Gijs MAM. Photonic nanojet array for fast detection of single nanoparticles in a flow. Nano Lett. 2015;15(3):1730–5. https://doi.org/10.1021/nl5044067.
32. Krivitsky LA, Wang JJ, Wang Z, Luk'yanchuk B. Locomotion of microspheres for super-resolution imaging. Sci Rep. 2013;3:3501. https://doi.org/10.1038/srep03501.
33. Mora CAG, Hartelt M, Bayer D, Aeschlimann M, Ilin EA, Oesterschulze E. Microsphere-based cantilevers for polarization-resolved and femtosecond SNOM. Appl Phys B. 2016;122:86. https://doi.org/10.1007/s00340-016-6359-2.
34. Li Y, Xin H, Lei H, Liu L, Li Y, Zhang Y, Li B. Manipulation and detection of single nanoparticles and biomolecules by a photonic nanojet. Light Sci Appl 2016;5:e16176. https://doi.org/10.1038/lsa.2016.176.
35. Zhang T, Li P, Yu H, Wang F, Wang X, Yang T, Yang W, Li WJ, Wang Y, Liu L. Fabrication of flexible microlens arrays for parallel super-resolution imaging. Appl Surf Sci. 2020;504:144375. https://doi.org/10.1016/j.apsusc.2019.144375.
36. Li J, Liu W, Li T, Rozen I, Zhao J, Bahari B, Kante B, Wang J. Swimming microrobot optical nanoscopy. Nano Lett. 2016;16(10):6604–9. https://doi.org/10.1021/acs.nanolett.6b03303.
37. Li J, Gao W, Dong R, Pei A, Sattayasamitsathit S, Wang J. Nanomotor lithography. Nat Commun. 2014;5:5026. https://doi.org/10.1038/ncomms6026.
38. Mcleod E, Arnold CB. Subwavelength direct-write nanopatterning using optically trapped microspheres. Nat Nanotechnol. 2008;3:413–7. https://doi.org/10.1038/nnano.2008.150.
39. McLeod E, Arnold CB. Array-based optical nanolithography using optically trapped microlenses. Opt Express. 2009;17(5):3640–50. https://doi.org/10.1364/oe.17.003640.
40. Bezryadina A, Li J, Zhao J, Kothambawala A, Ponsetto J, Huang E, Wang J. Zhaowei Liu, Localized plasmonic structured illumination microscopy with an optically trapped microlens. Nanoscale. 2017;9:14907–12. https://doi.org/10.1039/c7nr03654j.
41. Jin G, Bachman H, Naquin TD, Rufo J, Hou S, Tian Z, Zhao C, Huang TJ. Acoustofluidic scanning nanoscope with high resolution and large field of view. ACS Nano. 2020;14(7):8624–33. https://doi.org/10.1021/acsnano.0c03009.
42. Wang Z, Guo W, Li L, Luk'yanchuk B, Khan A, Liu Z, Chen Z, Hong M. Optical virtual imaging at 50 nm lateral resolution with a white-light nanoscope. Nat Commun. 2011;2:218. https://doi.org/10.1038/ncomms1211.
43. Yan Y, Li L, Feng C, Guo W, Lee S, Hong M. Microsphere-coupled scanning laser confocal nanoscope for sub-diffraction-limited imaging at 25 nm lateral resolution in the visible spectrum. ACS Nano. 2014;8(2):1809–16. https://doi.org/10.1021/nn406201q.
44. Chen L, Zhou Y, Wu M, Hong M. Remote-mode microsphere nano-imaging: new boundaries for optical microscopes. Opto Electron Adv 2018;1(1):170001. https://doi.org/10.29026/oea.2018.170001.
45. Yang H, Moullan N, Auwerx J, Gijs MAM. Super-resolution biological microscopy using virtual imaging by a microsphere nanoscope. Small. 2014;10(9):1712–8. https://doi.org/10.1002/smll.201302942.
46. Li L, Guo W, Yan Y, Lee S, Wang T. Label-free super-resolution imaging of adenoviruses by submerged microsphere optical nanoscopy. Light Sci Appl 2013;2:e104. https://doi.org/10.1038/lsa.2013.60.
47. Wenger J, Gérard D, Aouani H, Rigneault H. Disposable microscope objective lenses for fluorescence correlation spectroscopy

using latex microspheres. Anal Chem. 2008;80(17):6800–4. https://doi.org/10.1021/ac801016z.

48. Zhi Y, Yu X, Gong Q, Yang L, Xiao Y. Single nanoparticle detection using optical microcavities. Adv Mater. 2017;29(12):1604920. https://doi.org/10.1002/adma.201604920.
49. Liang L, Teh DB, Dinh N-D, Chen W, Chen Q, Wu Y, Chowdhury S, Yamanaka A, Sum TC, Chen C-H, Thakor NV, All AH, Liu X. Upconversion amplification through dielectric superlensing modulation. Nat Commun. 2019;10:1391. https://doi.org/10.1038/s41467-019-09345-0.
50. Guo J, Wu Y, Gong Z, Chen X, Cao F, Kala S, Qiu Z, Zhao X, Chen J, He D, Chen T, Zeng R, Zhu J, Wong KF, Murugappan S, Zhu T, Xian Q, Hou X, Ruan YC, Li B, Li Y, Zhang Y, Sun L. Photonic nanojet-mediated optogenetics. Adv Sci. 2022;9(12):2104140. https://doi.org/10.1002/advs.202104140.
51. Li H, Chen X, Wu T, Gong Z, Guo J, Bai X, Li J, Zhang Y, Li Y, Li B. Stimulation and imaging of neural cells via photonic nanojets. Photonics Res. 2024;12(8):1604–9. https://doi.org/10.1364/PRJ.523587.
52. Humar M, Yun SH. Intracellular microlasers. Nat Photonics. 2015;9:572–6. https://doi.org/10.1038/nphoton.2015.129.
53. Schubert M, Steude A, Liehm P, Kronenberg NM, Karl M, Campbell EC, Powis SJ, Gather MC. Lasing within live cells containing intracellular optical microresonators for barcode-type cell tagging and tracking. Nano Lett. 2015;15(8):5647–52. https://doi.org/10.1021/acs.nanolett.5b02491.
54. Schubert M, Woolfson L, Barnard IRM, Dorward AM, Casement B, Morton A, Robertson GB, Appleton PL, Miles GB, Tucker CS, Pitt SJ, Gather MC. Monitoring contractility in cardiac tissue with cellular resolution using biointegrated microlasers. Nat Photonics. 2020;14:452–8. https://doi.org/10.1038/s41566-020-0631-z.
55. Padgett MJ, Molloy JA, McGloin D. Optical tweezers: methods and applications. New York: CRC Press; 2010. https://doi.org/10.1080/00107514.2011.564309
56. Grier DG. A revolution in optical manipulation. Nature. 2003;424:810–6. https://doi.org/10.1038/nature01935.
57. Zhang Y, Min C, Dou X, Wang X, Urbach HP, Somekh MG, Yuan X. Plasmonic tweezers: for nanoscale optical trapping and beyond. Light Sci Appl 2021;10:59. https://doi.org/10.1038/s41377-021-00474-0.
58. Mandal S, Serey X, Erickson D. Nanomanipulation using silicon photonic crystal resonators. Nano Lett. 2010;10(1):99–104. https://doi.org/10.1021/nl9029225.
59. Yang AHJ, Moore SD, Schmidt BS, Klug M, Lipson M, Erickson D. Optical manipulation of nanoparticles and biomolecules in sub-wavelength slot waveguides. Nature. 2009;457:71–5. https://doi.org/10.1038/nature07593.
60. Li Y, Xin H, Liu X, Zhang Y, Lei H, Li B. Trapping and detection of nanoparticles and cells using a parallel photonic nanojet array. ACS Nano. 2016;10(6):5800–8. https://doi.org/10.1021/acsnano.5b08081.
61. Münzer HJ, Mosbacher M, Bertsch M, Zimmermann J, Leiderer P, Boneberg J. Local field enhancement effects for nanostructuring of surface. J Microsc. 2001;202(1):129–35. https://doi.org/10.1046/j.1365-2818.2001.00876.x.
62. Miccio L, Behal J, Mugnano M, Memmolo P, Mandracchia B, Merola F, Grilli S, Ferraro P. Biological lenses as a photomask for writing laser spots into ferroelectric crystals. ACS Appl Bio Mater. 2019;2(11):4675–80. https://doi.org/10.1021/acsabm.9b00569.
63. Chien Y-H, Wang C-H, Liu C-C, Chang S-H, Kong KV, Chang Y-C. Large-scale nanofabrication of designed nanostructures using angled nanospherical-lens lithography for surface enhanced infrared absorption spectroscopy. ACS Appl Mater Interfaces. 2017;9(29):24917–25. https://doi.org/10.1021/acsami.7b08994.
64. Jacassi A, Tantussi F, Dipalo M, Biagini C, Maccaferri N, Bozzola A, Angelis FD. Scanning probe photonic nanojet lithography. ACS Appl Mater Interfaces. 2017;9(37):32386–93. https://doi.org/10.1021/acsami.7b10145.

8 Lanthanide Ion-Doped Nanoparticles-Based Optical Trapping and Sensing

Fan Wang, Xuchen Shan, Tiange Zhang, and Dajing Wang

Abstract

Accurate force measurement is of great importance in both biological detection and physical processes. Optical tweezers have emerged as a vital tool for non-contact force detection, with nanometer-sized trapping probes playing a critical role in achieving high spatial resolution and force sensitivity. However, smaller nanoparticles with low refractive indices often result in reduced trapping forces, making optical trapping of nanoparticles for sensing a significant challenge. Recently, upconversion nanoparticles (UCNPs) have emerged as a promising new class of nanoprobes. These particles exhibit much higher trap stiffness compared to traditional gold nanoparticles of the same size. Furthermore, UCNPs offer stable and bright fluorescence, enabling fluorescence-based video trapping and precise force sensing. Leveraging their unique properties, such as surface modification and lanthanide ion emission, trapped UCNPs can also function as sensors for temperature and pH, broadening their application potential. These distinctive advantages make UCNPs an exceptional multifunctional probe for optical tweezers. This chapter provides an in-depth discussion on nanoparticle trapping, the mechanisms and benefits of UCNPs, and their applications in UCNP-based optical tweezers, which are poised to create novel opportunities in scientific research.

Keywords

Optical tweezers · Upconversion nanoparticles · Machine learning · Photonic force microscopy · Single nanoparticle tracking

F. Wang (✉) · X. Shan (✉) · T. Zhang · D. Wang
School of Physics, Beihang University, Beijing, China
e-mail: fanwang@buaa.edu.cn

X. Shan
e-mail: shanxuchen@buaa.edu.cn

8.1 Introduction

8.1.1 Nanoscale Trapping by Optical Tweezers

In 1970, Arthur Ashkin discovered that continuous lasers exert radiation pressure on micron-sized latex spheres [1]. By 1986, his team demonstrated that a single laser beam could generate a gradient force capable of stably trapping individual micron-sized particles, leading to the invention of the first single-beam optical tweezers [2]. The discovery that lasers can exert forces on particles has profoundly impacted two major areas of scientific research. On the one hand, it paved the way for the laser cooling of atoms, driving advancements in ultracold atom technology. On the other hand, optical tweezers have become an indispensable non-contact force-sensing tool, enabling groundbreaking studies in both physical processes and biological applications.

With the development of materials science and biological technology, the size requirement of probes has gradually shifted from the micron size to the nanometer level. However, for nanoparticles within 100 nm, the trapping force is usually insufficient to stably manipulate these nanoparticles. These demands have given rise to many new technologies for enhancing the optical trapping force of nanoparticle probes. For example, upconversion nanoparticles enhance the optical force due to the ion resonance effect formed by the high lanthanide ions doping nanoparticles [3]. These technologies for enhancing optical force have significantly contributed to fields such as physics, materials, and biological applications.

The probe can be modelled as a dipole for nanometer-sized optical tweezer probes, which belong to the Rayleigh regime. Therefore, the time-averaged force acting on the nano-scale probe can be expressed as:

B. Li et al. (eds.), *Biophotonic Manipulation*, Advances in Optics and Optoelectronics,
https://doi.org/10.1007/978-981-96-4982-2_8

$$\langle \boldsymbol{F} \rangle = \frac{1}{4} Re(\alpha_p) \nabla |\boldsymbol{E}|^2 + \frac{\sigma_{Ext}}{2c} Re(\boldsymbol{E} \times \boldsymbol{H}^*) + \frac{\sigma_{Ext} c \varepsilon_0}{4\omega i} \nabla \times E \times \boldsymbol{E}^* \quad (8.1)$$

here α_p is the polarisability of the particle. $\boldsymbol{E}$ is the electronic field, σ_{Ext} is the extinction cross-section, c is the light speed in vacuum, $\boldsymbol{H}$ is the magnetic field, ω is the angular frequency, Re stands for the real part, ε_0 is the vacuum dielectric permittivity [4]. The first term is the gradient force, the second term is the radiation pressure, and the third term is the polarization gradient.

8.1.2 Nanoscale Trapping Probe

In the field of nanotechnology, optical tweezers have demonstrated the ability to trap a wide variety of particles [5]. In 1986, the original paper on optical tweezers first showed the tapping of the dielectric microspheres, laying a solid foundation for trapping microscopic particles. Further, Ashkin and Dziedzic achieved the trapping of nanorod-like biomaterials with a diameter of tobacco mosaic virus, opening up the application of optical tweezers in the field of bioscience.

In the early work, dielectric particles were considered the best candidates for optical trapping due to their lower scattering interface. Svoboda and Block demonstrated that in the nanoscale range, although metal particles and dielectric particles have similar scattering characteristics, metal particles, with much higher polarizability, exhibit stronger stability and controllability during optical trapping, providing a new perspective on the diversity of material selection for optical trapping technology [6].

In the following decades, a large number of nanoparticles made from different materials emerged and became the subject of optical trapping research, including common materials such as polymers, glass, semiconductors, and graphene, as well as new generation low dimensions nanomaterials such as nanodiamond, quantum dots, and upconversion materials [7]. The trapping of these nanoparticles has been successfully verified in liquid media, vacuum, and air media. The trapping process of nanoparticles has also promoted the formation of particle clusters, enriching the research system and application scenarios of nanotechnology.

With the continuous development of technology, nanoscale optical trapping technology has developed rapidly. In addition to tobacco mosaic virus, it can also trap a variety of biological nanoparticles, including viruses, biological protein aggregates, etc., and through optical tweezers (Fig. 8.1), it has realized the research on the mechanism of virus infecting host cells, the dynamic process of virus nucleic acid injection into cells, the precise tracking of biomolecular signal transduction, and targeted drug delivery, opening up new avenues for disease diagnosis and treatment [8].

8.1.3 Challenge for Nano-OT

Since Ashkin developed optical tweezers in 1986, various advancements have significantly expanded their capabilities. Digital holographic optical tweezers [9] enable precise control of particle positions in three dimensions. Acoustic-optic deflector-based optical tweezers [10] allow for rapid manipulation of trapped particles at rates up to the megahertz range. Near-field optical trapping techniques, including plasmonic nano-tweezers [5, 11, 12], have enhanced precision, enabling the trapping of nanoparticles such as intracellular bioparticles, nanowires, and quantum dots. These innovations in nanoscale optical tweezers have opened new avenues for analyzing biomolecular interactions and advanced imaging. Hybrid approaches, such as combining optical and thermal forces [13], further improve trapping efficiency. Additionally, nanoscale lithography has been employed to optimize particle fabrication [14], achieving stronger trapping forces, better orientation control, and improved particle uniformity.

However, several challenges remain. Nanostructures small polarizability [15, 16] reduces trap stiffness, and biological nanoparticles with lower refractive indices (≈ 1.4–1.5) require high near-infrared laser power (~ 100 mW), causing potential overheating and reduced trapping efficiency. Metallic nanoparticles (~ 50 nm) perform better, but their utility in biological contexts is limited by heat sensitivity. Near-field optical trapping techniques [17, 18] show promise for low-index materials but often restrict manipulation to 2D and fixed positions, losing 3D capabilities. Additionally, low polarizability leads to weak scattering

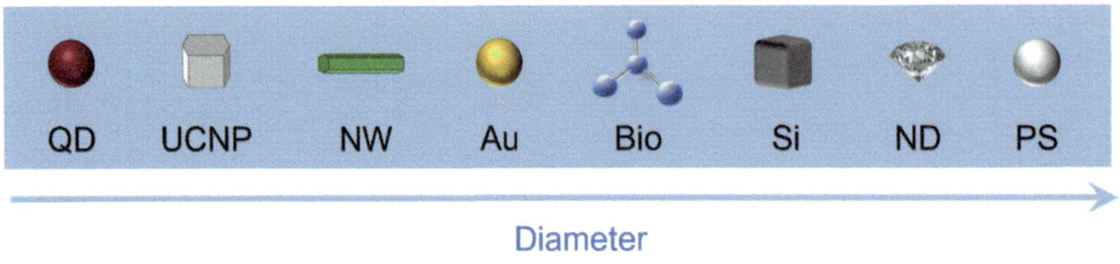

Fig. 8.1 Optically trappable particles, including Quantum dots (QD), Upconversion nanoparticles (UCNP), Nanowires (NW), Gold nanoparticles (Au), Biological nanoparticles (Bio), Silicon nanoparticles (Si), Nanodiamond (ND), Polystyrene sphere (PS)

cross-sections, complicating real-time position tracking, trap stiffness measurements, and single-particle detection.

Intracellular trapping poses further challenges [19]. Nanostructures encapsulated via endocytosis may aggregate, altering their properties and functionality. The higher refractive index and viscosity of cytoplasm decrease trapping efficiency, and the high laser power needed risks damaging organelles. Orientation control of nanoscale particles is limited; multi-spot holographic optical tweezers work for micrometer-scale Reconfigurable multi-component micromachines driven by optoelectronic tweezers nanowires but rely on polarization effects for nanoscale asymmetric particles, restricting control to 2D. The diffraction limit also creates trapping spots about ten times larger than nanoparticles, reducing efficiency and confinement. With the design and modulation of structured illumination, multi-component micromachines can be driven by optical tweezers [20].

Trapping ability is another key factor when applying this technology to more complicated scenarios, such as the delivery of specimens. Traditional plasmonic one can translate nanoscale objects, but the throughput is not high enough to conduct the analysis systematically. Mie resonances can be integrated into all-dielectric nano-tweezers, which feature low photo heating and stable trapping performance to address these challenges above. Researchers making use of this effect successfully achieve stable trapping of 20 nm dielectric objects in silicon-based nano-antennas [21]. Another way to bypass this obstacle is by using the thermophoresis effects and electrophoresis, generated by optical and electric fields on metallic substrates. This can also minimize the photodamage to the species [22, 23].

In perspective, there are many directions promising to advance nano-optical tweezers. The higher refractive index mismatch brings geometric resonances, which facilitate the trapping efficiency. The candidates to be trapped with higher polarizability are efficient in improving the trapping performance with lower laser power, leading to lower photoheating. Much research focused on the polarizability property finds that zeta potential, super-radiance, ion resonance, and other effects can engineer the gradient force of nano-OT. Besides, the design of particles with certain shapes that are easier to trap is an efficient way to generate stronger optical forces. Another famous technology, beam shaping, is introduced to generate the optimal optical beam to enhance the trap stiffness in three dimensions.

8.1.4 History of UCNPs

Impurity doping represents a technique capable of altering material properties. The incorporation of rare-earth ions into nanomaterials enables the modulation of the crystallographic phases, morphologies, sizes, and electronic configurations within inorganic nanomaterials. Moreover, the doping of rare-earth ions endows materials with abundant optical, electrical, magnetic, and catalytic properties, rendering the doped nanomaterials highly appealing for a multitude of technological applications. The rare-earths constitute a group of 17 elements, encompassing scandium, yttrium, and 15 lanthanides (extending from lanthanum to lutetium) [24]. These elements share numerous similar physical and chemical characteristics and are frequently found together in geological deposits. The history of rare-earth elements can be dated back to 1751, when the first rare-earth mineral was unearthed in Sweden [25]. However, this mineral was initially misidentified as calcium iron silicate. In 1787, Carl Arrhenius discovered another analogous mineral in Ytterby, Sweden. In 1794, a Finnish chemist named Johan Gadolin was the first to isolate a novel oxide from the Ytterby ore and named it ytterbia. Subsequently, in 1842, Mosander further divided the ytterbia oxide into yttria, erbia, and terbia. Nevertheless, later investigations disclosed that these three oxides were also mixtures of rare-earth elements. Owing to the similarity in the chemical properties of rare-earth elements, the isolation of all 17 rare-earth was not achieved until 1908–1909, when a repetitive fractional crystallization technique was employed. In 1914, Henry Moseley utilized X-ray spectroscopy to confirm all rare-earth elements, with the exception of the radioactive element promethium (Pm). Promethium was not characterized until 1945 at Clinton Laboratories, which is now Oak Ridge National Laboratory.

Since the 1990s, with the rapid development of nanoscience and nanotechnology, the potential of rare-earth-doped nanomaterials for a diverse array of applications has been progressively acknowledged. The early efforts were predominantly centred around rare-earth-doped luminescent nanoparticles, spurred by the invention of colloidal quantum dots and other inorganic nanophosphors [26]. The primary impetus was to diminish the size of conventional rare-earth-doped luminophores, rendering them dispersible in a colloidal suspension for novel applications like biolabeling and biosensing. One of the initial endeavors was documented in 1990 by Beverloo and colleagues, who employed 0.1–0.3 μm Y_2O_2S:Eu microparticles as luminescent biolabels in immune assays [27]. The merits of using rare-earth-doped luminescent materials encompass high photochemical stability, a narrow emission bandwidth, a long luminescence lifetime, and the remarkable ability to attain UV/visible emission under near-infrared (NIR) excitation through the distinctive process of upconversion [28]. Nevertheless, in the beginning, these nanophosphors could only be fabricated by pulverizing bulk phosphors via ball milling. This constrained the minimum size of the phosphors to several hundred nanometers, thereby impeding the broader application of these materials. Not until the dawn of the twenty-first century did novel synthetic methods for generating high-quality crystals smaller than 100 nm in size fuel the growth in this domain [29].

8.1.5 Energy Level Theory of UCNPs

Lanthanide-doped conversion nanoparticles comprise an inorganic crystalline host matrix with trivalent lanthanide ions incorporated within the host. UCNPs exhibit a nonlinear relationship between low pump intensity and high emission intensity, known as large anti-Stokes shifts. Other characteristics such as complex and sharp emission spectra and long tunable excited-state lifetimes suggest a wide range of applications. In this part, we explain the mechanism of upconversion emission using the commonly employed rate-equations-based approach [30].

The lanthanide elements possess highly abundant energy levels due to their electronic spin–orbit coupling, enabling photon upconversion [31]. Conventionally, lanthanide-doped upconversion mechanisms, depending on the lanthanide energy levels, are divided into four types: excited-state absorption (ESA), energy transfer upconversion (ETU), cooperative sensitization upconversion (CSU), and photon avalanche (PA).

Some of lanthanides like Er^{3+}, Tm^{3+}, Ho^{3+}, which have long-lived intermediate states to store energy, can achieve photon upconversion via ETU or ESA (Fig. 8.2a,b). Single-doped UCNPs suffer from low converted efficiency due to the small absorption section of Ln^{3+}. Most lanthanides doped UCNPs can achieve upconversion emission in an efficient manner via ETU. The photon with lower energy is firstly absorbed by the sensitizer, typically Yb^{3+}, and the energy is transferred to the activator. The transition from the high level to the ground state generates upconversion luminescence. Also, the co-doping system also improves the efficiency of the upconversion process. For the lanthanides without such energy level like Tb^{3+}, the upconversion process need cooperative sensitization (Fig. 8.2c). Another mechanism called EMU can also effectively excite the activator without intermediate state (Fig. 8.2d). The photon is absorbed by the sensitizer and accumulated to excite the ladder element, generally Tm^{3+}, to the high level and then the energy migration occurs between the migratory ion and finally the activator ion is excited.

In the mechanism of PA (Fig. 8.2e), there are some key requirements: low ground state absorption (GSA), strong excited state absorption (ESA), and effective cross-relaxation. When the pumping intensity is beyond the threshold of PA, the population are in the excited state; then the effective cross relaxation makes the accumulation of the population in the intermediate state. The ground-state absorption event initiates a chain reaction of excited-state absorption and cross-relaxation events between lanthanide ions, resulting in great nonlinearity.

Here, we take a commonly used ETU-based carrier rate equation as an example. In this example, Yb^{3+} serves as sensitizer of infrared photons and transfer the energy to the activator, here is Er^{3+}, for upconverted emission. As a simplified modal, Yb^{3+} has two levels, n4 and n5, representing ${}_2F^{7/2}$ and ${}_2F^{5/2}$, respectively. Er^{3+} has three-levels with n1, n2 and n3 representing ${}_4I^{15/2}$, ${}_4I^{11/2}$, and ${}_4F^{7/2}$ (${}_2H^{11/2}$), respectively. The differential rate equations are express as:

$$\frac{dn_1}{dt} = -W_1 n_5 n_1 + \frac{n_2}{\tau_2} + \frac{n_3}{\tau_3}$$
$$\frac{dn_2}{dt} = W_1 n_5 n_1 - W_2 n_5 n_2 - \frac{n_2}{\tau_2}$$
$$\frac{dn_4}{dt} = -\frac{\delta P}{hv} n_4 + \frac{n_5}{\tau_5} + W_1 n_5 n_1 + W_2 n_5 n_2$$
$$n_1 + n_2 + n_3 = n_a$$
$$n_4 + n_5 = n_s$$

which τ_i is the life time of n_i, δ is the absorption cross section of Yb^{3+}, P is the excitation intensity, h is the Planck's constant, v is the frequency of pump laser, W_1 and W_2 represent the energy transfer between Yb^{3+} and Er^{3+}, n_a and n_s are total

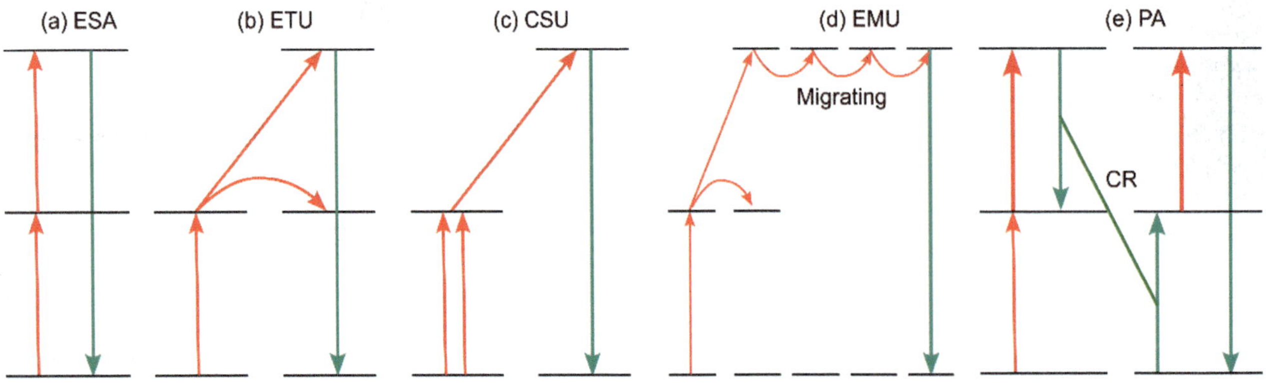

Fig. 8.2 Schematics of traditional upconversion mechanisms

number of populations of Er^{3+} and Yb^{3+}, respectively. Reader can get the steady numerical solution by solving these equations and can also obtain the time-dependent populations via mathematical method like Runge-Kuta method. The related parameter can be found at [32].

8.1.6 Advantages of UCNPs

UCNPs provide a wide array of exceptional advantages, making them highly promising for biomedical applications that integrate both diagnostic and therapeutic functions into a single platform. These attributes also underscore the significance of using UCNPs as optical trapping probes. The fluorescent nonlinearity also allows for the integration of numerous advanced microscopic techniques. The NIR super-resolution imaging for deep mouse liver tissue is achieved using the nonlinearity and stimulated emission depletion (STED) approach, as Fig. 8.3a,b show. When Structured Illumination Microscopy (SIM) is combined with UCNPs, video rate, and wide-field super-resolution can be witnessed in biological specimens. The parallel detection approach of Image Scanning Microscopy (ISM) exhibits a two-fold resolution enhancement with UCNPs, offering a novel outlook for imaging in a simple, cost-effective, and straightforward manner, as Fig. 8.3c shows. By capitalizing on the nonlinear fluorescent signals, the higher-order signal components can be isolated, leading to improved resolution. Thanks to the multiple and long-lived energy levels, the multi-photon process can be expanded via the Taylor series, and super-linearity can be adjusted under varying excitation laser powers.

The emerging optical multiplexing technology has drawn attention to diverse applications like imaging, information encryption, and disease diagnosis. With pre-engineered UCNPs containing different ions and structures, the fluorescence lifetime at specific wavelengths can be monitored and adjusted, presenting a unique optical signature for each particle type, as Fig. 8.3d shows. Using differently designed UCNPs, additional multiplexing aspects such as color, lifetime, optical nonlinearity, and polarization can be achieved.

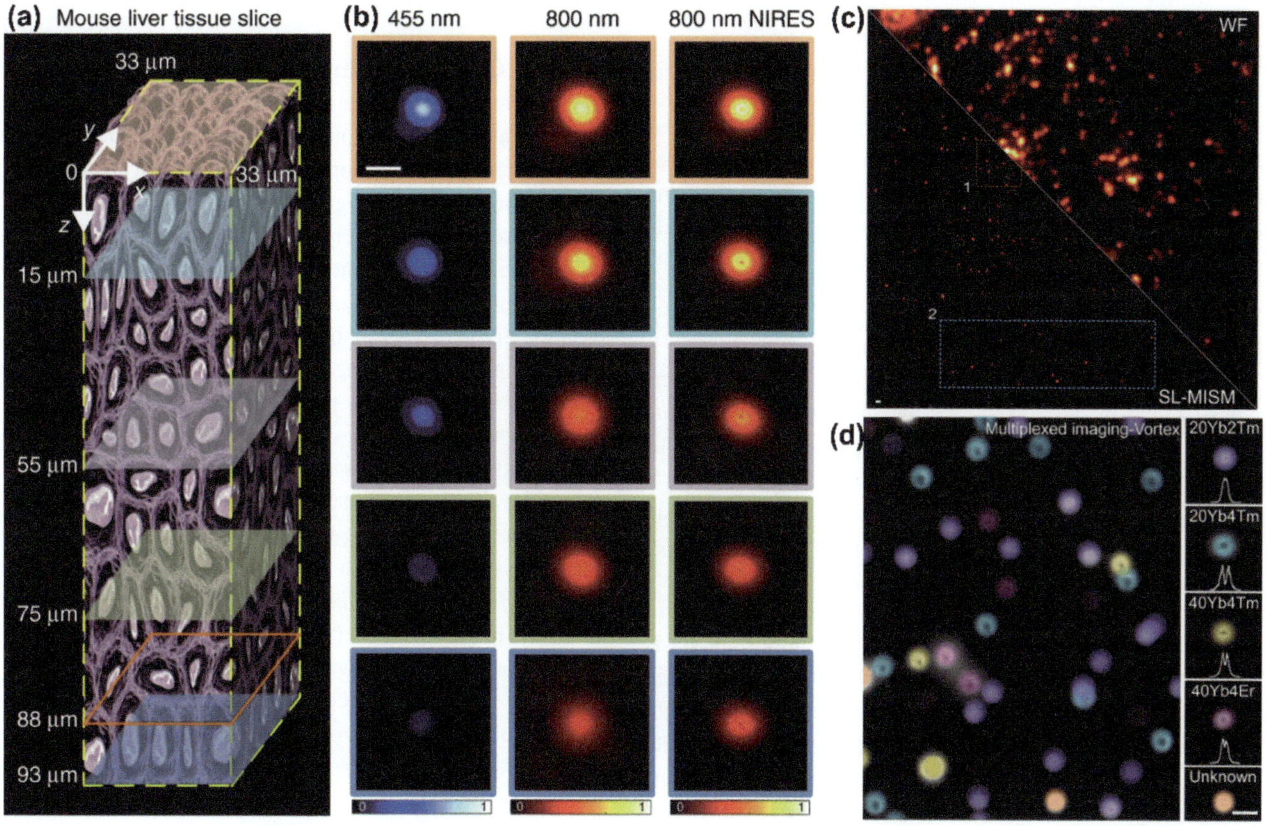

Fig. 8.3 Examples of application of UCNPs on super-resolution imaging. **a-b**. Multiphoton NIR imaging for deep mouse liver tissue using UCNPs. (Image adapted with permission from Elisa et al. [33] under Copyright license CC BY) **c** Resolution enhancement of image scanning miscopy based on the nonlinearity of UCNPs. (Image adapted with permission from Elisa et al. [34] under Copyright license CC BY) **d** Multiplexing microscopy enabled by optical nonlinearity. (Image adapted with permission from Elisa et al. [35] under Copyright license CC BY)

Moreover, nanocrystals with a high density of lanthanide ion doping can modify the permittivity and, as a result, substantially boost the optical force exerted on the nanoparticles due to the ion resonance effect. When the laser illumination frequency coincides with the dipole oscillation frequency of the lanthanide ions, a large number of lanthanide ions can significantly augment the amplitude of their electric susceptibility, permittivity, and consequently the overall polarizability of the nanocrystal. Additionally, doping lanthanide ions can modify the scattering characteristics of low-refractive-index nanoparticles, leading to a remarkable enhancement in the polarizability and scattering cross-section of the nanoparticles.

Furthermore, UCNPs can be combined with other imaging techniques, such as magnetic resonance imaging (MRI) [36, 37], computed tomography (CT) [38], and positron emission tomography (PET) [39], to offer multimodal imaging. This integration allows for a more comprehensive understanding of disease status, progression, and treatment response. UCNPs' biocompatibility further supports their clinical potential, as they are generally non-toxic and safe for use in medical diagnostics and therapy.

8.1.7 Applications of UCNPs

Scientific interest in nanocomposites based on UCNPs has been growing rapidly, and they have found applications in several newly emerging fields, such as bioimaging and treatment, photocatalysis, lighting and displays, and optical sensing (Fig. 8.4).

8.1.7.1 Bioimaging and Treatment

As an ideal probe for bioimaging, relying on the characteristics of near-infrared excitation and visible light emission, UCNPs can effectively avoid the interference of autofluorescence in biological tissues and achieve high-resolution imaging [40]. For example, in cell imaging, they help observe the distribution of molecules in cells and changes in physiological states, and in in vivo animal imaging, they are used to monitor the growth and metastasis processes of organs, tissues, and tumors, providing an important basis for disease diagnosis. In terms of disease treatment, in photodynamic therapy, they can convert near-infrared light into visible light that activates photosensitizers to generate singlet oxygen to kill tumor cells; in photothermal therapy, they can absorb near-infrared light and convert it into heat energy to achieve thermal ablation of tumors; they can also serve as drug carriers to accurately deliver drugs and control drug release through external stimuli, improving drug efficacy and reducing side effects [41].

8.1.7.2 Photocatalysis

UCNPs can be combined with semiconductor photocatalysts (such as TiO_2, CdS, etc.) to convert infrared light into ultraviolet or visible light, and promote the degradation of pollutants such as methylene blue and rhodamine B under 980 nm laser irradiation. In addition, by introducing specific rare-earth ions, the semiconductor bandgap can be adjusted, the recombination of photogenerated carriers can be inhibited, and the photocatalytic performance can be improved [42]. They also have application potential in reactions such as water splitting for hydrogen production, providing a new way for the development of the photocatalysis field.

8.1.7.3 Lighting and Displays

The optical device field also benefits from UCNPs. In all-optical communication, they can act as optical amplifiers or wavelength converters to improve signal processing capabilities and optimize the performance of optical communication systems, meeting the growing communication needs. Display technology helps to achieve display effects with high brightness, high resolution, and high colour saturation, providing strong support for the innovation and development of displays [43].

8.1.7.4 Optical Sensing

Rare-earth-doped upconversion nanoparticles play an important role in the field of optical sensing [44]. In temperature sensing, nanomaterials doped with rare-earth ions such as Er^{3+}, relying on the change of their optical properties with temperature, can accurately measure temperature by using the changes in fluorescence intensity ratio, wavelength or lifetime, and are widely used in biomedical and materials research. Some materials show excellent sensitivity and resolution. In pressure sensing, when specific rare-earth-doped crystals are stimulated by pressure, they change their luminescence characteristics based on the piezoelectric and crystal field effects and can be used for stress sensing, having great potential value in fields such as electronic recording. For pH sensing, by combining with pH-sensitive substances, the detection is achieved with the emission of rare-earth as an internal reference. In molecular detection, they can be used as sensors to detect a variety of biomolecules through energy transfer and can also achieve multi-modal detection in various ways. The single-particle counting form has high sensitivity, but the problem of high working temperature in gas sensing remains to be solved.

8.1.8 Background of Optical Trapped UCNPs

With advancements in material technology, a diverse array of novel optical force-enhanced nanoprobes has emerged.

Fig. 8.4 Applications of UCNP

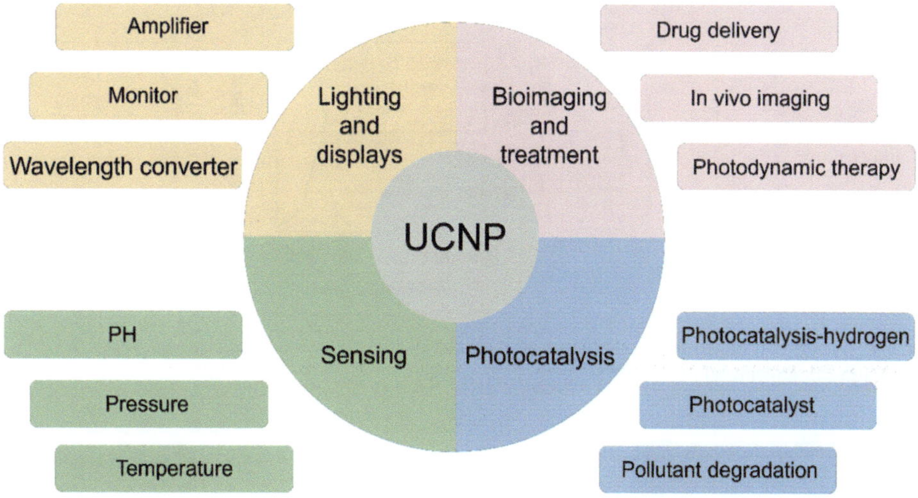

Among these, UCNPs stand out due to their unique properties, including a low imaginary refractive index, which minimizes photothermal effects, and highly stable fluorescence, facilitating reliable image detection. Moreover, UCNPs exhibit a resonance effect that enhances the polarizability of the nanocrystals, leading to a significant increase in optical force compared to gold nanoparticles of the same size.

This resonance effect is attributed to the resonance effect generated in the electromagnetic field by doping a high concentration of lanthanide ions in the nanoparticles, which in turn enhances the optical force. This approach to increasing the optical force provides a new method for enhancing the optical force of far-field optical tweezers of nanoparticles. Owing to this resonance effect, UCNPs can be used as optical force dyes and bound to the surface of other probes to induce an optical force enhancement phenomenon while also significantly reducing the requirement for the refractive index difference of the surrounding environment.

8.1.9 Theory of Resonance Effect

The manipulation of upconversion nanoparticles by optical tweezers can be regarded as the interaction between electromagnetic fields and matters. As a resonant substance, the charges of UCNPs can be polarized by the electromagnetic field. Based on the Lorentz model, the expression for the relationship between the polarization intensity and the electric field of lanthanide ion-doped $NaYF_4$ nanocrystals is shown as follows.

$$\frac{d^2\vec{\mathcal{P}}}{dt^2} + \sigma\frac{d\vec{\mathcal{P}}}{dt} + \omega_0^2\vec{\mathcal{P}} = \omega_0^2\varepsilon_0\chi'\vec{E} \quad (8.2)$$

$$\chi' = \frac{e^2N}{m\varepsilon_0\omega_0^2} = \frac{\omega_p^2}{\omega_0^2} \quad (8.3)$$

where $\vec{\mathcal{P}}$ is polarisation density, σ is the damping coefficient, ω_0 is atom resonance angular frequency, χ' is the electric susceptibility, m is the effective mass of an electron, N is the number of charges per unit volume, ω_p is plasma angular frequency.

For example, the dielectric constant of Yb^{3+}-ion-doped UCNPs is related to the susceptibility introduced by the resonant transition of Yb^{3+} ions, with resonance angular frequencies of ω_1, ω_2 and ω_3. This creates three susceptibilities χ_1, χ_2 and χ_3. The electric permittivity ε of $NaYF_4$ doped with Yb^{3+} ions is expressed as:

$$\varepsilon_{Ln_{NP}} = \varepsilon_0(1 + \chi_0 + \chi_1 + \chi_2 + \chi_3) \quad (8.4)$$

Based on the energy levels of the Yb^{3+} ions, we know that when the incident wavelength is around 920 nm, 940 nm and 972 nm, the susceptibilities and permittivity of these nanoparticles will be significantly increased (Fig. 8.5). This also implies an enhancement in the polarizability of the nanoparticles, which further leads to an increase in the optical force.

8.1.10 Force Enhancement by Resonance-OT

The theory of enhancing optical tweezers with the resonance effect has been described in the previous section. In this part, the applications of ion-resonant optical tweezers will be presented. Upconversion nanoparticles can serve as optical force enhancement probes and be linked with cells or other probes to act as optical force enhancement dyes.

For instance, optical tweezers in water do not easily manipulate large HeLa cells. If the method of increasing the trapping laser power is used to manipulate the cells forcibly, it

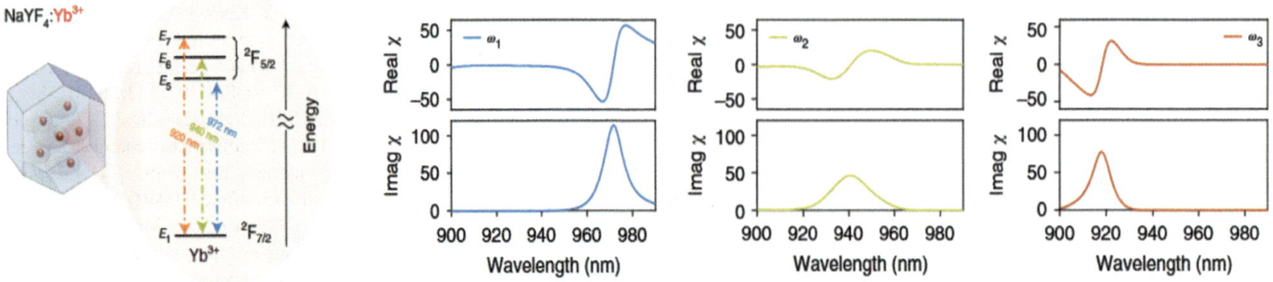

Fig. 8.5 Illustration of the energy levels of Yb^{3+} doped in $NaYF_4$ nanocrystals. (Image reprinted from Shan et al. [3] Copyright (2021), with permission from Springer)

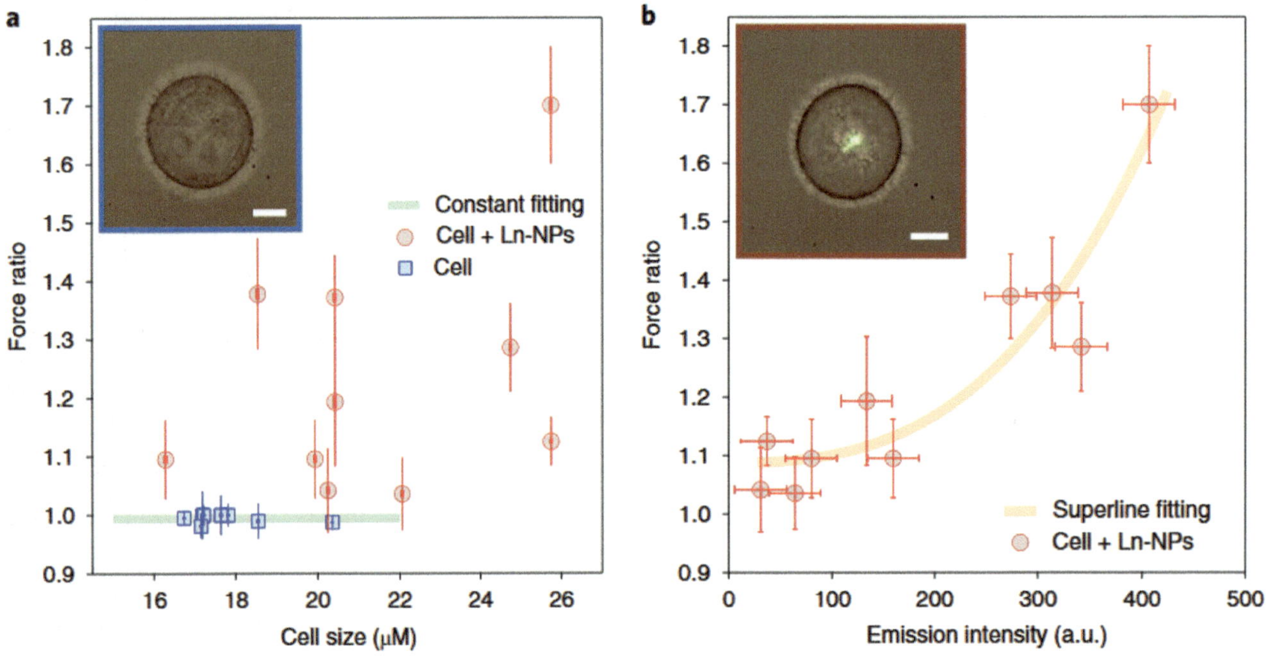

Fig. 8.6 Escape velocity measurements to quantify the trap stiffness for HeLa cells. (Image reprinted from Shan et al. [3] Copyright (2021), with permission from Springer)

will significantly impact the cell activity. When UCNPs are used as optical force dyes to coat the cell surface, the optical force of the cells is enhanced. Data analysis shows that the more optical force dyes there are, the more obvious the optical force enhancement is.

In the previous section, it was explained in detail that the upconversion nanoparticles doped with Yb^{3+} ions have an ion resonance effect near 972 nm, which can enhance the optical force of optical tweezers. In the experiment shown in Fig. 8.6, cells were subjected to rapid dragging experiments under two lasers with wavelengths of 976 nm and 808 nm, respectively and at almost the same laser power. The experimental results showed that under the manipulation of the 976 nm laser, the trapping force received by the cells was significantly increased compared with that under the 808 nm laser. Moreover, the more nanoparticles there were, the more obvious the characteristic of optical force enhancement was.

In another work, UCNPs also played a good role in enhancing the optical force. When researchers want to know the stiffness of cell membranes, they can obtain the hardness of cell membranes by squeezing the cell membranes with micro-probes. This method requires that the optical force probes have strong optical force characteristics. In the experiment, the probes coated with UCNPs obviously have stronger optical force than polymer spheres, the results shown in Fig. 8.7. Such optical force enhancement probes can explore the mechanical properties of cell membranes more accurately, thus making the experimental results more reliable.

8.1.11 Sub-femtonewton Force Sensing

Detecting weak forces in physiological solutions with high spatial resolution is important for exploring the biomechanics

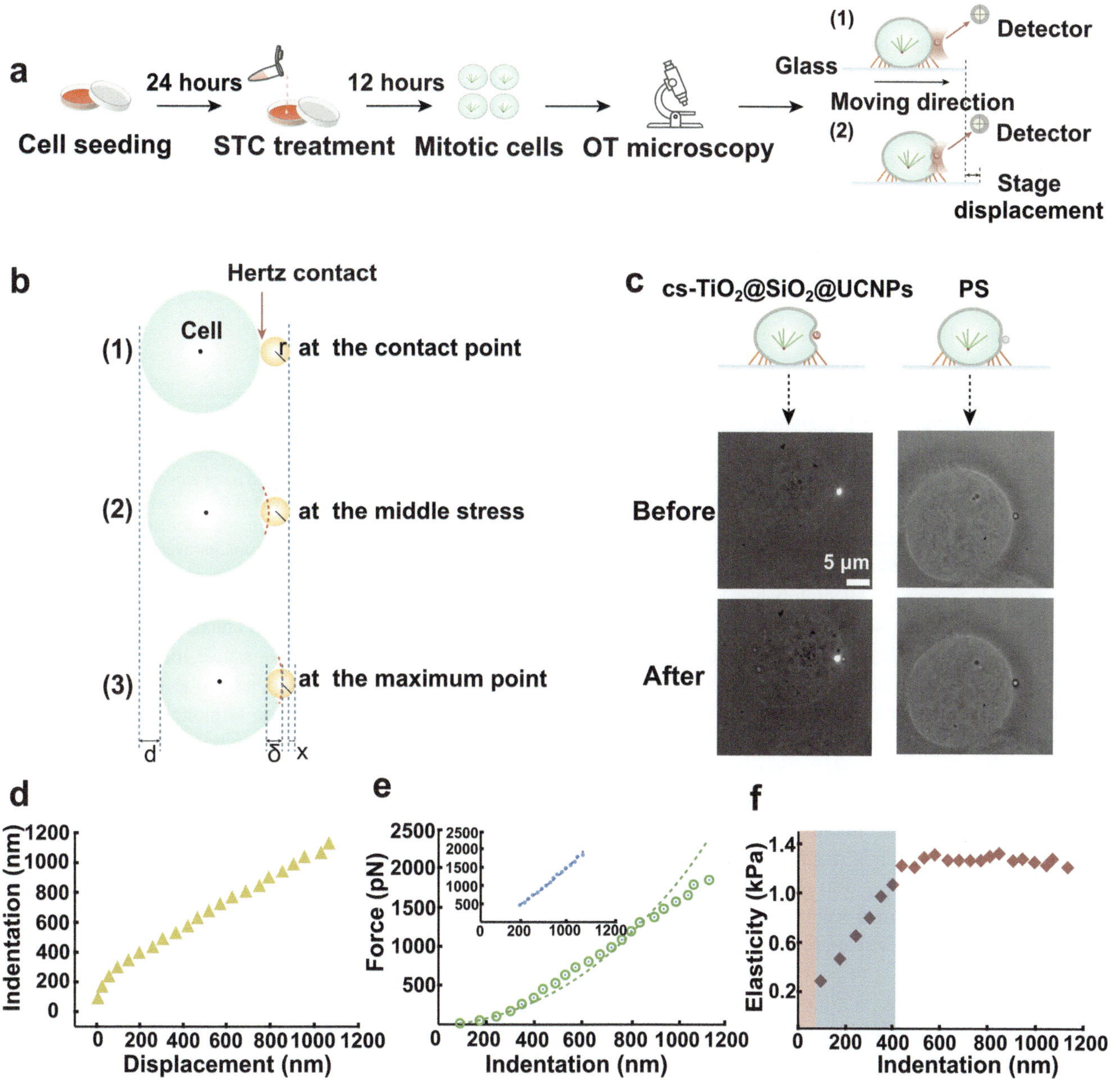

Fig. 8.7 Cell elasticity measured by optical tweezers. (Image reprinted with permission from Di et al. [45] Copyright (2024) American Chemical Society)

of biological events [46]. Forces generated by tissues and cells, such as contractile forces, can drop to the nanonewton (nN) magnitude. Traction forces exerted during single-cell migration fall within the piconewton level, while the string tension of a single DNA molecule and the organization of chromosomes reduces to the sub-piconewton order.

Cantilever-based force microscopy and microtube force spectroscopy are commonly used tools for detecting weak forces, yet they have limitations in spatial mapping dynamics. Molecular force microscopy is powerful but has a restricted dynamic force range. Magnetic tweezers are capable of three-dimensional (3D) force sensing, yet are constrained by the size of magnetic probes and can measure forces as low as 100 fN. A photonic force microscope (PFM), which relies on a trapped probe by optical tweezers, enables 3D dynamic control over force sensing, can achieve femtonewton (fN) sensitivities, and detect a large dynamic range (from nN to fN) in solution. However, surpassing the fN sensitivity while attaining temporal and spatial resolution remains a challenge.

The PFM makes use of the statistics of position fluctuations, and its sensitivity can be enhanced by increasing the sampling number. Nevertheless, the maximum efficient data sampling rate is limited by the inverse of the characteristic timescale (autocorrelation time, τ_{ot}), resulting in a thermal limit on force sensing. A larger force probe size leads to a higher drag coefficient and poorer force sensitivity. When using the scattering field as the signal, the best-reported force sensitivity in solution is 10 fN/Hz$^{1/2}$ for a 0.5 μm probe, and the weakest force detected in solution is below 3 fN. Using a nanoparticle as the trapping probe has the potential to yield a smaller drag coefficient and thus higher sensitivity. However, this approach comes with two challenges. Firstly, the scattering signal of nanoparticles becomes too weak to be used for measuring the trap stiffness. Secondly, the reduced trapping efficiency due to the smaller size requires high trapping power, which in turn causes a heating effect. More specifically, the minimum trap stiffness needed to maintain a nanoparticle within a $10k_BT$ trapping potential depth is $k_{min} = 40k_BT/w_0^2$. For the gold (Au) nanoparticle, which is a common traditional trapping nanoprobe, around 142 mW is still required to maintain stable trapping. This high power leads to a temperature increase of about 40 °C, which can damage biological samples and degrade the detection accuracy. Hence, the nanoscale thermal limit has not been reached.

Due to the ion resonance effect, UCNPs (upconversion nanoparticles) can be stably trapped by the laser beam with the resonant wavelength, which endows them with extraordinary trap stiffness. As a result, lanthanide-doped nanoparticles (Ln-NPs), a novel type of fluorescence probe characterized by long-term stable and bright emission and already applied in bioimaging, are employed as the trapping nanoprobe for the development of a super-resolved photonic force microscope (SRPFM) to achieve nanoscale thermally limited force sensing [4] (Fig. 8.8).

These probes possess multiple advantages when utilized for force sensing. They exhibit strong emission signals, have a weak photothermal effect, are of smaller size, and demonstrate good bio-compatibility. By tracking the 3D position of the trapped Ln-NP from its diffraction-limited emission spot, a solution to overcome the problem of the weak scattering signal can be identified. Thanks to leveraging the ion resonance effect in Ln-NPs, only a power of 10.22 mW is required to reach a potential depth of 10 kBT, leading to a merely modest temperature increase of 0.7 °C. When 25-nm-radius Ln-NPs are used as the trapping probes, the thermal limit is calculated to be 1.8 fN/Hz$^{1/2}$. It is important to emphasize that realizing this limit demands the development of both the detection method and the entire system.

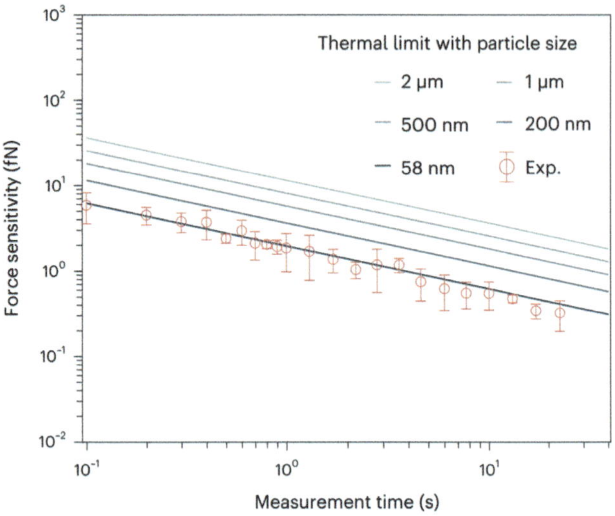

Fig. 8.8 Nanoscale thermally limited force sensitivity demonstrated by using a 58 nm Ln-NP without external force. The error bars represent s.d. from the four-time calculation. (Image reprinted from Shan et al. [4] Copyright (2024), with permission from Springer)

The sensitivity can be further enhanced through the application of more accurate nanoparticle localization methods. By using UCNPs as the trapped probe, the detected force range can be extended to the sub-femtonewton level, enabling the sensing of biological interactions. Moreover, UCNPs can be functionalized and linked to various bioparticles like exosomes and integrins to detect biological behaviors.

8.1.12 Laser Cooling

Laser cooling is a phenomenon in which the temperature of materials is reduced through laser excitation. Using UCNPs as trapping probes enables the simultaneous realization of laser cooling and optical trapping. UCNPs operate via an anti-Stokes emission mechanism, where the energy of emitted photons exceeds that of absorbed photons. This process effectively removes heat from the system. This cooling effect has great potential for applications ranging from microfluidic systems to biological studies, where precise thermal control is required at the micro- and nanoscale [47–49].

Recent research has focused on materials like ytterbium-doped yttrium lithium fluoride (YLF) nanocrystals and sodium yttrium fluoride ($NaYF_4$) microspinners, which exhibit significant cooling when exposed to near-infrared (NIR) light [43]. These materials absorb low-energy photons and emit higher-energy photons, leading to a net cooling effect. The efficiency of this cooling can be controlled by adjusting laser power and wavelength, making it a versatile

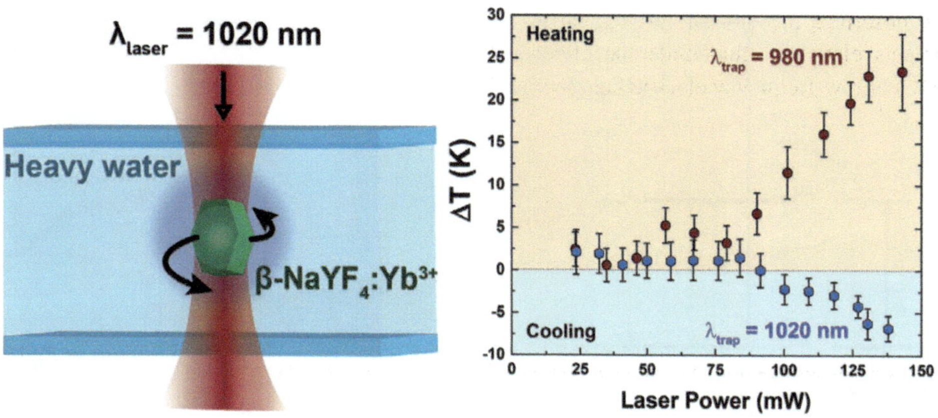

Fig. 8.9 a. A disk-shaped particle is trapped and rotated under the laser beam of 1020 nm. b. The relation between the temperature change and power under 980 nm and 1020 nm laser. (Image adapted with permission from Elisa et al. [47] under Copyright license CC BY)

technique for various applications, including biological and photonic systems (Fig. 8.9).

Optical tweezers play a critical role in laser cooling by providing precise control over the positioning and manipulation of individual particles. Using optical trapping, researchers can isolate nanocrystals or microspinners in a focused laser beam, enabling detailed studies of their cooling behavior. For example, in experiments with Yb^{3+}-doped YLF nanocrystals, optical tweezers were used to trap individual particles and analyze their Brownian motion. Since Brownian motion is temperature-dependent, this technique allows for accurate measurements of the temperature reduction in response to laser cooling. Researchers demonstrated temperature drops of over 10 °C in physiological media using Cold Brownian Motion (CBM) analysis, a significant achievement for biological applications [48].

In addition to providing precise trapping, optical tweezers enable the measurement of cooling effects through the manipulation of particle dynamics. For instance, $NaYF_4$ microspinners were trapped using optical tweezers, and their spinning velocity was monitored under laser irradiation. The cooling effect was observed as a reduction in the spinning velocity of the micro-spinners when illuminated with a 1020 nm laser, confirming the occurrence of local cooling. This method, where the rotational motion of the particles serves as a thermometer, offers a simple yet effective way to quantify the cooling effect.

By combining optical trapping with laser cooling, researchers can isolate individual particles, minimize external disturbances, and accurately measure temperature changes. This synergy is crucial for understanding the behavior of materials under laser irradiation and opens new possibilities for applications in temperature-sensitive biological research, microfluidics, and photonics. The ability to manipulate and measure the thermal properties of particles with high precision enhances both fundamental research and practical applications, making optical tweezers an indispensable tool in the field of laser refrigeration.

8.1.13 Temperature Sensing

Temperature is an important physical index for evaluating biological, physiological processes and detecting temperature changes is beneficial for the early prevention and treatment of diseases [50]. However, traditional contact temperature measurement methods are difficult to achieve temperature detection in many biological microenvironments. In view of the rapid development of optical tweezer technology and UCNPs, temperature detection can be achieved by laser trapping and exciting UCNPs. The temperature sensing ability of UCNPs is based on its specific energy level diagram (51). After exciting Yb^{3+} with infrared (980 nm) radiation, the energy is transferred to Er^{3+}. Subsequently, the excited Er^{3+} ions absorb 980 nm photons, filling their high-energy $^4F_{7/2}$ level, and then filling the $^4S_{3/2}$ and $^2H_{11/2}$ states. The radiative de-excitation from these states to the ground state is responsible for the green emission bands at 520 nm and 540 nm ($^2H_{11/2} \rightarrow {}^4I_{15/2}$ and $^4S_{3/2} \rightarrow {}^4I_{15/2}$, respectively). These two states are so close in energy that they are thermally coupled, that is, their relative populations follow the Boltzmann statistics (Fig. 8.10a). Therefore, the temperature dependence of the relative intensities of the emissions at 520 nm and 540 nm makes it possible to determine the temperature by a ratio analysis of the emission generated by Er^{3+}–Yb^{3+} co-doped UCNPs. In fact, $NaYF_4$:Er^{3+},Yb^{3+} UCNPs are the most commonly used lanthanide-doped nanothermometers because of their excellent luminescence brightness, good chemical and physical stability, and low toxicity.

In 2017, S. Drobczyński et al. carried out an examination of the thermal gradient in the proximity of diffuse large B-cell lymphoma (DLBCL) cells as a consequence of locally induced hyperthermia [52]. In this particular case, the authors made use of optical microthermometers (OμTs) that were fabricated by enclosing $NaYF_4$:Er^{3+}, Yb^{3+} UCNPs within amorphous silica shells approximately 6 μm in diameter. For the purpose of heating, mesoporous silicon microparticles (SiμPs) were employed. The DLBCL cells, microthermome-

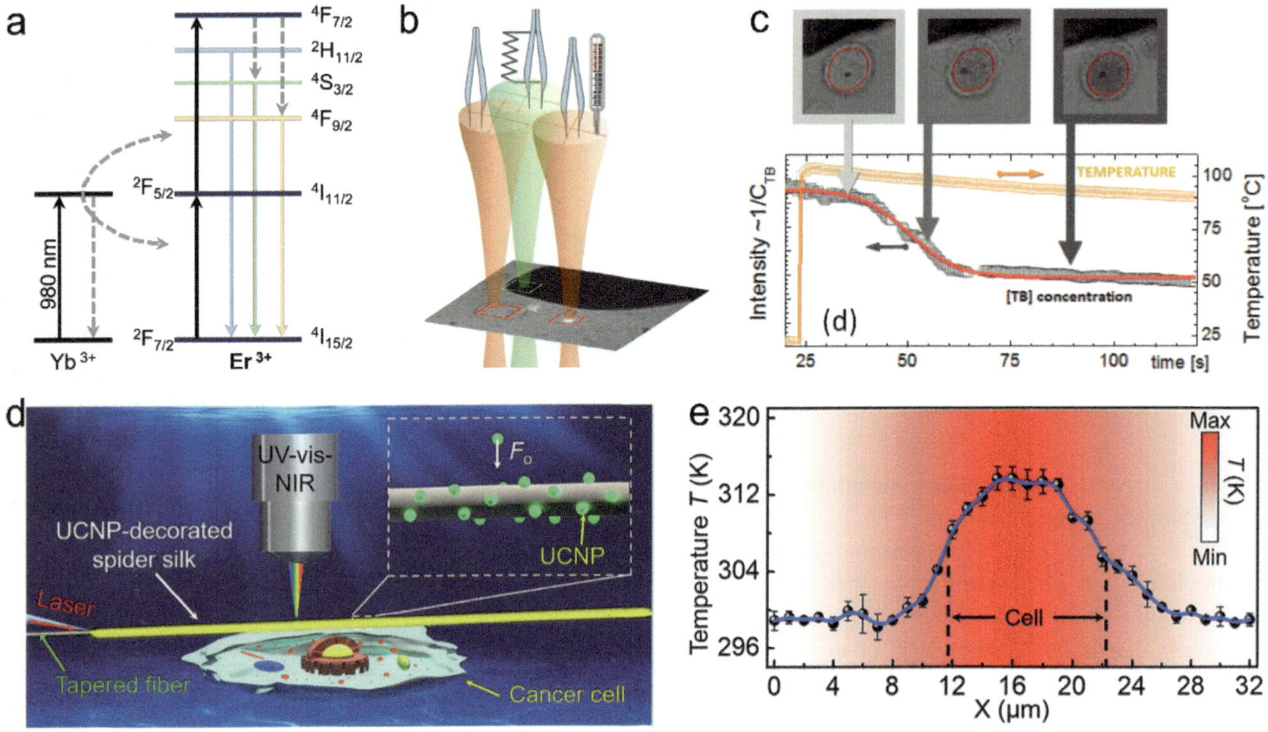

Fig. 8.10 **a** Energy scheme of Er^{3+}–Yb^{3+} co-doped UCNP. **b** The operational schematic of the active dual-wavelength optical multitrap configuration. The 980 nm beams (depicted as orange cones) are utilized for the trapping of the cell and the trapping/reading of the optical microthermometers, whereas the 1064 nm optical trap (shown as a green cone) is employed for optical trapping and heating. **c** Time-lapse snapshots of the cells being heated at 10 s, 30 s, and 65 s following the activation of the heating process. In the bottom-right corner, there are graphs presenting the relative Trypan Blue (TB) accumulation within the dying cell and the corresponding temperature (on the right axis). (b-c reprinted with permission from Drobczyński et al. [52] Copyright (2017) American Chemical Society) **d** Illustration of the temperature measurement process for a single living cell utilizing UCNP—decorated spider silk. **e** The scan of the FIR values surrounding the cell was acquired through the spectra at different points along the UCNP—decorated spider silk positioned on the cell. In the background gradient, the reddest shade indicates the maximum temperature of 312 K, while the whitest shade represents the minimum temperature of 298 K. (d-e image adapted from Gong et al. [53] Copyright (2021) American Chemical Society)

ters, and SiμPs particles had the capability to be entrapped simultaneously by an active dual-wavelength optical multitrap (Fig. 8.10b). In this setup, two separate 980 nm laser beams (represented as orange cones) were utilized to capture the DLBCL cells and OμTs, while a 1064 nm laser beam (illustrated as a green cone) was used to trap the SiμPs particles, concurrently converting light into heat. As can be seen in Fig. 8.10c, the temperature in the cells' environment rose to nearly 100 °C, thereby thermally harming them. This was also evidenced by the accumulation of TB in the damaged cells 15 s after the temperature increase. It took an additional 15 s for the TB to reach saturation in the damaged cells, suggesting that the damaged cells respond in an effort to mitigate the damage caused by heating.

In 2021, Gong et al. introduced a 980 nm laser into spider silk with natural optical fiber characteristics and decorated the spider silk with UCNPs, successfully constructing a thermometer for single-cell temperature measurement [53]. They brought the UCNP-decorated spider silk close to a single cancer cell (Fig. 8.10d). Since the ratio of the fluorescence emission intensities at 525 nm and 548 nm generated by the energy level transition of UCNPs changes due to the effect of temperature on the thermal coupling between the two thermal equilibrium states when excited by a 980 nm laser, they achieved accurate detection of the membrane temperature of the cancer cell during apoptosis by calculating the changes in the fluorescence intensity ratio. It is worth mentioning that the absolute sensitivity of this thermometer is from 3.3×10^{-3} K^{-1} to 4.5×10^{-3} K^{-1}, which provides reliable accuracy for temperature measurement at the single-cell level.

8.2 Conclusions and Perspectives

This chapter has reviewed the emergence and applications of upconversion nanoparticle-based optical trapping technology. The size and refractive index of optical tweezer probes are critical parameters that influence the optical force. Recent advancements in material technology and biotechnology have driven the development of smaller probes with lower refractive indices, highlighting the need for stable trapping techniques for such nanoprobes. We have primarily demonstrated how upconversion nanoscale fluorescent probes, with their ability to enhance ion-resonant optical forces, have significantly advanced the field of nanoscale optical tweezers. Furthermore, the unique optical properties of upconversion nanoparticles enable their application in laser cooling, temperature measurement, and weak force detection at the subcellular level. Nanoscale optical tweezers with force-enhanced capabilities are already playing a pivotal role in research areas such as materials science, physical processes, and biological applications. Looking ahead, high-precision nanoscale optical tweezers are poised to make even greater contributions to scientific progress.

Acknowledgements We acknowledge financial support from the National Natural Science Foundation of China (62275010, U23A20481), the Analysis & Testing Center in Beihang University, and the Beijing Natural Science Foundation (1232027).

References

1. Ashkin A. Acceleration and trapping of particles by radiation pressure. Phys Rev Lett. 1970;24(4):156–9.
2. Ashkin A, Dziedzic JM, Bjorkholm JE, Chu S. Observation of a single-beam gradient force optical trap for dielectric particles. Opt Lett. 1986;11(5):288–90.
3. Shan X, Wang F, Wang D, Wen S, Chen C, Di X, et al. Optical tweezers beyond refractive index mismatch using highly doped upconversion nanoparticles. Nat Nanotechnol. 2021;16(5):531–7.
4. Shan X, Ding L, Wang D, Wen S, Shi J, Chen C, et al. Sub-femtonewton force sensing in solution by super-resolved photonic force microscopy. Nat Photonics. 2024;18(9):913–21.
5. Grier DG. A revolution in optical manipulation. Nature. 2003;424(6950):810–6.
6. Svoboda K, Block SM. Optical trapping of metallic Rayleigh particles. Opt Lett. 1994;19(13):930–2.
7. Maragò OM, Jones PH, Gucciardi PG, Volpe G, Ferrari AC. Optical trapping and manipulation of nanostructures. Nat Nanotechnol. 2013;8(11):807–19.
8. Xin H, Li Y, Liu YC, Zhang Y, Xiao YF, Li B. Optical Forces: From fundamental to biological applications. Adv Mater. 2020;32(37):2001994–2018.
9. Liesener J, Reicherter M, Haist T, Tiziani HJ. Multi-functional optical tweezers using computer-generated holograms. Optics Communications. 2000;185(1):77–82.
10. Bola R, Treptow D, Marzoa A, Montes-Usategui M, Martín-Badosa E. Acousto-holographic optical tweezers. Opt Lett. 2020;45(10):2938–41.
11. Kwak E-S, Onuta T-D, Amarie D, Potyrailo R, Stein B, Jacobson SC, et al. Optical trapping with integrated near-field apertures. J Phys Chem B. 2004;108(36):13607–12.
12. Crozier KB. Quo vadis, plasmonic optical tweezers? Light Sci Appl. 2019;8(1):35–40.
13. Jones S, Andrén D, Karpinski P, Käll M. Photothermal heating of plasmonic nanoantennas: influence on trapped particle dynamics and colloid distribution. ACS Photonics. 2018;5(7):2878–87.
14. Desgarceaux R, Santybayeva Z, Battistella E, Nord AL, Braun-Breton C, Abkarian M, et al. High-resolution photonic force microscopy based on sharp nanofabricated tips. Nano Lett. 2020;20(6):4249–55.
15. Draine BT, Goodman J, editors. Beyond Clausius-Mossotti—wave propagation on a polarizable point lattice and the discrete dipole approximation. (electromagnetic scattering and absorption by interstellar grains) 1992.
16. Knöner G, Parkin S, Nieminen TA, Heckenberg NR, Rubinsztein-Dunlop H. Measurement of the index of refraction of single microparticles. Phys Rev Lett. 2006;97(15):157402–5.
17. Huang Q, Lee J, Arce FT, Yoon I, Angsantikul P, Liu J, et al. Nanofibre optic force transducers with sub-piconewton resolution via near-field plasmon–dielectric interactions. Nat Photonics. 2017;11(6):352–5.
18. Berthelot J, Aćimović SS, Juan ML, Kreuzer MP, Renger J, Quidant R. Three-dimensional manipulation with scanning near-field optical nanotweezers. Nat Nanotechnol. 2014;9(4):295–9.
19. Volpe G, Maragò OM, Rubinsztein-Dunlop H, Pesce G, Stilgoe AB, Volpe G, et al. Roadmap for optical tweezers. J Phys Photonics. 2023;5(2):022501.
20. Zhang S, Elsayed M, Peng R, Chen Y, Zhang Y, Peng J, et al. Reconfigurable multi-component micromachines driven by optoelectronic tweezers. Nat Commun. 2021;12(1):5349–57.
21. Andres-Arroyo A, Gupta B, Wang F, Gooding JJ, Reece PJ. Optical manipulation and spectroscopy of silicon nanoparticles exhibiting dielectric resonances. Nano Lett. 2016;16(3):1903–10.
22. Lin L, Wang M, Peng X, Lissek EN, Mao Z, Scarabelli L, et al. Opto-thermoelectric nanotweezers. Nat Photonics. 2018;12(4):195–201.
23. Hong C, Yang S, Ndukaife JC. Stand-off trapping and manipulation of sub-10 nm objects and biomolecules using opto-thermo-electrohydrodynamic tweezers. Nat Nanotechnol. 2020;15(11):908–13.
24. Dong H, Sun L-D, Yan C-H. Energy transfer in lanthanide upconversion studies for extended optical applications. Chem Soc Rev. 2015;44(6):1608–34.
25. Piguet C. Extricating erbium. Nat Chem. 2014;6(4):370.
26. Grundmann M, Christen J, Ledentsov NN, Böhrer J, Bimberg D, Ruvimov SS, et al. Ultranarrow luminescence lines from single quantum dots. Phys Rev Lett. 1995;74(20):4043–6.
27. Beverloo HB, van Schadewijk A, van Gelderen-Boele S, Tanke HJ. Inorganic phosphors as new luminescent labels for immunocytochemistry and time-resolved microscopy. Cytometry. 2005;11(7):784–92.
28. Auzel F. Upconversion and anti-stokes processes with f and d ions in solids. Chem Rev. 2004;104(1):139–74.
29. Mai H, Zhang Y, Si R, Yan Z, Sun L, You L, et al. High-quality sodium rare-earth fluoride nanocrystals: controlled synthesis and optical properties. J Am Chem Soc. 2006;128(19):6426–36.
30. Zhou B, Shi B, Jin D, Liu X. Controlling upconversion nanocrystals for emerging applications. Nat Nanotechnol. 2015;10(11):924–36.
31. Chen G, Qiu H, Prasad PN, Chen X. Upconversion nanoparticles: design, nanochemistry, and applications in theranostics. Chem Rev. 2014;114(10):5161–214.
32. Bergstrand J, Liu Q, Huang B, Peng X, Würth C, Resch-Genger U, et al. On the decay time of upconversion luminescence. Nanoscale. 2019;11(11):4959–69.
33. Chen C, Wang F, Wen S, Su QP, Wu MCL, Liu Y, et al. Multi-photon near-infrared emission saturation nanoscopy using upconversion nanoparticles. Nat Commun. 2018;9(1):3290–5.
34. Ding L, Chen C, Shan X, Liu B, Wang D, Du Z, et al. Optical nonlinearity enabled super-resolved multiplexing microscopy. Adv Mater. 2024;36(2):2308844–51.

35. Wang Y, Liu B, Ding L, Chen C, Shan X, Wang D, et al. Multi-photon super-linear image scanning microscopy using upconversion nanoparticles. Laser Photonics Rev. 2024;18(12):2400746–53.
36. Park YI, Kim JH, Lee KT, Jeon K-S, Na HB, Yu JH, et al. Nonblinking and nonbleaching upconverting nanoparticles as an optical imaging nanoprobe and T1 magnetic resonance imaging contrast agent. Adv Mater. 2009;21(44):4467–71.
37. Chen Y, Wang D, Yang H, Li K, Wang Y, Li Z. Polydopamine-coated manganese-doped upconversion nanoparticles as potential UCL/MRI dual-mode probes. J Phys Chem C. 2023;127(40):20041–8.
38. Liu Y, Li L, Guo Q, Wang L, Liu D, Wei Z, et al. Novel Cs-based upconversion nanoparticles as dual-modal CT and UCL imaging agents for chemo-photothermal synergistic therapy. Theranostics. 2016;6(10):1491–505.
39. Gallo J, Alam IS, Jin J, Gu Y-J, Aboagye EO, Wong W-T, et al. PET imaging with multimodal upconversion nanoparticles. Dalton Trans. 2014;43(14):5535–45.
40. Kumar B, Malhotra K, Fuku R, Van Houten J, Qu GY, Piunno PAE, et al. Recent trends in the developments of analytical probes based on lanthanide-doped upconversion nanoparticles. TrAC Trends Anal Chem. 2021;139–64.
41. Chen H, Ding B, Ma Pa, Lin J. Recent progress in upconversion nanomaterials for emerging optical biological applications. Adv Drug Deliv Rev. 2022;188–212.
42. Zheng B, Fan J, Chen B, Qin X, Wang J, Wang F, et al. Rare-earth doping in nanostructured inorganic materials. Chem Rev. 2022;122(6):5519–603.
43. Wang X, Valiev RR, Ohulchanskyy TY, Ågren H, Yang C, Chen G. Dye-sensitized lanthanide-doped upconversion nanoparticles. Chem Soc Rev. 2017;46(14):4150–67.
44. Wu J, Wu J, Wei W, Zhang Y, Chen Q. Upconversion nanoparticles based sensing: from design to point-of-care testing. Small. 2024;20(29):2311729–58.
45. Di X, Wang D, Shan X, Ding L, Zhong Z, Chen C, et al. Probing the nanonewton mitotic cell deformation force by ion-resonance-enhanced photonics force microscopy. Nano Lett. 2024;24(44):14004–11.
46. Liu L, Kheifets S, Ginis V, Capasso F. Subfemtonewton force spectroscopy at the thermal limit in liquids. Phys Rev Lett. 2016;116(22):228001-5.
47. Ortiz-Rivero E, Prorok K, Martín IR, Lisiecki R, Haro-González P, Bednarkiewicz A, et al. Laser refrigeration by an ytterbium-doped NaYF4 microspinner. Small. 2021;17(46):2103122–32.
48. Roder PB, Smith BE, Zhou X, Crane MJ, Pauzauskie PJ. Laser refrigeration of hydrothermal nanocrystals in physiological media. Proc Natl Acad Sci. 2015;112(49):15024–9.
49. Rahman ATMA, Barker PF. Laser refrigeration, alignment and rotation of levitated Yb3+:YLF nanocrystals. Nat Photonics. 2017;11(10):634–8.
50. Zhou J, Rosal B, Jaque D, Uchiyama S, Jin D. Advances and challenges for fluorescence nanothermometry. Nat Methods. 2020;17(17):967–80.
51. Zhang F, Camarero P, Haro-González P, Labrador-Páez L, Jaque D. Optical trapping of optical nanoparticles: fundamentals and applications. Opto Electron Sci. 2023;2(9):230019–42.
52. Drobczyński S, Prorok K, Tamarov K, Duś-Szachniewicz K, Lehto V-P, Bednarkiewicz A. Toward controlled photothermal treatment of single cell: optically induced heating and remote temperature monitoring in vitro through double wavelength optical tweezers. ACS Photonics. 2017;4(8):1993–2002.
53. Gong Z, Wu T, Chen X, Guo J, Zhang Y, Li Y. Upconversion nanoparticle decorated spider silks as single-cell thermometers. Nano Lett. 2021;21(3):1469–76.

Nanoscale Thermometry for Biophotonic Detection

9

Zhiyong Gong, Yongshi Ye, and Jinghui Guo

Abstract

In order to measure and monitor the temperature of various nanoscale substances or biological structures such as subcellular, especially to monitor local temperature changes such as living cells and organelles in vivo, it is of great significance to develop available nanoscale temperature measurement techniques for single cells. In recent years, nanothermometers have shown excellent performance with higher precision and higher sensitivity than earlier thermometers, providing a variety of multifunctional, high-durability nano-temperature sensors that make it possible to perform at higher time and space. Higher resolution enables monitoring of temperature changes in intracellular organelles during various physiological activities and metabolisms. With rapidly development, recent studies have shown that in some directions, nanothermometers still have some unavoidable limitations to overcome. Here, we present various design strategies and device developments used to construct nanothermometers in recent years, as well as their application prospects. As a new direction in the interdisciplinary field, possible opportunities and challenges are analyzed and discussed, and improvements to this cross-cutting field are proposed.

Keywords

Nanoscale thermometry · Temperature measurement · Biophotonics · Optical detection · Optical tweezer

Z. Gong · J. Guo (✉)
School of Medicine, The Chinese University of Hong Kong, Shenzhen, China
e-mail: guojinghui@cuhk.edu.cn

Z. Gong · Y. Ye
Guangdong Provincial Key Laboratory of Nanophotonic Manipulation, Institute of Nanophotonics, College of Physics and Optoelectronic Engineering, Jinan University, Guangzhou, China

9.1 Introduction

The development of nanoscience and nanotechnology has increased the demand for temperature measurement in a smaller size range and in a limited time. Especially the electronic components per unit area and the high precision calculations required to measure the temperature of substances in small size ranges make the problem of local overheating very prominent. Therefore, measuring the temperature of small-sized substances with high spatiotemporal resolution has become one of the keys to improving the temperature measurement accuracy. In addition, design strategies and guidelines for nanodevices can be improved [1–3].

To improve the subwavelength spatial resolution of temperature measurements, the most promising is to use high-precision scanning transmission electron microscopes and atomic force microscopes, such as scanning thermal microscopes, which can achieve local temperature measurements of substances with 10 nm precision [4]. However, they are often invasive and irreversible damage, often require the help of complex instruments and equipment, and are difficult to carry out practical applications in biomedicine and other fields [5]. In addition, the way they acquire data point by point inevitably limits the speed and efficiency of temperature imaging. Another way is to use non-contact far-field optical temperature measurement technology, such as infrared thermal imaging, optical interference, Raman spectroscopy, etc., to achieve high-precision, large-scale temperature measurement to a certain extent. However, the relationship between its temperature measurement accuracy and spatial resolution leads to its further development: the higher the accuracy, the lower the spatial resolution. And often a very complex calibration process is required to achieve temperature measurement in the smaller size range [6, 7].

To overcome the shortcomings of scanning transmission electron microscopy and non-contact far-field thermometry,

B. Li et al. (eds.), *Biophotonic Manipulation*, Advances in Optics and Optoelectronics,
https://doi.org/10.1007/978-981-96-4982-2_9

chemical synthesis and optical techniques have long been used to develop various temperature-sensitive materials to measure temperature [8–10]. A series of nanothermometers with temperature sensing properties have been developed, showing very bright prospects in biological research and in vivo diagnosis. Using the different temperature-sensitive characteristics of nanothermometers, the temperature response characteristics of materials at molecular, atomic and even electronic scales can be measured by monitoring various indicators such as intensity, peak shift, intensity ratio, and fluorescence lifetime [2]. Here, we focus on introducing and summarizing the development and research of nanothermometers in recent years, which are mainly different from fluorescence-based nanothermometers and non-fluorescence nanothermometers. Among them, fluorescence-based nanothermometers mainly utilize the temperature-sensitive properties of fluorescence generated by nanomaterials such as organic dyes, rare earth elements, quantum dots, and nanodiamonds to achieve nanoscale temperature measurement [11–16]. The non-fluorescent nanometer thermometer adopts advanced technical equipment to improve the scale of temperature measurement in terms of detection accuracy and detection methods, mainly including scanning thermal microscopy, Raman spectroscopy, thermocouples and thermal expansion [4, 17, 18]. They each have their advantages and disadvantages, but as nanothermometers, they still have considerable potential in fields such as bioimaging and temperature sensing (Fig. 9.1).

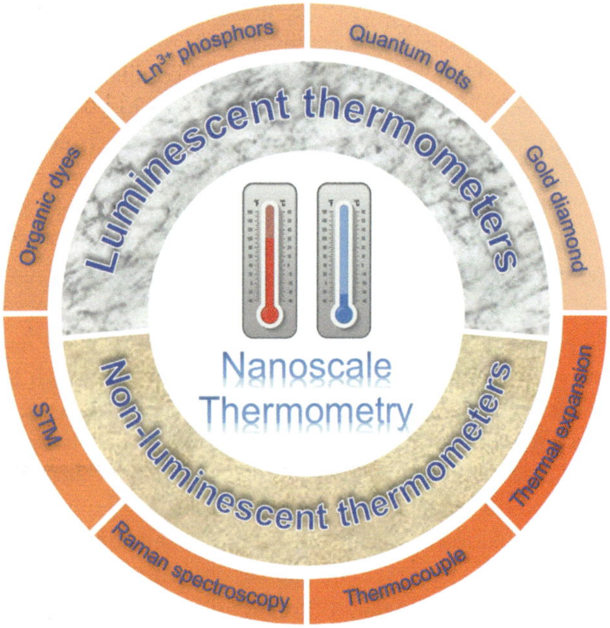

Fig. 9.1 Different types and applications of nanothermometers

9.2 Luminescent Thermometers

Nanotechnology has provided new tools for the measurement of temperature, such as nanomaterials including silver nanoparticles, quantum dots, upconversion nanoparticles and carbon nanotubes. One of the methods of temperature sensing is to use fluorescent nanomaterials or nanoparticles to detect temperature, because the characteristics of the material itself: the shape of the emission band, the position of the emission peak, the fluorescence intensity and the fluorescence lifetime are affected by temperature [19]. Based on the luminescence time lifetime change, after each excitation luminescence, the luminescence level exhibits an exponential decay with time, which is characterized by the dependence of its fluorescence lifetime on temperature [20]:

$$\tau = \frac{1}{W_r + W_{nr}(T)} = \frac{1}{\tau_0^{-1} + k\exp\left(-\frac{\Delta E}{k_B T}\right)} \tag{9.1}$$

where W_r and W_{nr} are the emissivity and non-emissivity, respectively, τ_0 is the radiative lifetime at absolute temperature, k is a constant, ΔE is the energy difference between the emission level and the excitation level, and k_B is the Boltzmann constant. Because the fluorescence lifetime is measured in the time or frequency domain, it is hardly affected by light scattering and reflection, fluorescence intensity fluctuations, and material distribution inhomogeneities, which are its key advantages as nanothermometers. However, the complexity of calculation and measurement of its lifetime makes it easier to overcome the intensity-based approach. The fluorescence intensity based on energy transitions is related to the population of excited states at temperature T [21]:

$$I \propto gAh\nu \exp\left(-\frac{E}{k_B T}\right) \tag{9.2}$$

where g is the degeneracy of the state, A is the spontaneous luminescence rate, h is Planck constant, ν is the frequency, and E is the energy at the energy level. However, in general, the change of fluorescence intensity through a single emission peak is often easily disturbed by external conditions, and temperature calibration needs to be carried out under very uniform conditions. Therefore, it is difficult to carry out in practical application. Therefore, based on the transition of two levels of adjacent emission levels, the ratio between their emission intensities (I_1 and I_2) is called the fluorescence intensity ratio FIR as a measure of the absolute temperature [22]:

$$\frac{I_1}{I_2} = \frac{g_1 A_1 h\nu_1}{g_2 A_2 h\nu_2}\exp\left(-\frac{\Delta E_{12}}{k_B T}\right) = B \times \exp\left(-\frac{\Delta E_{12}}{k_B T}\right) \tag{9.3}$$

This formula is a function of temperature, expressed as the temperature-dependent intensity ratio between two different emission spectra. The inherent reason is the law of energy conservation. The energy of one photon is limited, and the ratio of the energy allocated to the two energy levels will also change depending on the temperature, so temperature sensing can be achieved. Nanomaterials composed of fluorescent nanomaterials are very suitable for the preparation of nanothermometers because they cannot only provide high spatial resolution, but also realize temperature sensing in extremely small-scale areas such as organisms or cells [23]. And because of the additional visualization advantages provided by fluorescence, fluorescent nanothermometers are increasingly of interest to researchers in the biomedical field [24]. At present, there are many types of fluorescent nanothermometers to choose from. According to their luminescent materials, they can be divided into nanothermometers composed of materials such as organic dyes, rare earth materials, quantum dots and diamonds [25].

9.2.1 Organic Dyes

In the 1940s, organic dyes were first used as temperature indicators [26]. Until now, organic dyes such as rhodamine, fluorescein isothiocyanate, and cyanine dyes have become very popular for fluorescence imaging and temperature sensing in biological environments as the most readily available and used probes [27]. It once provided very favorable conditions for researchers to study various physiological activities in the biological field. A large number of organic dyes have been researched and sold on the market. People can choose a wide range of organic dyes, including their desired emission wavelength, emission range, dye solubility, composite luminescence, etc., and can be used for the construction of almost all kinds of temperature indicators. As fluorescent nanothermometers, materials such as rhodamine and fluorescein are common temperature-sensitive dyes commonly used in intracellular or microfluidic temperature sensing. However, the fluorescence quenching problem caused by molecular aggregation in these traditional organic fluorescence thermometers in concentrated solutions or solid environments will reduce the yield and accuracy of fluorescence quantum [28].

For higher sensitivity, the fluorescence intensity should be characterized by a significant change with temperature. Therefore, new dyes designed to have efficient nonradiative high temperature quenching were developed [29]. Molecules with twisted intramolecular charge-transfer state emission are one of the options for designing organic fluorescent nanothermometers.

An important example is reported using a high quantum yield organic dye AIEgen, 2-([1,1′-biphenyl]-4-yl)-3-(4-((E)-4-(diphenylamino)styryl) phenyl) fumaronitrile (TBB) with a rotatable electron donor–Π–acceptor (D − Π–A) structure was used for temperature sensing [30]. The dye fluoresces with high intensity and can be quenched at high temperatures, changing the fluorescent color. Thus, a fluorescent nanothermometer with switching response can be realized. By using F127 as a polymer matrix, two groups of the temperature-sensitive probe TBB and the reference dye rhodamine 110 can be encapsulated into a whole to construct a ratiometric fluorescence thermometer based on TBB & R110@F127 nanoparticles. The structure exhibits a good temperature sensitivity of 2.37% $°C^{-1}$, a wide temperature response range of 25–65 °C, and excellent temperature-sensitive emission reversibility. When monitoring the intracellular temperature, it is shown that the structure can monitor the temperature change of the cell from 25 to 53° in real time during the heating process of the photothermal therapy reagent, indicating that it has considerable potential applications in the field of biometrics.

The development of ratiometric fluorescent nanothermometers based on organic dyes in recent years has been limited by improvements in accuracy and anti-interference. The main reason is that the luminescence characteristics of single fluorescence are easily disturbed by the surrounding environment and the stability of temperature calibration. Thus, an important example is reported for a series of nanothermometers with temperature-dependent ratiometric bifluorescence of trimeric boron compounds [31]. The range and color of this molecular thermometer can be easily tuned by changing the substituents and medium, which allows the development of a wider range of nanotemperature probes that are more suitable for practical use. In liquid organic solutions, usable in the range of up to − 30 °C–110 °C, the fluorescent color of this thermometer varies from green–yellow to yellow–red, green to green–yellow and blue to green with decreasing temperature. Fluorescence spectra and colors can be correlated with temperature values with high accuracy (Fig. 9.2).

9.2.2 Ln^{3+} Phosphors

Many literature reports indicate that the energy level transitions of rare earth elements and the lifetimes of excited states of specific transitions are closely related to temperature [32]. Luminescent nanoprobes containing lanthanide ions Ln^{3+} have received special attention due to their unique electronic structure and optical properties. The unique electronic structure of lanthanum ions makes it rich in transition energy levels, which can cover the visible range from ultraviolet to near-infrared. Among the lanthanide ions, there are several ions such as Yb^{3+}, Tm^{3+}, Er^{3+}, Nd^{3+}, and Pr^{3+} which can generate fluorescence in the near-infrared II region [33].

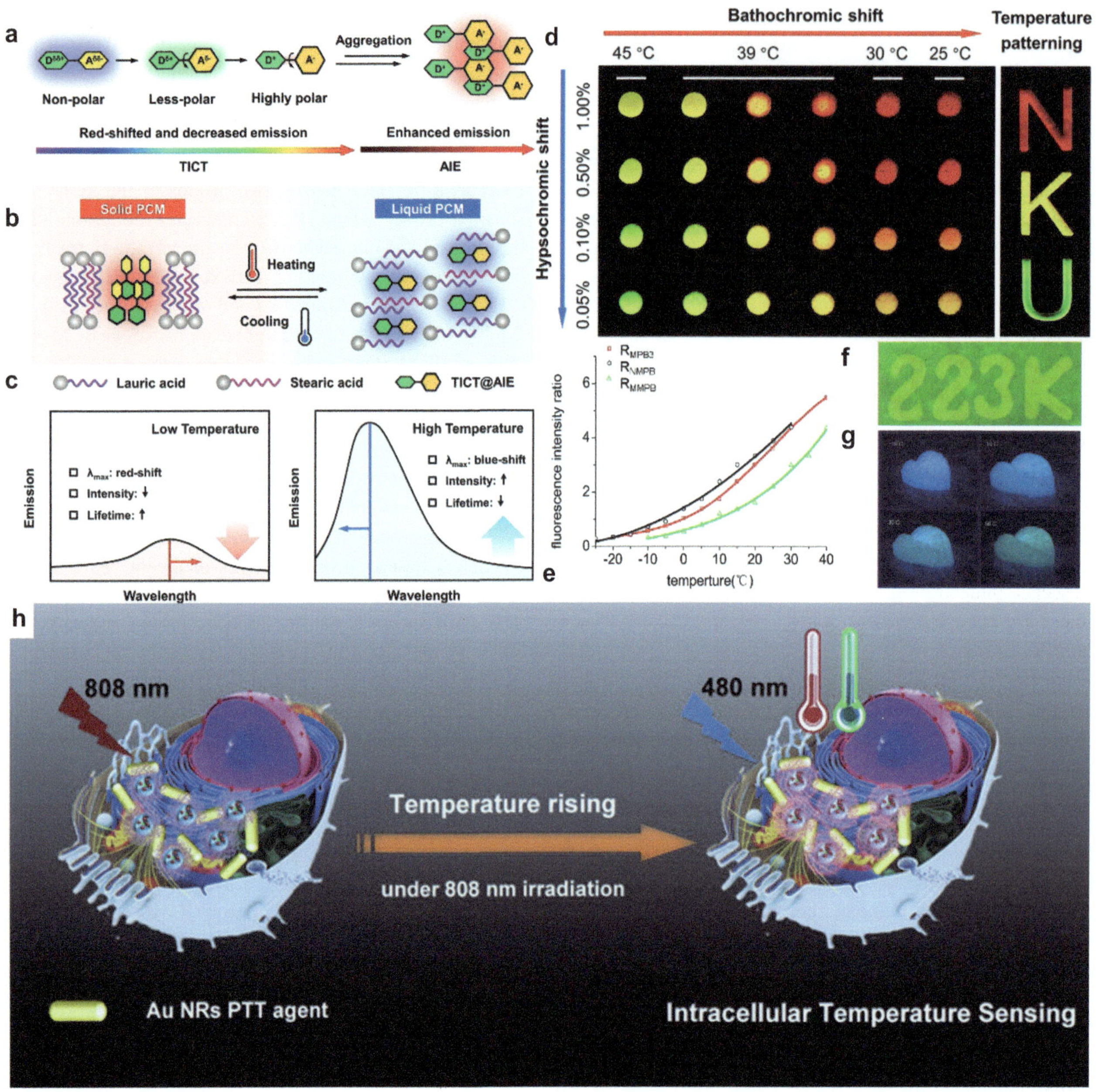

Fig. 9.2 Schematic illustrations of temperature sensing mechanism of organic dyes. **a** TICT@AIE organic dyes with simultaneous temperature-responsive and aggregation-responsive luminescence properties. **b** The mechanism of PCM modulating the fluorescence properti©of TICT@AIE. **c** Multi-parameter temperature sensing based on fluorescent nanothermometers. **d** The color changes of temperature patterning with temperature. Copyright © 2021 American Chemical Society [29]. **e** Temperature-dependent curves of several polymer systems. **f** Characters written on film at low temperature: 223 K. **g** Fluorescence pictures of polymers at different temperatures. Copyright © 2019 Elsevier B.V [31]. **h** Schematic diagram of temperature sensing of fluorescent probes with TICT properties. Copyright © 2020 American Chemical Society [30]

And this is very important for imaging in the near-infrared region II, which can achieve higher resolution, deeper penetration depth and lower biological background noise. Due to the insensitivity of lanthanides to the external environment, the generation of higher-energy and more stable fluorescence has great potential in multiplexing, high-resolution imaging, and high-precision detection.

So far, great efforts have been devoted to the research of lanthanide ion-mediated fluorescent nanothermometers, mainly including improving the absolute temperature sensitivity and relative temperature sensitivity of thermometers. A large number of lanthanide-based fluorescent nanothermometers have been reported [8]. An important example reports the development and application of a typical ratiometric lanthanide-based fluorescence nanothermometer $YVO_4:Yb^{3+},Er^{3+}$ operating in a wide spectral range (VIS–NIR) [34]. The developed fluorescent nanothermometer can display green fluorescence (anti-Stokes) luminescence that can be resolved by the human eye, as well as good temperature dependence. The energy level bands of the selected fluorescence intensity ratios are the thermally coupled and non-thermally coupled energy levels of Er^{3+} and Yb^{3+}. The ion attachment rate of the band intensities at 550 nm from Er^{3+} decreased significantly with increasing temperature, while the intensities of the bands above 900 nm from Yb^{3+} increased with increasing temperature [35]. Further, as a temperature measurement parameter, the ratio of the fluorescence intensity of the two is calculated. demonstrated that when a fluorescent nanothermometer exhibits nonlinear luminescence (e.g., upconversion luminescence), energy-level transitions, in which different numbers of photons participate in the relevant energy-level transitions, and changes in the optical power of the excitation light affect the sensor performance and temperature calibration parameters.

Due to their near-infrared luminescence properties and biocompatible properties, most lanthanide-based nanothermometers are used in bioimaging, biodetection, and clinical disease diagnosis and treatment. A typical in vivo nanotemperature measurement by a lanthanide-based fluorescent nanothermometer is reported [36]. The luminescence range of the fluorescent nanothermometer was changed to near-infrared II region, and the luminescence lifetime of thermally sensitive lanthanide elements was measured to realize temperature detection and sensing. To achieve higher luminescence intensity and less environmental interference, a ternary structure based on an inert core, an active shell, and an inert shell was developed to realize the encapsulation and assembly of lanthanide nanoparticles. This ternary structure maximizes the thermally sensitive properties of a given nanoparticle. Fluorescent nanoprobes were injected into living mice by injection, thereby realizing the imaging of the temperature distribution of mice. Not only that, but also can accurately monitor the pathological process of inflammation in mice [7]. The results of its temperature detection are consistent with those obtained using standard thermal imagers, making the lanthanide-based fluorescent nanothermometers ideal nanothermometers for temperature measurement in cells and even living mammals [2] (Fig. 9.3).

9.2.3 Quantum Dots

Nanoscale quantum dots have been used for small-scale thermometry because they exhibit temperature-dependent photoluminescence characteristics, such as changes in fluorescence intensity and red or blue shifts in fluorescence peak positions. Depending on the size of quantum dots, their luminescent properties such as emission frequency, quantum yield, and photostability also change significantly, making quantum dots an important reference for temperature sensing and imaging in many fields one [37]. Because of the bioconjugation of quantum dots, quantum dot nanothermometers have attracted the most attention in the fields of optoelectronics, biomedicine, and small-scale sensing. Compared with other nanomaterials, quantum dots have better biocompatibility and photochemical stability [38].

These features indicate its great potential in biological imaging. In recent years, two-photon excited nano-quantum dots have been attracting attention due to their long-wavelength excitation characteristics that help achieve deeper penetration depth and higher image resolution in photonics and biophotonics. An important example reports a surface S-rich CdS quantum dot synthesized using a green hydrothermal method [39]. When quantum dots were co-cultured with biological cells, the high uptake and high cellular activity exhibited by the cells demonstrated the good biocompatibility and low toxicity of quantum dot nanomaterials. The S-rich surface makes the quantum dots exhibit a smaller band gap at the characteristic wavelength. Thus, two-photon fluorescence is generated.

It is expected to be a candidate for cell bioimaging applications. An important example reports the non-contact local temperature measurement using wavelength shifting of a single CdSe quantum dot [40]. In contrast to other work, most have used a set of quantum dots to probe their spectral changes as a function of temperature. Notably, the spectral shift of a single quantum dot is characterized as 0.1 nm/°C in the room temperature range, and reconstruction of the temperature distribution of the microheater can be achieved using a quantum dot fluorescent nanothermometer.

Both experiments and simulations show good spatial resolution. Since the use of quantum dots alone for temperature sensing and biological imaging has great drawbacks, it is mainly manifested in the large uptake of biological cells and low temperature resolution. Therefore, an important example is reported using temperature probes based

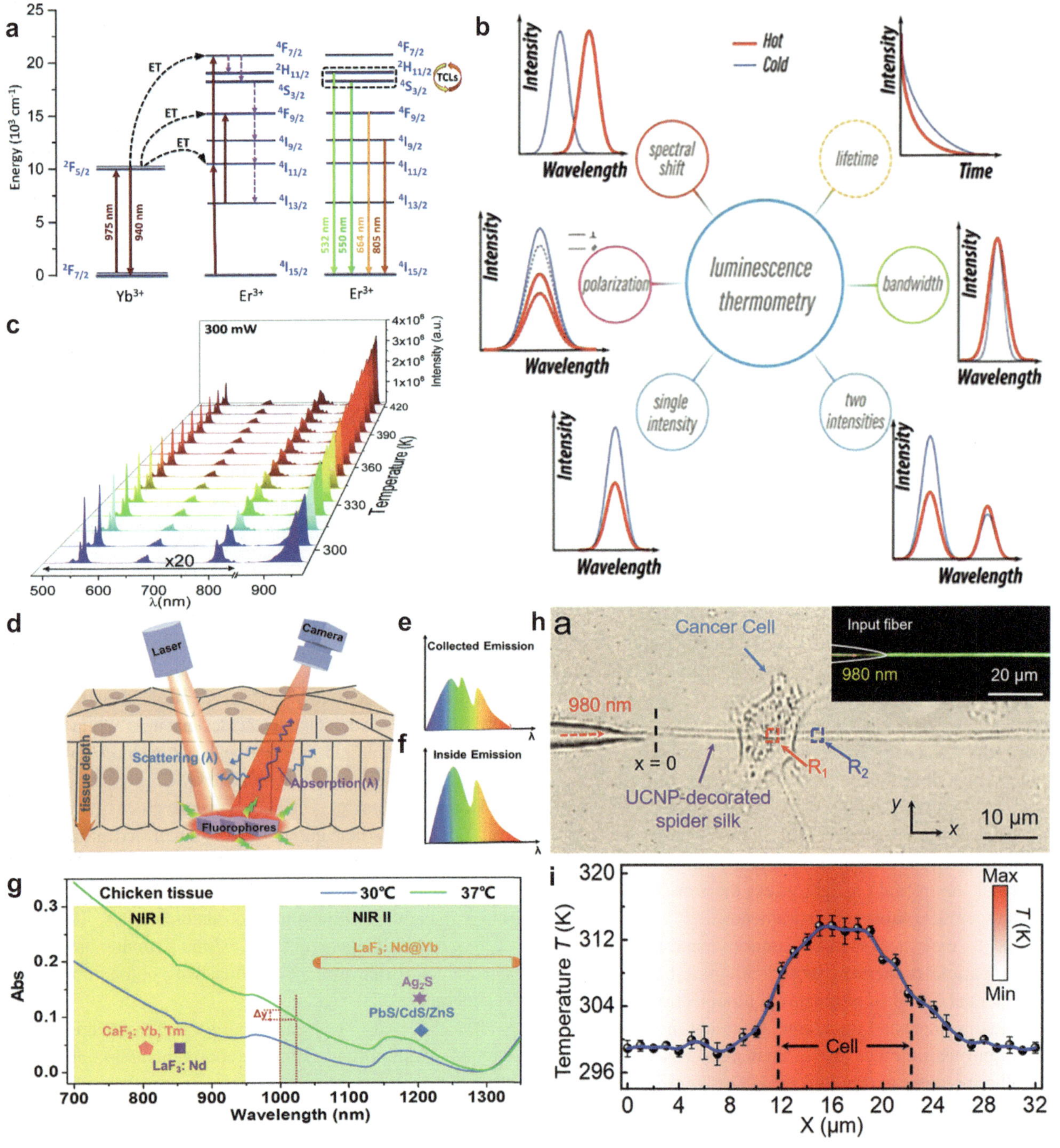

Fig. 9.3 Schematic illustrations of temperature sensing mechanism of Ln^{3+} phosphors. **a** Energy level diagram of potential radiative and nonradiative processes for Ln^{3+} phosphors. Copyright © The Royal Society of Chemistry 2021 [35]. **b** Elevated temperature affects the luminescence of fluorescent materials [8]. **c** Emission spectra of Ln^{3+} phosphors at different temperatures (laser power: 100 mW). Copyright © The Royal Society of Chemistry 2021 [35]. **d** Schematic diagram of thermal luminescence imaging of biological problems. **e** Luminescence spectrum in vitro. **f** Luminescence spectrum in vivo. **g** Nano-temperature labeling of biological windows in NIR-I and NIR-II regions. Copyright © 1999–2024 John Wiley & Sons, Inc or related companies [7]. **h** Ln^{3+} phosphors-modified natural spider silk acts as a nanothermometer to measure single-cell temperature. **i** Single-cell spatial temperature distributions measured by Ln^{3+} phosphors-modified spider silk. Copyright © 2021 American Chemical Society [36]

on lanthanide-based nanoparticles complexed with nano-quantum dots, while using gold nanoparticles as autonomous heaters, thereby realizing the luminescence signals of the two probes at different positions and Measurement of the temperature of nanoparticles [41]. When the gold nanoparticles located in the mesoporous silica nanoparticles are illuminated and heated, the heat generated can be detected by the temperature probe of the composite structure. The temperature difference reported by two probes 55 nm apart can be as high as 30 °C.

These results suggest that excellent nanoscale temperature-sensitive fluorescent probes can serve as powerful tools for probing biological cell temperature and higher image resolution. However, the photobleaching, fluorescence instability and blue emission of quantum dots under long-term illumination have led to their further development in the field of nanothermometry. In particular, these reasons will lead to the fact that the solubility, agglomeration, instability and biotoxicity of quantum dots in different environments during the synthesis process have become a major limitation of quantum dot fluorescent nanothermometers. Fluorescence imaging of biology using quantum dot nanomaterials has become a widely used method [42, 43]. However, the use of quantum dots for temperature measurement of biological cells is limited by the problems of massive cellular uptake and biological toxicity. Therefore, a lot of research is still needed to realize quantum dots as a reproducible and safe powerful tool in clinical medicine [44] (Fig. 9.4).

9.2.4 Gold Diamond

In recent years, fluorescent nanothermometers constructed with diamond nanomaterials are considered to have great prospects in the field of nanoscale biological thermometry. Diamond-mediated photoluminescence defect nitrogen-vacancy (NV) centers is one of the characteristics of diamond as a temperature-sensitive material [45]. Compared with other fluorescent nanothermometers, this electron spin-dependent photoluminescence associated with atomic defects has excellent photostability and can realize various optical detections, not only temperature, but also magnetic fields, electric fields and so on. Moreover, NV center-based nanoscale thermometers are based on quantum effects and thus have unprecedented sensitivity and precision, enabling temperature sensing and bioimaging at the single-molecule level [46]. An important example reports a typical nanodiamond-based nanothermometer. The method uses coherent manipulation of electron spins associated with nitrogen vacancy color centers in diamond [5]. Its nanothermometer can detect temperature changes as small as 1.8 mK, and its sensitivity can reach 9 $\mathrm{mKHz}^{-1/2}$. On the thermometric scale, we can realize the temperature measurement of the local thermal environment on the length scale as short as 200 nm, which greatly improves the research of life science on a smaller scale and realizes the temperature gradient at the subcellular level. control and mapping. Not only that, based on high-precision fluorescent nanothermometers, combined with gold nanoparticle-mediated thermal ablation therapy, can achieve site-specific selective identification and killing or killing of malignant cells without damaging surrounding tissues and cells, due to its high precision at the subcellular scale.

Not only that, the localization, quantification and identification of specific organelles in cells can be achieved by labeling specific free radicals specifically with nanodiamonds [47]. An important example reports quantum monitoring and temperature sensing of the metabolic activity of individual mitochondria with a fluorescent nanodiamond-based nanothermometer [48]. In this process, a specific modality: relaxation measurements were used to perform magnetic experiments on diamond, which allowed the detection of the noise of the spin due to free radicals in fluorescent nanodiamonds with nitrogen vacancy color centers. In the experiment, the temperature of the mitochondria measured by the fluorescent nanodiamond will slightly decrease with the increase of the actual temperature of the mitochondria, which is mainly caused by factors such as the surrounding background fluorescence, but the effect is small enough to be ignored.

In the process of nanothermometry, the temperature has an effect on the magnetic resonance frequency of the NV color center of the nanodiamond [48]:

$$\frac{\mathrm{d}\upsilon_{\pm}}{\mathrm{d}T} = \frac{\mathrm{d}D}{\mathrm{d}T} = -74.2\,\mathrm{kHzK}^{-1} \tag{9.4}$$

As the temperature increases, the resonant frequency shifts to a lower frequency. This transition is a result of the expansion of the defective diamond crystal leading to a reduction in electron spin and spin coupling, that is, a reduction in the interaction between the two electrons of the NV color center [49]. Although it will cause changes in other properties of the diamond, it will also allow the diamond to achieve subcellular level temperature measurement. An important example reports the measurement of temperature changes in human embryonic kidney cells by recording the spectral shift of the zero-phonon line of negatively charged nitrogen vacancy color centers in fluorescent nanodiamonds [50]. By incorporating a nanohybrid of fluorescent nanodiamonds and citrate-terminated gold nanorods as a two-in-one nanodevice, this approach overcomes the limitation of temperature sensing at the individual particle level. Using this composite material, the respective temperatures of cell membrane rupture and intercellular nanotube rupture were measured, and it was found that the temperature of membrane rupture was 15 °C higher than that of intercellular nanotube rupture. curvature or a higher degree of deformation. This is the first time

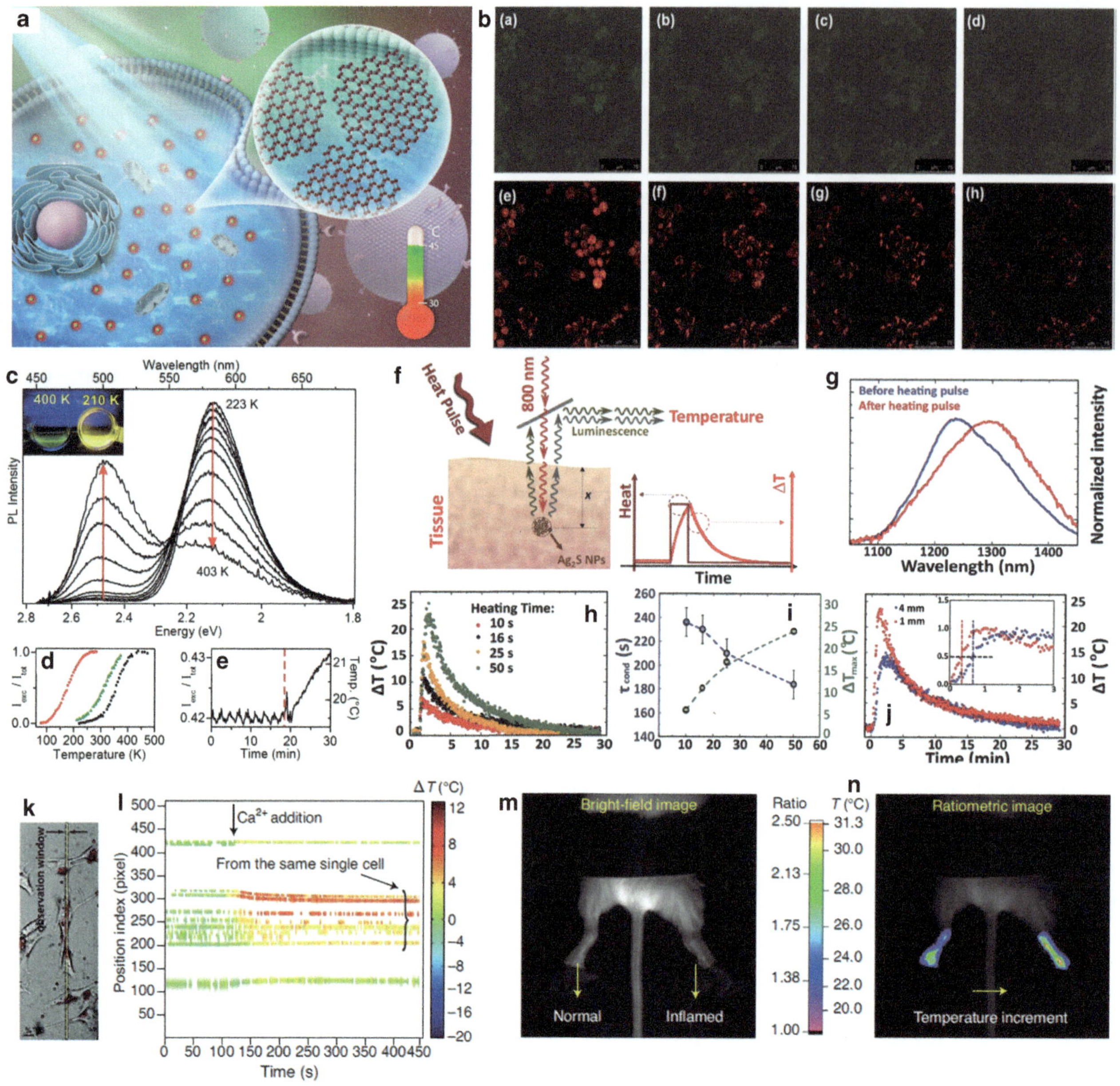

Fig. 9.4 Schematic illustrations of temperature sensing mechanism of nanoscale quantum dots. **a** Efficient ratiometric nanothermometer based on colloidal carbon quantum dots. **b** Temperature sensing micrographs of quantum dots at different temperatures. Copyright © 2021 Royal Society of Chemistry [44]. **c** PL spectra of colloidal quantum dots at different temperatures. **d** Response curves of three colloidal quantum dots. **e** PL response curves of colloidal quantum dots under temperature shock. **f** Colloidal quantum dots as subtissue temperature sensors. **g** Subtissue Fluorescence Emission Spectra under Pulse. **h** Incremental changes in temperature of sub-tissues under different pulse heating. **i** Maximum subtissue temperature increment as a function of heating time. **j** Subtissue temperature increments at different depths as a function of time. Copyright © 2020 The Authors. Published by WILEY–VCH Verlag GmbH & Co. KGaA, Weinheim [42]. **k** Brightfield micrographs of cells overlaid with fluorescent images of quantum dots. **l** Graph of position-dependent temperature versus time. **m** Bright-field images of normal and injured hind legs of mice. **n** Quantum dot-based ratiometric images of normal and injured hind legs of mice. Copyright © 2011 American Chemical Society [41]

that thermal events in living cells have been revealed at the nanoscale, and this work highlights the importance of selective killing and controlled thermal treatment of cancer cells for single-cell laser surgery and precise cancer hyperthermia as a powerful new tool (Fig. 9.5).

9.3 Non-luminescent Thermometers

Compared with fluorescent nanothermometers, the thermometric properties of non-fluorescent nanothermometers are not directly related to temperature. To a certain extent, some of these non-fluorescent nanothermometers cannot even be called thermometers because they cannot give a specific temperature value. The driving force for the development of non-fluorescent nanothermometers is mainly due to the vigorous development of nanolithography and micro-nano optoelectronics. It is urgent to improve the non-fluorescent nano-temperature measurement system, to realize the temperature measurement at the cellular level and the drawing of the large-scale and high-precision temperature distribution map of biological tissues from high spatial resolution and monitoring sensitivity [51].

9.3.1 Scanning Thermal Microscopy

After the development of scanning electron microscope (SEM) and confocal scanning electron microscope (FSEM), people urgently want to further improve the imaging accuracy of scanning microscopy. Therefore, for example, cryo-scanning electron microscope (Cryo-SEM) that can realize atoms and atomic force microscope (AFM) that can realize biological cells, etc., have been studied and developed vigorously. However, simple sample imaging can no longer meet people's needs, and the functionalization of scanning microscopy has been proposed, so scanning thermal microscopy (SThM) for temperature sensing has attracted attention [52]. Scanning thermal microscopy is developed on the basis of scanning probe microscopy (SPM), and can be called one of the important experimental techniques for studying nanoscale thermal phenomena. As with other scanning microscopy techniques, higher temporal and spatial resolution (milliseconds to microseconds, micronanometers to nanometers) is required [53]. A generalized scanning thermal microscope uses an atomic force microscope probe combined with a thermal probe or thermal resistance to realize the interaction between the tip and the sample and the temperature measurement of the target sample. In this process, due to the technology of atomic force microscopy, in During the measurement process, the integrity of the target sample can be well maintained. The measurement process can be divided into two categories: contact mode and non-contact mode [54].

In the contact mode, a certain mechanical force will act on the surface of the target sample, the detection accuracy is higher, the sensitivity is higher, and it will have higher spatial resolution; in the non-contact mode, the temperature measurement probe is suspended in the sample. Working tens to hundreds of nanometers above the surface allows for higher temporal resolution, with measurements several times faster than in contact mode. In traditional scanning thermal microscopes, due to the large size of the tip-sample contact area (about 50 nm) generated by the liquid surface tension film at the tip-sample interface of the temperature measurement probe, the accuracy of scanning thermal microscopy is limited to ~ 50 nm. nm. Therefore, how to reduce the contact area becomes the best choice to improve the resolution of scanning thermal microscope. An important example reports an ultra-high vacuum scanning thermal microscope (UHV-SThM) that can be used for quantitative thermometry at nanometer resolution [55]. The system overcomes some of the major limitations of conventional SThM through additional sweeps and additional convolutional mathematical models, allowing quantitative measurements of temperature fields with excellent thermal resolution (~ 15 mK) and spatial resolution (~ 10 nm). The system overcomes the generation of liquid surface at the tip-sample interface in principle, enabling solid–solid contact (~ 10 nm) to be applied. Despite greatly improving the thermal and spatial resolution of scanning thermal microscopy, UHV scanning thermal microscopy techniques are difficult to use on rough and uneven sample surfaces due to edge effects, and due to the probe-to-probe tip geometry There are differences in the models, so it is necessary to separately calculate the relationship between the collected thermal signal and the applied force for each probe, which has certain limitations.

In addition to ultra-high vacuum scanning thermal microscopy, in order to further expand the application of scanning thermal microscopy in the fields of biological materials, energy regeneration and photoelectrochemical catalysis, some scanning thermal microscopy based on other principles, such as liquid-based scanning thermal microscopy, combined with Scanning thermal microscopy of shear force and scanning thermal microscopy combined with infrared spectroscopy have been developed and applied [56]. Liquid-based scanning thermal microscopy uses liquid immersion to monitor the light intensity of fluorescent particles attached to the probe. Its structure is very stable, eliminating the constraints of expensive and harsh working environments. The scanning thermal microscope combined with the shear force is to realize the lateral movement of the probe when it is in contact with the sample, so as to measure the normal force and the transverse shear force through the cantilever deflection measurement method, and calculate the change of the sample temperature. Scanning thermal microscopy combined with infrared spectroscopy generates local temperature differences on the sample or probe through IR heating, so as to

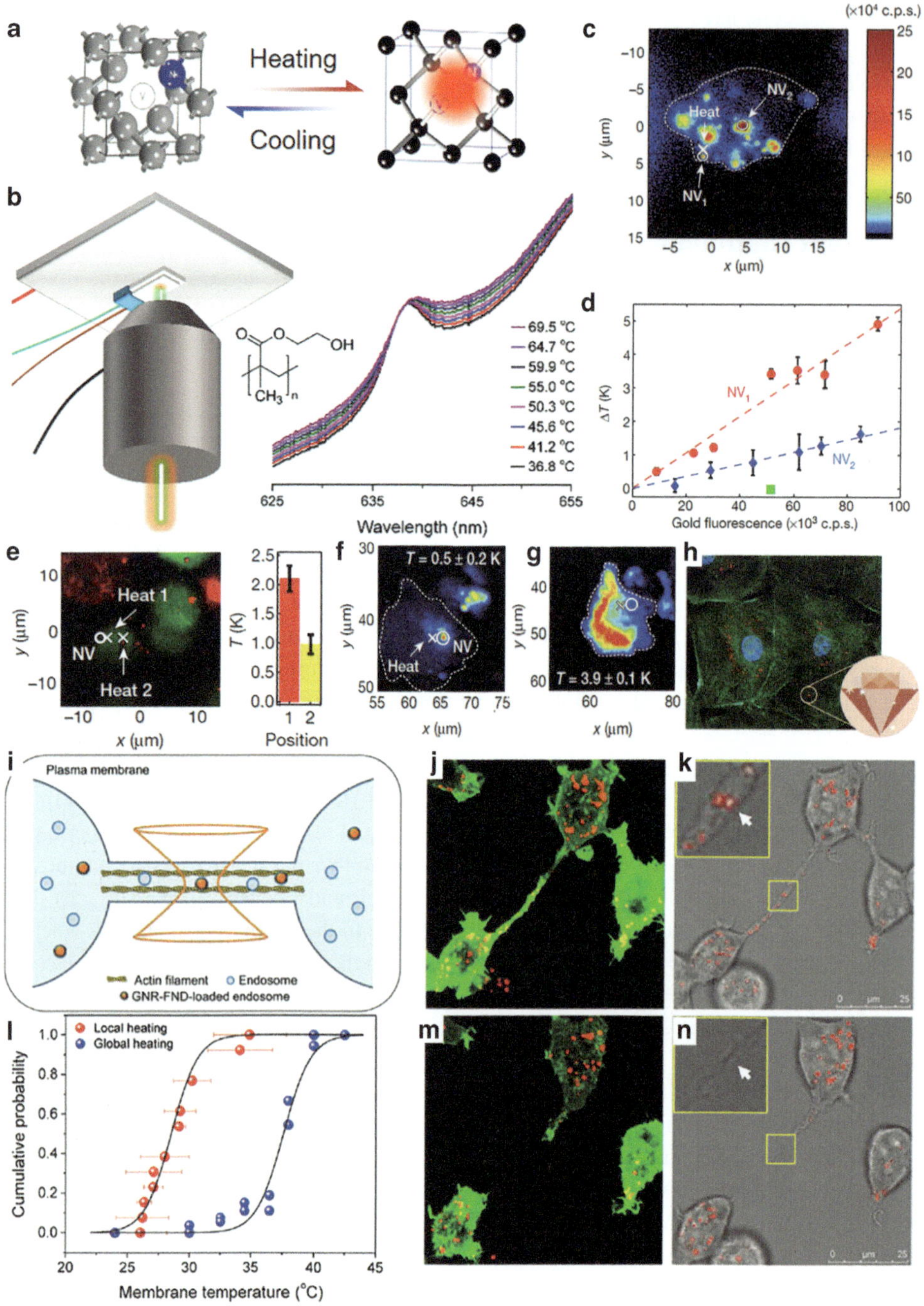

Fig. 9.5 Schematic illustrations of nanodiamond-based temperature sensing. **a** Diamond crystal structure changes with temperature. **b** Schematic diagram of all-optical thermometry based on nitrogen vacancies. Copyright © 2019 American Chemical Society [49]. **c** Temperature imaging of single cells based on nanodiamonds and heated gold nanoparticles. **d** Temperature change images of NV_1 and NV_2 at different locations. **e** Fluorescence Scanning Stained Cells (Live and Dead Assay). **f** Single-cell fluorescence image at low illumination power. **g** Single-cell fluorescence image at high illumination power. Copyright © 2013, Springer Nature Limited. [5] **h** Nanodiamond-based photoluminescence image of Hela cells. Copyright © 2018 WILEY–VCH Verlag GmbH & Co. KGaA, Weinheim [47]. **i** Temperature sensing of diamonds in cellular nanotubes. **j** Fluorescence pictures of GFP labeling (green) and diamond labeling (red). **k** Bright-field images of (**j**). **l** Images of membrane temperature changes during local and global heating. **m** Fluorescence image after 6 s of laser irradiation with laser power of 330 μW. **n** Bright-field images of **m**. Copyright © 2017 Wiley–VCH Verlag GmbH & Co. KGaA, Weinheim [50]

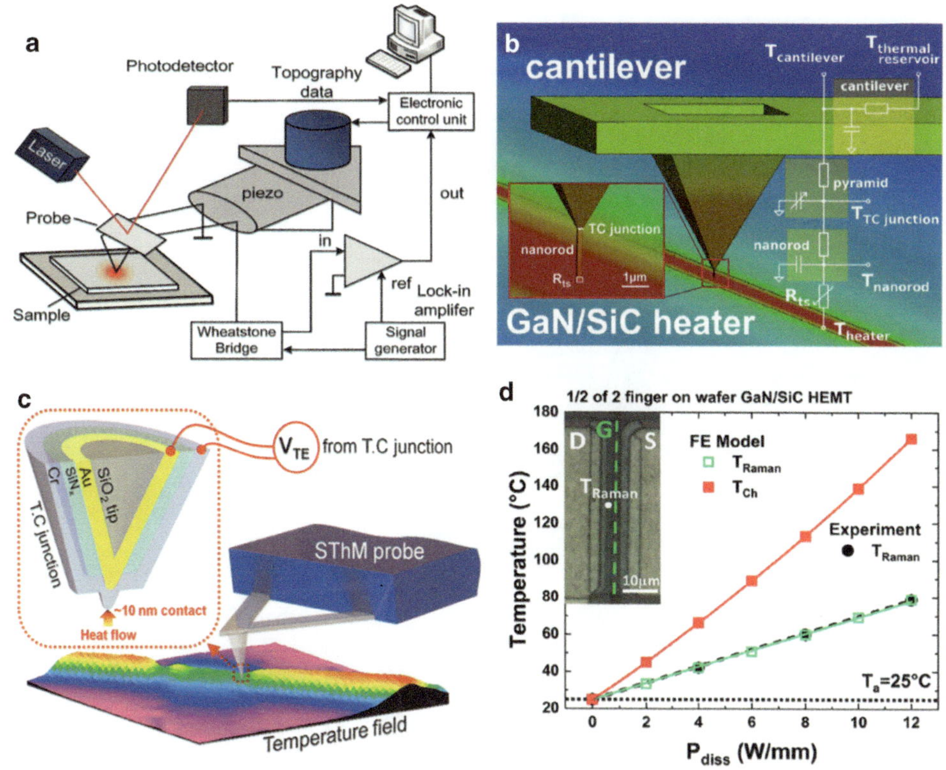

Fig. 9.6 Scanning thermal microscopy (SThM). **a** Equipment diagram of SThM. Copyright © 2019 The Authors. Published by WILEY–VCH Verlag GmbH & Co. KGaA, Weinheim [52]. **b** Schematic illustration of quantitative temperature measurement by SThM thermocouple tip. Copyright © 2012 American Chemical Society [55]. **c** Transient thermal simulations were performed using the VTP-200 sThM probe to match the probe's thermal time constant. **d** Comparison curves of experimental and simulated Raman temperature. Copyright © 2021 Elsevier Ltd. [57]

measure the thermal conductivity of the sample and obtain a temperature distribution map. Future challenges of scanning thermal microscopy will focus on extending its applicability and the physical study of heat transfer mechanisms during thermometry. Greater progress in more areas is expected [57–60] (Fig. 9.6).

9.3.2 Raman Spectroscopy

When light passes through a transparent medium, light scattered by molecules undergoes a frequency change, a phenomenon known as Raman scattering. The spectral line whose frequency is symmetrically distributed on both sides of the incident light frequency is the Raman spectrum. Among them, when the frequency of the scattered light is higher than the frequency of the incident light, the generated spectral line is the anti-Stokes Raman line; when the frequency of the scattered light is lower than the frequency of the incident light, the generated spectral line is the Stokes Raman line [61].

The use of nanoparticles, such as gold nanoparticles, to achieve nanoscale heating research and applications has grown tremendously in recent years. However, due to the lack of effective nanoscale thermometry, further development is limited. Especially in some applications of nanoplasmonics, the side effects of nanoplasma heating on nanoscale temperature measurement have been neglected [62]. Traditional methods of measuring temperature using Raman spectroscopy require calibration and may produce temperatures that are not the same as the material being measured. Therefore, an important example reports the use of label-free, insensitive to other physical parameters, the anti-Stokes Raman thermometry method that can directly obtain the temperature to achieve nano-scale temperature measurement [63]. Since the initial state of the particle to achieve anti-Stokes Raman scattering must be an excited state energy level. But this energy filling is difficult to achieve at room temperature. Therefore, at room temperature, according to the Boltzmann distribution, the ratio of the anti-Stokes signal to the Stokes signal is relatively weak. As the temperature increases, the initial state of the excited state gradually increases, and the ratio of the two gradually increases accordingly. Based on this temperature dependence, the temperature measurement of the material can be achieved by monitoring the ratio of the two [64, 65].

The development of current methods to measure temperature using the anti-Stokes to Stokes ratio is limited by the fact that nanoparticle heating is required to generate surface plasmons, and more experiments and theory to achieve further development and research. It is worth noting that it has always been crucial to measure temperatures above 1000 degrees Celsius. Because the research on high temperature structural

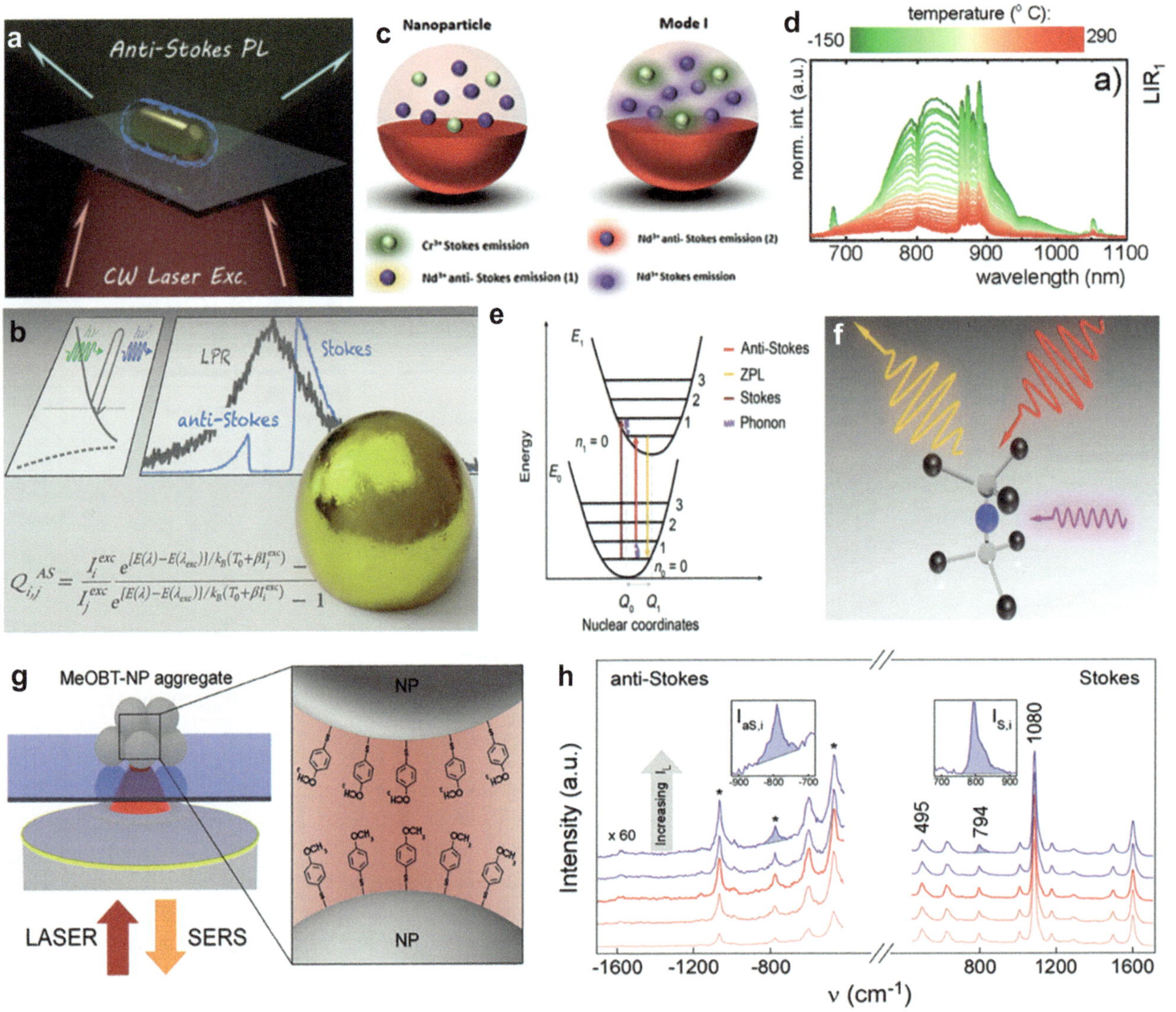

Fig. 9.7 Raman spectroscopy as nanothermometer. **a** Schematic illustration of anti-Stokes Raman emission of hot carriers in gold nanorods. Copyright © 2019 American Chemical Society [64]. **b** Anti-Stokes Raman thermometry in nanoplasmas. Copyright © 2021 American Chemical Society [65]. **c** Schematic representation of the readout of Mode I temperature using nanoparticles. **d** Thermal evolution curve of the emission spectrum in Mode I. Copyright © 2020 Elsevier B.V [62]. **e** Diamond gains extra energy through phonon absorption in the anti-Stokes case. **f** Schematic illustration of the anti-Stokes Raman mechanism at the rhombohedral center of diamond. Copyright © 2019, The American Association for the Advancement of Science [67]. **g**, **h** Schematic illustration of SERS-based Stokes and anti-Stokes thermometry of metal surface molecules. Copyright © 2022, American Chemical Society [63]

changes of various industrial materials and minerals is within this temperature range. However, the traditional visible light excited Raman spectrum has a limited measurement range, so how to use an effective thermometry method to measure the temperature in the range of 1000–1500 is very important. As another application of Raman spectroscopy, an important example is reported using UV-excited Raman scattering spectroscopy, using anti-Stokes Raman scattering to achieve a wide range of temperature measurements from 500 °C to 1500 °C The calibration greatly improves the upper limit of the experimental temperature using Raman spectroscopy [66, 67] (Fig. 9.7).

9.3.3 Thermocouple

Like scanning thermal microscopes and atomic force microscopes, probes using thermistors have been developed for high-precision area temperature measurement. With access to the size, sensitivity, and stability of suitable nanoprobes, temperature measurements with high resolution and localization accuracy can be greatly achieved [68, 69]. In addition, the introduction of cantilever beams and the preparation of probes based on nanolithography technology enable higher-precision (20 nm) and even single-point-assisted area temperature measurement to be realized [70].

The preparation of the thermistor probe and the design selection of the temperature measurement method are one of the most important factors affecting the temperature measurement accuracy of the thermistor nanothermometer. An important example reports the use of well-established Focused Electron Beam Induced Deposition (FEBID) to fabricate free-standing, multi-legged 3D bridge architectures that are electrically manipulated to enable temperature-dependent resistance change measurements [71]. In addition to greatly improving predictability and reliability, this approach can also be used for 3D nanoscale applications, including electromagnetism, plasmonics, and nanosensing. In addition to its 3D capabilities, focused electron beam-induced deposition allows direct write preparation on almost any given material without requiring any additional processing in specific areas. This approach provides a very small effective volume, so there is no reduction in resolution and heat transfer effects of thermometry convection. Compared with traditional substituted resistors, 3D nanothermistors based on focused electron beam-induced deposition: (1) The effective contact area of the probe tip is reduced to within 10 nm, thus achieving higher temperature imaging resolution; (2) The method of using focused electron beam induced deposition in the early stage greatly reduces the processing time of the sample (~ 10 min), eliminating more and more complicated step procedures [72]. In addition to being used alone to achieve nanometer temperature sensing, thermistors are often combined and assembled with other technologies to achieve higher precision and higher resolution temperature sensing functions, such as scanning thermal microscopy [73] (Fig. 9.8).

9.3.4 Thermal Expansion

To achieve temperature measurement with subwavelength spatial resolution, the most promising are of course those methods such as scanning thermal microscopy based on highly sensitive thermistors, but this technology is limited by complex equipment, imaging temperature measurement speed and high invasiveness It is difficult to use in daily practical applications. Therefore, the long-range detection of surface deposition of temperature-sensitive materials based on surface optics has attracted much attention. The phenomenon of thermal expansion of temperature-sensitive phase change materials (PCMs) caused by temperature changes has attracted much attention, especially its superiority in thermal sensitivity, thermometric properties, and temperature counting. As a promising PCM material, photothermal conversion induced by molecular spin-intersecting nanoparticles is often used for surface temperature measurement and thermal imaging [74].

An important example is reported of a molecularly self-selected cross-complex [$Fe(HB(tz)_3)_2$] (tz = 1,2,4-triazol-1-yl)1, deposited in high-quality vacuum form A type of film that overcomes the obstacles of repeated mechanical fatigue problems of common SCO materials [75]. This material exhibits exceptionally strong spin stability and durability after repeatable thermal cycling, as well as excellent spin transition properties. After probing by optical reflectivity, the material was used for temperature visualization and surface nanotemperature measurements. Although limited by its submicron spatial resolution, this SCO material has a low signal-to-noise ratio and remains the most traditional thermal imaging modality. In recent years, through the combination of atomic force microscopy and thermistor, the temporal and spatial resolution has been further improved, so that local temperature measurement at the nanometer scale can be performed. Undeniably, this discovery constitutes a crucial step towards mature practical application of SCO materials.

Not only that, the spin-crossover phenomenon of materials induced by temperature change itself has attracted attention due to the long lifetime of the photo-induced state and the inability to return to the initial energy level [76]. Under such constraints, it is urgent to achieve high spin state transition tolerance and stability by changing the composition of molecular complexes. An important example has been reported using the SCO composite ([$Fe(HB(1,2,4\text{-triazol-1-yl})_3)_2$] (1)) to achieve exceptionally high spin state transition tolerance, resulting in Reliable, quantitative pump-probe measurements can be achieved in unique single-crystal thermal hysteresis lines [77]. Among them, an important way is to study the further effects of smaller size on thermodynamically driven switching phenomena at the micro-nano scale. So far, in thermal expansion nanothermometers, it remains the ultimate challenge to study the temperature-dependent dynamics of the spin-crossover phenomenon (Fig. 9.9).

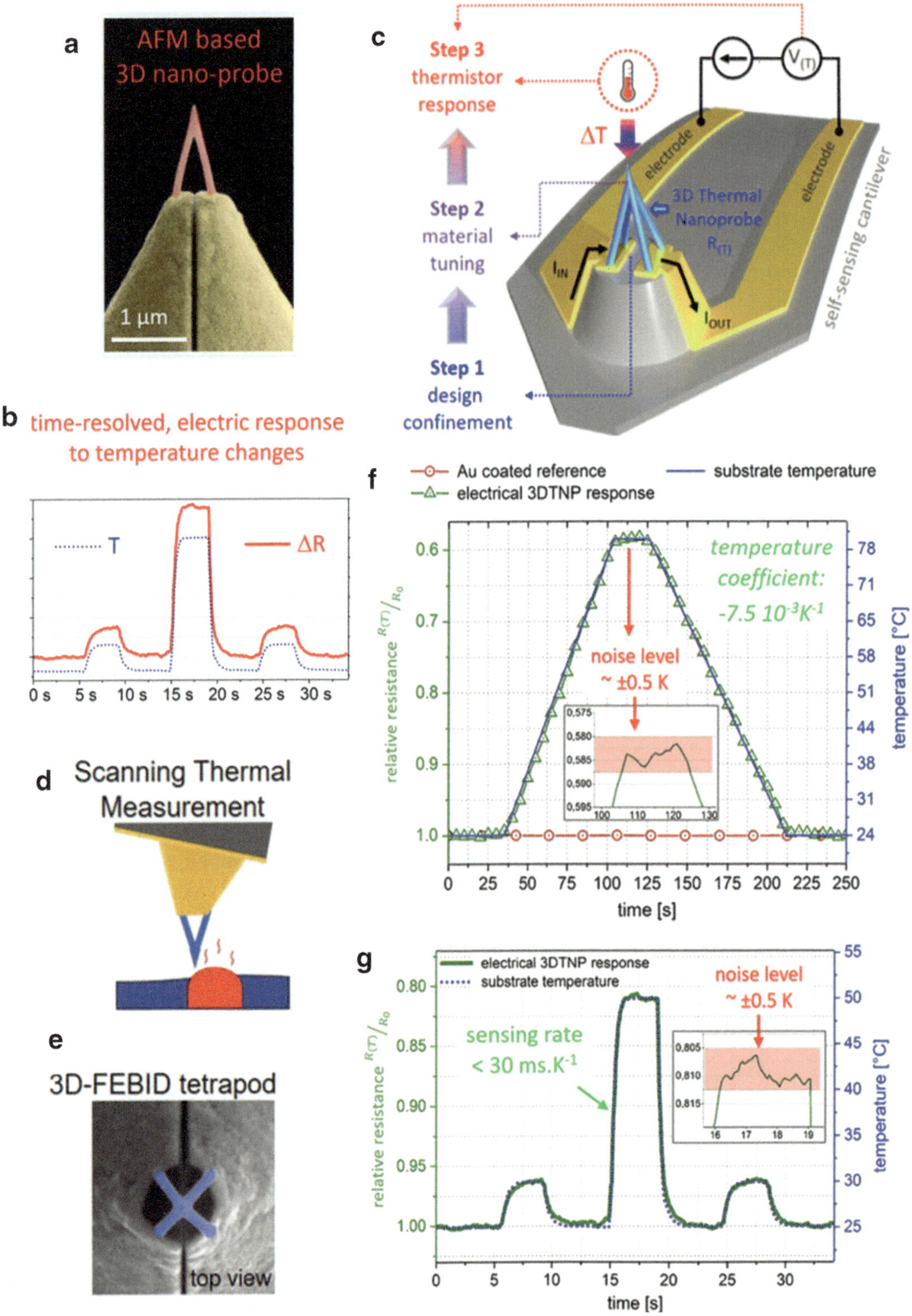

Fig. 9.8 Thermocouple as nanothermometer. **a** 3D Nano-thermocouple Scanning Probe for Atomic Force Microscopy. **b** Time-resolved thermocouple thermometry curve in response to temperature. Copyright © 2019, American Chemical Society [71]. **c** Schematic diagram of thermocouple-based nanometer temperature detection. **d** Thermocouple-based scanning thermal detection technology. **e** 3D SEM image of a thermocouple probe. Copyright © 2009, American Chemical Society [72]. **f** Electrical response curves of a three-dimensional thermocouple probe to temperature increase. **g** Evaluation curves for temperature measurements but with the highest heating efficiency. Copyright © 2017, The Author(s) [73]

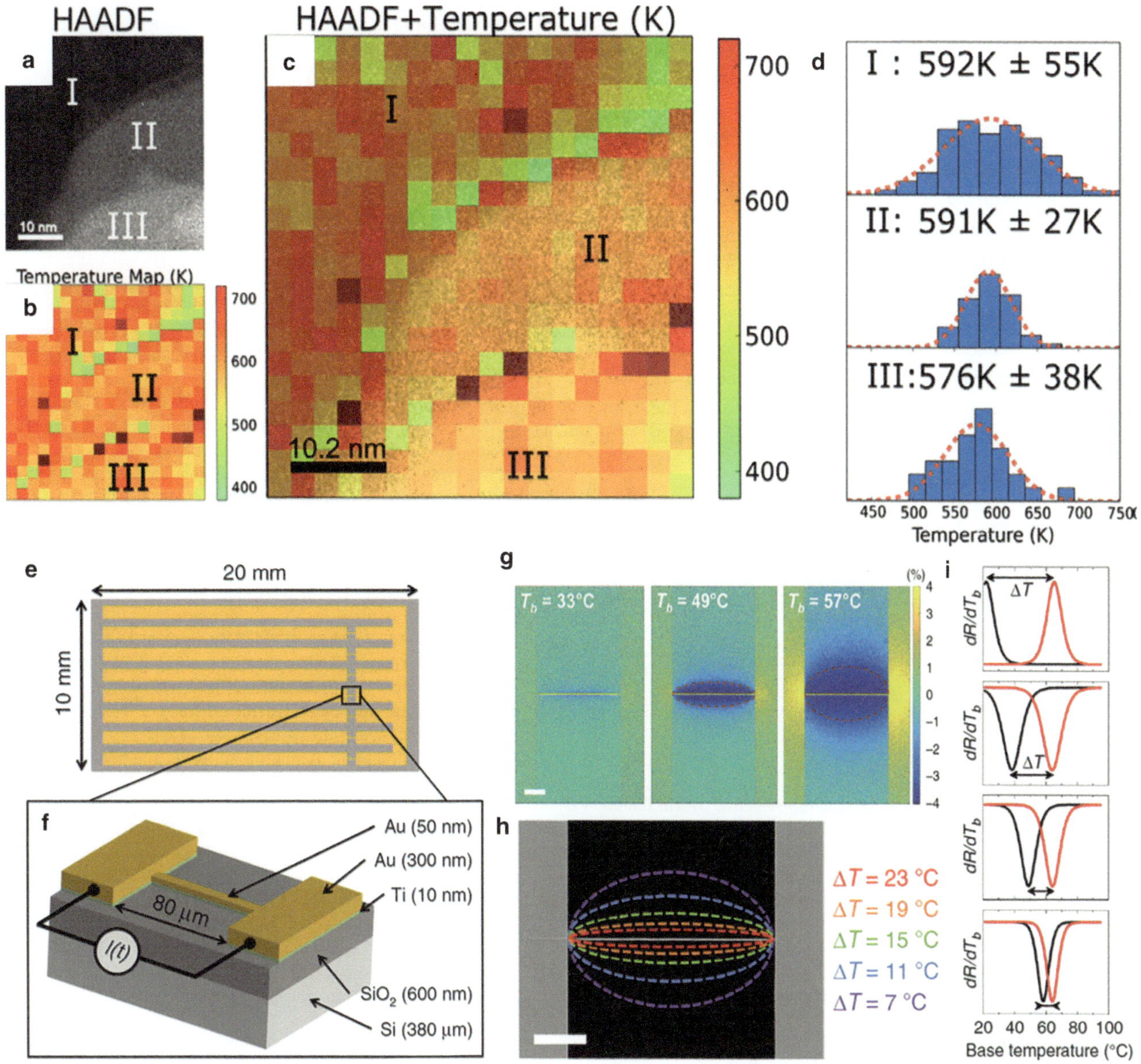

Fig. 9.9 Nanothermometers based on thermal expansion. **a** HAADF image of $MoSe_2$ nanosheets. **b** Corresponding temperature map of $MoSe_2$ nanosheets at a sample temperature of 573 K. **c** Overlaid plots of (**a**) and (**b**). **d** Temperature distribution map for each area. Copyright © 2018 American Physical Society [76]. **e** Schematic diagram of the chip array for thermal expansion temperature sensing. **f** Partial schematic diagram of thermal expansion nano-temperature thermometry. **g** Optical reflectance images of nanowires at selected temperatures. **h** Surface temperature contour plots obtained at different reference temperatures. **i** Graph showing the measured temperature difference as the substrate temperature changes. Copyright © 2021 Wiley–VCH GmbH [77]

9.4 Conclusion

This review systematically presents various approaches to achieve nanoscale temperature measurement using both fluorescent and non-fluorescent modes. For example, the construction of nanometer thermometers is realized by using fluorescent materials such as organic dyes, lanthanides, quantum dots, and diamonds. Using scanning thermal microscopy, anti-Stokes scattering, thermistors, and surface material spin-crossing properties, etc., can also construct nanothermometers for temperature visualization and sensing. As a hot field, nanoscale temperature measurement has experienced very rapid and huge development in recent years, and will continue to change rapidly for many years to come with the innovation of micro-nano miniaturization technology. Although very gratifying progress has been made in related fields, the rapid development has also brought many problems, including the mechanism research of various thermometry and the combination of various cross-technologies, etc., which limit the further development of nano thermometry. Until now, the field is still in the early stages of development, and there are many unknowns to learn before nanothermometers become commercial products.

To sum up, there are still many research topics and research directions for nanothermometers worthy of study. The numerous challenges and obstacles encountered in the development process require the joint efforts of experts in various fields and disciplines. As a new interdisciplinary subject, nanothermometer will surely shine in the fields of nanomaterials, chemical synthesis, biomedicine, optoelectronics, micro-nano microscopy and other fields in the future.

Acknowledgements This work was supported by National Natural Science Foundation of China (No. 12372317), Shenzhen Science and Technology Program (No. 202308073001816) and the China Postdoctoral Science Foundation (No. 2024M753104). Z.G. and Y.Y. contributed equally to this work.

References

1. Liu CF, Leong WH, Xia K, et al. Ultra-sensitive hybrid diamond nanothermometer. National Sci Rev. 2021;8(5): nwaa194.
2. Zhou J, Del Rosal B, Jaque D, et al. Advances and challenges for fluorescence nanothermometry. Nat Methods. 2020;17(10):967–80.
3. Cheng Q, Rajauria S, Schreck E, et al. Precise nanoscale temperature mapping in operational microelectronic devices by use of a phase change material. Sci Rep. 2020;10(1):20087.
4. Liu H, Fan Y, Wang J, et al. Intracellular temperature sensing: An ultra-bright luminescent nanothermometer with non-sensitivity to pH and ionic strength. Sci Rep. 2015;5(1):14879.
5. Kucsko G, Maurer PC, Yao NY, et al. Nanometre-scale thermometry in a living cell. Nature. 2013;500(7460):54–8.
6. Wu X, Zhan S, Han J, et al. Nanoscale ultrasensitive temperature sensing based on upconversion nanoparticles with lattice self-adaptation. Nano Lett. 2020;21(1):272–8.
7. Tan M, Li F, Cao N, et al. Accurate in vivo nanothermometry through NIR-II lanthanide luminescence lifetime. Small. 2020;16(48):2004118.
8. Nexha A, Carvajal JJ, Pujol MC, et al. Lanthanide doped luminescence nanothermometers in the biological windows: strategies and applications. Nanoscale. 2021;13(17):7913–87.
9. Mohammed LJ, Omer KM. Carbon dots as new generation materials for nanothermometer. Nanoscale Res Lett. 2020;15(1):182.
10. Xiang G, Xia Q, Liu X, et al. Upconversion nanoparticles modified by Cu_2S for photothermal therapy along with real-time optical thermometry. Nanoscale. 2021;13(15):7161–8.
11. Fujiwara M, Shikano Y. Diamond quantum thermometry: From foundations to applications. Nanotechnology. 2021;32:482002.
12. Fischer LH, Harms GS, Wolfbeis OS. Upconverting nanoparticles for nanoscale thermometry. Angew Chem Int Ed. 2011;50(20):4546–51.
13. Bradac C, Lim SF, Chang HC, et al. Optical nanoscale thermometry: from fundamental mechanisms to emerging practical applications. Adv Opt Mater. 2020;8(15):2000183.
14. Bednarkiewicz A, Drabik J, Trejgis K, et al. Luminescence based temperature bio-imaging: status, challenges, and perspectives. Appl Phys Rev. 2021;8(1):011317.
15. Kniec K, Piotrowski W, Ledwa K, et al. Spectral and thermometric properties altering through crystal field strength modification and host material composition in luminescence thermometers based on Fe3+ doped AB2O4 type nanocrystals (A= Mg, Ca; B= Al, Ga). J Mater Chem C. 2021;9(2):517–27.
16. Sadat S, Tan A, Chua YJ, et al. Nanoscale thermometry using point contact thermocouples. Nano Lett. 2010;10(7):2613–7.
17. Hartman T, Geitenbeek RG, Wondergem CS, et al. Operando nanoscale sensors in catalysis: all eyes on catalyst particles. ACS Nano. 2020;14(4):3725–35.
18. Shimizu T, Ding W, Kameta N. Soft-matter nanotubes: a platform for diverse functions and applications. Chem Rev. 2020;120(4):2347–407.
19. Su X, Wen Y, Yuan W, et al. Lifetime-based nanothermometry in vivo with ultra-long-lived luminescence. Chem Commun. 2020;56(73):10694–7.
20. Ji Z, Cheng Y, Cui X, et al. Heating-induced abnormal increase in Yb 3+ excited state lifetime and its potential application in lifetime luminescence nanothermometry. Inorg Chem Front. 2019;6(1):110–6.
21. Hemmer E, Acosta-Mora P, Mendez-Ramos J, et al. Optical nanoprobes for biomedical applications: shining a light on upconverting and near-infrared emitting nanoparticles for imaging, thermal sensing, and photodynamic therapy. J Mater Chem B. 2017;5(23):4365–92.
22. Vetrone F, Naccache R, Zamarrón A, et al. Temperature sensing using fluorescent nanothermometers. ACS Nano. 2010;4(6):3254–8.
23. Qin T, Liu B, Zhu K, et al. Organic fluorescent thermometers: highlights from 2013 to 2017. TrAC Trends Anal Chem. 2018;102:259–71.
24. Liu J, Zhang H, Selopal GS, et al. Visible and near-infrared, multiparametric, ultrasensitive nanothermometer based on dual-emission colloidal quantum dots. ACS Photonics. 2019;6(10):2479–86.
25. Yuexuan Y, Daocheng W. Research shortcomings of fluorescent nanothermometers in biological and medical fields. Nanomedicine. 2020;15(08):735–8.
26. Wan Y, Cui Y, Yang Y, et al. Dye confined in metal-organic framework for two-photon fluorescent temperature sensing. Microporous Mesoporous Mater. 2018;268:202–6.
27. Zhao D, Yue D, Jiang K, et al. Ratiometric dual-emitting MOF⊃ dye thermometers with a tunable operating range and sensitivity. J Mater Chem C. 2017;5(7):1607–13.
28. Cui Y, Song R, Yu J, et al. Dual-emitting MOF⊃ dye composite for ratiometric temperature sensing. Adv Mater. 2015;27(8):1420–5.
29. Xue K, Wang C, Wang J, et al. A sensitive and reliable organic fluorescent nanothermometer for noninvasive temperature sensing. J Am Chem Soc. 2021;143(35):14147–57.
30. Meng L, Jiang S, Song M, et al. TICT-based near-infrared ratiometric organic fluorescent thermometer for intracellular temperature sensing. ACS Appl Mater Interfaces. 2020;12(24):26842–51.
31. Liu X, Liu J, Zhou H, et al. Ratiometric dual fluorescence tridurylboron thermometers with tunable measurement ranges and colors. Talanta. 2020;210:120630.
32. Brites CDS, Balabhadra S, Carlos LD. Lanthanide-based thermometers: at the cutting-edge of luminescence thermometry. Adv Opt Mater. 2019;7(5):1801239.
33. Jia M, Sun Z, Zhang M, et al. What determines the performance of lanthanide-based ratiometric nanothermometers? Nanoscale. 2020;12(40):20776–85.

34. Kolesnikov IE, Mamonova DV, Kalinichev AA, et al. Construction of efficient dual activating ratiometric YVO 4: Nd 3+/Eu 3+ nanothermometers using co-doped and mixed phosphors. Nanoscale. 2020;12(10):5953–60.
35. Stopikowska N, Runowski M, Skwierczyńska M, et al. Improving performance of luminescent nanothermometers based on non-thermally and thermally coupled levels of lanthanides by modulating laser power. Nanoscale. 2021;13(33):14139–46.
36. Gong Z, Wu T, Chen X, et al. Upconversion nanoparticle decorated spider silks as single-cell thermometers. Nano Lett. 2021;21(3):1469–76.
37. Molaei MJ. Carbon quantum dots and their biomedical and therapeutic applications: a review. RSC Adv. 2019;9(12):6460–81.
38. Lin X, Kong M, Wu N, et al. Measurement of temperature distribution at the nanoscale with luminescent probes based on lanthanide nanoparticles and quantum dots. ACS Appl Mater Interfaces. 2020;12(47):52393–401.
39. Zhang N, Liu X, Wei Z, et al. Cell imaging using two-photon excited CdS fluorescent quantum dots working within the biological window. Nanomaterials. 2019;9(3):369.
40. Li S, Zhang K, Yang JM, et al. Single quantum dots as local temperature markers. Nano Lett. 2007;7(10):3102–5.
41. Yang JM, Yang H, Lin L. Quantum dot nano thermometers reveal heterogeneous local thermogenesis in living cells. ACS Nano. 2011;5(6):5067–71.
42. Zhao H, Vomiero A, Rosei F. Tailoring the heterostructure of colloidal quantum dots for ratiometric optical nanothermometry. Small. 2020;16(28):2000804.
43. Cao S, Zheng J, Zhao J, et al. Robust and stable ratiometric temperature sensor based on Zn–In–S quantum dots with intrinsic dual-dopant ion emissions. Adv Func Mater. 2016;26(40):7224–33.
44. Han Y, Liu Y, Zhao H, et al. Highly efficient ratiometric nanothermometers based on colloidal carbon quantum dots. J Mater Chem B. 2021;9(20):4111–9.
45. Zhang T, Pramanik G, Zhang K, et al. Toward quantitative biosensing with nitrogen–vacancy center in diamond. ACS sensors. 2021;6(6):2077–107.
46. Casola F, Van Der Sar T, Yacoby A. Probing condensed matter physics with magnetometry based on nitrogen-vacancy centres in diamond. Nat Rev Mater. 2018;3(1):17088.
47. Chipaux M, van der Laan KJ, Hemelaar SR, et al. Nanodiamonds and their applications in cells. Small. 2018;14(24):1704263.
48. Sotoma S, Zhong C, Kah JCY, et al. In situ measurements of intracellular thermal conductivity using heater-thermometer hybrid diamond nanosensors. Sci Adv. 2021;7(3):eabd7888.
49. Hui YY, Chen OY, Azuma T, et al. All-optical thermometry with nitrogen-vacancy centers in nanodiamond-embedded polymer films. J Phys Chem C. 2019;123(24):15366–74.
50. Tsai PC, Epperla CP, Huang JS, et al. Measuring nanoscale thermostability of cell membranes with single gold–diamond nanohybrids. Angew Chem Int Ed. 2017;56(11):3025–30.
51. Chmielewski A, Ricolleau C, Alloyeau D, et al. Nanoscale temperature measurement during temperature controlled in situ TEM using Al plasmon nanothermometry. Ultramicroscopy. 2020;209:112881.
52. Zhang Y, Zhu W, Hui F, et al. A review on principles and applications of scanning thermal microscopy (SThM). Adv Func Mater. 2020;30(18):1900892.
53. Gomès S, Assy A, Chapuis PO. Scanning thermal microscopy: a review. physica status solidi. 2015;212(3):477–94
54. Pylkki RJ, Moyer PJMPJ, West PEWPE. Scanning near-field optical microscopy and scanning thermal microscopy. Jpn J Appl Phys. 1994;33(6S):3785.
55. Kim K, Jeong W, Lee W, et al. Ultra-high vacuum scanning thermal microscopy for nanometer resolution quantitative thermometry. ACS Nano. 2012;6(5):4248–57.
56. Tovee PD, Pumarol ME, Rosamond MC, et al. Nanoscale resolution scanning thermal microscopy using carbon nanotube tipped thermal probes. Phys Chem Chem Phys. 2014;16(3):1174–81.
57. Gucmann F, Pomeroy JW, Kuball M. Scanning thermal microscopy for accurate nanoscale device thermography. Nano Today. 2021;39:101206.
58. Nguyen TP, Lemaire E, Euphrasie S, et al. Microfabricated high temperature sensing platform dedicated to scanning thermal microscopy (SThM). Sens Actuators A. 2018;275:109–18.
59. Barker A, Sapkota B, Oviedo JP, et al. Automated plasmon peak fitting derived temperature mapping in a scanning transmission electron microscope. AIP Adv. 2021;11(3):035330.
60. Umatova Z, Zhang Y, Rajkumar R, et al. Quantification of atomic force microscopy tip and sample thermal contact. Rev Sci Instrum. 2019;90(9):095003.
61. Prochazkova AJ, Mayr F, Gugujonovic K, et al. Anti-Stokes photoluminescence study on a methylammonium lead bromide nanoparticle film. Nanoscale. 2020;12(31):16556–61.
62. Maciejewska K, Marciniak L. Multimodal Stokes and Anti-Stokes luminescent thermometers based on GdP_5O_{14} co-doped with Cr^{3+} and Nd^{3+} ions. Chem Eng J. 2020;402:126197.
63. Park S, Yeon GJ, Lee H, et al. Self-Referenced SERS Thermometry of Molecules on a Metallic Nanostructure. J Phys Chem C. 2021;126(1):451–8.
64. Cai YY, Sung E, Zhang R, et al. Anti-Stokes emission from hot carriers in gold nanorods. Nano Lett. 2019;19(2):1067–73.
65. Baffou G. Anti-Stokes thermometry in nanoplasmonics. ACS Nano. 2021;15(4):5785–92.
66. Fujimori H, Kakihana M, Ioku K, et al. Advantage of anti-Stokes Raman scattering for high-temperature measurements. Appl Phys Lett. 2001;79(7):937–9.
67. Tran TT, Regan B, Ekimov EA, et al. Anti-Stokes excitation of solid-state quantum emitters for nanoscale thermometry. Sci Adv. 2019;5(5):eaav9180.
68. Basov M. Schottky diode temperature sensor for pressure sensor. Sens Actuators A. 2021;331:112930.
69. Pligovka A, Lazavenka A, Gorokh G. Anodic niobia column-like 3-D nanostructures for semiconductor devices. IEEE Trans Nanotechnol. 2019;18:790–7.
70. Winkler R, Fowlkes JD, Rack PD, et al. 3D nanoprinting via focused electron beams. J Appl Phys. 2019;125(21):210901.
71. Sattelkow J, Fröch JE, Winkler R, et al. Three-dimensional nanothermistors for thermal probing. ACS Appl Mater Interfaces. 2019;11(25):22655–67.
72. Strelcov E, Lilach Y, Kolmakov A. Gas sensor based on metal–insulator transition in VO2 nanowire thermistor. Nano Lett. 2009;9(6):2322–6.
73. Yang F, Li G, Yang J, et al. Measurement of local temperature increments induced by cultured HepG2 cells with micro-thermocouples in a thermally stabilized system. Sci Rep. 2017;7(1):1721.
74. Hu Y, Picher M, Tran NM, et al. Photo-Thermal Switching of Individual Plasmonically Activated Spin Crossover Nanoparticle Imaged by Ultrafast Transmission Electron Microscopy. Adv Mater. 2021;33(52):2105586.
75. Ridier K, Bas AC, Zhang Y, et al. Unprecedented switching endurance affords for high-resolution surface temperature mapping using a spin-crossover film. Nat Commun. 2020;11(1):3611.
76. Hu X, Yasaei P, Jokisaari J, et al. Mapping thermal expansion coefficients in freestanding 2D materials at the nanometer scale. Phys Rev Lett. 2018;120(5):055902.
77. Ridier K, Nicolazzi W, Salmon L, et al. Sequential activation of molecular and macroscopic spin-state switching within the hysteretic region following pulsed light excitation. Adv Mater. 2022;34(6):2105468.

Raman Detection and Machine Learning in Biomedical Applications

10

Chenghong Lin, Qian Zhang, and Yuchao Li

Abstract

Raman spectroscopy, with its ability to detect substances at the molecular level and its non-labeling, high-precision, and non-invasive characteristics, is widely applied within the biomedical field. Compared with non-biological samples, biological samples exhibit a higher complexity in terms of material composition and distribution, coupled with substantial individual variances. Moreover, the Raman scattering signals emitted by these biological samples are inherently weak. Therefore, difficulties such as faint signals, data collection obstacles, and analysis complexities are often encountered during their Raman detection. With the advancement of computer technology, machine learning has played a significant role in various fields such as biomedicine, optics, and chemistry in recent years. It has performed well in tasks like classification, data analysis, and data prediction. In response to the challenges associated with Raman spectroscopy examination of biological samples, many researchers in recent years have employed a combination of Raman spectroscopy and machine learning for the analysis of biological samples at various scales, which has realizesd functions such as cancer diagnosis, cell sorting, bacteria identification, and virus monitoring. The combination of Raman detection and machine learning thus opens up a new research avenue in the realm of biomedicine.

Keywords

Raman spectroscopy · Machine learning · Biomedicine · Cancer diagnosis · Cell identification

C. Lin · Q. Zhang · Y. Li (✉)
Guangdong Provincial Key Laboratory of Nanophotonic Manipulation, Institute of Nanophotonics, College of Physics and Optoelectronic Engineering, Jinan University, Guangzhou, China
e-mail: liyuchao@jnu.edu.cn

10.1 Introduction

10.1.1 Raman Spectroscopy

In 1928, Indian scientist Raman first observed the phenomenon of frequency changes in light scattering when it passed through a transparent medium [1], which later came to be known as Raman scattering. Subsequently, in 1934, Pracheck refined the fundamental theory of Raman scattering [2]. Due to the low intensity of Raman scattering, coupled with the rapid development of infrared technology, research and applications of Raman spectroscopy once fell into a developmental trough. Not until the advent of the ruby laser in 1960, with its characteristics of high power density, monochromaticity, and strong directivity, was the issue of low Raman scattering efficiency resolved. This crucial development made the research of Raman spectroscopy finally ushered in a peak. In 1974, the discovery of Surface Enhanced Raman Scattering (SERS) by Fleischmann et al. further propelled the development of Raman spectroscopy research [3]. From the end of the twentieth century to the present, the advent of various advanced Raman spectrometers and Raman microscopes has increasingly highlighted the importance of Raman spectroscopy. It plays a significant role in numerous fields such as biology, medicine, chemistry, and industry.

Raman spectroscopy, commonly dubbed the "molecular fingerprint" of substances, operates on the principle that when a portion of the photons from the excitation light collide inelastically with molecules, an energy exchange occurs, thereby altering the frequency and direction of the light. The majority of photons from the excitation light will undergo elastic collisions with molecules, leaving the photon frequency unchanged. This type of scattering is known as Rayleigh scattering. The frequency difference between Raman and Rayleigh scattering is referred to as the Raman shift (Fig. 10.1a). The magnitude of the Raman shift depends on the change in the vibrational energy level of the molecule,

B. Li et al. (eds.), *Biophotonic Manipulation*, Advances in Optics and Optoelectronics,
https://doi.org/10.1007/978-981-96-4982-2_10

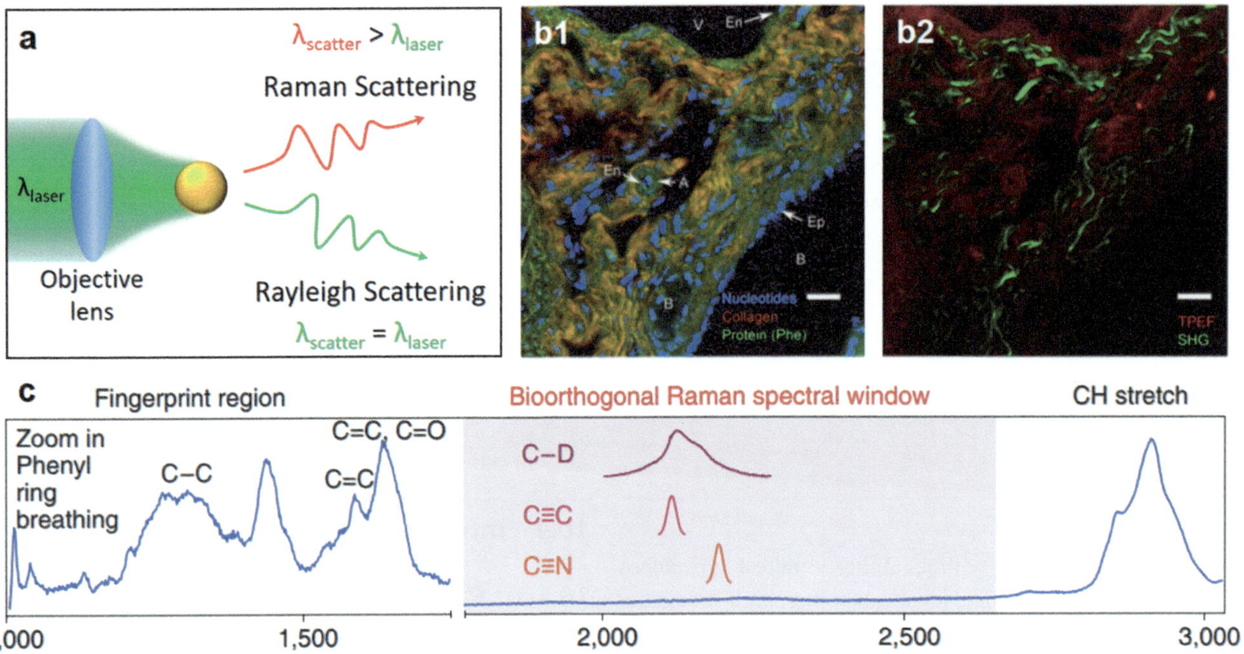

Fig. 10.1 Principles and applications of Raman spectroscopy **a** Basic principle of Raman spectroscopy. **b** Raman imaging of portal veins in mouse liver tissue, with nuclei in blue, collagen in orange, and protein content in green. (Image adapted the Open Access paper Nitta et al. [10], under Copyright (2020) the Authors). **c** A typical cellular Raman spectrum with a spectrally silent region of the cell. (Image adapted with permission from Hu et al. [7], Copyright (2019), with permission from Springer Nature)

each substance has different chemical bonds and vibration modes, which produce distinctive Raman shifts during Raman scattering, forming a Raman spectrum. These spectra reflect the molecular composition of the substances. Therefore, by analyzing the Raman spectra, we can obtain information about the molecular composition of the substances.

Raman spectroscopy is a technique that is non-invasive, label-free, and requires no sample preparation. It has less interference from water compared to infrared spectroscopy. The Raman peaks of water mainly appear beyond 3000 cm^{-1}, while the characteristic peaks of biological samples are primarily within the range of 1000–3000 cm^{-1} (Fig. 10.1c). This makes Raman spectroscopy particularly suitable for detecting biological samples [4–6]. The diverse molecular composition in biological samples, along with the distribution of characteristic molecules in Raman spectra, provides the possibility to distinguish different samples. Raman imaging, which involves the collection of Raman spectra according to the spatial intensity distribution of specific Raman peaks (Fig. 10.1b), provides valuable spatial information regarding the material composition of biological samples [7–9]. This unique capability gives Raman spectroscopy a significant edge over other detection methods. The simplicity of sample preparation, cost-effective testing, the ability to deliver rapid and reliable results render Raman spectroscopy particularly suited for use in clinical settings.

10.1.2 Machine Learning

Machine learning, first proposed in the mid-twentieth century, set the stage for the development of artificial neural networks. This network evolved from the mathematical neuron model proposed by McCulloch and Pitts in 1943 [11]. In 1957, Rosenblatt et al. proposed the concept of the perceptron [12], utilizing the equipment available at that time to tune parameters according to training data. However, the perceptron was inherently flawed in that it could not tackle nonlinear problems. It couldn't even perform simple logic functions such as XOR. Moreover, Efforts to resolve nonlinear problems by stacking multiple perceptrons were impeded by the complexities of model construction and parameter tuning. In order to solve this problem, in 1982, Hopfield et al. proposed the first complete artificial neural network model in the true sense [13], and perfected its theoretical basis. In 1986, Rumelhart et al. proposed the back-propagation algorithm [14], a method that gained widespread use in neural network training and yielded significant results. Consequently, the two major issues hindering the development of neural networks were finally resolved. This subsequently ushered in an upsurge in the field. However, due to the local optima encountered with the gradient descent method and the sluggish training of multi-layer neural networks owing to inadequate hardware performance, the progression of neural network technology soon hit another slump. In 2006, the Boltzmann machine and deep belief network, as proposed by Hinton et al., effectively addressed the issue of local optima [15]. Specifically, these models simplified the convergence of training results to the target extremum through the process of parameter initialization. Along with the advancement of computer technology, there has been a gradual enhancement in computational capacities. This progression has facilitated the emergence of increasingly complex neural networks, giving birth to the concept of "machine learning". Consequently, its research and application have progressively ushered in a new wave of interest. The proposal of the ReLU function, the groundbreaking success of the Convolutional Neural Network (CNN) AlexNet in the ImageNet Challenge, and AlphaGo's victory over Go champion Lee Sedol all pointed to the fact that machine learning technology was increasingly gaining prominence across diverse fields. Concurrently, this advancement drew an unprecedented level of interest from researchers in numerous disciplines.

Machine learning, as one of the core technologies in the realm of artificial intelligence, primarily functions to extract patterns from intricate data via algorithms. These algorithms need both high-quality data and significant computational power from the systems involved. In this era of advanced computer technology, machine learning technology has made significant progress, and has already yielded notable achievements in various research fields such as optics, material science, and chemistry. In recent years, owing to the intricate molecular composition of biological samples and the inherent variability amongst individual specimens, the data obtained through various measurements often not intuitive (Fig. 10.2a1, a2). Moreover, many in vivo studies require the samples to be undamaged (Fig. 10.2b1, b2, c). These issues often pose challenges to their analysis. To overcome this, numerous biological studies now employ machine learning methods for sample analysis (Fig. 10.2d). This includes automatic segmentation and classification of sample images [16], predicting brain activation regions during reading [17], and capturing gesture information via wearable devices [18], among other applications. After combined with Raman spectroscopy, functionalities like diabetes detection [19], cellular metabolite gradient detection [20], and identification of different edible oil types [21] have been accomplished, demonstrating its vast potential in the field of biomedicine.

10.1.3 Raman Spectroscopy Combined with Machine Learning in Other Fields

The Raman spectra of biological samples contain abundant molecular information. Traditional analysis methods often struggle to fully exploit this information, whereas machine learning techniques can effectively analyze both Raman spectra and Raman imaging. Therefore, the majority of current techniques combining Raman spectroscopy with machine learning are employed within biological research. However, these methods are not confined to this domain. Raman spectroscopy and Raman imaging can also elucidate the molecular information of non-biological samples. By combining with machine learning, the accuracy and readability of Raman information can be improved. For example, Usant Kajendirarajah and others have developed two machine learning networks, which can quickly analyze the Raman imaging of carbon nanotubes and improve the spatial contrast of Raman imaging. These two networks can perform point-by-point analysis of Raman imaging according based on the background, the number of vibrational bands, and the ID/IG ratio, respectively (Fig. 10.3a), to obtain more accurate and intuitive carbon nanotube defect density distribution information [22]. Machine learning is also capable of addressing the instrumental noise and spectral anomalies in Raman spectroscopy. Brandt et al. employed Raman spectroscopy for detecting the concentration of microplastic particles in samples, with machine learning being used to process the spectral data. This approach simultaneously accomplished two functions: correcting distortions in the spectrum and enhancing its signal-to-noise ratio [23].

Similarly, this technique can be applied within the realm of information technology. Chen et al. crafted a suspended array utilizing Raman spectral coding (Fig. 10.3b1). This design leverages micro-quartz chips to bind with diverse molecular probes, enabling the Raman spectrum of the array to present distinct characteristic peaks (Fig. 10.3b2), thereby accomplishing Raman spectral coding. In this study, machine learning is utilized as a decoder. It decodes the Raman spectrum of the analyte using the ResNet network to extract the information contained in the array [24]. Beyond the function of the decoder, the integration of Raman spectroscopy and machine learning also enables comprehensive material characterization. Zhang et al. utilized high-speed Raman imaging methods in combination with machine learning to detect the location, quantity, and metallic properties of carbon nanotubes. They cultivated the carbon nanotubes on a fork-shaped growth substrate (Fig. 10.3c), enabling them to collect the Raman spectroscopy data using a line scanning method, thereby optimizing the collection speed. Subsequently, they applied convolutional neural networks to analyze the Raman data, obtaining information related to the carbon nanotubes [25]. It is proved that the method of Raman spectroscopy combined with machine learning also has certain potential in the field of industrial production. Within the realm of food safety, Berghian-Grosan et al. employed various machine learning networks to analyze the Raman spectra of fruit distillates, aiming to discern the fruit varieties and producers [26]. Hu et al. used three distinct

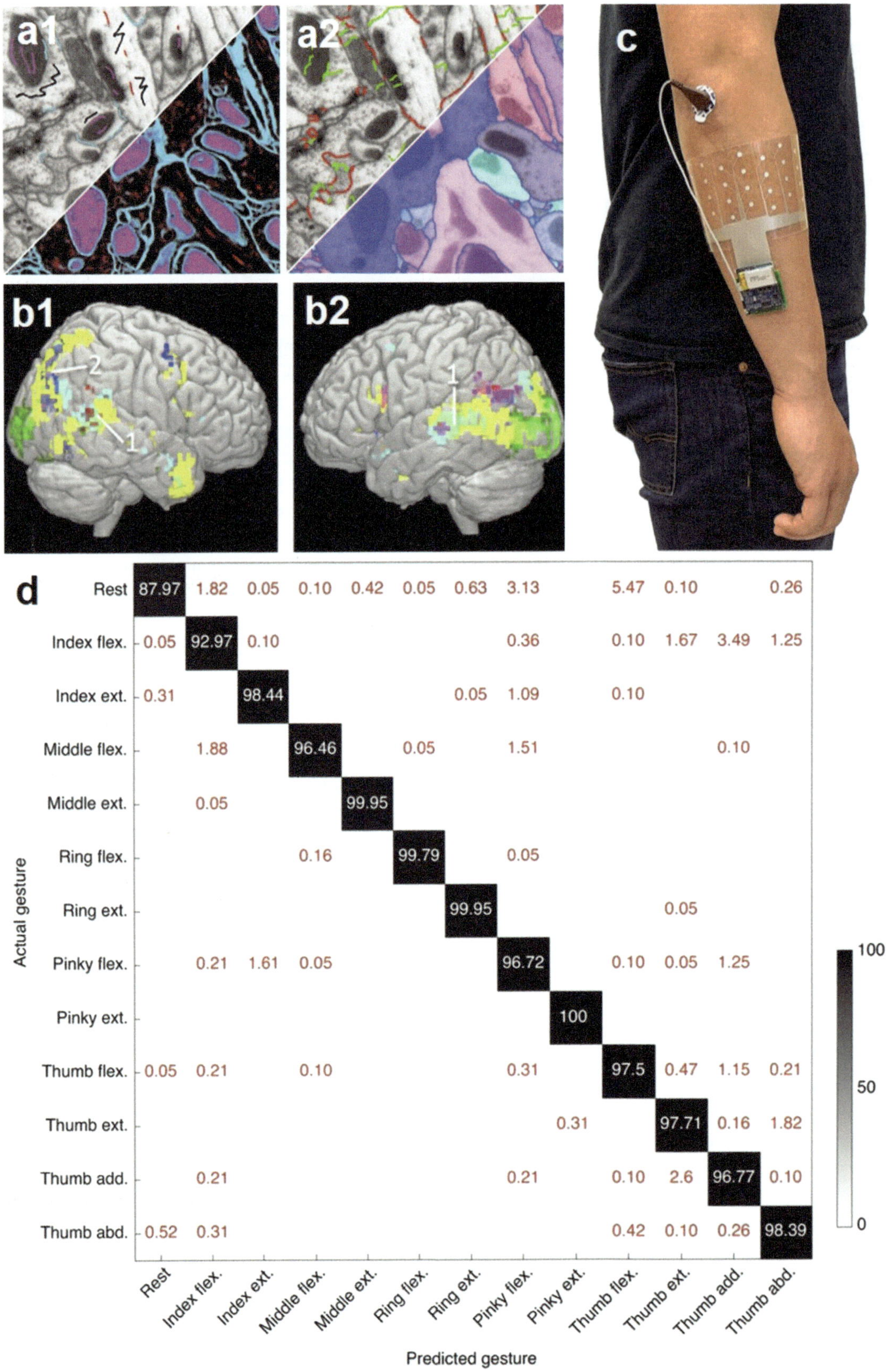

Fig. 10.2 Several applications of machine learning. **a1** Pixel classification, colored labels are used to predict which class all pixels belong to (mitochondria in magenta; membrane in blue; cytoplasm in black; microtubules in red). **a2** Automatic segmentation, clicking on the correct edge (green, false edge; red, true edge) on the cell edge to segment the correct cell region. (Image adapted with permission from Berg et al. [16], Copyright (2019), with permission from Springer Nature). **b** Judging the brain regions corresponding to neural activity when humans process various content while reading. (Image adapted the Open Access paper Wehbe et al. [17], under Copyright (2014) the Authors). **c** Arm sensor for gesture recognition. **d** Machine learning gesture recognition result. (Image adapted with permission from Moin et al. [18], Copyright (2020), with permission from Springer Nature)

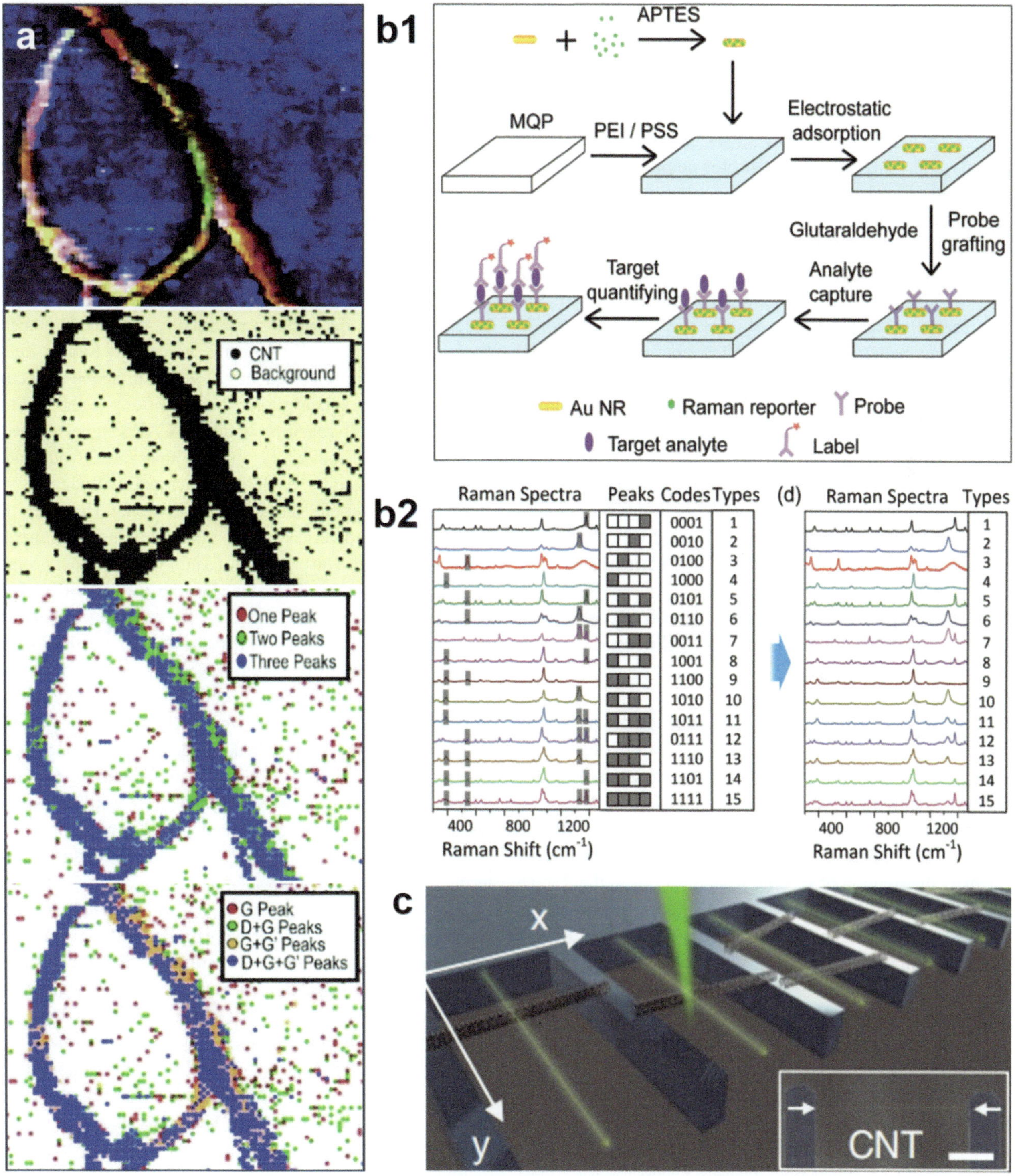

Fig. 10.3 Several applications of machine learning combined with Raman spectroscopy. **a** Machine learning performs point-by-point classification of Raman images of carbon nanotubes based on Raman spectral peak positions. (Image adapted with permission from Kajendirarajah et al. [22], Copyright (2020), with permission from Royal Society of Chemistry). **b1** Preparation of Raman spectrally encoded suspended arrays. **b2** Decoding of Raman spectrally encoded SAs using conventional methods. (Image adapted with permission from Chen et al. [24], Copyright (2019), with permission from Royal Society of Chemistry). **c** Schematic diagram of high-speed Raman imaging of carbon nanotubes on forked growth substrates. (Image adapted the Open Access paper Zhang et al. [25], under Copyright (2022) the Authors)

machine learning models—support vector machines, probabilistic neural networks, and convolutional neural networks—to classify pure and adulterated honey samples with high accuracy [27]. In the field of analytical chemistry, Fan et al. developed a convolutional neural network to identify mixed compounds. They utilized anol, acetonitrile, and distilled water to prepare 94 samples with varied proportions, subsequently collecting Raman spectra for component identification via machine learning. At the same time, they assessed the recognition performance of various neural network models and constructed a convolutional neural network that yielded the best performance [28]. In the industrial field, Baliyan and colleagues proposed an analytical framework (AF) for automatically identifying Raman spectral characteristics in lithium-ion battery electrodes. They constructed a neural network capable of accurately marking these spectral features. This approach is poised to facilitate real-time quality control and the development of professional analytical technologies [29].

10.2 Application of Raman Spectroscopy and Machine Learning in Bioanalysis

10.2.1 Application of Raman Spectroscopy and Machine Learning at Biological Tissue Level

Biological tissues consist of cells and interstitial constituents. Serving as the structural layer within living systems that bridges cells and organs, these tissues, due to the specificity of their cellular population structures, are frequently employed for pathological analysis. Raman imaging is a microscopy technique based on the spatial intensity distribution of specific peak positions in the Raman spectrum, which can specifically display the distribution of substances corresponding to these characteristic Raman peaks within tissue samples. With its ability to display the spatial distribution of different substances on the same sample based on distinct peak positions, Raman imaging offers a more targeted approach for molecular substance detection compared to traditional optical imaging. Raman imaging, displaying a unique texture of material distribution, is particularly suited to machine learning analysis. Its efficient and rapid data acquisition process facilitates the collection of substantial datasets necessary for robust machine learning training. The combination of these two methods can implement specific functions such as prediction and classification.

In recent years, the fusion of Raman imaging with machine learning has frequently been employed for label-free diagnosis of cancer tissue samples. There exists a variety of Raman imaging techniques such as conventional Raman imaging (Fig. 10.4a), Coherent anti-Stokes Raman scattering (CARS) (Fig. 10.4b), and Stimulated Raman Scattering (SRS) (Fig. 10.4c). Among them, conventional Raman imaging, which involves sequentially illuminating a sample with a laser and recording the Raman spectra at each point, generates an image composed of these spectra. This method enables the display of spatial intensity distribution at a specific peak position of the Raman spectrum, providing information about the spatial distribution of corresponding molecules. However, it is also characterized by certain limitations such as low precision and slow acquisition speed. In contrast, CARS utilizes two that are overlapped in both time and space to identify specific molecular vibrations, generating direction-specific signals via a four-wave mixing process. When the frequency difference between the pump and Stokes fields matches the vibration frequency of a Raman-active molecule in the sample, resonant molecular transitions to the excited state occur. These transitions generate CARS signals that are several orders of magnitude stronger than those from traditional Raman imaging, enabling higher quality imaging with lower laser power and shorter acquisition time. However, the CARS spectrum diverges from its corresponding spontaneous Raman spectrum. The presence of a non-resonant background complicates spectral allocation, leading to the difficulty of image interpretation and limiting the detection sensitivity. Compared with CARS, SRS can provide detection results that are free from background interference and easier to interpret. Moreover, it achieves higher detection sensitivity through the use of megahertz phase-sensitive detection.

Upon acquiring Raman imaging data, it can be analyzed using machine learning. With substantial training on large datasets, machine learning can efficiently perform various tasks such as judgment, classification, and evaluation. Helal et al. and his team proposed a Simple Linear Iterative Clustering algorithm for pixel segmentation in three-channel images, leading to an improved Signal-to-Noise Ratio (SNR). This methodology more effectively preserves the spatial characteristics of Raman images. In addition, they utilized Rate Distortion Theory (RDT) for unsupervised clustering of the spectrum, which performs well under conditions of noise and cluster overlap. Based on this, cellular-level information in liver tissue samples was evaluated [30]. CNN is applicable for the classification of various characteristic images and is commonly used in Raman imaging processing. Weng et al., for instance, used CNN in their Raman imaging studies to distinguish various lung cancer cells from their normal cells [31]. Similarly, Hollon et al. utilized CNN to differentiate brain tumor cell types and classify cancerous regions [32]. These deep learning networks exhibited high precision in their respective classification tasks, demonstrating the potential of combining Raman spectroscopy and machine learning for applications within histology.

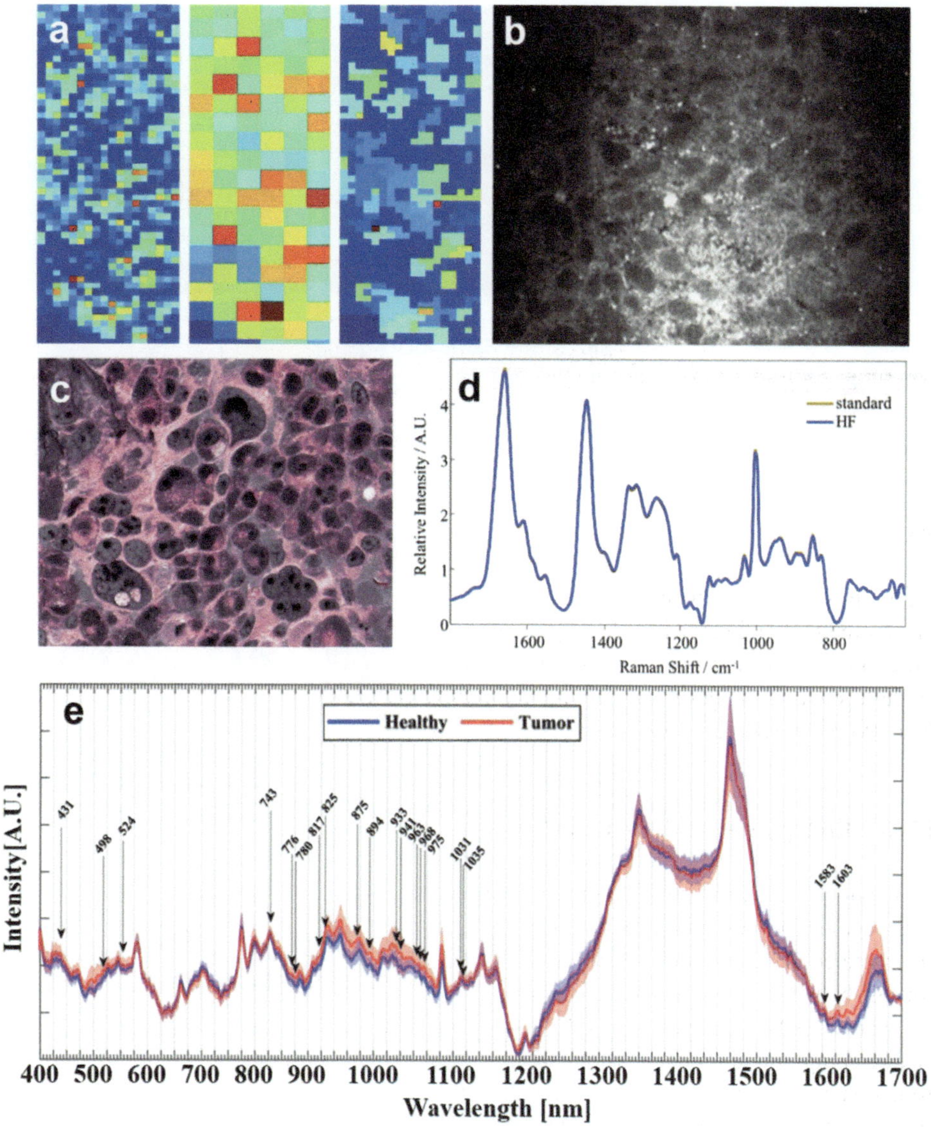

Fig. 10.4 Several commonly used Raman spectroscopy techniques at the histological level. **a** Raman imaging of liver tissue collected by traditional Raman imaging technology (left). (Image adapted with permission from Helal et al. [30], Copyright (2019), with permission from John Wiley and Sons). **b** Raman imaging of lung adenocarcinoma tissue collected by coherent anti-Stokes Raman imaging technology. (Image adapted the Open Access paper Weng et al. [31], under Copyright (2017) the Authors). **c** Raman imaging of brain tissue acquired by stimulated Raman spectroscopy. (Image adapted with permission from Hollon et al. [32], Copyright (2020), with permission from Springer Nature). **d** Comparison of Serum Raman spectra of rats fed standard and high-fat diets (with Alzheimer's disease). (Image adapted with permission from Ralbovsky et al. [33], Copyright (2021), with permission from Elsevier). (**e**) Comparison of Raman spectra of gliomas and normal brain tissues. (Image adapted the Open Access paper Riva et al. [34], under Copyright (2021) the Authors)

Simultaneously, while single-point Raman spectrum overlooks spatial distribution information, it retains its significance in detection as it contains the information of molecular composition of the samples (Fig. 10.4d, e), and Combined with artificial neural networks, it can also demonstrate excellent analytical performance. Riva et al. successfully differentiated between healthy and cancerous tissues by applying the Random Forest (RF) and Gradient Boosting Trees (GB) algorithms to the analysis of Raman spectra collected from both tumor and healthy tissues [34]. Additionally, Shu et al. utilizing a fiber-optic Raman endoscope system in combination with deep learning, developed a fiber-optic Raman diagnostic platform that achieved real-time diagnosis of nasopharyngeal carcinoma in vivo and post-treatment review of patients with nasopharyngeal carcinoma [35].

10.2.2 Application of Raman Spectroscopy and Machine Learning at the Single Cell Level

With continuous breakthroughs at the microscopic scale, the detection precision of Raman spectroscopy-related instruments has also improved correspondingly. As a result, the focus on single-cell detection has become increasingly prominent [36–38]. Diverse types of machine learning networks, both powerful and versatile. Recent studies frequently employ Raman spectroscopy in conjunction with these networks to facilitate various functions, including cell imaging, automatic counting, classification, and the detection of radiation resistance, differentiation, and metabolism.

Compared to tissue-level detection, cell-level detection demands higher accuracy. Utilizing Raman spectra, the "molecular fingerprint" of materials, for Raman imaging

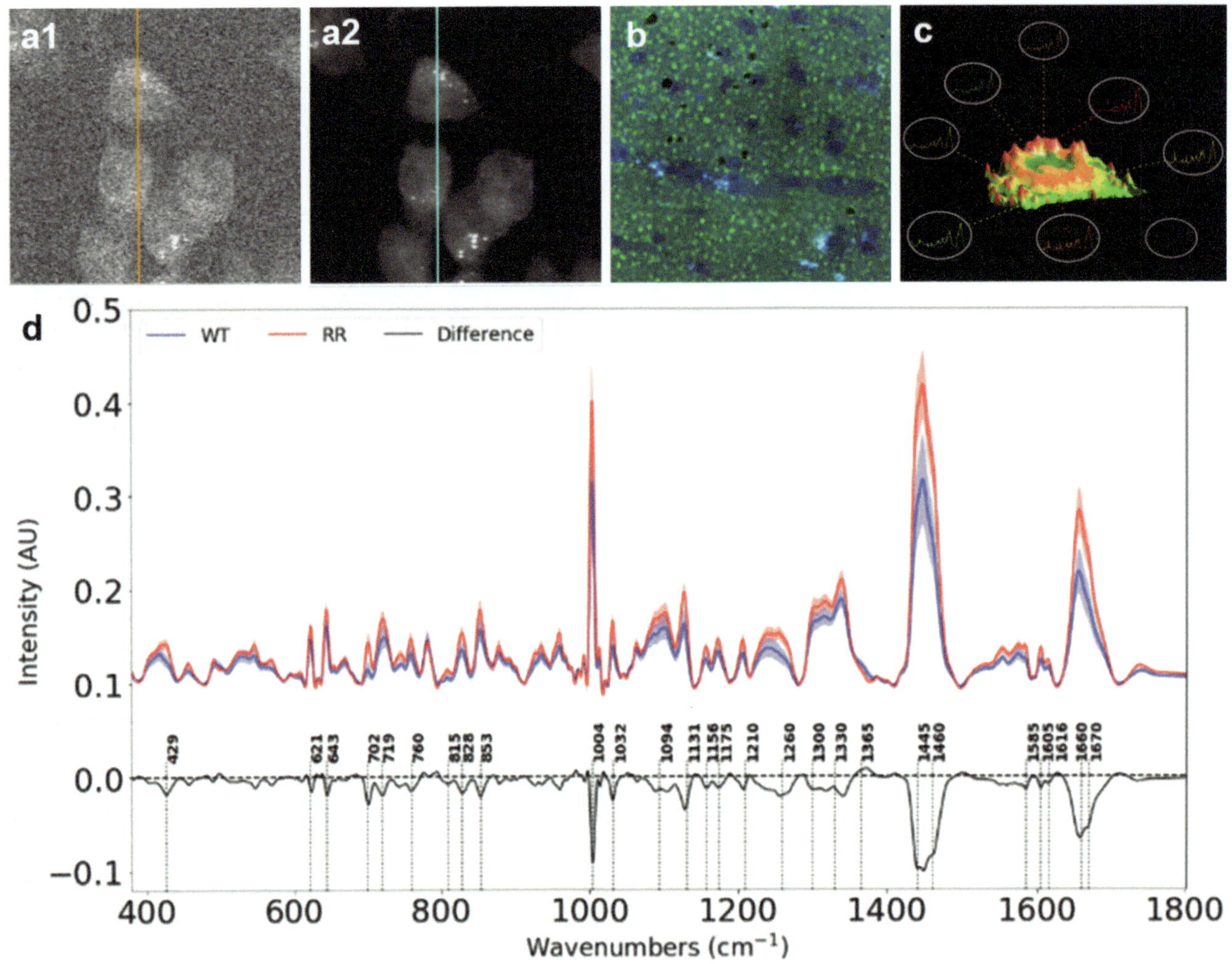

Fig. 10.5 Several common applications of Raman techniques combined with machine learning at the single-cell level. **a** Raman imaging of Hela cells collected by SRS (top) and processed by convolutional neural network denoising algorithm (bottom). (Image adapted the Open Access paper Manifold et al. [39], under Copyright (2019) the Authors). **b** Raman imaging of cell bodies embedded in the background of adipose tissue collected by SRS. (Image adapted the Open Access paper Zhang et al. [40], under Copyright (2021) the Authors). **c** Schematic diagram of distribution density of thyroid cell material collected by stimulated Raman spectroscopy. (Image adapted the Open Access paper Taylor et al. [41], under Copyright (2019) the Authors). **d** Comparison of mean Raman spectra of radiation-tolerant (red) and common breast cancer cells (blue) and their difference (black). (Image adapted the Open Access paper Tipatet et al. [42], under Copyright (2021) the Authors)

of cells allows for not only the observation of cellular morphology but also the detection of intracellular material distribution. However, due to the relatively weak intensity of the Raman signal, increasing laser power to enhance this signal could damage the cell sample. Even the currently fastest and clearest imaging technique, SRS, often struggles to provide Raman spectra with a high signal-to-noise ratio. To address this issue, Manifold et al. designed a de-noising algorithm based on U-NET convolutional neural network, and de-noised Raman imaging collected by SRS, achieving remarkable results (Fig. 10.5a) [39]. Through the analysis of Raman imaging, a wealth of useful information can usually be obtained. Zhang utilized the U-NET algorithm to analyze SRS Raman images (Fig. 10.5b), which enabled automatic determination of the cell count in the Raman images [40]. Beyond displaying cell morphology and count, Raman imaging contains a wealth of material information, ripe for comprehensive exploitation through machine learning. Taylor et al. integrated Raman imaging with Rate Distortion Theory (RDT) (Fig. 10.5c). RDT, often employed as an information theory-based clustering method, maximizes compression based on the goodness of fit. RDT excels in scenarios where there is significant overlap between clustered objects, enabling effective classification of healthy thyroid cells and thyroid cancer cells, along with the analysis on the relevant components of these two distinct cell types [41].

Given the wealth of material information inherent in Raman spectra, single-point Raman spectra can be collected for analysis when clear Raman imaging is not promptly available from the instrument (Fig. 10.5d). Tipatet et al. used Raman spectroscopy of breast cancer cells combined with linear discriminant analysis (LDA) and principal component analysis to determine whether they were radioactive resistant [42]. Zhang et al. adopted five feature extraction algorithms: Partial Least Squares (PLS), Principal Component Analysis (PCA), Kernel Principal Component Analysis (KPCA), Isometric Feature Mapping (Isomap), and Local Linear Embedding (LLE). These were combined with six classification models: KNN, EXTREME Learning Machine (ELM), Decision Tree (DT), Back Propagation Neural Network (BP), Genetic Optimization Back Propagation Neural Network (GA-BP), and Linear Discriminant Analysis (LDA), to create a total of 30 classification models. The aim was to distinguish Tumor Regenerative Cells (TRCs) from parental control cells, with the performance of these models subsequently compared. Ultimately, the eight combinations with the best performance were selected [43]. Unnan Geng collected Raman spectra from neural stem cells ranging from undifferentiated to highly differentiated states. An artificial neural network was employed to analyze their types and differentiation rates, enabling the precise analysis of eight Raman spectral data points with the highest accuracy [44].

10.2.3 Application in Detection of Bacteria and Viruses

Bacteria and viruses, posing significant threats to human life and health, have been the focus of extensive research over the years, leading to the development of numerous detection methods. Compared to traditional bacterial detection methods such as morphological observation, immunodiagnostic technology [45], and PCR technology [46], the detection technology with single bacterial accuracy offers advantages like rapid speed, high precision, and the elimination of the need for bacterial community cultivation. When utilizing Raman spectroscopy for the detection of individual bacteria, the typically small size of bacteria results in suboptimal imaging outcomes that often fall short of meeting various testing requirements, compounded by lengthy acquisition times. Hence, current research predominantly utilizes single or multi-point bacterial Raman spectra as the primary subjects of study [47–49].

In nature, bacteria exhibit vast diversity. Due to the extensive overlap in molecular composition among different bacterial species, their Raman spectral peak positions are almost indistinguishable (Fig. 10.6a). Therefore, conventional analysis methods often fail to achieve satisfactory accuracy. In addition to convolutional neural networks, Yu et al. designed a Long Short-Term Memory network (LSTM) for pathogen identification in seafood, which achieved a higher recognition accuracy rate [50]. In a comparative study between the kpCA-DT model and the PCA-SVM model for foodborne pathogen classification, Yan et al. found that the kpCA-DT model demonstrated superior performance over the PCA-SVM model in terms of bacterial classification and serotype level identification [51].

Because of the global spread of the coronavirus, there is an urgent need for rapid and accurate virus detection techniques. However, due to the nanoscale size of viruses, the Raman signal from a single virus is often overwhelmed by environmental noise. Consequently, recent studies on Raman detection of viruses have typically struggled to achieve single-virus precision, focusing instead on determining whether a sample is infected with a virus. Huang et al. developed a sensitive, rapid, and on-site SARS-CoV-2 antigen detection technology by integrating surface-enhanced Raman spectroscopy with deep learning. This technique utilizes residual neural networks to analyze the Raman spectra of patient samples, capable of detecting SARS-CoV-2 antigens in patients' throat swabs or sputum with an accuracy of 87.7% [52]. In addition to the novel coronavirus, Hepatitis B virus (HBV), which has long posed a threat to human health, also often draws the attention of researchers. Khan et al. combined Raman spectroscopy with the classical Support Vector Machine (SVM) model to analyze HBV infection in human serum (Fig. 10.6b1, b2), achieving an accuracy rate of 98% [53].

10.3 Conclusions

All in all, the combination of deep learning with Raman spectroscopy allows for rapid, label-free detection of biological samples, demonstrating substantial potential applications in biology. This approach is versatile across various scales, from macroscopic to microscopic. In recent years, amidst the global spread of COVID-19, this method of rapid and non-contact detection has increasingly showcased its advantages. Moreover, it also holds significant potential for application in several important areas closely related to human life and health, including cancer diagnosis, cell imaging, bacterial identification, and viral infections. However, Raman spectroscopy is incapable of performing clear Raman imaging on nano-scale samples, nor can it detect samples with extremely weak signals and complex backgrounds. The integration of deep learning methods can

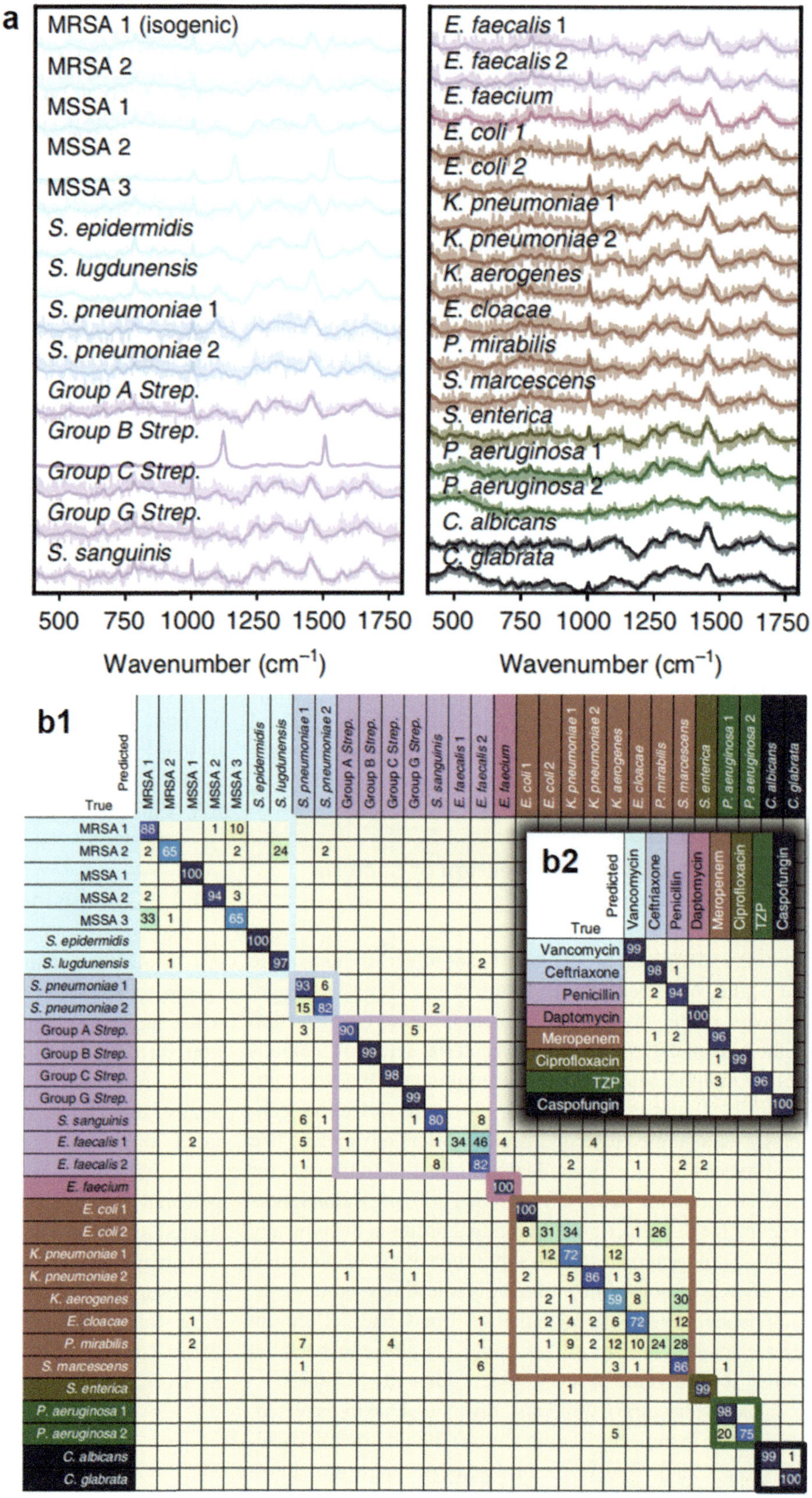

Fig. 10.6 Raman spectra of bacteria and viruses. **a** Raman spectra of 30 different bacteria. (Image adapted with permission from Ho et al. [54], Copyright (2019), with permission from Springer Nature). **b** Raman spectra of serum samples from normal (solid green line) and HBV infection (dotted red line). **b1** classification results of 30 species of bacteria and fungi with 100 different types using trained convolutional neural network, **b2** and prediction results of antibiotic use. (Image adapted with permission from Ho et al. [54], Copyright (2019), with permission from Springer Nature)

help mitigate these shortcomings, yet a substantial scope for enhancement remains. On the other hand, deep learning typically requires a large amount of data, while acquiring sufficient data can be notably difficult in certain application scenarios. How to construct high-performance machine learning models with a small amount of Raman spectroscopy or Raman imaging data is also a significant problem that researchers need to solve currently.

Acknowledgements This work was supported by National Natural Science Foundation of China (No. 62422505). C.L. and Q.Z. contributed equally to this work.

References

1. CV Raman KS Krishnan 1928 A new type of secondary radiation Nature 121 3048 501 502 https://doi.org/10.1038/121501c0
2. Placzek G. Rayleigh-streuung und Raman-effekt (Akademische Verlagsgesellschaft. 1934.
3. M Fleischmann PJ Hendra AJ McQuillan 1974 Raman spectra of pyridine adsorbed at a silver electrode Chem Phys Lett 26 2 163 166 https://doi.org/10.1016/0009-2614(74)85388-1
4. Camp Jr CH, Lee YJ, Heddleston JM, Christopher MH, Hight Walker AR, Rich JN, Lathia JD, Cicerone MT. High-speed coherent Raman fingerprint imaging of biological tissues. Nat Photonics. 2014;8:627–34. https://doi.org/10.1038/nphoton.2014.145
5. Ji M, Arbel M, Zhang L, Freudiger CW, Hou SS, Lin D, Yang X, Bacskai BJ, Xie XS. Label-free imaging of amyloid plaques in Alzheimer's disease with stimulated Raman scattering microscopy. Sci Adv. 2018;4:eaat7715. https://doi.org/10.1126/sciadv.aat7715
6. Y Zhang Y Gu J He BD Thackray J Ye 2019 Ultrabright gap-enhanced Raman tags for high-speed bioimaging Nat Commun 10 3905 https://doi.org/10.1038/s41467-019-11829-y
7. F Hu L Shi W Min 2019 Biological imaging of chemical bonds by stimulated Raman scattering microscopy Nat Methods 16 830 842 https://doi.org/10.1038/s41592-019-0538-0
8. Kang JW, Park YS, Chang H, Lee W, Singh SP, Choi W, Galindo LH, Dasari RR, Nam SH, Park J, So PTC. Direct observation of glucose fingerprint using in vivo Raman spectroscopy. Sci Adv. 2020;6(4):eaay5206. https://doi.org/10.1126/sciadv.aay5206
9. G Bodelón V Montes-García V López-Puente EH Hill C Hamon MN Sanz-Ortiz S Rodal-Cedeira C Costas S Celiksoy I Pérez-Juste L Scarabelli AL Porta J Pérez-Juste I Pastoriza-Santos LM Liz-Marzán 2016 Detection and imaging of quorum sensing in pseudomonas aeruginosa biofilm communities by surface-enhanced resonance Raman scattering Nat Mater 15 1203 1211 https://doi.org/10.1038/nmat4720
10. Nitta N, Iino T, Isozaki A, Yamagishi M, Kitahama Y, Sakuma S, Suzuki Y, Tezuka H, Oikawa M, Arai F, Asai T, Deng D, Fukuzawa H, Hase M, Hasunuma T, Hayakawa T, Hiraki K, Hiramatsu K, Hoshino Y, Inaba M, Inoue Y, Ito T, Kajikawa M, Karakawa H, Kasai Y, Kato Y, Kobayashi H, Lei C, Matsusaka S, Mikami H, Nakagawa A, Numata K, Ota T, Sekiya T, Shiba K, Shirasaki Y, Suzuki N, Tanaka S, Ueno S, Watarai H, Yamano T, Yazawa M, Yonamine Y, Carlo DD, Hosokawa Y, Uemura S, Sugimura T, Ozeki Y, Goda K. Raman image-activated cell sorting. Nat Commun. 2020;11:3452. https://doi.org/10.1038/s41467-020-17285-3
11. WS McCulloch W Pitts 1943 A logical calculus of the ideas immanent in nervous activity Bull Math Biol 5 115 133 https://doi.org/10.1007/BF02478259
12. Rosenblatt F. The perceptron—a perceiving and recognizing automaton. Cornell Aeronautical Laboratory; 1957.
13. JJ Hopfield 1982 Neural networks and physical systems with emergent collective computational abilities Proc Natl Acad Sci USA 79 8 2554 2558 https://doi.org/10.1073/pnas.79.8.2554
14. D Tank J Hopfield 1986 Simple'neural'optimization networks: an A/D converter, signal decision circuit, and a linear programming circuit IEEE Trans Circuits Syst 33 5 533 541 https://doi.org/10.1109/TCS.1986.1085953
15. GE Hinton RR Salakhutdinov 2006 Reducing the dimensionality of data with neural networks Science 313 5786 504 507 https://doi.org/10.1126/science.1127647
16. S Berg D Kutra T Kroeger CN Straehle BX Kausler C Haubold M Schiegg J Ales T Beier M Rudy K Eren JI Cervantes B Xu F Beuttenmueller A Wolny C Zhang U Koethe FA Hamprecht A Kreshuk 2019 Ilastik: interactive machine learning for (bio) image analysis Nat Methods 16 1226 1232 https://doi.org/10.1038/s41592-019-0582-9
17. L Wehbe B Murphy P Talukdar A Fyshe A Ramdas T Mitchell 2014 Simultaneously uncovering the patterns of brain regions involved in different story reading subprocesses PLoS ONE 9 11 e112575https://doi.org/10.1371/journal.pone.0123148
18. A Moin A Zhou A Rahimi A Menon S Benatti G Alexandrov S Tamakloe J Ting N Yamamoto Y Khan F Burghardt L Benini AC Arias JM Rabaey 2021 A wearable biosensing system with in-sensor adaptive machine learning for hand gesture recognition Nat. Electron. 4 54 63 https://doi.org/10.1038/s41928-020-00510-8
19. E Guevara JC Torres-Galván MG Ramírez-Elías C Luevano-Contreras FJ González 2018 Use of Raman spectroscopy to screen diabetes mellitus with machine learning tools Biomed Opt Express 9 10 4998 5010 https://doi.org/10.1364/boe.9.004998
20. F Lussier D Missirlis JP Spatz JF Masson 2019 Machine-learning-driven surface-enhanced Raman scattering optophysiology reveals multiplexed metabolite gradients near cells ACS Nano 13 2 1403 1411 https://doi.org/10.1021/acsnano.8b07024
21. C Berghian-Grosan DA Magdas 2020 Raman spectroscopy and machine-learning for edible oils evaluation Talanta 218 121176https://doi.org/10.1016/j.talanta.2020.121176
22. U Kajendirarajah MO Avilés F Lagugné-Labarthet 2020 Deciphering tip-enhanced Raman imaging of carbon nanotubes with deep learning neural networks Phys Chem Chem Phys 22 17857 17866 https://doi.org/10.1039/D0CP02950E
23. J Brandt K Mattsson M Hassellöv 2021 Deep learning for reconstructing low-quality FTIR and Raman spectra–a case study in microplastic analyses Anal Chem 93 49 16360 16368 https://doi.org/10.1021/acs.analchem.1c02618
24. X Chen L Xie Y He T Guan X Zhou B Wang G Feng H Yu Y Jie 2019 Fast and accurate decoding of Raman spectra-encoded suspension arrays using deep learning Analyst 144 4312 4319 https://doi.org/10.1039/C9AN00913B
25. J Zhang ML Perrin L Barba J Overbeck S Jung B Grassy A Agal R Muff R Brönnimann M Haluska C Roman C Hierold M Jaggi M Calame 2022 High-speed identification of suspended carbon nanotubes using Raman spectroscopy and deep learning Microsyst Nanoeng 8 19 https://doi.org/10.1038/s41378-022-00350-w
26. C Berghian-Grosan DA Magdas 2020 Application of Raman spectroscopy and machine learning algorithms for fruit distillates discrimination Sci Rep 10 21152 https://doi.org/10.1038/s41598-020-78159-8
27. S Hu H Li C Chen C Chen D Zhao B Dong X Lv K Zhang Y Xie 2022 Raman spectroscopy combined with machine learning algorithms to detect adulterated Suichang native honey Sci Rep 12 3456 https://doi.org/10.1038/s41598-022-07222-3
28. X Fan W Ming H Zeng Z Zhang H Lu 2019 Deep learning-based component identification for the Raman spectra of mixtures Analyst 144 1789 1798 https://doi.org/10.1039/C8AN02212G

29. A Baliyan H Imai 2019 Machine learning based analytical framework for automatic hyperspectral Raman analysis of lithium-ion battery electrodes Sci Rep 9 18241 https://doi.org/10.1038/s41598-019-54770-2
30. KM Helal JN Taylor H Cahyadi A Okajima K Tabata Y Itoh H Tanaka K Fujita Y Harada T Komatsuzaki 2019 Raman spectroscopic histology using machine learning for nonalcoholic fatty liver disease FEBS Lett 593 18 2535 2544 https://doi.org/10.1002/1873-3468.13520
31. S Weng X Xu J Li STC Wong 2017 Combining deep learning and coherent anti-stokes Raman scattering imaging for automated differential diagnosis of lung cancer J Biomed Opt 22 10 106017https://doi.org/10.1117/1.jbo.22.10.106017
32. TC Hollon B Pandian AR Adapa E Urias AV Save SSS Khalsa DG Eichberg RS D'Amico ZU Farooq S Lewis PD Petridis T Marie AH Shah HJL Garton CO Maher JA Heth EL McKean SE Sullivan SL Hervey-Jumper PG Patil BG Thompson O Sagher GM McKhann RJ Komotar ME Ivan M Snuderl ML Otten TD Johnson MB Sisti JN Bruce KM Muraszko J Trautman CW Freudiger P Canoll H Lee S Camelo-Piragua DA Orringer 2020 Near real-time intraoperative brain tumor diagnosis using stimulated Raman histology and deep neural networks Nat Med 26 52 58 https://doi.org/10.1038/s41591-019-0715-9
33. Ralbovsky NM, Fitzgerald GS, McNay EC, Lednev IK. Towards development of a novel screening method for identifying Alzheimer's disease risk: Raman spectroscopy of blood serum and machine learning. Spectrochim Acta Part A. 2021;254:119603. https://doi.org/10.1016/j.saa.2021.119603
34. M Riva T Sciortino R Secoli E D'Amico S Moccia B Fernandes MC Nibali L Gay M Rossi ED Momi L Bello 2021 Glioma biopsies classification using Raman spectroscopy and machine learning models on fresh tissue samples Cancers 13 5 1073 https://doi.org/10.3390/cancers13051073
35. C Shu H Yan W Zheng K Lin A James S Selvarajan CM Lim Z Huang 2021 Deep learning-guided fiberoptic Raman spectroscopy enables real-time in vivo diagnosis and assessment of nasopharyngeal carcinoma and post-treatment efficacy during endoscopy Anal Chem 93 31 10898 10906 https://doi.org/10.1021/acs.analchem.1c01559
36. CS McGinnis LM Murrow ZJ Gartner 2019 DoubletFinder: doublet detection in single-cell RNA sequencing data using artificial nearest neighbors Cell Syst 8 4 329 337 https://doi.org/10.1016/j.cels.2019.03.003
37. e Jong OG, Murphy DE, Mäger I, Willms E, Garcia-Guerra A, Gitz-Francois JJ, Lefferts J, Gupta D, Steenbeek SC, van Rheenen J, Andaloussi SE, Schiffelers RM, Wood MJA, Vader P. A CRISPR-Cas9-based reporter system for single-cell detection of extracellular vesicle-mediated functional transfer of RNA. Nat Commun. 2020;11:1113.https://doi.org/10.1038/s41467-020-14977-8
38. Y Cao J Feng L Tang C Yu G Mo B Deng 2020 A highly efficient introduction system for single cell-ICP-MS and its application to detection of copper in single human red blood cells Talanta 206 120174https://doi.org/10.1016/j.talanta.2019.120174
39. B Manifold E Thomas AT Francis AH Hill D Fu 2019 Denoising of stimulated Raman scattering microscopy images via deep learning Biomed Opt Express 10 8 3860 3874 https://doi.org/10.1364/boe.10.003860
40. Q Zhang KK Yun H Wang SW Yoon F Lu D Won 2021 Automatic cell counting from stimulated Raman imaging using deep learning PLoS ONE 16 7 e0254586https://doi.org/10.1371/journal.pone.0254586
41. JN Taylor K Mochizuki K Hashimoto Y Kumamoto Y Harada K Fujita T Komatsuzaki 2019 High-resolution Raman microscopic detection of follicular thyroid cancer cells with unsupervised machine learning J Phys Chem B 123 20 4358 4372 https://doi.org/10.1021/acs.jpcb.9b01159
42. KS Tipatet L Davison-Gates TJ Tewes EK Fiagbedzi A Elffick B Neu A Downes 2021 Detection of acquired radioresistance in breast cancer cell lines using Raman spectroscopy and machine learning Analyst 146 3709 https://doi.org/10.1039/d1an00387a
43. H Zhang C Chen R Gao Z Yan Z Zhu B Yang C Chen X Lv H Li Z Huang 2021 Rapid identification of cervical adenocarcinoma and cervical squamous cell carcinoma tissue based on Raman spectroscopy combined with multiple machine learning algorithms Photodiagn Photodyn Ther 33 102104https://doi.org/10.1016/j.pdpdt.2020.102104
44. J Geng W Zhang C Chen H Zhang A Zhou Y Huang 2021 Tracking the differentiation status of human neural stem cells through label-free Raman spectroscopy and machine learning-based analysis Anal Chem 93 30 10453 10461 https://doi.org/10.1021/acs.analchem.0c04941
45. van Hooij A, Fat EMTK, da Silva MB, Bouth RC, Messias ACC, Gobbo AR, Lema T, Bobosha K, Li J, Weng X, Salgado CG, Spencer JS, Corstjens PLAM, Geluk A. Evaluation of immunodiagnostic tests for leprosy in Brazil, China and Ethiopia. Sci Rep. 2018;8:17920. https://doi.org/10.1038/s41598-018-36323-1
46. JY Jung HK Yoon S An JW Lee E-R Ahn YJ Kim HC Park K Lee JH Hwang SK Lim 2018 Rapid oral bacteria detection based on real-time PCR for the forensic identification of saliva Sci Rep 8 10852 https://doi.org/10.1038/s41598-018-29264-2
47. Whang K, Lee JH, Shin Y, Lee W, Kim YW, Kim D, Lee LP, Kang T. Plasmonic bacteria on a nanoporous mirror via hydrodynamic trapping for rapid identification of waterborne pathogens. Light: Sci Appl. 2018;7:68. https://doi.org/10.1038/s41377-018-0071-4
48. SD Dryden S Anastasova G Satta AJ Thompson DR Leff A Darzi 2021 Rapid uropathogen identification using surface enhanced Raman spectroscopy active filters Sci Rep 11 8802 https://doi.org/10.1038/s41598-021-88026-9
49. Liu TY, Tsai KT, Wang HH, Chen Y, Chen YH, Chao YC, Chang HH, Lin CH, Wang JK, Wang YL. Functionalized arrays of Raman-enhancing nanoparticles for capture and culture-free analysis of bacteria in human blood. Nat Commun. 2011;2:538. https://doi.org/10.1038/ncomms1546
50. S Yu X Li W Lu H Li YV Fu F Liu 2021 Analysis of Raman spectra by using deep learning methods in the identification of marine pathogens Anal Chem 93 32 11089 11098 https://doi.org/10.1021/acs.analchem.1c00431
51. Yan S, Wang S, Qiu J, Li M, Li D, Xu D, Li D, Liu Q. Raman spectroscopy combined with machine learning for rapid detection of food-borne pathogens at the single-cell level. Talanta 2021;226:122195. https://doi.org/10.1016/j.talanta.2021.122195
52. J Huang J Wen M Zhou S Ni W Le G Chen L Wei Y Zeng D Qi M Pan J Xu Y Wu Z Li Y Feng Z Zhao Z He B Li S Zhao B Zhang P Xue S He K Fang Y Zhao Y Zhao K Du 2021 On-site detection of sars-cov-2 antigen by deep learning-based surface-enhanced Raman spectroscopy and its biochemical foundations Anal Chem 93 26 9011 9298 https://doi.org/10.1021/acs.analchem.1c01061
53. S Ali M Hassan M Saleem SF Tahir 2021 Deep transfer learning based hepatitis B virus diagnosis using spectroscopic images Int J Imaging Syst Technol 31 1 94 105 https://doi.org/10.1002/ima.22462
54. CS Ho N Jean CA Hogan L Blackmon SS Jeffrey M Holodniy N Banaei AAE Saleh S Ermon J Dionne 2019 Rapid identification of pathogenic bacteria using Raman spectroscopy and deep learning Nat Commun 10 4927 https://doi.org/10.1038/s41467-019-12898-9

Single-Molecule Manipulation with Optical Tweezers

11

Yu-Xuan Ren

Abstract

The optical manipulation of biomacromolecules allows the study of the dynamics and interaction of molecules under mechanical tension. More importantly, the ability to manipulation single molecule at a time offers unique mechanical features of a specific molecule distinguishable to the ensemble study using the average over huge number of molecules. In this chapter, we introduce the basics of single molecule optical tweezers, the most successful examples of single molecule optical tweezers instrument, and the representative applications of the single molecule manipulation. This chapter will be reference for researchers who want to study the single molecule manipulation with optical tweezers. We also hope the chapter will excite more applications of the single molecule optical tweezers.

Keywords

Optical tweezers · Single-molecule biophysics · Protein folding · Motor protein · Fluorescence microscopy

11.1 Introduction

Light photons carry momentum, and the transfer of momentum on absorbing particles effectively creates a forward scattering force which can be used to levitate microparticles [1, 2]. Optical levitation was also studied in high vacuum [3], and further used to cold atoms. Moreover, the single beam can be focused by an objective with a high numerical aperture (NA), and attract the microparticles towards the diffraction-limited focus. This is nowadays well known as the optical tweezers which was first invented by Ashkin [4].

In Ashikin's earlier demonstrations, he used the optical tweezers to study the biological system, to trap, move, and guide cells [5, 6]. Back to 1989, S. Block et al. had determined the torsional compliance of flagella in tethered *Escherichia coli* and a motile *Streptococcus* [7]. They identified that the flagella behave as linear torsion springs for roughly half a revolution, but became much more rigid when turned beyond this point in either direction [7]. Most recent reports on manipulating the cells and particles in vivo involve either single and multiple trapping in the zebrafish larva [8, 9], as well as study of the bacterial motility [10]. The optical torque can be induced on the microscopic birefringent particles of calcite [11]. More generally, the asymmetric particle also experience optical torque under the act of even linearly polarized beam [12]. The most challenging part to use optical tweezers in biophysics lies in the characterization of single molecule mechanical behavior. The most representative one is the identification of the kinesin motor movement [13], which has been the demonstration for biological application with optical tweezers during the announcement of awarding A. Ashkin the 2018 Nobel prize in physics. F. Fazal and S. Block at Stanford University wrote in their commentary [14],

> Optical tweezers have become one of the primary weapons in the arsenal of biophysicists, and have revolutionized the new field of single-molecule biophysics.

In this chapter, we intend to briefly introduce fundamental concepts and some representative tweezer modules widely used in single molecule studies and the related techniques that impowers this optical tool, such as single-molecule fluorescence (SMF), single-molecule foster resonance energy transfer (smFRET) [15], and single-molecule Raman spectrum (smRaman), in addition to the well-studied single molecule systems.

Y.-X. Ren (✉)
Jinshan Hospital, and Institute for Translational Brain Research, Fudan University, Shanghai, China
e-mail: yxren@fudan.edu.cn

B. Li et al. (eds.), *Biophotonic Manipulation*, Advances in Optics and Optoelectronics,
https://doi.org/10.1007/978-981-96-4982-2_11

11.2 Detection Principle of Optical Tweezers

The optical force on a particle exerted by a focused light beam can be understood by the interaction of optical field with the particle. In Mie scattering regime, the particle geometry is much greater than the light wavelength, ray optics evaluates the optical force. In Rayleigh regime, the particle approximates a dipole, and the optical force is the consequence of the light-field dipole interaction. In general, the electromagnetism evaluates the light-particle interaction considering the electric dipole, magnetic dipole, and the interaction of both. In this section, we focus on the detection of the optical force in practical experiments.

The trapped microparticle experiences confined Brownian motion, and the particle position keeps fluctuating. The particle motion can be monitored by a camera, and be evaluated by the video analysis. More precisely, the positional signal of the microsphere can be detected by a high-resolution position sensitive detector (PSD) through back focal plane interferometry [16]. Figure 11.1 shows the optical schematics of the position detection in a single beam optical trap.

Once the optical trap catches a single micro-/nano- particle, the forward scattering light interfere with the unscattered light, and form an interference pattern at the back focal plane of the detection objective/condenser. A PSD captures the intensity centroid of the interference pattern and the position signal is digitalized by a data acquisition card (DAQ). The back-focal-plane interferometry allows the simultaneous measurement of the displacement and force [17]. The power spectrum for the positional signal of the trapped microsphere reads [16],

$$|\tilde{x}(f)|^2 = \frac{k_b T}{c\pi^2\beta\left[\left(\frac{\alpha}{2\pi\beta}\right)^2 + f^2\right]}, \tag{11.1}$$

where k_b is Boltzman constant, T is the temperature of the fluid, and the conversion coefficient c is normally within $0.1-3\,\text{mW/nm}$, and the trap stiffness α is usually within $0.01-0.8\,\text{pN/nm}$, $\beta = 6\pi r\eta$ is the drag coefficient. The corner frequency can be fitted by the measured power spectrum, and determines the final trap stiffness $\alpha = 2\pi\beta f_c$. With these two parameters, the positional signal can be readily converted to both the displacement and force.

Figure 11.2 illustrates a typical power spectrum density of the positional signal for a 1.9 μm bead (red), and in the absence of microsphere when the laser is on (blue) and off (rose red). This is very sensitive to environment noise, including drift of the microscope stage, laser pointing fluctuation, and Brownian motion of the trapped bead. In order to achieve a bp resolution for single molecule applications, the optical tweezers usually operate in a vibration-isolated environment with sophisticated temperature and air flow control. All components of the 'dumbbell' force clamp assay are optically levitated in the buffer and decoupled with any mechanical motion from the sample chamber. To further inhibit the air fluctuation, all the optical element external to the microscope can be enclosed in a seal box filled with helium gas with a refractive index much closer to unity. Other efforts include the sound-proof isolation, and differential detection [19–21].

Figure 11.3 demonstrates the resolution test using double stranded DNA under the dual-trap optical tweezers. Each end of the dsDNA is attached to a microsphere through either digoxigin-antidigoxigin or biotin-streptavidin linkage. The molecule was stretched step by step with a step size of 2

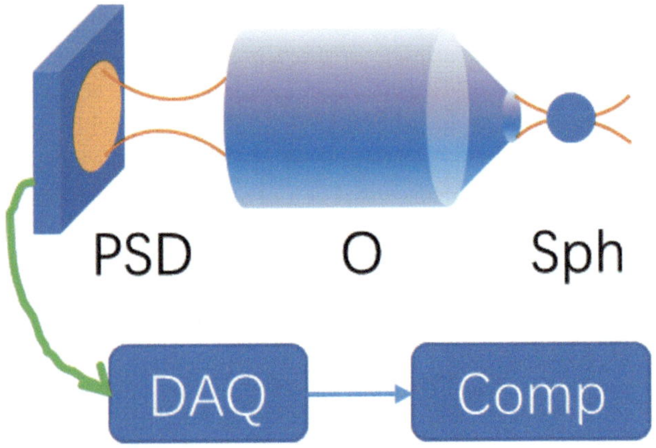

Fig. 11.1 Detection of optical tweezers. PSD, position sensitive detector; O, objective; Sph, microsphere; DAQ, data acquisition card; Comp, control computer

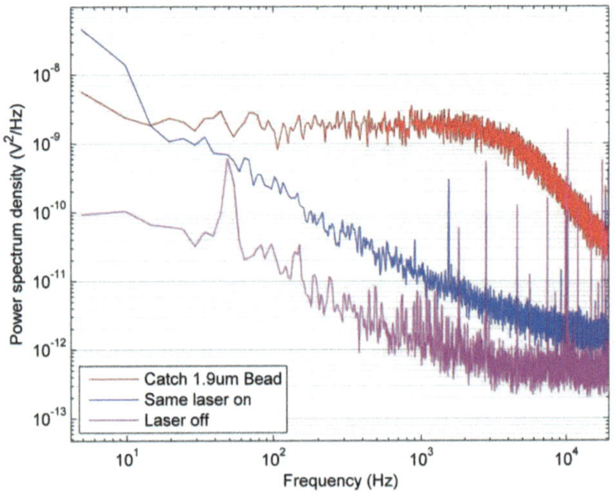

Fig. 11.2 Power spectrum density of positional signal for a 1.9 μm bead (red), and the PSD in the absence of microsphere when the laser is on (blue) and off (rose red). Reproduced from Chinese Laser Press [18]

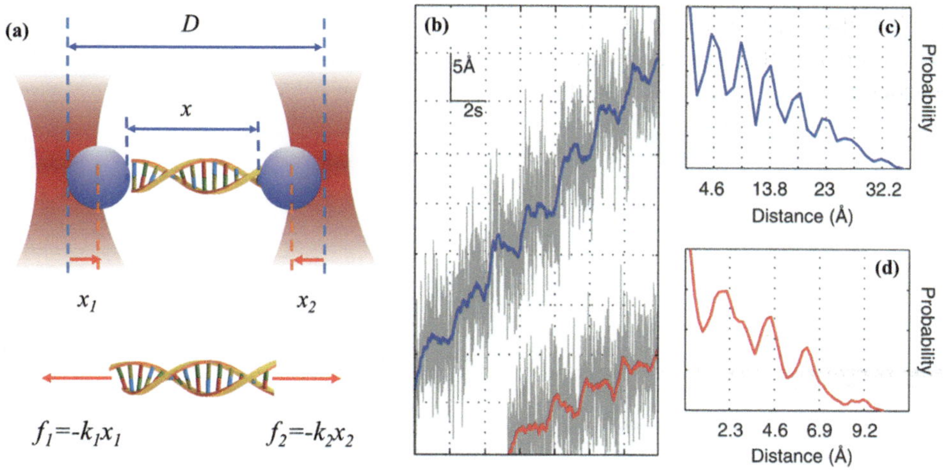

Fig. 11.3 Resolution test of single-molecule optical tweezers. **a** Schematics of the resolution test. **b** Resolution measurement under the sampling frequency of 100 kHz. Pairwise distance distribution of **b** suggests a step size of **c** 4.6 Å and **d** 2.3 Å [20]. Reproduced with permission from American Institute of Physics [20]

and 1 bps, and the end-to-end distance is sampled at 100 kHz. The pairwise distance distribution suggests a step size of (c) 4.6 Å and (d) 2.3 Å [20]

11.3 Optical Tweezers Revolutionized Single-Molecule Biophysics

Optical tweezers, with high-spatial and temporal resolution, can be applied to measure the transient force of single protein molecules under mechanical tension.

11.3.1 Modules of Optical Tweezers Used in Single Molecule Biophysics

Since the invention in 1986, single molecule biophysicists have kept improving the optical tweezers tailored for monitoring the mechanical behavior of a single molecule. Basic quantities involve the force, displacement, torque, and fluorescence. The molecular activity in modern molecular biology often takes place on the DNA template, and the basic length unit for the DNA is base pair (1 bp = 0.34 nm, basic length unit for double stranded DNA).

Chronologically, the tweezers instruments evolve from single beam to dual-trap optical tweezers. Some popular modules used in excellent labs are shown in Fig. 11.4. The first module applies a single beam optical trap to catch and hold a polymer sphere as probe to detect the force and displacement (Fig. 11.4a). The polymer tether was formed between the trapped sphere and glass slide. The extension of the molecule can be changed by traversing the glass slide with a piezo-stage. As the extension increases, the force on the tether molecule also increases and pulls the trapped sphere outwards in the linear region of the trap. Once the sphere moves to a critical position but still in the linear region, the feedback loop is activated to clamp the position by ramping up the laser power.

The single-beam trap can also be used to trap birefringent particles and dual-beam detection with orthogonal polarizations enables the measurement of the torque exerted on the particle (Fig. 11.4b) [11, 23]. Experimental configuration to measure supercoil torque on a DNA molecule with Angular Optical Tweezers is shown in Fig. 11.4b [24, 28]. Such optical wrench can be used to study the chromatin torsional mechanics and topoisomerase activity.

As the stage traverses, the trapped sphere experiences both translational and axial motions, and the combined motion complicates the data analysis. A possible solution is to hold one sphere with a micropipette, while the second sphere is held in the optical trap as force and displacement probe (Fig. 11.4c). This module is well known as the mini-tweezers instrument (http://tweezerslab.unipr.it) with a highly folded optics and compact size. The minitweezers can be hang over the top of the ceiling to minimize air fluctuation and mechanical noise [27]. In both the glass-trap and micropipette-trap modules, the mechanical drift of the stage couples to the detection system, thus degrades the resolution.

To eliminate the noise from the stage drift, a dual-trap optical tweezers with differential detection was developed (Figs. 11.3a and 11.4c) [19]. This is achieved by splitting the laser beam into orthogonal polarizations with equal power and forming two traps. The captured spheres are attached to DNA molecules via chemical linking. Due to air fluctuation, each of the single trap was not able to resolve ~ bp change, while the differential detection minimizes the noise from air fluctuation, and resolves ~ bp displacement change (Fig. 11.3b). This revolving ability is also verified in the pairwise distance histogram (Fig. 11.3c, d). In fact, the dual-trap optical tweezers can also be configured into many modes, including the passive, constant force, and position clamp modes [20, 29]. As an example, a DNA hairpin 20R55/T4 (where 20 indicates the stem length in bp, R represents random sequence, 55 is the percentile GC content in stem, and T4 indicates the thymidine tetraloop sequence)

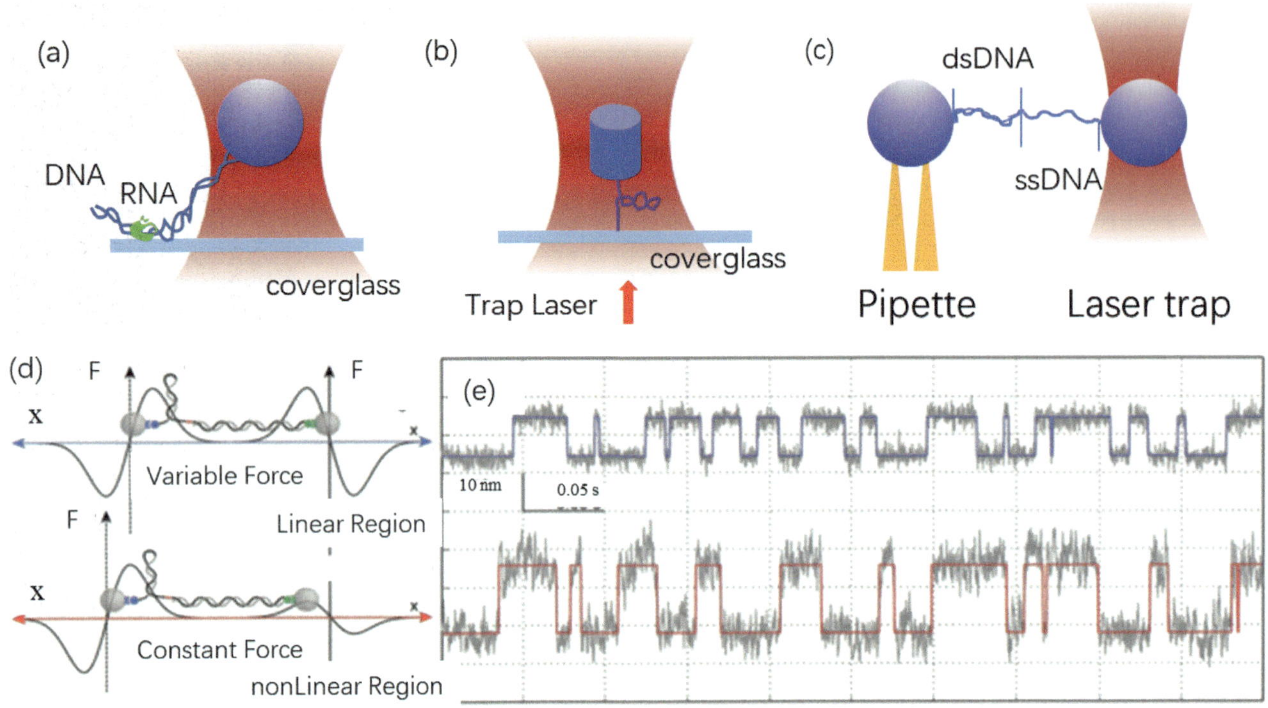

Fig. 11.4 Configurations of optical trapping setup used in single-molecule biophysics. **a** single-beam optical trap with position clamp [22]. A single DNA molecule is tethered between a trapped bead and the bottom glass slide. **b** Angular optical trapping configuration [23, 24]. **c** Micropipette-tweezers configuration [25–27]. **d** Schematic of variable and constant force mode using dual-trap optical tweezers. Both traps have identical intensity for variable mode, while one trap is significantly weaker than the other in the constant force mode [20]. **e** temporal extension traces of the DNA hairpin recorded in variable (top) and constant (bottom) force mode show the hairpin dynamic folding and unfolding under mechanical force [20]. Reproduced with permission from AIP [20]

is stretched using the dual-trap geometry (Fig. 11.3a). In passive mode with fixed trap separation, when the molecules folds and unfolds, both the force and extension hop between the two states (top row in Fig. 11.4d, e). While one of the traps is configured with a weak power, the sphere is easily moved out of the linear region. There exists a considerably long range, within which the force fluctuation keeps small, before the sphere escapes (bottom of Fig. 11.4e). In this case, only the extension hops with the force kept constant (bottom of Fig. 11.4e). Some unconventional single molecule tools include the quadruple-trap optical tweezer that can manipulate dual DNA molecules [30].

11.3.2 Combining Single Molecule Fluorescence

The strongly focused laser has a much higher power density than that for the fluorescence signal emitted by a single molecule. Less successful reports were found on single molecule fluorescence at the early stage of single molecule studies. Until 2003, M. Lang et al. demonstrated the first coincidence between single molecule fluorescence with optical trapping [31]. They adopt the evanescent wave to excite the fluorescence over a region of several square micrometer. The optical-trapping and position-detection laser beams are focused to diffraction-limited spots and overlapping with each other.

In Fig. 11.5(a), a 1010-bp DNA duplex with an overhanging segment was attached at one end via biotin-avidin linkage to a 500-nm diameter polystyrene sphere. The distal end of the DNA is annealed to a 15-bp oligonucleotide, which is anchored directly to glass surface via a digoxygenin-antidigoxygenin antibody linkage [31]. One Tetramethylrhodamine (TAMRA) dye molecule was placed on the 3′ end of the shorter 15-mer and one on the 5′ end of the long strand, such that the dye proximity cause fluorescence quench. The rhodamine self-quenching leads to approximately 30-fold decrease in fluorescence due to the formation of noncovalent dimers based on stacked xanthine rings [32]. The distance sensitivity probed by dye self-quenching is exceedingly short (about 1–2 nm), and smaller than those by fluorescence resonance energy transfer (FRET). The DNA-bead complex was stretched at constant velocity unitl rupture (‘unzipping’) of the 15 base-pair duplex occurred [31]. TAMRA dyes attached at the ends of the DNA strands provide a fluorescence signal (red dots). The fluorescence signal was monitored by silicon avalanche photodiode (APD) under confocal mode with detection area limited to a region of approximately 950 nm diameter in the specimen plane. In Fig. 11.5(b), prior to rupture, force increased monotonically, and the low fluorescence counts indicated the quenched dye state. At rupture (t ~ 2 s), the force dropped from 9 to 0 pN and the fluorescent counts increased, indicating that the two fluorophores are separated, unquenching the dye bound to the 15-mer kept on the slide. The dye photobleached in a single step, and the light signal returned to the background levels at t ~ 9 s.

Comstock et al. demonstrated an ultrahigh-resolution optical trap with single fluorophore sensitivity, which is showcased by detecting the single-molecule fluorescence change during the binding/unbinding of a fluorophore labeled ssDNA oligonucleotide [21]. Subsequently, George et al. reported a dual-trap optical tweezers instrument combined with single molecule fluorescence and single-molecule visualization [20]. Figure 11.5c shows the schematics of the time-shared dual-trap optical tweezers with single molecule fluorescence detection. The hairpin conjugates to one dsDNA handle through a disulfide bond and to the other through a streptavidin molecule. The complex DNA construct is further tethered to two anti-digoxigenin coated microsphere. The hairpin 20R55/T4 molecule unfolds under optical force and refolds due to thermal fluctuation. A small fragment of Cy3-labeled oligonucleotide with complementary sequence to a region of the DNA hairpin labels with a single fluorophore and is diffusing in the buffer. The oligonucleotide binding and unbinding events can be read from the extension and fluorescence count curves. When the DNA hairpin molecule is folded, the swimming ssDNA fragment has a probability to bind on the DNA hairpin molecule and clamp the hopping dynamics. Large gap on the extension curve indicates the pause of hopping accompanied by the fluorescence increase, which implies that the fluorophore is binding on the

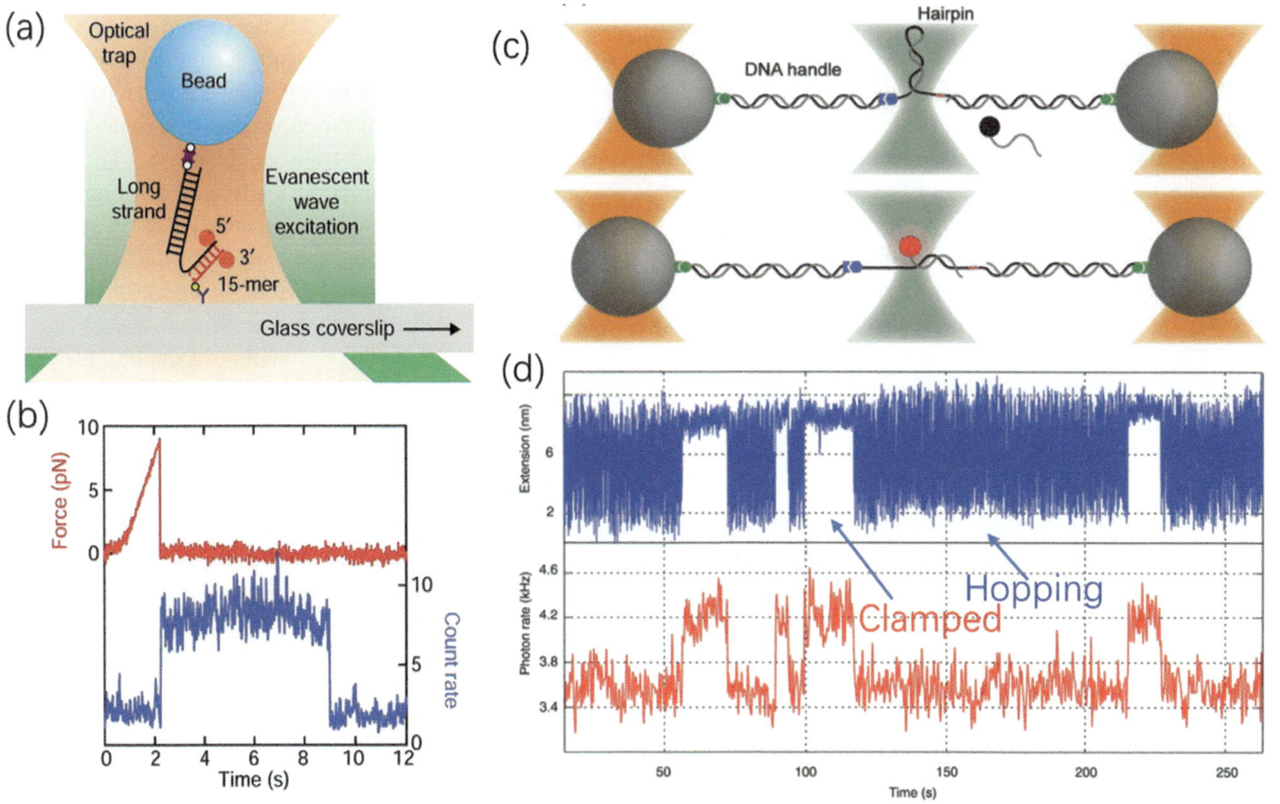

Fig. 11.5 Combining optical tweezers and single molecule fluorescence. **a** The experimental geometry for single-beam optical trap combined with total internal reflection fluorescence microscopy (not to scale). **b** simultaneous traces of force (red) and fluorescent photon counts (blue). **c** DNA hairpin is attached to beads via dsDNA handles. The hairpin unfolds due to applied force and refolds under thermal fluctuation. **d** Simultaneous monitor of the oligonucleotide binding and unbinding events by both extension and fluorescence counts. Reproduced with permission from Nature and APS [20, 31]

hairpin molecule (Fig. 11.5d). Apart from single-molecule epi-fluorescence [33], total internal reflection fluorescence (TIRF) [34], smFRET [35] and smRaman spectrum are two more advanced techniques that help provide more information on the molecules [36].

11.3.3 Biological Applications

Single molecule optical tweezers have successfully studied a diverse of protein and nucleic acid systems. The well-understood biomacromolecules ranges from simple hairpin, formed in RNA or double-stranded DNA, to the bacterial ribosome, a macromolecular machine comprised of over 50 protein subunits, molecular motors, i.e., polymerase and helicase, and three structured RNA molecules [14]. This section intends to introduce some of the representative examples in the single molecule studies.

The DNA can pack and fold into chromosomes or serve as a template during transcription and replication, and the function of DNA depends on the elastic properties of the molecules as well as the modification by local interactions [37]. The study of DNA stretching and mechanical flexibility has been the most fundamental experiment in single-molecule biophysics. In 1992, S. Smith et al. reported the direct measurement of mechanical elasticity of a single-molecule λ DNA by tethering the molecule between the glass surface and a magnetic bead [37].

In 1996, Smith et al. applied optical tweezers to stretch single-molecule λ-phage dsDNA [25]. Under a longitudinal stress of ~ 65 pN, dsDNA molecule undergoes a highly cooperative transition into a stable form with 5.8 angstroms rise per bp. The molecules rapidly and reversibly contracted to the normal contour lengths when the stress was released below 65 pN. The force-extension curve shows hysteresis during relaxation [25].

Subsequently in 1997, M. Wang et al. reported the Force-extension relation of a single DNA molecule under various buffer conditions using a surface-trap geometry [38]. One end of the single DNA molecule was attached to the cover-glass surface with a stalled RNA polymerase complex, and the other end was attached to a microsphere. The microsphere was clamped at a certain trapping position by dynamically altering the trap stiffness using an AOD [38]. The DNA molecule has a persistence length of $\sim$ 47 μm in a buffer containing 10 mM Na^{+}. An elasticity theory incorporating both enthalpic and entropic contributions fits the experimental results with an elastic modulus of ~ 1100 pN [38].

In 2001, Liphardt et al. reported single-molecule study of three types of RNA molecules representing major structural units of large RNA assemblies [26]. They are P5ab, P5absΔA, and P5abc RNAs [26]. P5ab is a simple A-form double helix RNA hairpin that typifies the basic unit of RNA. P5abcΔA has an additional helix and thus a three-helix junction. Finally, P5abc is comparatively complex and contains an A-rich bulge, enabling P5abc to pack into a stable tertiary structure in the presence of Mg^{2+} ions [26]. RNA molecules were attached between two 2-μm beads with ~ 500 bp RNA–DNA hybrid handles. One bead was held in a force-measuring optical trap, and the other bead was controlled by a piezo-electric actuator through a micropipette [25]. The length versus time traces for P5abc in EDTA at various force levels shows the folding/unfolding dynamics mediated by mechanical force [26].

Some ssDNA can also fold to hairpin structure due to complementary sequence. By attaching the hairpin molecule to two microspheres through dsDNA handles, Woodside et al. studied the folding dynamaics under mechanical force for hairpin molecules with different stem length and GC content (Fig. 11.6b) [39]. Figure 11.6b shows the force-extension curve (FEC) for hairpin 20R55/T4 with traps separated at a speed of 10–15 nm/s. The hairpin unfolds abruptly at $F \approx 13\,pN$, displaying a characteristic sawtooth pattern as the extension (force) increases (drops) [39]. A record segment for hairpin 20R55/T4 under a 13.2-pN load shows the two-state behavior, with rapid switching between folded and unfolded states (Fig. 11.6c). The extension histogram displays two peaks with separation between the two $\Delta x = 18.0 \pm 0.5$ nm. The energy landscape of the molecule is dominated by two potential wells separated by a single barrier. The applied mechanical force tilts the energy landscape, and consequently, alters the free energy difference between unfolded and folded states [39]. The unfolding force $F_{1/2}$ at which the hairpin molecule has a 50% probability of being unfolded can be determined by measuring the force dependence of population of the two states. The probability of the unfolded state as function of force (Fig. 11.6d) fits the Boltzmann relation, $P_u(F) = \left(1 + \exp\left[\left(F_{1/2} - F\right) \cdot \Delta x\right]\right)^{-1}$ with fitting parameters $F_{1/2} = 13.2 \pm 0.7pN$ and $\Delta x = 18 \pm 2\,\text{nm}$. The lifetimes of the folded and unfolded states also depend on the applied force, which changes the energy barrier height. Assuming that the position of the energy barrier is independent on force, the lifetime of the folded state is $\tau_f = \tau_{f,0}\exp\left(F\Delta x_f^{\ddagger}/k_BT\right)$, where $\Delta x_f^{\ddagger}$ is the distance from the folded to the transition state and $\tau_{f,0}$ is the lifetime of the folded state at $F = 0$. The slopes of the logarithmic average lifetime of the folded and unfolded states as function of force shows that the transition barrier located 12 ± 2 nm from the folded state and 5.3 ± 0.6 nm from the unfolded state (Fig. 11.6e). The unfolding rate at zero force was derived from the intercept of the folded-state lifetime, $k_{u,0} = 10^{-15\pm2}\,\text{s}^{-1}$. The unfolding rate and free energy supply measure of hairpin stability.

The biological activities are driven by energy and forces with the action of a variety of molecular motors. Kinesin, as a mechanoenzyme, couples ATP hydrolysis to movement along

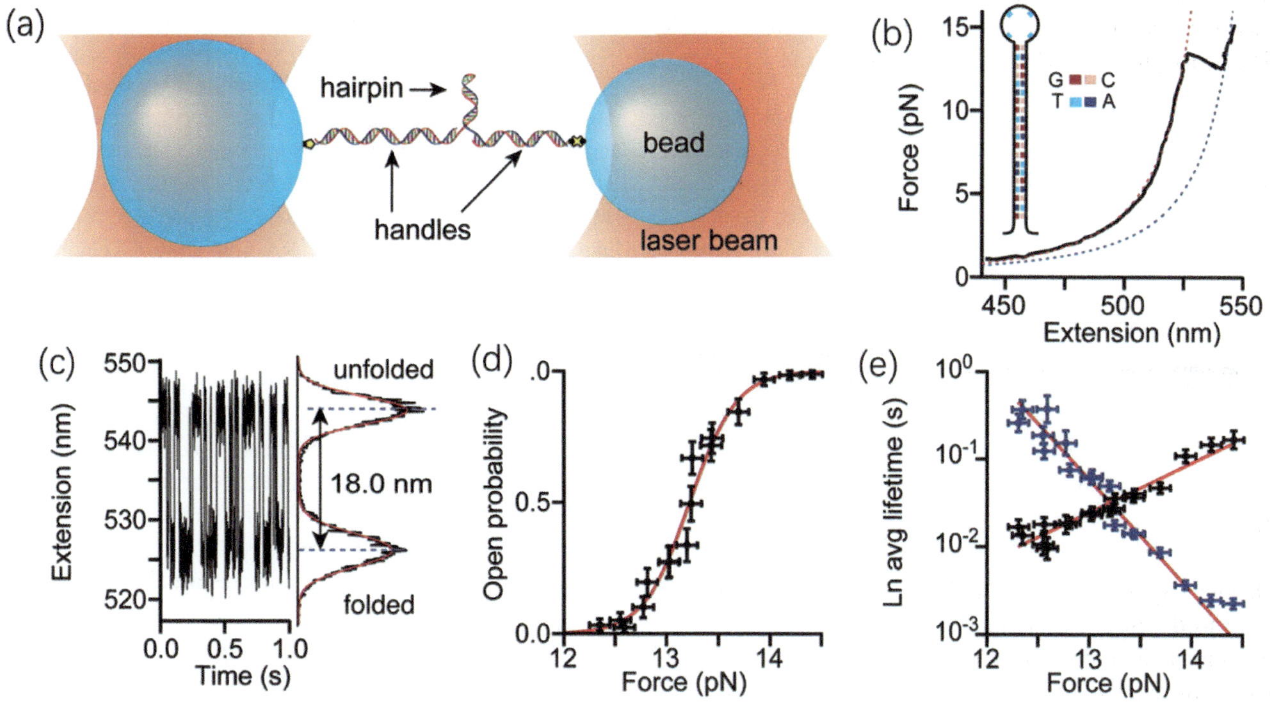

Fig. 11.6 Single molecule study of nucleic acid. **a** Geometry for DNA hairpin streching experiment. A DNA hairpin is attached via dsDNA handles to beads held in dual traps [39]. **b** FEC for hairpin 20R55/4 T shows WLC behavior of handles at low F (dotted red line) followed by hairpin unfolding at ~ 13 pN. **c** Temporal extension traces at constant F reveals two-state behavior with an opening distance of 18.0 ± 0.5 nm. **d** Unfolded state probability (black) as a function of F follows two-state Boltzmann statistics (red) [39]. **e** Lifetimes of folded (blue) and unfoled (black) states vary exponentially with force [39]. Reproduced with permission from PNAS [39]

microtubules. It powers vesicle transport and other forms of microtubule-based motility [13]. In 1990, S. Block et al. used optical tweezers to manipulate the microsphere coated with carrier protein to move onto the microtubules. They found that the bead with excess kinesin moved smoothly along the microtubule for many micrometers, but those carry 0.17–3 kinesin molecules moved only about 1.4 μm and then spontaneously released from the microtubule. In 1993, Svoboda et al. reported the direct observation of the 8-nm steps in kinesin motor [40]. In 1994, K. Svoboda, and S. Block measured the force and velocity of a single kinesin molecule [41]. In 2003, S. Block et al. further identified that the kinesin moves by asymmetric hand-over-hand mechanism [42].

RNA polymerase (RNAP) is another molecular motor that moves along supercoiled DNA while carrying out transcription. Wang et al. measured the transcriptional velocities for single molecules of Escherichia coli RNAP as progressively larger forces by a feedback-controlled optical trap [22]. The shapes of RNAP force–velocity curves are distinct from those of the motor enzymes myosin or kinesin, which implies that the biochemical steps limiting transcription rates at low loads do not generate movement.

The backtracking and recovery by single molecules of RNAP was observed using ultrastable optical trapping with two-bead assay to record the transcriptional elongation [43, 44]. The key feature for RNA polymerase proofreading mechanism is a short backtracking motion of the enzyme along the DNA template. The transcription bubble shifts and the DNA-RNA duplex remains in register during backtracking. The RNAP cleaves off and discard the most recently added base(s) by endonucleolysis, and short RNA segments carrying misincorporated bases can be replaced, leading to the correction of transcriptional errors [44].

John et al. monitored the transcription by single molecules of Escherichia coli RNA polymerase in real time over long template distances [45]. The reversible pausing is a kinetic intermediate between normal elongation and the arrested state. RNA polymerase (RNAP) transcribes supercoiled DNA, whose torsional state is constantly varying, but how RNAP deals with DNA supercoiling remains unexplored. Ma and colleagues directly measured the individual Escherichia coli RNAPs as they transcribed supercoiled DNA [46]. They found that a resisting torque slowed RNAP and increased its pause frequency and duration [46].

11.3.4 Recent Progresses and Future Directions

Noble metal nanoparticles can absorb resonant light and produce a huge amount of heat to vaporize surrounding liquid. Such plasmonic bubbles are gaining increasing attention and induce a plethora of applications, such as, micromanipulation, robot propulsion, and clinical therapies [47]. The microrobots can be manipulated through the optical modulation [48]. Moreover, the metallic nanoparticles can be used to enhance the signal strength of Raman spectroscopy. Combined with optical tweezers, the in situ Raman spectroscopy can control the hotspot between two Ag nanoparticle-coated silica spheres, generating tunable and reproducible SERS enhancements with single-molecule sensitivity [49]. The hybrid instrument directly identifies the structural features of the transient species of alpha-synuclein among its predominant monomers by reducing the ensemble averaging [49].

The optical plasmonic trap built a dynamic nanocavity that reduces the diffraction-limited detection volume and provides reproducible electromagnetic field enhancements to achieve high-throughput single-molecule surface-enhanced Raman spectroscopy (SERS) characterizations [50]. Based on time-dependent SERS spectra, two types of low-populated transient species of human Islet Amyloid Polypeptide (amylin, hIAPP) containing either turn or β-sheet structure are characterized during the early-stage incubation at neutral condition [50]. These are pivotal to drive irreversible amyloid fibril developments even after a subsequent adjustment of pH.

The magnetic tweezers offers much greater detection range than optical tweezers, and allows for the study of multi-domain interaction mediated strength-building [51]. The force-dependent lifetimes of homo- and hetero-dimers of human α-actinins, reveals an ultra-high mechanical stability of the dimers associated with > 100 s lifetime within 40 pN forces under shear-stretching geometry [51]. The strong dimer stability is arisen from much weaker sub-domain pair interactions, suggesting the existence of distinct dimerized functional states of the dimer, spanning a spectrum of mechanical stability [51].

The Combined Optical and Magnetic BIomolecule TWEEZers (COMBI-Tweez) integrates optical trapping, time-resolved electromagnetic tweezers, and fluorescence microscopy, and can controllably form and visualize higher order structural motifs including plectonemes [52]. Such hybrid setup is crucial to determine the biopolymer topology inside cell environments, and the biochemical properties is important to cellular roles. The dynamic structure of chiral biopolymers depends on the extension and torsion, and the correlation of force, torque, and spatial location are crucial to determine the structural motifs upon physiological twisting and stretching [52].

Figure 11.7a shows the setup of the hybrid instrument combining optical tweezers, magnetic tweezers and fluorescence microscopy, for which the optical tweezers is collocated with a pair of orthogonal Helmholtz coils [52]. Figure 11.7b shows the schematics of the working principle of the hybrid instrument. One end of a DNA molecule is chemically conjugated to a surface-immobilised "anchor" bead with the other tethered to an optically trapped magnetic microbead. Figure 11.7c shows the snapshots of video of the continuous stretch-release of one representative tether. Figure 11.7d shows the representative force-extension data compiled from $n = 8$ consecutive stretch-release half cycles obtained from same tether by WLC fits (red curve). Mean persistence length is 53 ± 15 nm ($n = 9$ separate tethers), comparable with that obtained for unlabelled DNA construct [52].

Toehold-mediated strand displacement (TMSD) is extensively studied in dynamic DNA nanotechnology and a wide range of DNA or RNA-based reaction circuits [53]. Single-molecule force spectroscopy offers tool to dissect the dynamics of strand displacement processes at the single-molecule level. The TMSD proceeds very rapidly under load with single step times of 1 μs. Tuning invasion efficiency by introducing mismatches allows studying thousands of forward/backward invasion events on a single molecule and analyze the kinetics of the invasion process [53].

Correct mitotic spindle size is required for accurate chromosome segregation during cell division [54]. It is controlled by mechanical forces generated by molecular motors and non-motor proteins acting on spindle microtubules. Single molecule experiment reveals that microtubules are essential force generators for establishing spindle size. The microtubule plus-end binding proteins act at microtubule tips synergistically with minus-end directed motors to produce a system that can generate both pushing and pulling forces [54].

Extended synaptotagmins (E-Syts) mediate lipid exchange between the endoplasmic reticulum (ER) and the plasma membrane (PM) [55]. Single-molecule optical tweezers can detect the Ca^{2+}- and lipid-dependent membrane-binding kinetics of both E-Syt1 and E-Syt2, and revealed the stepwise membrane binding of extended synaptotagmins [55].

Combining confocal fluorescence microscopy and optical tweezers, the association and dissociation rates of the protein-DNA complex can be uncovered at the single molecule level. Distinct mechanisms for oligomer growth via the binding of protein to adjacent dsDNA or direct interaction with bound protein, resembles primary and secondary nucleation.

With dual-trap optical tweezers, the DNA molecule can be tethered by two optically trapped microspheres. An excitation beam (and superimposed STED beam) that is scanned over the DNA can perform the STED nanoscopy on optically stretched

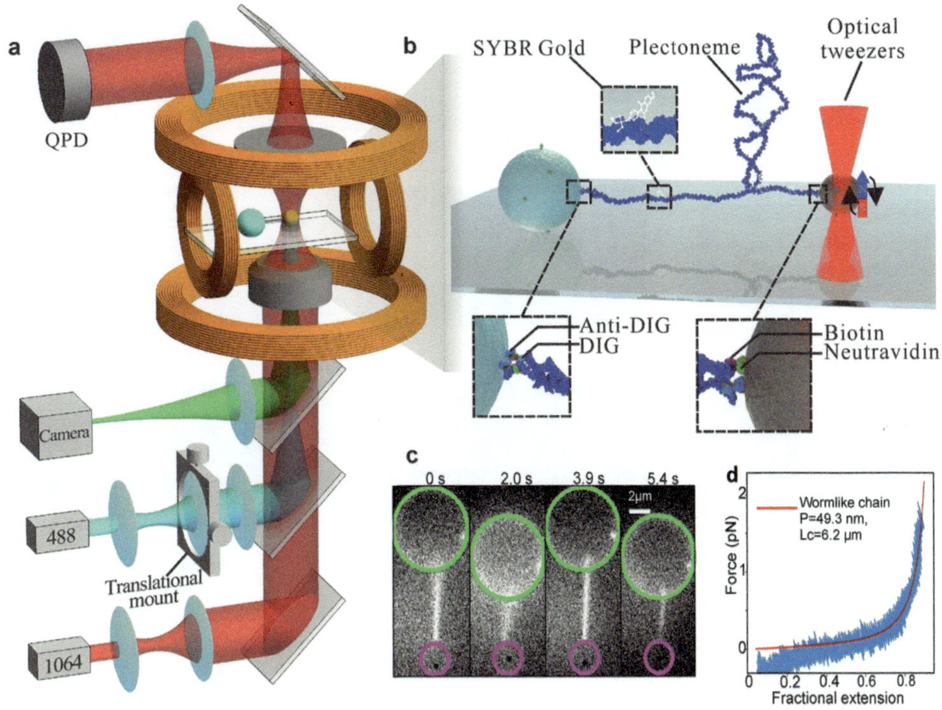

Fig. 11.7 **a** Setup and **b** schematic of optical tweezers combined with magnetic tweezers and fluorescence microscopy [52]. **c** One representative tether showing snapshots extracted from continuous stretch-release experiment. **d** Representative force-extension data compiled from $n = 8$ consecutive stretch-release half cycles obtained from same tether by WLC fits (red). Reproduced under the Creative Commons Attribution 4.0 International License [52]

DNA [56]. Recent progress suggests that this module can be used to visualize the dynamic binding of the protein molecule on the DNA double helix.

Xu and colleagues revealed that the polymerase undergoes rapid and autonomous exchange during replication not coordinated by other proteins [57]. They also observed a 'memory effect' in DNA polymerase rebinding, i.e., the enzyme tends to preserve its prior activity upon reassociation. The single molecule observation provides evidence to support an autonomous replication model that includes rapid protein exchange, burst of activity, and a 'memory effect' while moving processively forward.

Pathogenic dsDNA prompts AIM2 assembly leading to the formation of the inflammasome, which triggers the inflammatory response. The recognition of foreign dsDNA involves AIM2 self-assembly concomitant with dsDNA binding [58]. The kinetic formation and propagation of the assembly sheds light on innate immunity's time response and specificity. The single AIM2 molecules do not diffuse along the DNA, suggesting that oligomerization depends on stochastic encounters with DNA and/or DNA-bound AIM2 [58].

Oligomer growth is detected by analysis of the change in photon counts as a function of time in kymographs of single AIM2 molecules and oligomers bound to dsDNA (Fig. 11.8). The large oligomer triplicates its size in 0.85 s growing from 11 to 35 molecules [58]. The decrease in intensity for oligomers larger than 3 molecules (Fig. 11.8) reflects photobleaching and does not indicate AIM2-DNA dissociation. However, 2-molecule clusters can detach from the dsDNA (Fig. 11.8). Kymographs of single AIM2 molecules and oligomers bound to tethered dsDNA (left) and the corresponding heat maps (right) showing an increase in fluorescence intensity (Fig. 11.8). The AIM2 monomer grows at a rate of $5.8\,s^{-1}$, whereas the dimer at the bottom kymograph forms a hexamer at a rate of $94\,s^{-1}$ (middle kymograph of Fig. 11.8). Both kymographs were acquired at 1 nM AIM2. The starting number of molecules likely influences the oligomerization rate.

The thymine DNA glycosylase (TDG) exhibits multiple modes of linear diffusion, including hopping and sliding [59]. Using single molecule fluorescence, TDG active site variants and truncated N-terminus, reveals these variants alter base modification search and recognition mechanism of TDG. On DNA containing an undamaged nucleosome, TDG either bypass, colocalize with, or encounter but not bypass the nucleosome [59].

Many neurodegenerative diseases feature misfolded proteins that propagate via templated conversion of natively folded molecules [60]. The prion-like conversion in single molecules of superoxide dismutase-1 (SOD1) reveals that the SOD1 misfolding is linked to amyotrophic lateral sclerosis [60]. The mutants of SOD1 vastly increased misfolding of the wild-type molecule. Crucially, the pattern of misfolding was the same in the mutant and converted wild-type domains reflecting the templating effect expected for prion-like conversion [60].

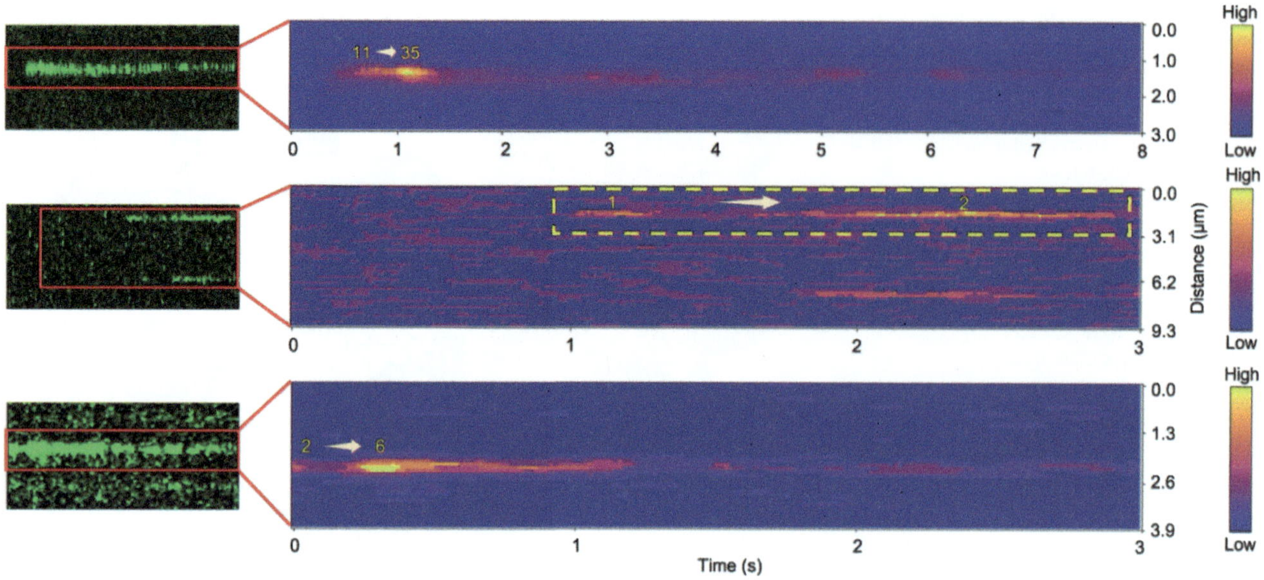

Fig. 11.8 Kymographs of single AIM2 molecules and oligomers bound to tethered dsDNA (left) and the corresponding heat maps (right) showing an increase in fluorescence intensity. AIM2 concentration: top (10 nM); middle and bottom (1 nM). Reproduced under the Creative Commons Attribution 4.0 International License [58]

11.4 Conclusion

In conclusion, single-molecule optical tweezers can manipulate single biomolecules with high spatio-temporal resolution, and monitor the dynamics of the molecule under mechanical tensor. With advancements in fluorescence detection, Raman spectroscopy, and magnetic manipulation, the single-molecule manipulation techniques are still developing and continue to offer deeper insights of the molecule behavior in single molecule level.

Acknowledgements This work was supported by National Natural Science Foundation of China (No. 62475047).

References

1. Ashkin A, Dziedzic JM. Optical levitation by radiation pressure. Appl Phys Lett. 1971;19:283–5. https://doi.org/10.1063/1.1653919.
2. Ashkin A, Dziedzic JM. Stability of optical levitation by radiation pressure. Appl Phys Lett. 1974;24:586–8. https://doi.org/10.1063/1.1655064.
3. Ashkin A, Dziedzic JM. Optical levitation in high vacuum. Appl Phys Lett. 1976;28:333–5. https://doi.org/10.1063/1.88748.
4. Ashkin A, Dziedzic JM, Bjorkholm JE, Chu S. Observation of a single-beam gradient force optical trap for dielectric particles. Opt Lett. 1986;11:288–90. https://doi.org/10.1364/OL.11.000288.
5. Ashkin A, Dziedzic JM. Optical trapping and manipulation of viruses and bacteria. Science. 1987;235:1517–20.
6. Ashkin A, Dziedzic JM, Yamane T. Optical trapping and manipulation of single cells using infrared laser beams. Nature. 1987;330:769–71.
7. Block SM, Blair DF, Berg HC. Compliance of bacterial flagella measured with optical tweezers. Nature. 1989;338:514. https://doi.org/10.1038/338514a0.
8. Johansen PL, Fenaroli F, Evensen L, Griffiths G, Koster G. Optical micromanipulation of nanoparticles and cells inside living zebrafish. Nat. Commun. 2016;7. https://doi.org/10.1038/ncomms10974
9. Hörner F, Meissner R, Polali S, Pfeiffer J, Betz T, Denz C, Raz E. Holographic optical tweezers-based in vivo manipulations in zebrafish embryos. J Biophotonics. 2017;10:1492–501. https://doi.org/10.1002/jbio.201600226.
10. Min TL, Mears PJ, Chubiz LM, Rao CV, Golding I, Chemla YR. High-resolution, long-term characterization of bacterial motility using optical tweezers. Nat. Methods. 2009;6:831. https://doi.org/10.1038/nmeth.1380. https://www.nature.com/articles/nmeth.1380#supplementary-information.
11. Friese MEJ, Nieminen TA, Heckenberg NR, Rubinsztein-Dunlop H. Optical alignment and spinning of laser-trapped microscopic particles. Nature. 1998;394:348–50.
12. Ren Y-X, Frueh J, Zhang Z, Rutkowski S, Zhou Y, Mao H, Kong C, Tverdokhlebov SI, Liu W, Wong KKY, Li B. Topologically protected optical pulling force on synthetic particles through photonic nanojet. Nanophotonics. 2024;13:239–49. https://doi.org/10.1515/nanoph-2023-0740.
13. Block SM, Goldstein LSB, Schnapp BJ. Bead movement by single kinesin molecules studied with optical tweezers. Nature. 1990;348:348–52.
14. Fazal FM, Block SM. Optical tweezers study life under tension. Nat Photon. 2011;5:318–21.
15. Comstock MJ, Whitley KD, Jia H, Sokoloski J, Lohman TM, Ha T, Chemla YR. Direct observation of structure-function relationship in a nucleic acid–processing enzyme. Science. 2015;348:352–4. https://doi.org/10.1126/science.aaa0130.

16. Berg-Sørensen K, Flyvbjerg H. Power spectrum analysis for optical tweezers. Rev Sci Instrum. 2004;75:594–612. https://doi.org/10.1063/1.1645654.
17. Gittes F, Schmidt CF. Interference model for back-focal-plane displacement detection in optical tweezers. Opt Lett. 1998;23:7–9. https://doi.org/10.1364/ol.23.000007.
18. Gong F, Ren Y-X. Advances in laser dual-trap optical tweezers in single-molecule biophysics. Chin J Lasers. 2023;50:1507402.
19. Moffitt JR, Chemla YR, Izhaky D, Bustamante C. Differential detection of dual traps improves the spatial resolution of optical tweezers. Proc Natl Acad Sci. 2006;103:9006.
20. Sirinakis G, Ren Y-X, Gao Y, Xi Z-Q, Zhang Y-L. Combined versatile high-resolution optical tweezers and single-molecule fluorescence microscopy. Rev Sci Instrum. 2012;83: 093708.
21. Comstock MJ, Ha T, Chemla YR. Ultrahigh-resolution optical trap with single-fluorophore sensitivity. Nat Methods. 2011;8:335. https://doi.org/10.1038/nmeth.1574. https://www.nature.com/articles/nmeth.1574#supplementary-information.
22. Wang MD, Schnitzer MJ, Yin H, Landick R, Gelles J, Block SM. Force and velocity measured for single molecules of RNA polymerase. Science. 1998;282:902.
23. Deufel C, Forth S, Simmons CR, Dejgosha S, Wang MD. Nanofabricated quartz cylinders for angular trapping: DNA supercoiling torque detection. Nature Methods. 2007;4:223. https://doi.org/10.1038/nmeth1013. https://www.nature.com/articles/nmeth1013#supplementary-information.
24. La Porta A, Wang MD. Optical torque wrench: angular trapping, rotation, and torque detection of quartz microparticles. Phys Rev Lett. 2004;92(190801): 190801. https://doi.org/10.1103/PhysRevLett.92.190801.
25. Smith SB, Cui Y, Bustamante C. Overstretching B-DNA: the elastic response of individual double-stranded and single-stranded DNA molecules. Science. 1996;271:795.
26. Liphardt J, Onoa B, Smith SB, Tinoco I, Bustamante C. Reversible unfolding of single rna molecules by mechanical force. Science. 2001;292:733.
27. Wuite GJL, Smith SB, Young M, Keller D, Bustamante C. Single-molecule studies of the effect of template tension on T7 DNA polymerase activity. Nature. 2000;404:103. https://doi.org/10.1038/35003614.
28. Le TT, Gao X, Park SH, Lee J, Inman JT, Lee JH, Killian JL, Badman RP, Berger JM, Wang MD. Synergistic coordination of chromatin torsional mechanics and topoisomerase activity. Cell. 2019;179:619-631.e615. https://doi.org/10.1016/j.cell.2019.09.034.
29. Greenleaf WJ, Woodside MT, Abbondanzieri EA, Block SM. Passive all-optical force clamp for high-resolution laser trapping. Phys Rev Lett. 2005;95: 208102. https://doi.org/10.1103/PhysRevLett.95.208102.
30. Heller I, Laurens N, Vorselen D, Broekmans OD, Biebricher AS, King GA, Brouwer I, Wuite GJL, Peterman EJG. Versatile quadruple-trap optical tweezers for dual DNA experiments. In: Gennerich, A, editors. Optical tweezers: methods and protocols. 2017. pp. 257–72. https://doi.org/10.1007/978-1-4939-6421-5_9.
31. Lang MJ, Fordyce PM, Block SM. Combined optical trapping and single-molecule fluorescence. J Biol. 2003;2:6. https://doi.org/10.1186/1475-4924-2-6.
32. Blackman MJ, Corrie JET, Croney JC, Kelly G, Eccleston JF, Jameson DM. Structural and biochemical characterization of a fluorogenic rhodamine-labeled malarial protease substrate. Biochemistry. 2002;41:12244–52. https://doi.org/10.1021/bi0263661.
33. Avellaneda MJ, Franke KB, Sunderlikova V, Bukau B, Mogk A, Tans SJ. Processive extrusion of polypeptide loops by a Hsp100 disaggregase. Nature. 2020;578:317–20. https://doi.org/10.1038/s41586-020-1964-y.
34. Sparkes I, White RR, Coles B, Botchway SW, Ward A. Using optical tweezers combined with total internal reflection microscopy to study interactions between the ER and Golgi in plant cells. In: Hawes C, Kriechbaumer V, editors. Plant endoplasmic reticulum: methods and protocols. 2018. pp. 167–78. https://doi.org/10.1007/978-1-4939-7389-7_13.
35. Ngo TTM, Zhang Q, Zhou R, Yodh JG, Ha T. Asymmetric unwrapping of nucleosomes under tension directed by DNA local flexibility. Cell. 2015;160:1135–44. https://doi.org/10.1016/j.cell.2015.02.001.
36. Bustamante CJ, Chemla YR, Liu S, Wang MD. Optical tweezers in single-molecule biophysics. Nat Rev Methods Primers. 2021;1:1–29. https://doi.org/10.1038/s43586-021-00021-6.
37. Smith SB, Finzi L, Bustamante C. Direct mechanical measurements of the elasticity of single DNA molecules by using magnetic beads. Science. 1992;258:1122.
38. Wang MD, Yin H, Landick R, Gelles J, Block SM. Stretching DNA with optical tweezers. Biophys J. 1997;72:1335–46. https://doi.org/10.1016/S0006-3495(97)78780-0.
39. Woodside MT, Behnke-Parks WM, Larizadeh K, Travers K, Herschlag D, Block SM. Nanomechanical measurements of the sequence-dependent folding landscapes of single nucleic acid hairpins. Proc Natl Acad Sci. 2006;103:6190–5. https://doi.org/10.1073/pnas.0511048103.
40. Svoboda K, Schmidt CF, Schnapp BJ, Block SM. Direct observation of kinesin stepping by optical trapping interferometry. Nature. 1993;365:721–7. https://doi.org/10.1038/365721a0.
41. Svoboda K, Block SM. Force and velocity measured for single kinesin molecules. Cell. 1994;77:773–84. https://doi.org/10.1016/0092-8674(94)90060-4.
42. Asbury CL, Fehr AN, Block SM. Kinesin moves by an asymmetric hand-over-hand mechanism. Science. 2003;302:2130.
43. Abbondanzieri EA, Greenleaf WJ, Shaevitz JW, Landick R, Block SM. Direct observation of base-pair stepping by RNA polymerase. Nature. 2005;438:460. https://doi.org/10.1038/nature04268. https://www.nature.com/articles/nature04268#supplementary-information.
44. Shaevitz JW, Abbondanzieri EA, Landick R, Block SM. Backtracking by single RNA polymerase molecules observed at near-base-pair resolution. Nature. 2003;426:684. https://doi.org/10.1038/nature02191. https://www.nature.com/articles/nature02191#supplementary-information.
45. John R, Davenport, Wuite GJL, Landick R, Bustamante C. Single-molecule study of transcriptional pausing and arrest by *E. coli* RNA polymerase. Science. 2000;287:2497.
46. Ma J, Bai L, Wang MD. Transcription under torsion. Science. 2013;340:1580.
47. Wang F, Wang Z, Dong L, Liu H, Yang L, Wang Y. Plasmonic bubbles: from fundamentals to applications. Adv Funct Mater. 2024;n/a:2403606. https://doi.org/10.1002/adfm.202403606.
48. Liu C, Huang Z, Huang S, Zhang Y, Li B, Nan F, Zheng Y. Robotic nanomanipulation based on spatiotemporal modulation of optical gradients. ACS Nano. 2024;18:19391–400. https://doi.org/10.1021/acsnano.4c06596.
49. Dai X, Fu W, Chi H, Mesias VSD, Zhu H, Leung CW, Liu W, Huang J. Optical tweezers-controlled hotspot for sensitive and reproducible surface-enhanced Raman spectroscopy characterization of native protein structures. Nat Commun. 2021;12:1292. https://doi.org/10.1038/s41467-021-21543-3.
50. Fu W, Chi H, Dai X, Zhu H, Mesias VSD, Liu W, Huang J. Efficient optical plasmonic tweezer-controlled single-molecule SERS characterization of pH-dependent amylin species in aqueous milieus. Nat Commun. 2023;14:6996. https://doi.org/10.1038/s41467-023-42812-3.

51. Zhang Y, Du J, Liu X, Shang F, Deng Y, Ye J, Wang Y, Yan J, Chen H, Yu M, Le S. Multi-domain interaction mediated strength-building in human α-actinin dimers unveiled by direct single-molecule quantification. Nat Commun. 2024;15:6151. https://doi.org/10.1038/s41467-024-50430-w.
52. Shepherd JW, Guilbaud S, Zhou Z, Howard JAL, Burman M, Schaefer C, Kerrigan A, Steele-King C, Noy A, Leake MC. Correlating fluorescence microscopy, optical and magnetic tweezers to study single chiral biopolymers such as DNA. Nat Commun. 2024;15:2748. https://doi.org/10.1038/s41467-024-47126-6.
53. Szalai AM, Ferrari G, Richter L, Hartmann J, Kesici M-Z, Ji B, Coshic K, Dagleish MRJ, Jaeger A, Aksimentiev A, et al. Single-molecule dynamic structural biology with vertically arranged DNA on a fluorescence microscope. Nat Methods. 2024. https://doi.org/10.1038/s41592-024-02498-x.
54. Chu L-Y, Stedman D, Gannon J, Cox S, Pobegalov G, Molodtsov MI. Force-transducing molecular ensembles at growing microtubule tips control mitotic spindle size. Nat Commun. 2024;15:9865. https://doi.org/10.1038/s41467-024-54123-2.
55. Ge J, Bian X, Ma L, Cai Y, Li Y, Yang J, Karatekin E, De Camilli P, Zhang Y. Stepwise membrane binding of extended synaptotagmins revealed by optical tweezers. Nat Chem Biol. 2021. https://doi.org/10.1038/s41589-021-00914-3.
56. Heller I, Sitters G, Broekmans OD, Farge G, Menges C, Wende W, Hell SW, Peterman EJG, Wuite GJL. STED nanoscopy combined with optical tweezers reveals protein dynamics on densely covered DNA. Nat Methods. 2013;10:910–6. https://doi.org/10.1038/nmeth.2599.
57. Xu L, Halma MTJ, Wuite GJL. Mapping fast DNA polymerase exchange during replication. Nat Commun. 2024;15:5328. https://doi.org/10.1038/s41467-024-49612-3.
58. Sharma M, de Alba E. Assembly mechanism of the inflammasome sensor AIM2 revealed by single molecule analysis. Nat Commun. 2023;14:7957. https://doi.org/10.1038/s41467-023-43691-4.
59. Alcón P, Kaczmarczyk AP, Ray KK, Liolios T, Guilbaud G, Sijacki T, Shen Y, McLaughlin SH, Sale JE, Knipscheer P, et al. FANCD2–FANCI surveys DNA and recognizes double- to single-stranded junctions. Nature. 2024;632:1165–73. https://doi.org/10.1038/s41586-024-07770-w.
60. Neupane K, Narayan A, Sen Mojumdar S, Adhikari G, Garen CR, Woodside MT. Direct observation of prion-like propagation of protein misfolding templated by pathogenic mutants. Nat Chem Biol. 2024;20:1220–6. https://doi.org/10.1038/s41589-024-01672-8.

Light-Driven Micro/Nanomotors in Biomedical Applications

12

Yanan Zhao, Xiaoxia Chen, and Yuchao Li

Abstract

Micro/nanomotors are power devices that convert various forms of external energy into their own mechanical energy at the micro/nanoscale. Constructed with different structures and driven by various methods, micro/nanomotors execute movements such as propulsion, rotation, and aggregation. Based on this, micro/nanomotors exhibit specific functionalities under the influence of external fields, including electricity, magnetism, acoustics, and light. Among these, light-driven micro/nanomotors have unique advantages, such as non-contact operation, high flexibility, and high precision. Combined with optical gradient force propulsion, bubble propulsion, thermophoresis, electrophoresis, and diffusiophoresis, light-driven micro/nanomotors have progressively gained functionality in environmental management and micro/nano processing, with particular significance in biomedicine. They exhibit promising development potential and can serve as high-precision therapeutic tools for targeted drug delivery, thrombolysis, and photothermal therapy. Building upon the use of light-driven micro/nanomotors in biological and biomedical applications, this paper first provides a brief overview of their classification. Subsequently, it delves into their applications in cell manipulation, cargo delivery, photodynamic therapy, and photothermal therapy. Finally, it summarizes and provides prospects for the application of light-driven micromotors in the field of biomedicine.

Y. Zhao · X. Chen · Y. Li (✉)
Guangdong Provincial Key Laboratory of Nanophotonic Manipulation, College of Physics and Optoelectronic Engineering, Institute of Nanophotonics, Jinan University, Guangzhou, China
e-mail: liyuchao@jnu.edu.cn

Keywords

Light-driven micromotors · Biomedicine · Cell regulation · Photothermal therapy

12.1 Introduction

As power devices at the micro/nanoscale, micro/nanomotors achieve active motion by harnessing various forms of external energy (such as light energy, electrical energy, magnetic energy, thermal energy, chemical energy, etc.) and converting it into their own mechanical energy [1–6]. The individual object driven by hydrogen peroxide was discovered by Whitesides and co-authors, marking a significant milestone in motor development in 2002 [7]. To enhance the adaptability of motors for tasks in micro-environments, micro/nanomotors with diverse structures and components have been successively developed. The controlled movements of these micro/nanomotors are driven by various driving mechanisms, including chemical driving, enzyme driving, electric field driving, optical field driving, etc. However, a notable limitation arises in the biomedical application of most chemically driven micro/nanomotors, as they commonly rely on hydrogen peroxide as fuel, which poses biological toxicity, and restricts their widespread use in the biomedical field [8, 9]. Micro/nanomotors powered by biological enzymes exhibit high biocompatibility, yet there is a need to enhance their movement speed and catalytic efficiency [10–12]. In contrast to the aforementioned micro/nanomotors, those propelled by external fields eliminate the need for fuel intervention, presenting significant advantages in the field of biomedicine. For instance, micro/nanomotors driven by magnetic and ultrasonic fields demonstrate outstanding active propulsion performance, energy penetration, and minimal restrictions in the working environment. However, their preparation process and

B. Li et al. (eds.), *Biophotonic Manipulation*, Advances in Optics and Optoelectronics,
https://doi.org/10.1007/978-981-96-4982-2_12

control parameter settings are intricate, and achieving long-term controllable motion poses a challenge [13–15]. The key issue in the application of micro/nanomotors is whether it is feasible to attain simpler preparation processes and more precise, flexible, and controllable operations while ensuring biocompatibility.

Functioning as an external control signal, parameters such as light intensity, frequency, polarization degree, and propagation direction are meticulously tuned and inputted into the micromotor. This precise adjustment enables remote wireless control of the micromotor with high spatial and temporal resolution [16–19]. Consequently, light-driven micro/nanomotors possess distinctive advantages, including non-contact operation, minimal damage, and the ability to flexibly control their movement speed and direction. The spatial precision achieved through remote wireless control of light-driven micro/nanomotors proves advantageous for active motion in micro-environments [20–25]. To enhance the performance of light-driven micro/nanomotors in autonomous motion under diverse environmental conditions, continuous research focuses on micro/nanomotors based on optical gradient force propulsion, light-driven bubble driving, light-driven autothermophoresis, and light-driven self-electrophoresis. Gradually, light-driven micro/nanomotors are being functionalized through alterations in motor preparation materials, construction of different light-responsive structures, and drug loading on their surfaces. This functionalization expands their applications in environmental management, micro/nano processing, and various fields. In recent years, there has been rapid development in light-driven micro/nanomotors, particularly in the field of biomedicine. These micro/nanomotors exhibit promising capabilities in drug-targeted delivery and tumor photothermal therapy, have the potential to significantly improve drug therapy and anticancer efficacy, and develop into powerful tools for biomedical applications.

This article reviews the light-driven micro/nanomotors applied to biological and biomedical, detailing the specific application of light-driven micro/nanomotor in the biomedical field, including cell regulation, cargo delivery, photodynamic treatment, and photothermal treatment. Furthermore, the article addresses the significant challenges that light-driven micro/nanomotors encounter in biological and biomedical applications.

12.2 Light-Driven Micro/Nanomotors for Cell Regulation

As the smallest units of life, cells play a crucial role in maintaining various physiological activities. Based on this, light-driven micro/nanomotors have been applied as tools for cell regulation, offering a non-contact and non-invasive approach to influence the physiological activities of cells. They play a significant role in regulating cell growth, apoptosis, eliminating dead cells, and facilitating cell transport [26, 27]. Axons are pivotal in the development, repair, and regeneration of the nervous system, with the growth characteristics of axon tips determining the direction of axon growth and migration [28, 29]. Consequently, light-driven micro/nanomotors present considerable advantages as a high-precision, flexible, and controllable control method. As shown in Fig. 12.1a, light-driven birefringent microspheres were applied to control the growth direction of individual axons (nerve fibers) by Wu and co-authors [30]. Birefringent microspheres were initially captured by one or two optical traps and positioned at the periphery of the axonal growth cones of cultured goldfish retinal ganglion cells. Utilizing the angular momentum of circularly polarized light, the trapped birefringent microspheres were induced to spin at approximately 1 Hz, creating localized microfluidics that generated a shear force of about 0.17 pN on the adjacent axonal growth cone. As depicted in Fig. 12.1a(I), prior to the manipulation with light-driven micro/nanomotors, the axons of the target neuron grew linearly (along the dotted arrow 1). The birefringent microspheres were positioned to the left of the growth cone by an optical trap with a counterclockwise circularly polarized light applied. Subsequently, after 540 s (Fig. 12.1a(II)), the target neuron axons were redirected to the lower right (along the dashed arrow 2) to promote growth. The light-driven micro/nanomotor precisely controlled the growth direction of a single axon.

Light-driven micro/nanomotors can be wirelessly controlled remotely by adjusting parameters such as optical intensity and power, thereby modifying the motion performance of the micro/motor. As shown in Fig. 12.1b(I), Wu and co-authors [31] proposed a polymer tubular light-driven micro/nanomotor comprised of gold nanoshells (AuNSs). The presence of AuNSs within the motor induces strong plasmonic resonance absorption in the near-infrared region, creating an asymmetric local heat distribution on the micro/nanomotor. This asymmetry results in different thermophoretic forces along the long axis of the micro/nanomotor structure. Consequently, micro/nanomotors driven by near-infrared light exhibit directed movement toward the target cell. Under near-infrared radiation at a condition of 4 mW μm^{-2}, the micro/nanomotors moved towards the target HeLa cell, achieving a maximum speed of approximately 11 μm^{-1} (Fig. 12.1b(II–IV)). Upon reaching the target HeLa cell action site, the near-infrared laser power was promptly increased to 30 mW μm^{-2}, and the micro/nanomotors were irradiated for 1 s. Subsequently, red fluorescence appeared in the nuclei of the targeted HeLa cells, while other cells exhibited no red fluorescence (Fig. 12.1b(V)). With the application of high-power near-infrared radiation, the surrounding medium's temperature increased due to the photothermal effect of the light-driven micro/nanomotors,

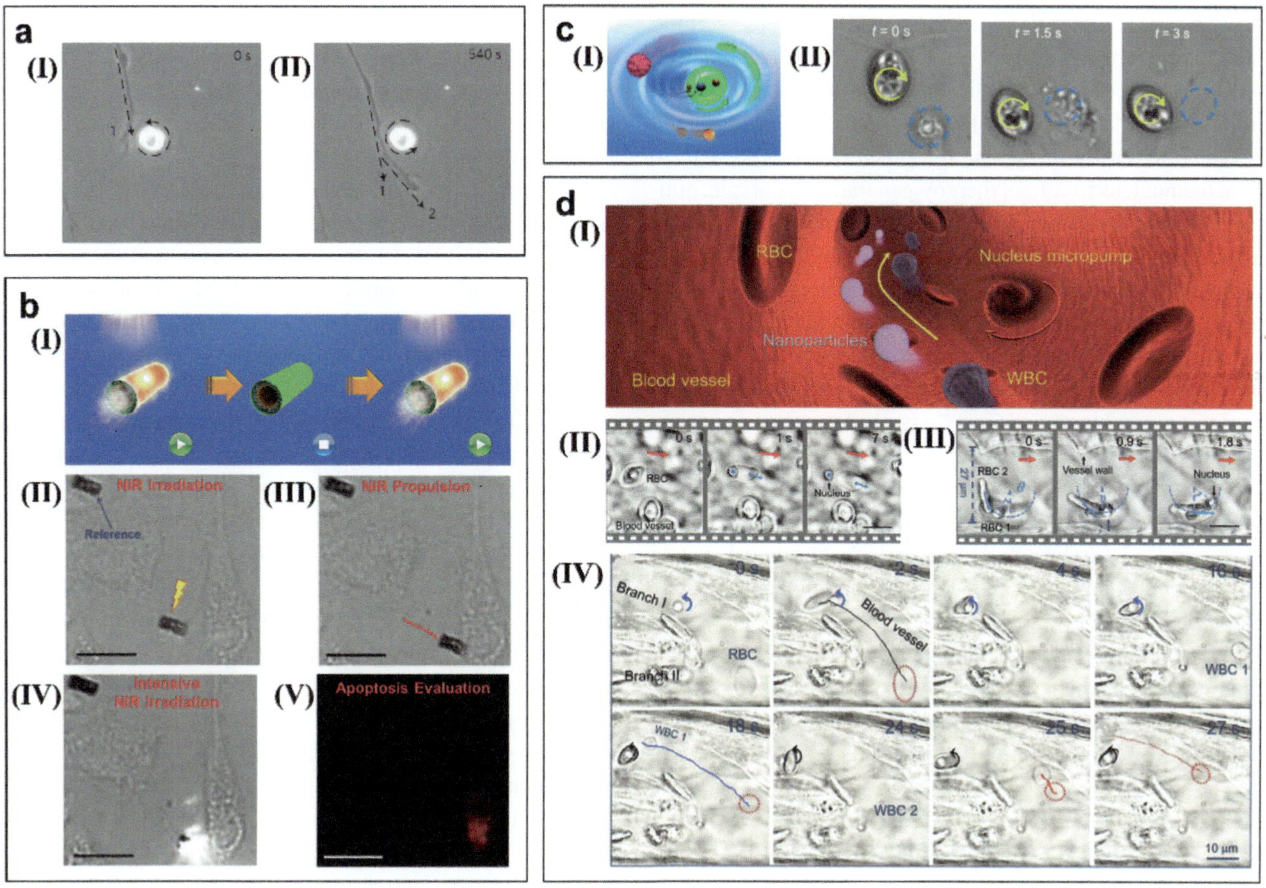

Fig. 12.1 Application of light-driven micro/nanomotors in cell regulation. **a** Optical microscopy images of nerve fiber growth before (I) and after (II) manipulation of light/driven micro/nanomotors. Reprinted from Wu et al. [30], Copyright (2012), with permission from Springer Nature. **b** Micro/nanomotors regulate cancer cell apoptosis. (I) Schematic diagram of the micro/nanomotor, and (II–IV) show the optical microscopy and fluorescence images of light-driven micro/nanomotors-induced apoptosis in cancer cells, respectively. Reprinted from Wu et al. [31], Copyright (2016), with permission from Wiley–VCH Verlag. **c** Micro/nanomotors clear dead cells. (I) Schematic diagram of the micro/nanomotor. (II) Optical microscopy image of the micro/nanomotor clearing dead cells. Reprinted from Xin et al. [32], Copyright (2020), with permission from American Chemical Society. **d** Light-driven micro/nanomotors transport cells. (I) Schematic diagram of the nuclear micropump. (II, III) Optical microscopy images of the dynamic extraction of nuclei by multiple optical traps and squeezing out of the nucleus by rotating red blood cells as a push hammer, respectively. (IV) Optical microscopy images of red blood cells transported by nucleus micropump. Reprinted from Gao et al. [33], Copyright (2021), with permission from Elsevier Ltd

inducing apoptosis in targeted cancer cells. Notably, other HeLa cells showed no red fluorescence, suggesting that the near-infrared enhancement had no apparent adverse effects on the cells. These results underscore the application of light-driven micro/nanomotors as a flexible and controllable photothermal agent, holding significant potential in cancer treatment.

Light-driven micro/nanomotors also play a crucial role in clearing dead cells and facilitating cell trafficking as a biocompatible approach. As shown in Fig. 12.1c(I), Xin and co-authors selected a single cell (green microalgae: *Chlamydomonas reinhardtii*) as a living, multifunctional micromotor driven by optical force [32]. When the optical traps provided by the optical tweezers stably capture randomly swimming cells, their flagella-induced swimming transforms into rotation around the focal point. The circular scanning trap sequence controlled by the acousto-optic deflection system was applied to the outer boundary of the micromotor, enabling a circular scanning optical trap to control the rotational motion of the micromotor. The local micro-flow generated by the rotation of the micro/nanomotors acts on different targets to accomplish various functions. The controllable rotation of the light-driven micro/nanomotor induces flagellar beating, effectively disrupting nearby biological aggregates (Fig. 12.1c(II)). Dead cells are completely fragmented and removed within 3 s. The light-driven micro/nanomotors efficiently dismantle biological aggregates and eliminate dead cells, offering a new avenue for cleaning the biological microenvironment. To inherently and fully ensure biocompatibility, Gao and co-authors designed a method for constructing light-driven micro/nanomotors in the form of the cell nucleus, serving as an endogenous biological micropump [33]. The construction principle of the nucleus micropump is illustrated in Fig. 12.1d(I). The nucleus of red blood cells were extracted and then driven by the scanning optical tweezers system to drive the nucleus to rotate, thereby generating local microfluidics. The multifunctional dynamics of the micropump control the transmission speed and direction. Among them, there are two ways to extract the erythrocyte nucleus. As shown in Fig. 12.1d(II), two discrete optical traps fix the nucleus (blue dots) and move the cell membrane (blue arrows), successfully isolating the nucleus. The alternative approach utilizes a spinning red blood cell as a "push hammer" to compress another red blood cell, thus separating the nucleus from the cell (Fig. 12.1d(III)). The micropump propels blood flow, prompting an erythrocyte to move towards Branch 1 (black curve) (Fig. 12.1d(IV)). Red blood cells synchronously rotate with the micropump ($t = 2$ s). Two white blood cells, WBC 1 and WBC 2, appear in the vessel ($t = 16$ s). Under the action of the nuclear micropump, WBC 1 and WBC 2 successively move toward Branch 1, following the blue and red curves, respectively. Thus, light-driven micro/nanomotors with full biocompatibility enable highly precise cell regulation in vivo.

12.3 Light-Driven Micro/Nanomotors for Cargo Delivery

Light-driven micro/nanomotors have demonstrated excellent capabilities in trapping, transporting, and releasing colloidal particles and living microorganisms without any magnetic guidance or additional surface functionalization for capturing/releasing microparticles. They have been developed as promising candidates for nanomedicine with the potential to enhance the delivery and anticancer effects of various drugs [34–36]. Visible-light-activated $BiVO_4$ micro/nanomotors were constructed by Villa and co-authors (Fig. 12.2a(I)), which consisted of a single-component $BiVO_4$. When placed in pure water and aqueous solutions under light irradiation (0.01 wt %), they demonstrated self-propulsion [37]. Since the surface of microorganisms is composed of a variety of biomolecules, their complex constituent molecules may exhibit different interactions than light-driven micro/nanomotors. As shown in Fig. 12.2a(II)–(V), yeast cells were the target microorganisms, and the interaction between $BiVO_4$ micro/nanomotors and yeast cells under visible light irradiation was explored. The initial motion direction of the $BiVO_4$ micro/nanomotor was indicated by the red arrow. In this biological microenvironment, it rotated about 45° to change the direction of motion and began to move toward the yeast cell (Fig. 12.2a(III)). After reaching the target site, the $BiVO_4$ micromotor captured and continued to transport the yeast cells (Fig. 12.2a(IV)). When the light was turned off, the self-propulsion of $BiVO_4$ micro/nanomotors and the generation of chemical gradients were hindered, resulting in the release of yeast cells (Fig. 12.2a(V)). The micromotor can achieve flexible and controllable cargo delivery under the driving of visible light, applicable to self-propulsion in different environments and even in highly viscous solutions. Near-infrared-driven micromotors can also easily and flexibly control the motion characteristics of the nanomotor by remotely modulating laser parameters, benefiting from lower optical damage. Near-infrared-driven Janus mesoporous silica micro/nanomotors with a diameter of 50 nm were constructed by Xuan and co-authors [38]. A Janus structure was formed by vacuum sputtering a 10 nm Au layer on one side of the mesoporous silica (Fig. 12.2b(I)). Under near-infrared laser irradiation, a thermal gradient was generated on the surface of the mesoporous silica due to the local photothermal effect of the Au layer, thus forming micro/nanomotors driven by autothermal electrophoresis (Fig. 12.2b(II)). In the absence of near-infrared illumination, most nanomotors remain in the left reservoir of the microfluidic chip due to their low diffusion rate (Fig. 12.2b(III)); and there are no green fluorescent dots appearing in the channel and on the right side of the microfluidic chip, indicating that drug delivery cannot be achieved. Under the irradiation of 3 W/cm^2 near-infrared light, green fluorescent dots representing fluorescein isothiocyanate appeared in the channel and the reservoir on the right side of the microfluidic chip, indicating that the irradiation of near-infrared light drove the active movement of the nanomotor to the target position. The drug-loaded micro/nanomotors driven by near-infrared light realize the active transport process and have great development potential in the field of biomedical drug delivery.

By adjusting the intensity of the incident light, the speed of the micro/nanomotors are controlled quickly and precisely. Combining this feature, Chen and co-authors took advantage of the limited penetration depth of light in semiconductor materials to construct an asymmetric surface chemical reaction on isotropic semiconductor particles, forming TiO_2 micro/nanomotors with a diameter of 3.5 μm (Fig. 12.2c(I)) [39]. Under illumination, a photocatalytic concentration gradient formed on its surface, promoting the movement of the TiO_2 micro/nanomotor. It is worth noting that, due to its isotropic structure, the motion direction of the micro/nanomotor was always consistent with the illumination direction and was not disturbed by its rotational Brownian diffusion or local flow. As shown in Fig. 12.2c(II), the TiO_2 micro/nanomotor was placed in a 0.66 wt% H_2O_2 environment. Three silica particles with a diameter of 1 μm were dispersed in H_2O_2 as a transport particle model and targeting particle, respectively. The TiO_2 micro/nanomotor

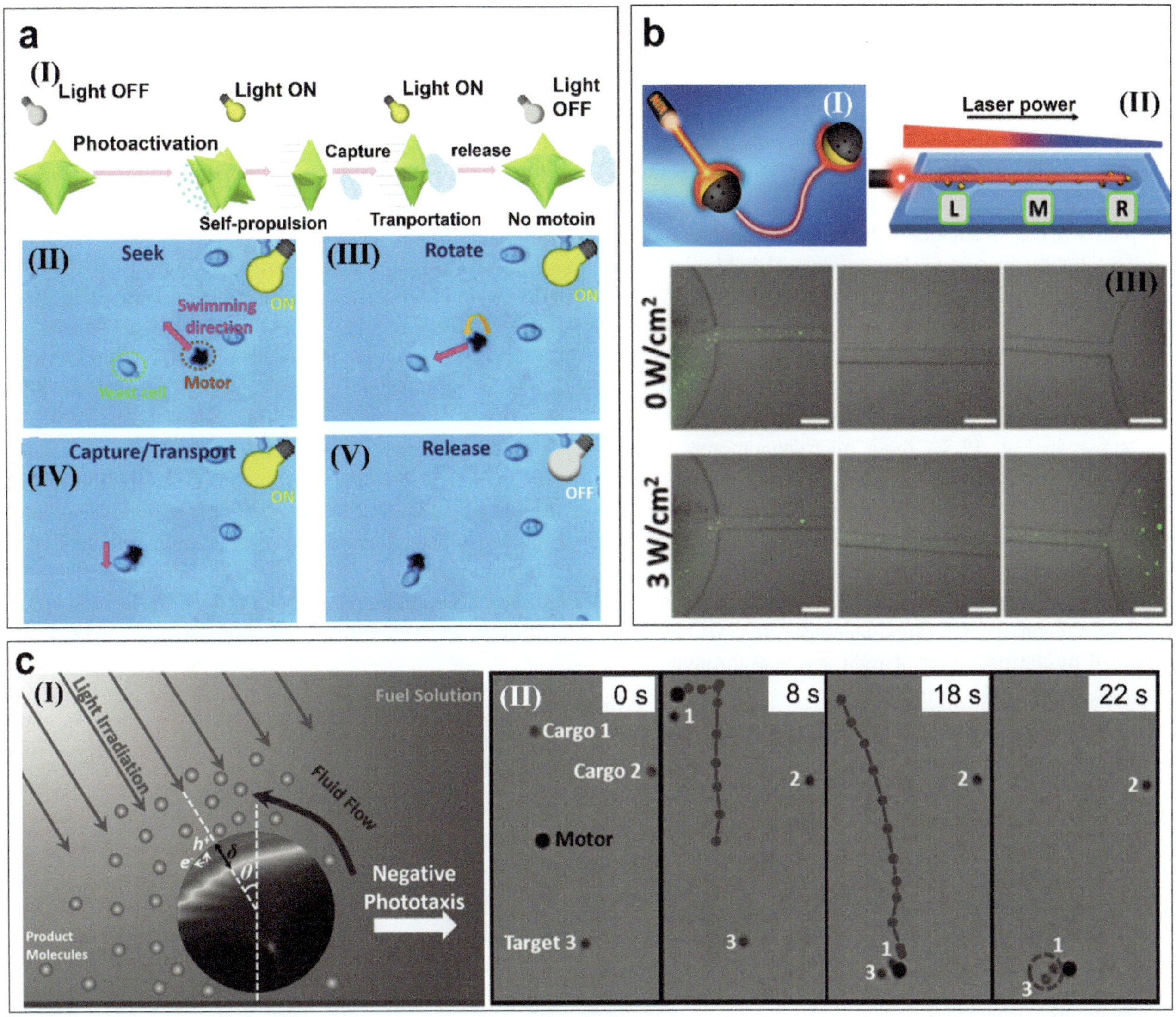

Fig. 12.2 Application of light-driven micro/nanomotors in cargo delivery. **a** Visible light photoactivation of $BiVO_4$ micro/nanomotors to control yeast delivery. (I) Schematic diagram of $BiVO_4$ micro/nanomotor construction. (II–V) Interaction of $BiVO_4$ micromotors with yeast under visible light activation. Reprinted from Villa et al. [37], Copyright (2019), with permission from American Chemical Society. **b** Janus mesoporous silica micro/nanomotors driven by near-infrared light for drug loading and delivery. (I) Schematic diagram of Janus micro/nanomotors construction. (II) Schematic diagram of Janus micro/nanomotors transporting drugs. (III) Relative fluorescence intensities of the drug-loaded mesoporous silica micro/nanomotors at three locations without and with near-infrared light irradiation, respectively; Scale bar = 200 μm. Reprinted from Xuan et al. [38], Copyright (2016), with permission from American Chemical Society. **c** Precisely controlled particle-targeted delivery of TiO_2 micro/nanomotors. (I) Schematic diagram of TiO_2 micro/nanomotors construction. (II) Optical microscopy images of TiO_2 micro/nanomotors for cargo delivery. Reprinted from Chen et al. [39], Copyright (2017), with permission from Wiley

was driven to the perimeter of particle 1 within 8 s by regulating the UV light driving in the X and Y directions respectively. When the distance between the micro/nanomotor and particle 1 was 2 μm, particle 1 was captured by the TiO_2 micro/nanomotor due to the effect of phoretic attraction. Driven by UV light in the Y direction, the micro/nanomotor transported the captured particle 1 to the target particle 3 within 18 s. At this point, the UV light was turned off, and particle 1 was released (22 s) due to the disappearing electrophoretic attraction and Brownian motion. In the same principle, the micro/nanomotor continued moving driven by ultraviolet light, realizing the continuous process of capturing, transporting, and releasing the particles 2. Based on this, the light-driven micro/nanomotor realizes precisely controlled particle-targeted transportation, providing the possibility for targeted drug delivery and treatment of biomedical drugs.

12.4 Light-Driven Micro/Nanomotors for Photodynamic Therapy

Photodynamic therapy (PDT) stands out as a promising cancer treatment strategy, typically killing cancer cells by converting non-toxic oxygen into reactive oxygen species through the utilization of photosensitizers [40, 41]. When coupled with the advantages of flexible, controllable, and non-invasive light-driven micro/nanomotors, it holds the potential to significantly enhance the efficacy of photodynamic therapy. Cao and co-authors proposed the construction of polymersomes comprised of biodegradable copolymers and encapsulated with asymmetric gold nanoshells to engineer micro/nanomotors tailored for photodynamic therapy [42]. As shown in Fig. 12.3a, these micro/nanomotors harness the synergies of two distinct materials, aggregation-induced emission (AIE) and Au, under the excitation of two-photon near-infrared light. The motors convert radiant energy into mechanical energy, initiating thermoelectrophoretic motion, triggering autonomous propulsion, and the generation of reactive oxygen species. Simultaneously, the Au layer was activated by the provision of light and heat through highly stable fluorescence, which promotes the micromotor motion behavior more vigorously. This enhancement leads to an improved photodynamic therapy effect and a more potent cytotoxic response. Based on this, AIE/Au nanomotors exhibit both propulsive motions to augment cellular interactions and the phototherapeutic generation of reactive oxygen species, furnishing an effective, direct, and highly localized cytotoxicity to target cells.

Light-driven micro/nanomotors manifest their active photodynamic properties as carriers in dynamic therapy. As shown on the left side of Fig. 12.3b, Ma and co-authors loaded the hydrophobic photosensitizer zinc phthalocyanine (ZnPc) onto adamantane (Ad)-modified mesoporous silica nanoparticle channels [43]. Notably, Ad dissociated ZnPc and prevented its aggregation within the motor channel. Thus, ZnPc maintains monomeric properties under light irradiation with low optical power output (675 nm), and the nanomotors combined with it efficiently generate cytotoxic singlet oxygen ($^1O_2$) in the cellular microenvironment. In the upper right was a fluorescence microscope image of HeLa cells cultured for 24 h under dark conditions. The bottom was a fluorescence

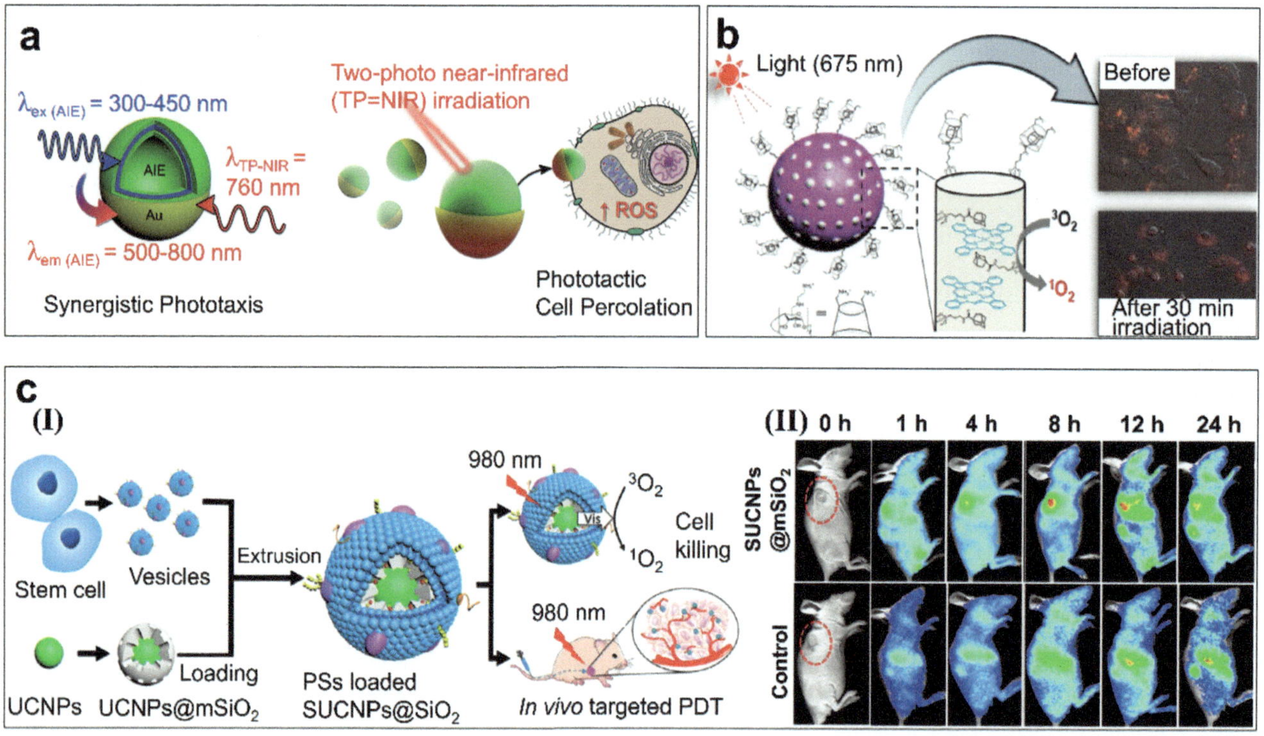

Fig. 12.3 Application of light-driven micro/nanomotors in photodynamic therapy. **a** Enhanced phototherapy via aggregation-induced emission of photoactivated nanomotors. Reprinted from Cao et al. [42], Copyright (2021), with permission from Springer Nature. **b** Degradation of ZnPc in micro/nanomotor channels and its photodynamic activity. Reprinted from Ma et al. [43], Copyright (2013), with permission from American Chemical Society. **c** Near-infrared light-activated stem cell membrane camouflaged upconversion nanomotors for remote photodynamic therapy in vivo. (I) The preparation process of micro/nanomotors and the principle of photodynamic therapy. (II) Fluorescence images in vivo and the control group at 1, 2, 4, 8, 12, and 24 h after injection of nanomotors into the tail vein of mice. Red circles indicate tumor sites. Reprinted from Gao et al. [44], Copyright (2016), with permission from American Chemical Society

microscope image of HeLa cells observed after 30 min of light irradiation at low optical power output (675 nm) and 24 h of culture. It can be clearly observed that under illumination, the nanomotors cause severe photodynamic damage to HeLa cells, eventually leading to cell apoptosis. To overcome the limitation of visible light penetration depth in photodynamic therapy, Gao and co-authors proposed preparing stem cell membrane-derived vesicles from mesenchymal stem cells and then encapsulating β-$NaYF_4$:Yb^{3+} with mesoporous silica by fusion synthesis. Two PSs (ZnPC and MC540) were loaded to prepare light-driven stem-cell-membrane-coated UCNPs (SUCNPs@mSiO_2) micro/nanomotors [44]. Upon near-infrared light irradiation, the motor produces cytotoxic reactive oxygen species that kill surrounding cells (Fig. 12.3c(I)). Combining the photodynamic effect of stem cell membranes and tumor-targeting capabilities, the micro/nanomotors were applied for in vivo photodynamic therapy. To verify the tumor-targeting ability of the SUCNPs@mSiO_2 micro/nanomotors in vivo, Cy7-labeled SUCNPs@mSiO_2 and UCNPs@mSiO_2 were intravenously injected with equal amounts into mice bearing HeLa tumors and then combined with fluorescence imaging to evaluate the real-time distribution of micro/nanomotors and tumor aggregation in vivo. Both SUCNPs@mSiO_2 and UCNPs@mSiO_2 were widely distributed throughout the body 1 h after intravenous injection (Fig. 12.3c(II)). However, the tumor fluorescence intensity of the SUCNPs@mSiO_2-injected group was significantly higher than that of the UCNPs@mSiO_2 group with the passage of time, indicating that the tumor aggregation degree of SUCNPs@mSiO_2 was higher. The transformed stem cell membranes play a role in enhancing the active targeting efficacy and tumor-promoting accumulation of SUCNPs@mSiO_2. Combined with the long-term blood circulation of stem cells and tumor specificity, light-driven SUCNPs@mSiO_2 micro/nanomotors further enhance the anti-tumor effect in photodynamic therapy.

12.5 Light-Driven Micro/Nanomotors for Photothermal Therapy

In photothermal therapy, light-driven micro/nanomotors also assume a crucial role, leveraging their active motion to enhance the cell-killing effect, presenting significant potential in cancer therapy [45–47]. The structure-controllable Au-bispyrene self-propulsion (Au-BP@SP) Janus micro/nanomotor was fabricated by Yang and co-authors, which demonstrated for the first time that the kinetic energy of the nanocomposite based on Janus thermophoresis is helpful for efficient cancer treatment [48]. As shown in Fig. 12.4a(I), the micro/nanomotors comprised inorganic gold stars and organic BP nano-aggregates, enveloped by a coalesced silica shell. Under near-infrared light irradiation, the BP nano-aggregates exhibited robust fluorescence due to the heat generated by the Au on the micro/nanomotor's surface. This resulted in a temperature gradient on both sides of the BP nano-aggregates, enabling active motion and the generation of photothermal effects. Notably, the Janus Au-BP@SP micro/nanomotor could convert kinetic energy into heat through collisions, leading to an increased perimeter temperature and further enhancing the photothermal effect. To validate the photothermal effect of the micro/nanomotor, equal concentrations (100 μg mL^{-1}, 50 μL) of Au-BP7@SP and Au@SP were injected into the left leg muscles of mice, respectively. The mouse tumor was irradiated with an 808 nm laser (2 W cm^{-2}, 5 min), and the temperature of the tumor during irradiation was recorded with a thermal imager after 1 h. The temperature of the tumor site treated by Au-BP7@SP Janus increased significantly to 55.7 °C within 5 min, a notable elevation compared to Au@SP nanoparticles (46.6 °C) (Fig. 12.4a(II)). Tumor cells have been effectively killed by a significant photothermal effect produced in a short time.

The mobility of light-driven micro/nanomotors in the near-infrared endows them with remarkable thrombolytic capabilities in photothermal therapy. Wan and co-authors designed and fabricated mesoporous/macroporous silica/platinum nanomotors coated with platelet membranes [49]. The thrombolytic drug urokinase was loaded onto the motor's surface due to its macroporous structure, and then the nanomotor was modified with a platelet membrane to facilitate its aggregation at the thrombus site (Fig. 12.4b(I)). Under near-infrared light irradiation, the non-uniform distribution of Pt nanoparticles on the surface of the nanomotor led to resonance absorption, creating a gradient thermal distribution, thereby advancing the motor thermophoresis into the thrombus site. Interestingly, the modified platelet membranes ruptured under near-infrared illumination, promoting the release of the drug with the greatest capacity by the micro/nanomotor, thereby averting the side effects of the drug reaching other parts of the body. The thermophoretic motion of the light-driven nanomotor under 808 nm near-infrared irradiation is depicted in Fig. 12.4b(II). Building upon this, the thrombolytic ability of the motor after entering the thrombus was explored. It can be clearly observed that, in the absence of NIR irradiation, the bionic motor alone has little thrombolytic effect. In contrast, the micro/nanomotors reduce the thrombus volume by about 74.7% after 3 h of light refraction.

Biomimetic optical micromotors play a pivotal role in medicine, and when combined with photothermal therapy, they exhibit unique advantages in actively traversing the blood–brain barrier. Deng and co-authors constructed natural killer (NK) cell-excited nanomotors with AIE characteristics (NK@AIEdots) [50]. As shown in Fig. 12.4c(I), the micromotors were constructed by encapsulating natural killer cells with active polymers to form the NK@AIEdots structure. They

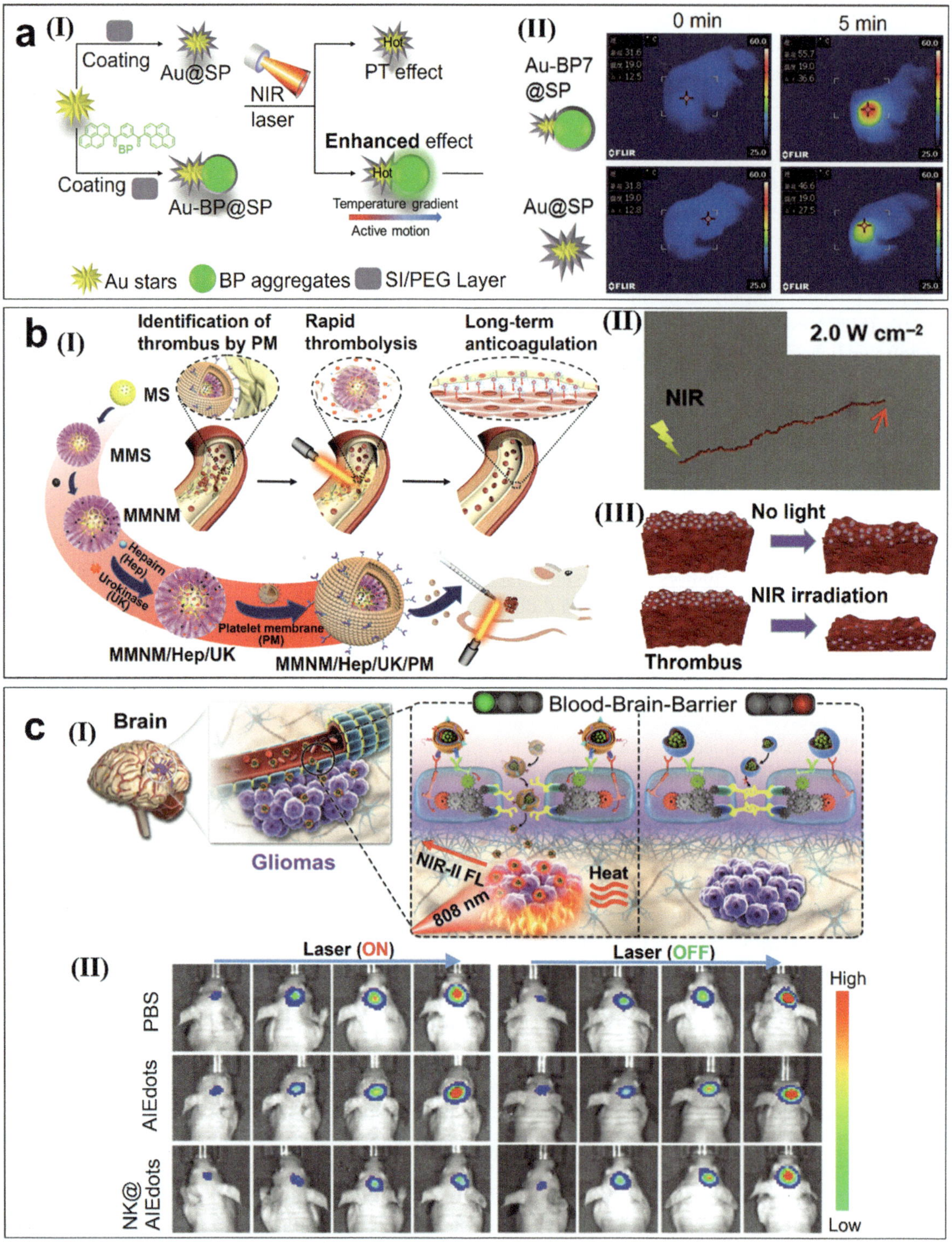

Fig. 12.4 Application of light-driven micro/nanomotors in photothermal therapy. **a** Janus micromotors for photothermal tumor therapy. (I) Schematic diagram of Au-BP@SP Janus micro/nanomotors. (II) Infrared thermal images of MCF-7 tumor mice injected with Au@SP and Au-BP7@SP under 808 nm laser irradiation at different time points. (Reprinted from Yang et al. [48], Copyright (2016), with permission from Wiley–VCH Verlag). **b** Platelet-derived light-driven porous nanomotor for thrombus therapy. (I) Schematic diagram of the preparation of nanomotors and their application in thrombus targeting and thrombolytic therapy. (II) Time-lapse images of the motion trajectories of micro/nanomotors in the near-infrared (2.0 W cm^{-2}). (III) Thrombolytic effect of nanomotors in vitro. Reprinted from Wan et al. [49], Copyright (2020), with permission from American Association for the Advancement of Science. **c** Biomimetic light-driven micro/nanomotors for near-infrared-II fluorescence-guided glioma therapy. (I) Schematic illustration of tight junction (TJ)-regulated blood–brain barrier penetration of NK@AIEdots micro/nanomotors. (II) Representative in vivo bioluminescence images of in situ gliomas at different time points after photothermal therapy. Reprinted from Deng et al. [50], Copyright (2020), with permission from American Chemical Society

autonomously traversed the blood–brain barrier by unraveling TJ structures, specifically accumulating in brain tumors within the complex brain stroma for in situ transcranial-scalp fluorescence imaging and photothermal therapy. As shown in Fig. 12.4c(II), intravenous injection of AIEdots, NK@AIEdots, and PBS served as experimental control groups to evaluate the therapeutic effect of the nanomotor on glioma. The mouse brain was irradiated with a laser with a power density of 0.3 W/cm^2 and 808 nm for 8 min 24 h after injection. Comparing the images detected by the three groups of infrared imagers, it can be clearly observed that after injection of NK@AIEdots and laser irradiation, the increase in bioluminescence intensity in mice is much slower, indicating that the growth of gliomas has been effectively inhibited. These results demonstrate that photothermal therapy based on NK@AIEdots micro/nanomotors effectively inhibits glioma.

12.6 Summary and Outlook

This article starts with the application of light-driven micro/nanomotors in biomedicine and systematically introduces the principles and applications of different types of light-driven micro/nanomotors in cell regulation, cargo delivery, photodynamic therapy, and photothermal therapy. Light-driven micro/nanomotors have the advantages of high manipulation accuracy, spatiotemporal resolution, and biocompatibility, which provide a versatile method for the manipulation of biological samples and can be used as specific therapeutic tools for biomedical diagnosis and treatment.

Nonetheless, the current research on light-driven micro/nanomotors encounters challenges related to biosafety, the feasibility of different light wavelengths, and the integration of complex functions. In light of these challenges, we propose recommendations for the continued advancement of light-driven micro/nanomotors in biological and biomedical applications.

Firstly, biosafety stands as a crucial prerequisite for the practical implementation of light-driven micro/nanomotors in biomedical clinical applications. The materials constituting these micro/nanomotors should ideally be biodegradable or controllably removable after completing their designated tasks. Secondly, the use of ultraviolet light, a common driver for micro/nanomotors, may pose risks to biological tissues. The near-infrared light located in the second region of the near-infrared spectrum (1000–1700 nm) has higher tissue penetration and lower photon absorption and scattering in the tissue. Developing near-infrared light-driven micro/nanomotors can enhance biosafety and efficiency in applications such as thrombus ablation and tissue repair. In addition, expanding research on light-driven micro/nanomotors with integrated multiple functions is essential for their tasks in intricate microscopic systems. By exploring efficient light-responsive micromotor preparation materials to improve their light conversion capabilities; by modifying fluorescent materials on the surface of micromotors, and by loading drug particles, to make them have biosensing, imaging, medical treatment, and other functions. The rapid and efficient movement of nanomotors in the complex in vivo flow environments opens avenues for improving drug delivery efficiency and enabling precisely targeted therapy using light-driven micro/nanomotors.

Acknowledgements This work was supported by the National Key Research and Development Program of China (No. 2022YFA1206300), and National Natural Science Foundation of China (No. 62422505).

References

1. Soler L, Magdanz V, Fomin VM, Sanchez S, Schmidt OG. Self-propelled micromotors for cleaning polluted water. ACS Nano. 2013;7:9611–20. https://doi.org/10.1021/nn405075d.
2. Fernández-Medina M, Ramos-Docampo MA, Hovorka O, Salgueiriño V, Städler B. Recent advances in nano-and micromotors. Adv Funct Mater. 2020;30:1908283. https://doi.org/10.1002/adfm.201908283.
3. Srivastava SK, Clergeaud G, Andresen TL, Boisen A. Micromotors for drug delivery in vivo: the road ahead. Adv Drug Delivery Rev. 2019;138:41–55. https://doi.org/10.1016/j.addr.2018.09.005.
4. Pourrahimi AM, Pumera M. Multifunctional and self-propelled spherical Janus nano/micromotors: recent advances. Nanoscale. 2018;10:16398–415. https://doi.org/10.1039/C8NR05196H.
5. Katuri J, Caballero D, Voituriez R, Samitier J, Sanchez S. Directed flow of micromotors through alignment interactions with micropatterned ratchets. ACS Nano. 2018;12:7282–91. https://doi.org/10.1021/acsnano.8b03494.
6. Mujtaba J, Liu J, Dey KK, Li T, Chakraborty R, Xu K, Makarov D, Barmin RA, Gorin DA, Tolstoy VP, Huang G, Solovev AA. Micro-bio-chemo-mechanical-systems: micromotors, microfluidics, and nanozymes for biomedical applications. Adv Mater. 2021;33:2007465. https://doi.org/10.1002/adma.202007465.
7. Ismagilov RF, Schwartz A, Bowden N, Whitesides GM. Autonomous movement and self-assembly. Angew Chem Int Ed. 2002;41:652–4. https://doi.org/10.1002/1521-3773(20020215)41:4%3c652::AID-ANIE652%3e3.0.CO;2-U.
8. Safdar M, Wani OM, Jänis J. Manganese oxide-based chemically powered micromotors. ACS Appl Mater Interfaces. 2015;7:25580–5. https://doi.org/10.1021/acsami.5b08789.
9. Brooks AM, Tasinkevych M, Sabrina S, Velegol D, Sen A, Bishop KJM. Shape-directed rotation of homogeneous micromotors via catalytic self-electrophoresis. Nat Commun. 2019;10:495. https://doi.org/10.1038/s41467-019-08423-7.
10. Tang S, Zhang F, Gong H, Wei F, Zhuang J, Karshalev E, Esteban-Fernández de Avila B, Huang C, Zhou Z, Li Z, Yin L, Dong H, Fang RH, Zhang X, Zhang L, Wang J. Enzyme-powered Janus platelet cell robots for active and targeted drug delivery. Sci Robot. 2020;5:eaba6137. https://doi.org/10.1126/scirobotics.aba6137.
11. Ma X, Hortelão AC, Patino T, Sanchez S. Enzyme catalysis to power micro/nanomachines. ACS Nano. 2016;10:9111–22. https://doi.org/10.1021/acsnano.6b04108.

12. Zhao X, Gentile K, Mohajerani F, Sen A. Powering motion with enzymes. Acc Chem Res. 2018;51:2373–81. https://doi.org/10.1021/acs.accounts.8b00286.
13. Zhou H, Mayorga-Martinez CC, Pané S, Zhang L, Pumera M. Magnetically driven micro and nanorobots. Chem Rev. 2021;121:4999–5041. https://doi.org/10.1021/acs.chemrev.0c01234.
14. Valdez-Garduño M, Leal-Estrada M, Oliveros-Mata ES, Sandoval-Bojorquez DI, Soto F, Wang J, Garcia-Gradilla V. Density asymmetry driven propulsion of ultrasound-powered Janus micromotors. Adv Funct Mater. 2020;30:2004043. https://doi.org/10.1002/adfm.202004043.
15. Xu H, Medina-Sánchez M, Schmidt OG. Magnetic micromotors for multiple motile sperm cells capture, transport, and enzymatic release. Angew Chem Int Ed. 2020;59:15029–37. https://doi.org/10.1002/anie.202005657.
16. Wang J, Xiong Z, Zheng J, Zhan X, Tang J. Light-driven micro/nanomotor for promising biomedical tools: principle, challenge, and prospect. Acc Chem Res. 2018;51:1957–65. https://doi.org/10.1021/acs.accounts.8b00254.
17. Zhan X, Zheng J, Zhao Y, Zhu B, Cheng R, Wang J. From strong dichroic nanomotor to polarotactic microswimmer. Adv Mater. 2019;31:1903329. https://doi.org/10.1002/adma.201903329.
18. Xu L, Mou F, Gong H, Luo M, Guan J. Light-driven micro/nanomotors: from fundamentals to applications. Chem Soc Rev. 2017;46:6905–26. https://doi.org/10.1039/C7CS00516D.
19. Dong R, Cai Y, Yang Y, Gao W, Ren B. Photocatalytic micro/nanomotors: from construction to applications. Acc Chem Res. 2018;51:1940–7. https://doi.org/10.1021/acs.accounts.8b00249.
20. Li Y, Liu X, Xu X, Xin H, Zhang Y, Li B. Red-blood-cell waveguide as a living biosensor and micromotor. Adv Funct Mater. 2019;29:1905568. https://doi.org/10.1002/adfm.201905568.
21. Zou X, Zheng Q, Wu D, Lei H. Controllable cellular micromotors based on optical tweezers. Adv Funct Mater. 2020;30:2002081. https://doi.org/10.1002/adfm.202002081.
22. Wang Y, Zhou C, Wang W, Xu D, Zeng F, Zhan C, Gu J, Li M, Zhao W, Zhang J, Guo J, Feng H, Ma X. Photocatalytically powered matchlike nanomotor for light-guided active SERS sensing. Angew Chem. 2018;130:13294–7. https://doi.org/10.1002/ange.201807033.
23. Liu C, Wang D, Zhang S, Cheng Y, Yang F, Xing Y, Xu T, Dong H, Zhang X. Biodegradable biomimic copper/manganese silicate nanospheres for chemodynamic/photodynamic synergistic therapy with simultaneous glutathione depletion and hypoxia relief. ACS Nano. 2019;13:4267–77. https://doi.org/10.1021/acsnano.8b09387.
24. Wan M, Wang Q, Li X, Xu B, Fang D, Li T, Yu Y, Fang L, Wang Y, Wang M, Wang F, Mao C, Shen J, Wei J. Systematic research and evaluation models of nanomotors for cancer combined therapy. Angew Chem Int Ed. 2020;59:14458–65. https://doi.org/10.1002/anie.202002452.
25. Wang D, Gao C, Wang W, Sun M, Guo B, Xie H, He Q. Shape-transformable, fusible rodlike swimming liquid metal nanomachine. ACS Nano. 2018;12:10212–20. https://doi.org/10.1021/acsnano.8b05203.
26. Zhu J, Thompson CB. Metabolic regulation of cell growth and proliferation. Nat Rev Mol Cell Biol. 2019;20:436–50. https://doi.org/10.1038/s41580-019-0123-5.
27. Wang Y, Feng L, Wang S. Conjugated polymer nanoparticles for imaging, cell activity regulation, and therapy. Adv Funct Mater. 2019;29:1806818. https://doi.org/10.1002/adfm.201806818.
28. Moser M, Murphy KM. Dendritic cell regulation of TH1–TH2 development. Nat Immunol. 2000;2000(1):199–205. https://doi.org/10.1038/79734.
29. Sleigh JN, Rossor AM, Fellows AD, Tosolini AP, Schiavo G. Axonal transport and neurological disease. Nat Rev Neurol. 2019;15:691–703. https://doi.org/10.1038/s41582-019-0257-2.
30. Wu T, Nieminen TA, Mohanty S, Miotke J, Meyer RL, Rubinsztein-Dunlop H, Berns MW. A photon-driven micromotor can direct nerve fibre growth. Nat Photonics. 2012;6:62–7. https://doi.org/10.1038/nphoton.2011.287.
31. Wu Z, Si T, Gao W, Lin X, Wang J, He Q. Superfast near-infrared light-driven polymer multilayer rockets. Small. 2016;12:577–82. https://doi.org/10.1002/smll.201502605.
32. Xin H, Zhao N, Wang Y, Zhao X, Pan T, Shi Y, Li B. Optically controlled living micromotors for the manipulation and disruption of biological targets. Nano Lett. 2020;20:7177–85. https://doi.org/10.1002/smll.201502605.
33. Gao Q, Wang W, Li X, Li Y, Ferraro P, Jiao X, Liu X, Zhang Y, Li B. Cell nucleus as endogenous biological micropump. Biosens Bioelectron. 2021;182: 113166. https://doi.org/10.1016/j.bios.2021.113166.
34. Alapan Y, Yasa O, Schauer O, Giltinan J, Tabak AF, Sourjik V, Siitti M. Soft erythrocyte-based bacterial microswimmers for cargo delivery. Sci Robot. 2018;3:eaar4423. https://doi.org/10.1126/scirobotics.aar4423.
35. He Q, Kiesewetter DO, Qu Y, Fu X, Fan J, Huang P, Liu Y, Zhu G, Liu Y, Qian Z, Chen X. NIR-responsive on-demand release of CO from metal carbonyl-caged graphene oxide nanomedicine. Adv Mater. 2015;27:6741. https://doi.org/10.1002/adma.201502762.
36. Zhang L, Chen Y, Li Z, Li L, Saint-Cricq P, Li C, Lin J, Wang C, Su Z, Zink JI. Tailored synthesis of octopus-type janus nanoparticles for synergistic actively-targeted and chemo-photothermal therapy. Angew Chem Int Ed. 2016;55:2118–21. https://doi.org/10.1002/anie.201510409.
37. Villa K, Novotný F, Zelenka J, Browne MP, Ruml T, Pumera M. Visible-light-driven single-component $BiVO_4$ micromotors with the autonomous ability for capturing microorganisms. ACS Nano. 2019;13:8135–45. https://doi.org/10.1021/acsnano.9b03184.
38. Xuan M, Wu Z, Shao J, Dai L, Si T, He Q. Near infrared light-powered Janus mesoporous silica nanoparticle motors. J Am Chem Soc. 2016;138:6492–7. https://doi.org/10.1021/jacs.6b00902.
39. Chen C, Mou F, Xu L, Wang S, Guan J, Feng Z, Wang Q, Kong L, Li W, Wang J, Zhang Q. Light-steered isotropic semiconductor micromotors. Adv Mater. 2017;29:1603374. https://doi.org/10.1002/adma.201603374.
40. Dolmans DE, Fukumura D, Jain RK. Photodynamic therapy for cancer. Nat Rev Cancer. 2003;3:380–7. https://doi.org/10.1038/nrc1071.
41. Niculescu AG, Grumezescu AM. Photodynamic therapy—an up-to-date review. Appl Sci. 2021;11:3626. https://doi.org/10.3390/app11083626.
42. Cao S, Shao J, Wu H, Song S, De Martino MT, Pijpers IAB, Friedrich H, Abdelmohsen LKEA, Williams DS, van Hest JCM. Photoactivated nanomotors via aggregation induced emission for enhanced phototherapy. Nat Commun. 2021;12:2077. https://doi.org/10.1038/s41467-021-22279-w.
43. Ma X, Sreejith S, Zhao Y. Spacer intercalated disassembly and photodynamic activity of zinc phthalocyanine inside nanochannels of mesoporous silica nanoparticles. ACS Appl Mater Interfaces. 2013;5:12860–8. https://doi.org/10.1021/am404578h.
44. Gao C, Lin Z, Wu Z, Lin X, He Q. Stem-cell-membrane camouflaging on near-infrared photoactivated upconversion nanoarchitectures for in vivo remote-controlled photodynamic therapy. ACS Appl Mater Interfaces. 2016;8:34252–60. https://doi.org/10.1021/acsami.6b12865.
45. Zhi D, Yang T, O'hagan J, Zhang S, Donnelly RF. Photothermal therapy. J Control Releas. 2020;325:52–71. https://doi.org/10.1016/j.jconrel.2020.06.032.

46. Liu Y, Bhattarai P, Dai Z, Chen X. Photothermal therapy and photoacoustic imaging via nanotheranostics in fighting cancer. Chem Soc Rev. 2019;48:2053–108. https://doi.org/10.1039/C8CS00618K.
47. Wu Z, Si T, Gao W, Lin X, Wang J, He Q. Superfast near-infrared light-driven polymer multilayer rockets. Small. 2015;12:577–82. https://doi.org/10.1002/smll.201502605.
48. Yang PP, Zhai YG, Qi GB, Lin YX, Luo Q, Yang Y, Xu A, Yang C, Li Y, Wang L, Wang H. NIR light propulsive janus-like nanohybrids for enhanced photothermal tumor therapy. Small. 2016;12:5423–30. https://doi.org/10.1002/smll.201601965.
49. Wan M, Wang Q, Wang R, Wu R, Li T, Fang D, Huang Y, Yu Y, Fang L, Wang X, Zhang Y, Miao Z, Zhao B, Wang F, Mao C, Jiang Q, Xu X, Shi D. Platelet-derived porous nanomotor for thrombus therapy. Sci Adv. 2020;6:eaaz014. https://doi.org/10.1126/sciadv.aaz9014.
50. Deng G, Peng X, Sun Z, Zheng W, Yu J, Du L, Chen H, Gong P, Zhang P, Cai L, Tang B. Natural-killer-cell-inspired nanorobots with aggregation-induced emission characteristics for near-infrared-II fluorescence-guided glioma theranostics. ACS Nano. 2020;14:11452–62. https://doi.org/10.1021/acsnano.0c03824.

Light-Controlled Microalgae Robots for Biomedical Applications

13

Ting Pan, Liliang Ye, and Hongbao Xin

Abstract

Bio-micro/nanorobots, which leverage the unique mobility and biological properties of natural organisms such as bacteria, cells and exosomes, hold exciting prospects for biomedical applications. Among these, microalgae exhibit excellent biocompatibility, phototaxis, and other unique features, rendering them promising candidates for the development of living bio-micro/nanorobots. As a new micromanipulation tool with the advantages of wireless manipulation with high precision and excellent biocompatibility with minimal invasiveness, light manipulation has become a research hotspot to precisely control the microrobot, especially living bio-microrobots. In this chapter, we discuss the advancements in the light-controlled microalgae robots, focusing on the unique features of microalgae, light-driven mechanisms and highlights their applications in nano-biothreat removal, targeted drug delivery, and the creation of light-deformable soft robots. These light-controlled microalgae robots exhibit high biocompatibility, minimal invasiveness, and show great promise for the biomedical applications in bio-microenvironments.

Keywords

Bio-microrobots · Optical manipulation · microalgae · Phototaxis · Biomedical applications

13.1 Introduction

The development of micro/nanorobots has brought remarkable progress and potential in the field of biomedical and even clinical applications because of their high controllability and tiny size [1]. Traditional design of micro/nanorobots often rely exclusively on artificial materials to convert chemical energy or external energy sources into propulsion forces. However, such synthetic micro/nanorobots face several limitations. One of the trickiest problems is that the toxicity of artificial materials to cells and tissues in the body. Therefore, there is a substantial demand for biocompatible micromotors with minimal invasiveness and negligible toxicity for effective task performance in biological environments.

To address the demands in micro/nanorobots for biomedical applications, there has been a growing focus on utilizing living cells (such as bacteria [2], microalgae [3], and sperm [4]) to develop biohybrid microrobotic systems with inherent biocompatibility. These living cells have advanced motility systems, enabling efficient self-propulsion capabilities and adaptive responses to versatile environments and external stimuli [5–7]. By integrating living entities with synthetic materials, biohybrid microrobots represent a promising approach for achieving biomimetic behaviors and enhanced functionalities with high biocompatibility [8]. Microalgae-based biohybrid microrobots, in particular, have gained significant interest for their long-lasting self-propulsion, adaptability to harsh environments, and robust cargo transportation abilities [9]. However, such biohybrid microrobots still inevitably employ synthetic materials (such as magnetic materials for magnetic actuation), which may induce undesirable toxicity or immune reactions.

The development of light-controlled micro/nanorobots has opened new pathways for creating living cell robots entirely free of synthetic components [10–12]. Light is a renewable, eco-friendly, and safe energy source, capable of

T. Pan · L. Ye · H. Xin (✉)
Guangdong Provincial Key Laboratory of Nanophotonic Manipulation, College of Physics and Optoelectronic Engineering, Institute of Nanophotonics, Jinan University, Guangzhou, P. R. China
e-mail: hongbaoxin@jnu.edu.cn

B. Li et al. (eds.), *Biophotonic Manipulation*, Advances in Optics and Optoelectronics,
https://doi.org/10.1007/978-981-96-4982-2_13

being transmitted to specific locations in a wireless and non-contact manner. Besides, light energy can be transformed into other energy forms like electrical and thermal energy with high efficiency. These advantages impart endow light-controlled micro/nanorobots with noninvasiveness, high-precision controllability and flexibility. Such properties are particularly valuable for biological and biomedical applications. Therefore, in this chapter, we focus on the advancements in the light-controlled microalgae robots, emphasizing the unique features of microalgae and light-control mechanisms. The potential biomedical applications of these light-controlled microalgae robots are also discussed.

13.2 The Unique Properties of Microalgae

Microalgae are a diverse group of organisms with photosynthetic organelles that synthesize energy and nutrients, which are widely distributed on the Earth. The bioactive compounds within algae cells, such as proteins, polysaccharides and pigments, have attract great interest treating diseases like cancer [13] and inflammation [14]. The size of microalgae spans from a few micrometers to hundreds of micrometers, and their motility varies depending on species and structural features. A typical microalgal cell structure comprises a flagellum, a cell wall and a cell body that contains an eyespot (Fig. 13.1a) [11]. The beating of flagellum can generate hydrodynamic force that propels the microalgae to swim freely in the water. For example, the flagellar beating of the freshwater green microalga *Chlamydomonas reinhardtii* (*C. reinhardtii*) shows a cyclic motion at 60–70 Hz (Fig. 13.1b) [15]. This motility allows algae to move toward nutrient-rich zones or escape harmful environments. This intrinsic motility shows great promise to develop self-propelled living microrobots. Besides, the rapid locomotion of algae, reaching velocities of several hundred micrometers per second, coupled with the beating motion of its flagella, can generate significant fluid dynamics in the surrounding environment (Fig. 13.1c) [16].

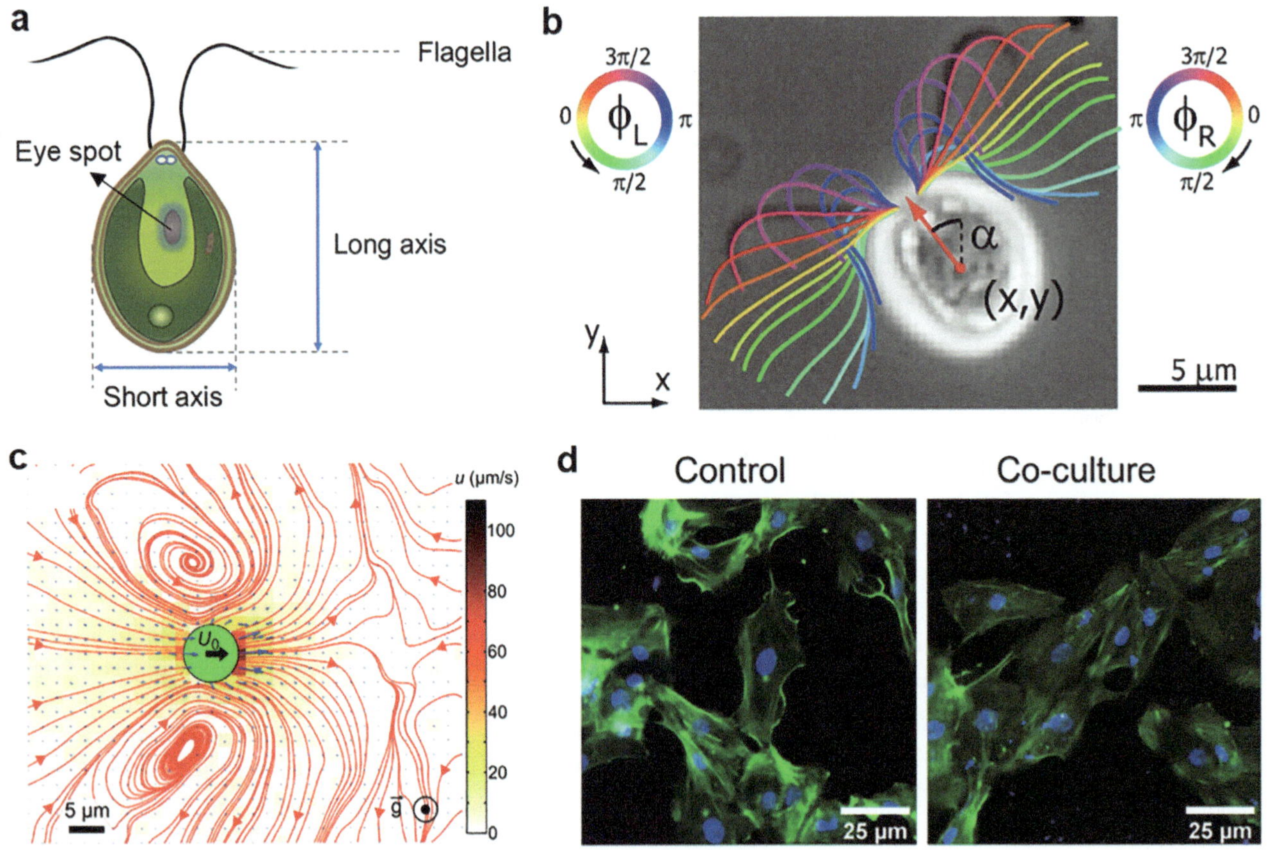

Fig. 13.1 **The unique properties of microalgae**. **a** Typical structure of a diflagellate green algae (Xin et al. [11], Copyright (2020), with permission from American Chemical Society). **b** The beating patterns of flagellum in *C. reinhardtii* (Geyer et al. [15], Copyright (2013), with permission from National Academy of Sciences). **c** The hydrodynamic flow generated by moving microalgae (Drescher et al. [16], Copyright (2010), with permission from American Physical society). **d** Co-culture of microalgae with endothelial cells demonstrating its good biocompatibility (Ehrenfeld et al. [17], Copyright (2023), with permission from Springer Nature)

The generation of flow fields, particularly in collective algal movements, holds potential for hydrodynamic manipulation of diverse micro-objects with arbitrary material and shape.

Microalgae are inherently biocompatible, unlike other microorganisms such as bacteria (e.g., *Escherichia coli*), which are harmful to animal tissues. Interestingly, the mechanical and physical properties of microalgae, such as *C. reinhardtii*, are found to closely resemble those of human erythrocytes [17]. When co-cultured with endothelial cells, *C. reinhardtii* shows no adverse effects on cell morphology or viability (Fig. 13.1d) [17]. Additionally, injecting microalgae into live mice has demonstrated their excellent compatibility in vivo [17]. With their extremely thin cell walls, microalgae can be naturally absorbed by the body, eliminating the need for synthetic modifications. Moreover, their ability to produce oxygen through photosynthesis, a function absents in animal cells that rely entirely on external oxygen sources, holds significant medical potential [18]. This feature makes microalgae valuable for addressing hypoxic conditions like organ transplantation, cerebral hypoxia, and cancer-related hypoxia.

13.3 Light-Control Mechanisms

Here, the driving mechanisms of light-controlled microalgae robots are mainly classified into two types: optical-tweezers-based manipulation and biologically based actuation. Optical tweezers (OT) serve as a powerful tool for the trapping and manipulation of cells with the advantages of minimal invasiveness, precise and flexible manipulation. The method of biologically based actuation typically employs the phototaxis characterization of microalgae to manipulate the navigation of bio-microrobots.

13.3.1 OT-Based Manipulation

The optical tweezers (OT) technique was initially demonstrated by Ashkin in 1986, when he successfully achieved the stable trapping of a dielectric microparticle using a single, tightly focused laser beam [19]. OT utilize "optical force" to trap and manipulate micro-objects. The basic configuration of OT involves focusing a laser beam to a diffraction-limited point with a high numerical aperture objective lens [20]. This highly focused laser beam can generate a light field with non-uniform intensity distribution, which creates an optical gradient force to trap the micro-objects interacted with the laser beam. By shifting the laser's focal point, OT can be used to trap and manipulate various micro-objects including dielectric particles, biological cells, and biomolecules. To further enable the simultaneous manipulation of multiple micro-objects, laser scanning techniques were introduced to develop scanning optical tweezers (SOT) [21]. SOT systems employ optical scanning devices, such as scanning mirrors and acousto-optic deflectors (AOD), to dynamically adjust the beam angle, allowing rapid scanning across the focal plane to create multiple optical traps [20]. Notably, AOD-based systems can operate at scanning frequencies exceeding 250 kHz, ensuring uniform laser power distribution across all scanning points, thereby producing high-quality optical traps for manipulating multiple targets simultaneously.

Two primary optical forces are responsible for the OT-based manipulation: optical scattering force and optical gradient force. The optical scattering force results from the conversion of photonic momentum. When light interacts with a micro-object and is scattered in a direction other than that of the incident beam, the light's momentum is altered and this alteration can be transmitted to the micro-object [22]. In contrast, the gradient force occurs in light fields with nonuniform intensity. For transparent micro-objects significantly larger than the laser wavelength, the gradient force can be understood as the collective effect of multiple scattering forces. The optical gradient force depends on the light intensity gradient, which can either pull or repel the micro-object, depending on the difference in refractive index between the micro-object and its medium. When applying a tightly focused Gaussian beam, micro-objects can be stably trapped at the focal point. In this situation, the optical gradient force and the optical scattering force synergistically govern the trapping process. The gradient force draws particles to the focal region and the scattering force drives them along the light propagation direction. The equilibrium position of the trapped micro-objects is established when these forces counterbalance each other, forming a dynamic system comparable to a harmonic spring oscillator.

13.3.2 Phototaxis-Driven

A key feature of microalgae is their photo-responsiveness, which is realized by the eyespot of microalgae to sense the light in the surrounding environment. The eyespot also functions as an ion channel to activate ion influx for triggering downstream signal transduction. This structure consists of carotenoid-rich granules, known as eyespot globules (Fig. 13.2a) [23]. In flagellated microalgae, the eyespot does not function solely as a photoreceptor; rather, it works in conjunction with another component that serves as the primary photoreceptor, together forming the photoreceptive

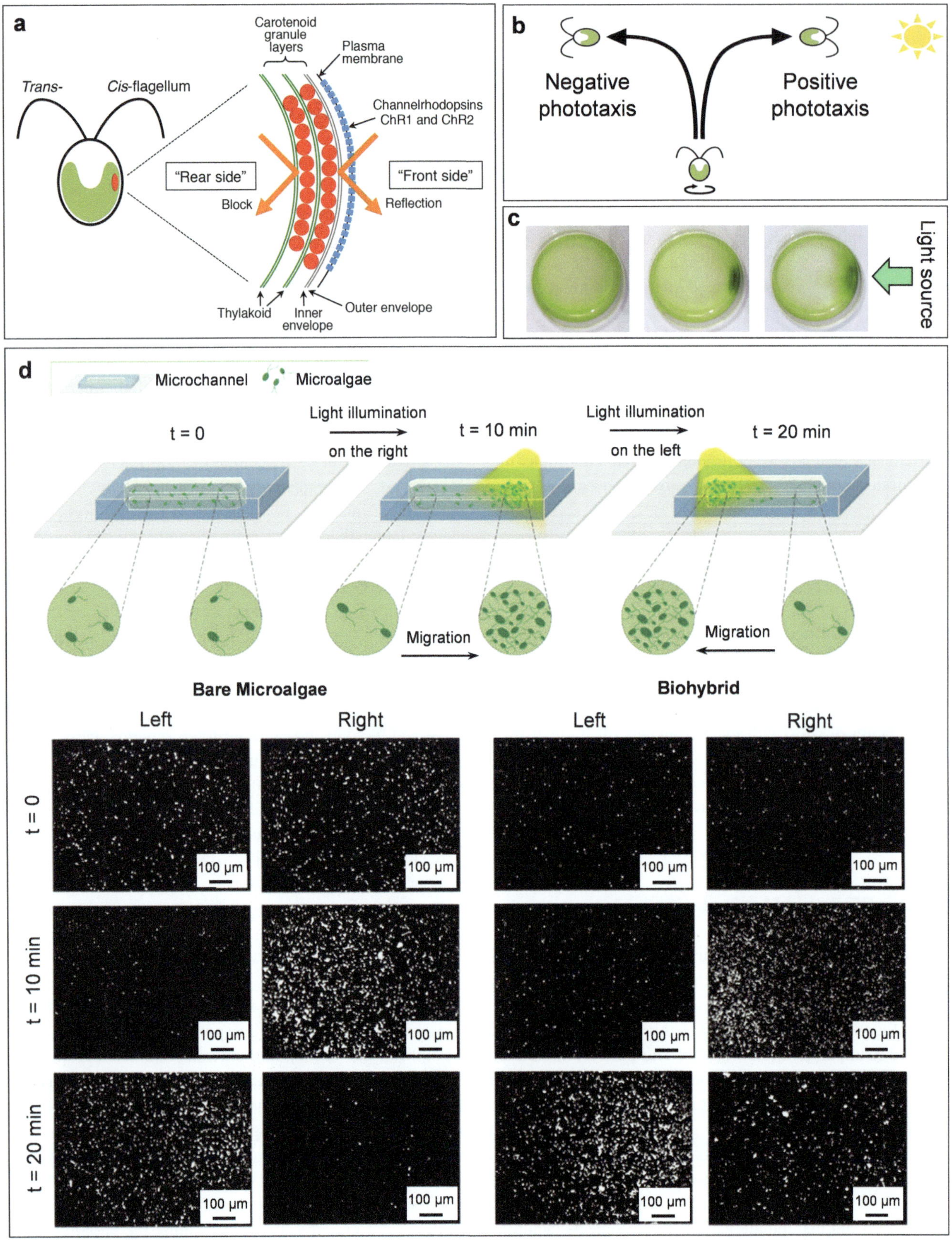

Fig. 13.2 **Phototactic behavior of microalgae**. **a** The structure of eyespot containing the carotenoid granule layers (red) and photoreceptor proteins. **b** The positive and negative phototaxis behavior of microalgae. **c** Phototactic behavior of collective microalgae swarm (Ueki et al. [23], Copyright (2016), with permission from National Academy of Sciences). **d** Phototaxis-driven microalgae robot with controlled navigation in microfluidic channels for cargo transportation (Akolpoglu et al. [25], Copyright (2020), with permission from John Wiley and Sons)

apparatus. Different microalgae groups have specific light-sensitive molecules serving as photoreceptors. For example, in two different algae species, *C. reinhardtii* and *Euglena gracilis* (*E. gracilis*), the photoreceptors for phototaxis are channel rhodopsins (ChRs) and photoactivated adenylyl cyclase (PAC), respectively, which differ significantly in their functions [24]. ChRs act as calcium ion channels responsive to light, while PAC is an enzyme that generates cAMP in response to light. The eyespot apparatus is ubiquitously found in different kinds of flagellated microalgae, which shows significant variation in different species.

Microalgae's phototactic behavior is primarily driven by their eyespot apparatus, which allows them to detect and respond to light. Under the control of the eyespot, microalgae can move toward a light source to enhance photosynthesis, a behavior known as positive phototaxis; or they can swim away from harmful light conditions to protect their photosynthetic machinery, a behavior called negative phototaxis (Fig. 13.2b) [23]. Figure 13.2c shows this behavior of a microalgae swarm, where microalgae in a petri dish congregate toward a light source [23]. This light-responsive behavior is highly beneficial for developing advanced light-controlled bio-microrobots, capable of performing specific tasks upon reacting to light stimuli. For instance, a phototaxis-driven microalgae robots can navigate in microfluidic channels under light control for cargo transportation, enabling efficient back-and-forth motion (Fig. 13.2d) [25].

13.4 Biomedical Applications

13.4.1 Removal of Bio-threats

The aggregation of biological threats (biothreats), such as thrombus formed from blood clotting, the development of microbial biofilms, and calculi formed by mineral crystallization, lead to many medical complications and endanger human health. Addressing these challenges requires precise disruption and removal of these aggregates. A novel solution involves a living microalgae robot, composed solely of *C. reinhardtii* cells and controlled by optical forces, has been demonstrated for in vitro disruption of biological aggregates (Fig. 13.3a) [11]. When using OT to trap a swimming *C. reinhardtii* cell, the flagella-driven translational motion of the microalgae turns into rotational motion around the optical focus (Fig. 13.3b). By further applying an annular scanning trap to the motor's outer boundary, the speed and direction of the rotation can be precisely manipulated. This micromotor demonstrates controlled movement in diverse biological fluids, including cell culture media, saliva, serum, plasma, blood, and bone marrow fluid (Fig. 13.3c). Moreover, the beating flagella of living microalgae serve as bio-powered arms capable of executing various biological tasks, including the effective disruption of biological aggregates like blood clots through targeted repetitive beating (Fig. 13.3d).

Nanoscale biological threats (nano-biothreats) like viruses and mycoplasmas are major contaminants in cell culture, which pose substantial risks to cell-based studies and bioanalysis. Although the use of antibiotics like penicillin and streptomycin for bacteria killing or the use of anti-mycoplasma reagent have been commonly applied in cell culture process, these chemical agents may also induce toxicity to the cultured cells. Therefore, eliminating these nano-biothreats non-invasively presents a major challenge, particularly for valuable cell (such as primary neurons and stem cells) cultures. To address this issue, a diatom-based microrobot, termed a diatombot, has been developed for trapping and removal of nano-biothreats in bio-microenvironments of cell cultures by integrating optical trapping with opto-hydrodynamic effects [26]. This system utilizes the triradiate *Phaeodactylum tricornutum Bohlin* (PTB), a naturally abundant diatom with a three-armed structure, displaying average arm dimensions of 7.9 μm in length and 1.8 μm in width. The diatombot achieves controlled rotational motion when an annularly scanning optical trap is applied using an optical tweezers system. These diatombots actively and selectively remove nano-biothreats, including viruses, mycoplasmas, and bacteria, from cell culture microenvironments without influencing the viability of cultured cells like primary neurons. Moreover, the surface modification of chitosan endows the diatombots with effective antibacterial properties after bacteria collection, further enhancing cell survival rates and gene transfection efficiency. This innovative approach offers a non-invasive, efficient, and versatile solution for nano-biothreat removal in cell culture process, presenting significant potential for single-cell studies and cell-based manufacturing applications.

13.4.2 Drug Delivery and Cargo Transportation

Targeted drug delivery and cargo transportation offer immense potential in biomedical applications, which are also essential capabilities of microrobotic systems. Micromanipulation techniques, such as optical and magnetic tweezers, offer powerful tools for the precise manipulation of micro-

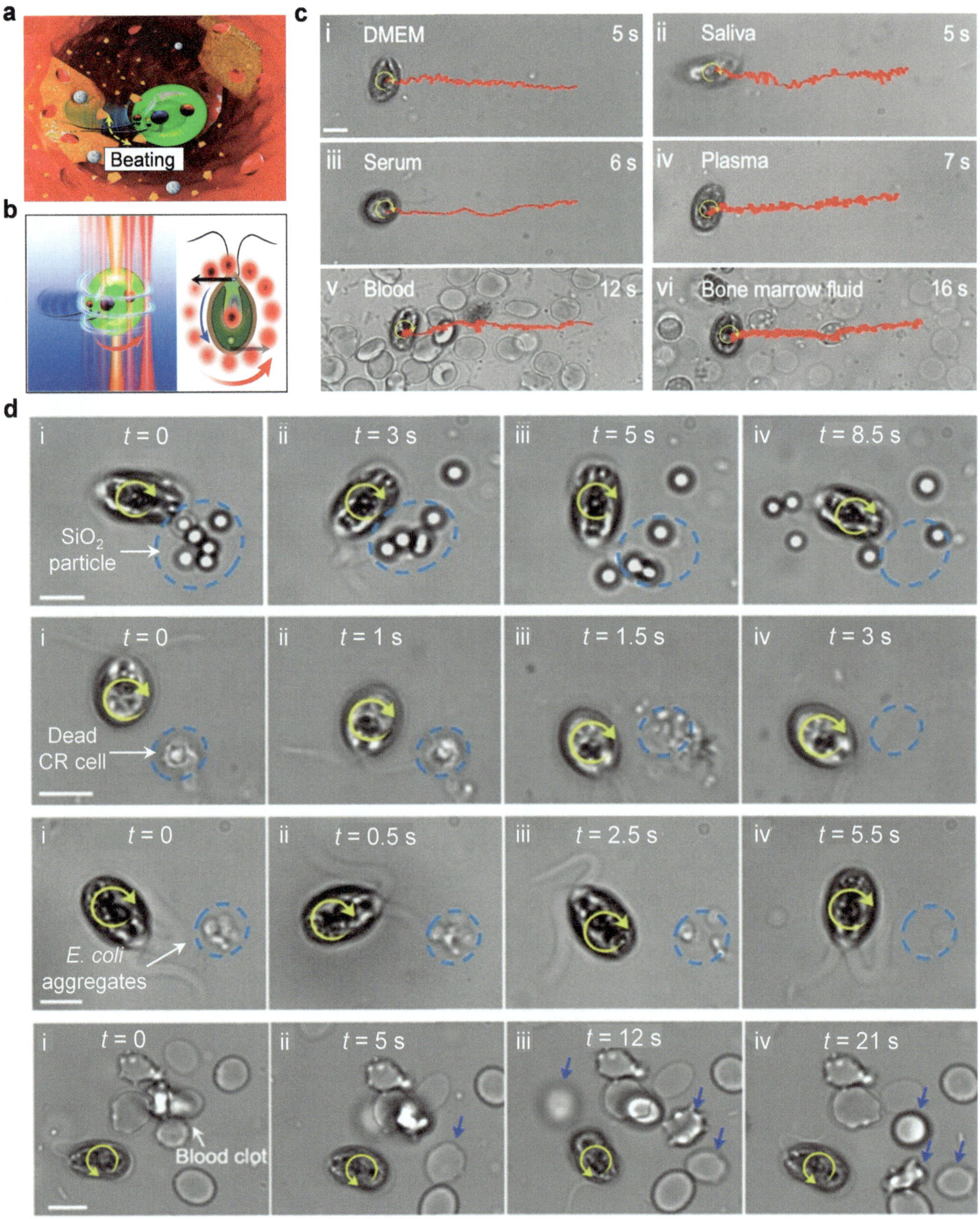

Fig. 13.3 Removal of biological aggregates by optically controlled microalgae robots (Xin et al. [11], Copyright (2020), with permission from American Chemical Society). **a** Schematic illustration of controllable disruption of biological aggregates. **b** The rotation of microalgae robot controlled by optical force. **c** Locomotion in different biological media. **d** Controllable disruption of different aggregates

objects for targeted drug delivery, cargo transportation and other biomedical tasks [27, 28]. While these micromanipulation techniques enable contactless trapping and manipulation of micro-objects, they still remain limitations for biomedical applications. For example, optical manipulation with optical tweezers needs to directly illuminate biological samples with high-intensity laser, which can lead to photodamage in biological samples and degrade drugs through photothermal or photochemical reactions. And these methods demand specific material properties of the targets. The micro-objects that can be trapped by optical tweezers need to have a higher refractive index than the surrounding medium, appropriate shape, and optical properties that allow for effective momentum transfer with the light field. Only magnetic materials can be trapped by magnetic tweezers. Dielectrophoresis relies on differences in dielectric properties between the micro-object and the surrounding medium. To overcome these limitations, a novel opto-hydrodynamic bio-micromotor tweezers system using living microalgae, *C. reinhardtii*, has been developed (Fig. 13.4a) [10]. In this system, two algae cells are trapped and manipulated with optical tweezers to form a pair of rotary micromotors, which can generate two adjacent microflows. By adjusting the two algae micromotors to rotate synchronously in the same direction at similar speeds, the generated microflows can produce hydrodynamic forces that noninvasively and stably "clamp" target objects in the middle of the bio-micromotor tweezers system without direct contact (Fig. 13.4b, c). This enables the trapping of diverse cargo types, including SiO_2 microparticles, and living cells like red blood cells and bacteria, within the localized fluid flow fields created by the rotating algae (Fig. 13.4d). This method can be further applied for noninvasive delivery of therapeutic cargos to single cell, realizing precision therapies at single cell level (Fig. 13.5). The algae-based opto-hydrodynamic micromotor tweezers pave the way for non-contact cargo transportation and drug delivery for many biomedical applications, such as precise diagnostics and minimally invasive therapies.

13.4.3 Light-Deformable Soft Microrobots

The biomedical applications of micro/nanorobots often faces the challenge of moving through narrow spaces within living tissues, such as the gaps in epithelial layers or the dense matrix of tumor microenvironments. While traditional rigid micro/nanorobots offer promising functionality, it is quite difficult to operate these micro/nanorobots in soft and fragile biological systems with minimal invasiveness, particularly in confined or complex spaces with limited accessibility. Thus, creating adaptive, flexible micro/nanorobots that can alter their shape and efficiently navigate tight spaces is critical for advancing their practical use in the complex biological environments. In an inventive study, Xiong et al. utilized *Euglena gracilis* (EG), a single-celled microalga, as a biologically powered, light-responsive, and deformable soft microrobot, termed the "Ebot" (Fig. 13.6a) [12]. The EG microalga naturally exhibits self-motility, autofluorescence, and a pronounced sensitivity to light stimuli, especially blue light. Its movement is regulated by the orientation of photoreceptors and the shading effects of its eyespot relative to the light source. The rotation of EG is controlled by periodically stimulating its photoreceptors, and variations in light intensity regulate its flagellar activity and structural deformation (Fig. 13.6b). Therefore, by adjusting the periodicity and intensity of light stimulation, the EG cell can be turned into a Ebot with precise and repeatable control over its movement, orientation, and deformation (Fig. 13.6c).

Although various microalgae have been utilized as microrobots, the Ebot stands out as a soft, deformable organism capable of undergoing repeatable, light-induced shape changes. This property enables it to maneuver effectively in complex and confined environments. For example, as shown in Fig. 13.6d, the Ebot successfully navigate through narrow microchannels containing obstructing structures and extremely thin pathways. Such unique adaptability is useful for drug delivery and therapeutic procedures that require reaching deep into complex tissues like epithelial layers and tumor microenvironments. Thanks to these characteristics, Ebots can perform a wide range of biomedical tasks, including delivering drugs to specific locations (Fig. 13.7a), and eliminating diseased cells from intestinal tissues (Fig. 13.7b). Moreover, the Ebots can also facilitate precise photodynamic therapy (PDT). The photosynthesis of chlorophyll in EG facilitate O_2 production when exposed to red light irradiation (Fig. 13.7c). Furthermore, chlorophyll serves as a natural photosensitizer, converting the generated O_2 into reactive oxygen species (ROS) under the 670-nm laser irradiation (Fig. 13.7d). Consequently, effective PDT is achieved through the application of Ebot and 670-nm light irradiation, resulting in a cancer-cell-killing efficiency up to 99% (Fig. 13.7e). The deformable structure and biological adaptability of Ebot set it apart from conventional rigid microrobots, making it a promising innovation for advanced medical applications in challenging environments.

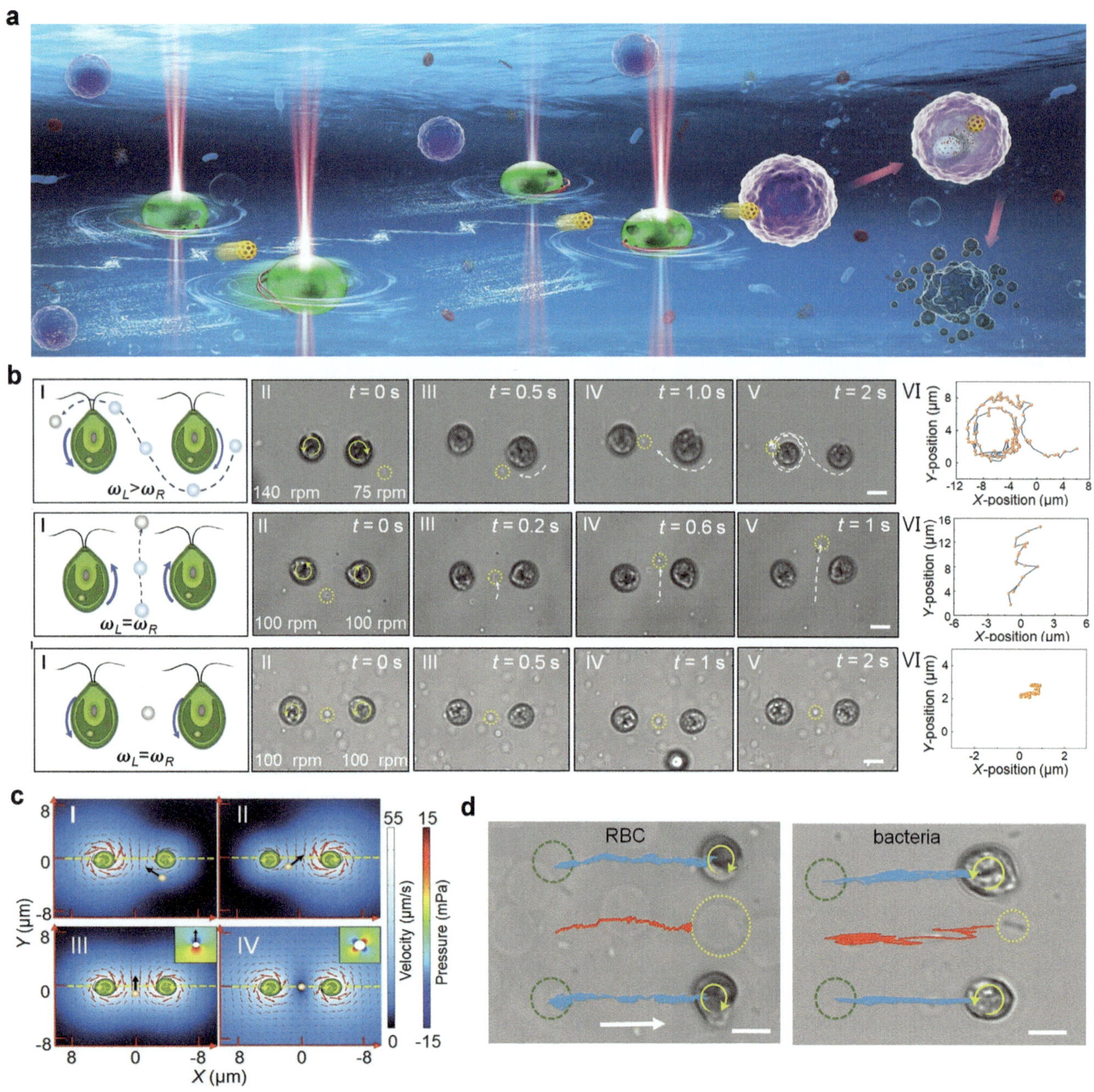

Fig. 13.4 Bio-micromotor tweezers for noncontact cargo transportation (Pan et al. [10], Copyright (2021), with permission from John Wiley and Sons). **a** Schematic illustration of the bio-micromotor tweezers controlled by optical trapping for noninvasive cargo delivery and precise therapy. **b** Hydrodynamic manipulation of microparticle. **c** Controllable delivery of living cells. **d** Targeted delivery of drug into single HL-60 cell for precise therapy

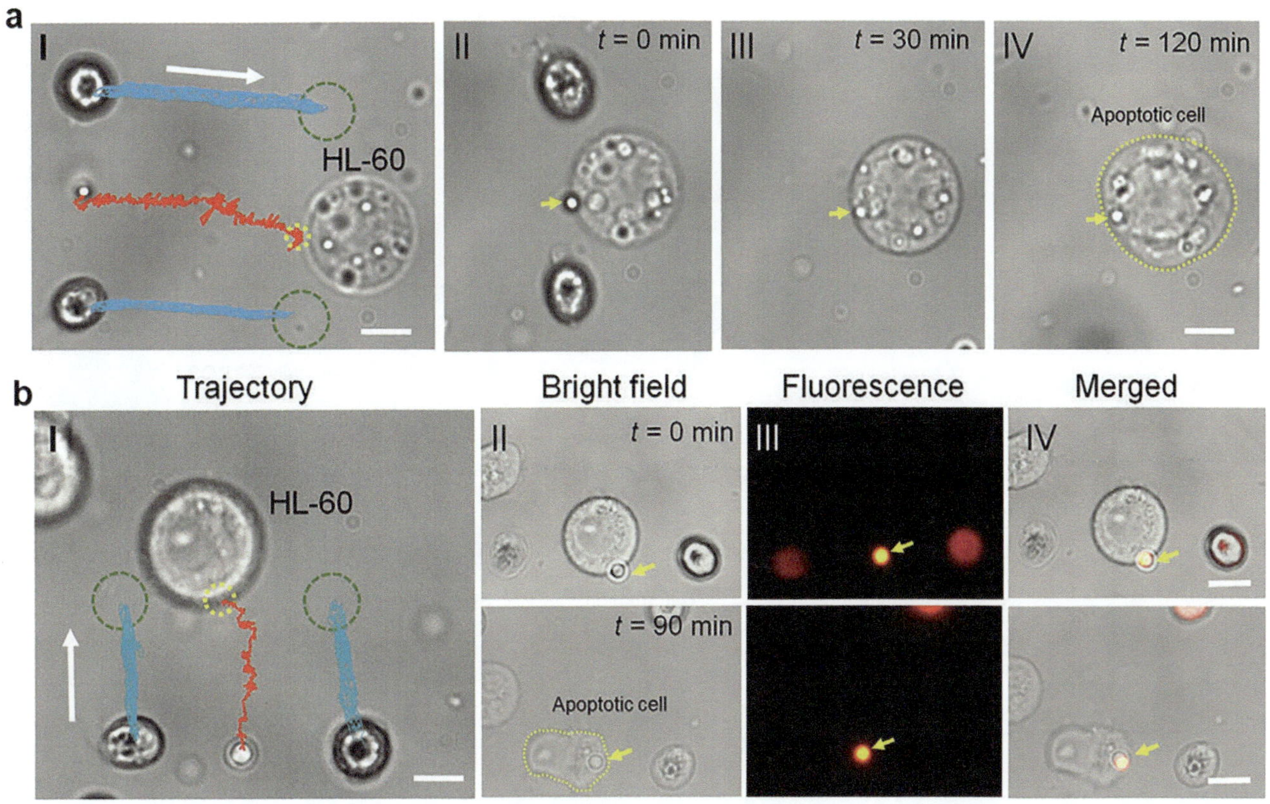

Fig. 13.5 **Bio-micromotor tweezers for precise therapy** (Pan et al. [10], Copyright (2021), with permission from John Wiley and Sons). **a** Targeted delivery of DOX-loaded mesoporous silica particles with diameter of **a** 1 μm and **b** 2.5 μm into single cancer cell HL-60 for precise cell killing

13.5 Conclusion

In this chapter, we reviewed the advancements in light-controlled microalgae robots, emphasizing their unique biological properties, mechanisms of light actuation, and their diverse biomedical applications. The remarkable self-propulsion, biocompatibility, phototaxis and adaptability of microalgae make them ideal candidates for bio-microrobotic systems. We explored optical tweezers-based and phototaxis-driven strategies for manipulating microalgae robots, demonstrating their precision and flexibility. Furthermore, we highlighted their potential in removing nano-biothreats, delivering drugs and cargos, and constructing light-deformable soft microrobots to navigate confined spaces for advanced biomedical applications. Light-controlled microalgae robots promise transformative opportunities in minimally invasive biomedical applications. Future efforts may focus on enhancing their functional versatility, integration with advanced imaging and control systems, and scaling their use for clinical applications.

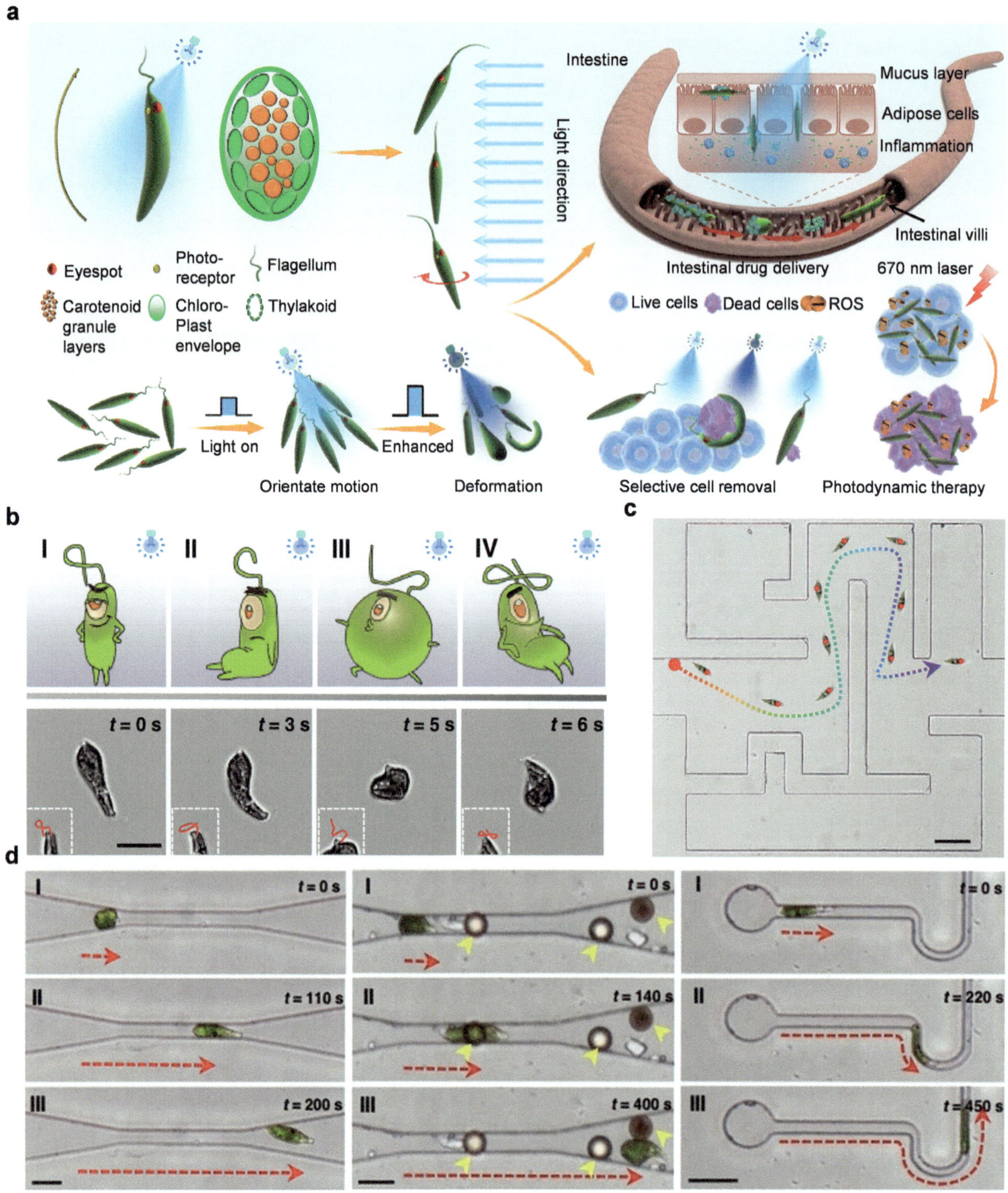

Fig. 13.6 Light-deformable soft algae robot, i.e. Ebots (Xiong et al. [12], Copyright (2024), with permission from Springer Nature). **a** Schematic illustration of EG phototaxis, deformation and multifunctional applications of Ebots. **b** Deformation progress of Ebot controlled by light. **c** Phototaxis of Ebots through complex channels. **d** Light-deformable Ebots to across micro-channels with obstacles and tight 5-μm channels

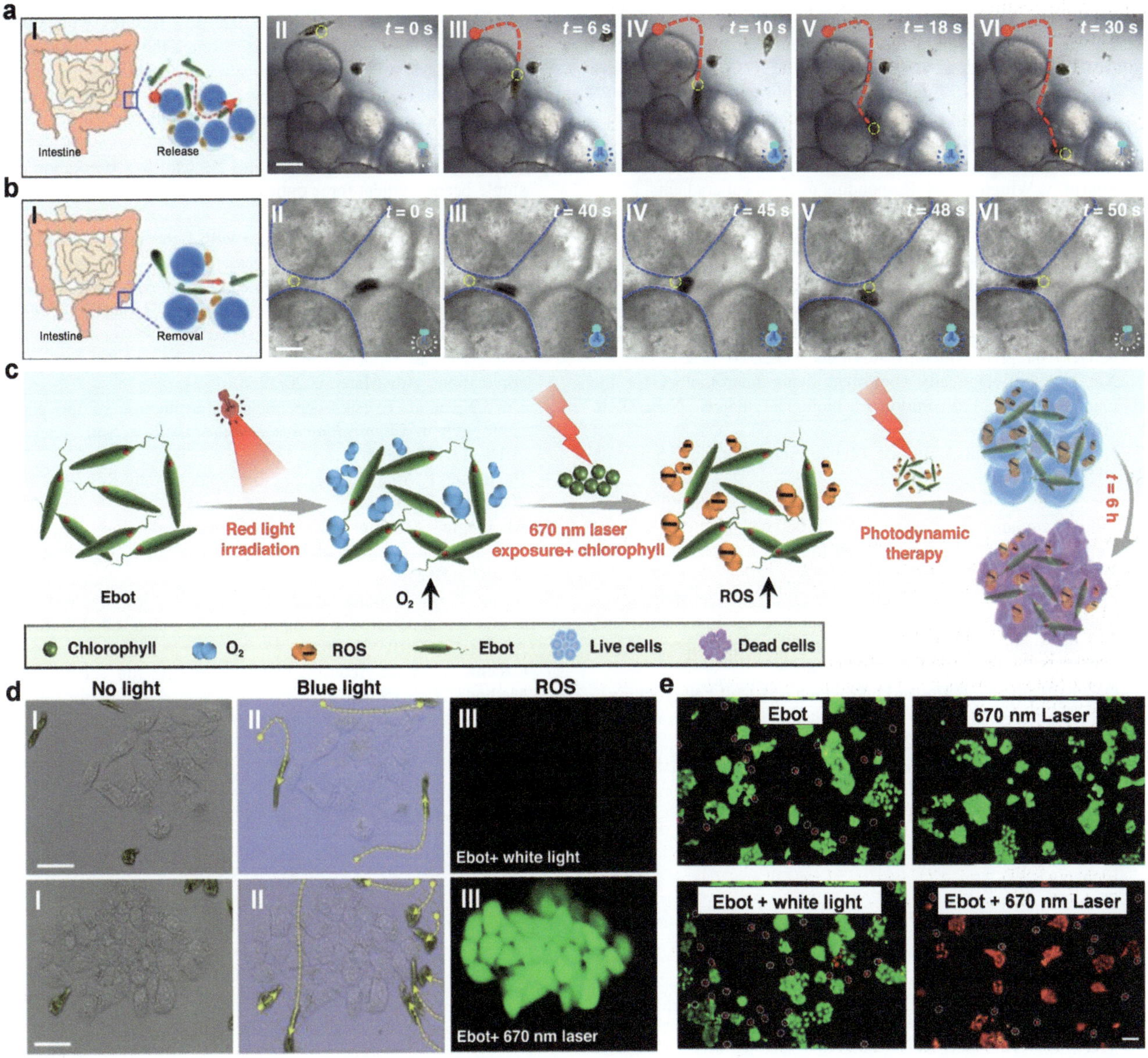

Fig. 13.7 **Ebots for multifunctional biomedical task execution** (Xiong et al. [12], Copyright (2024), with permission from Springer Nature). **a** Intestinal targeted drug delivery in an intestine in vitro. **b** Removal of a porous silica particle in the gut gap in vitro. **c** Schematic illustration of PDT using Ebot. **d** Ebot navigation and ROS generation in Hela cells. **e** Fluorescent images showing the results of PDT in Hela cells with different treatments of Ebot

Acknowledgements This work was supported by the National Natural Science Foundation of China (12374286, 32271405, and 12204196), the Natural Science Foundation of Guangdong Province (2024A1515012514), and the Fundamental Research Funds for the Central Universities (21624217).

References

1. Schmidt CK, Medina-Sánchez M, Edmondson RJ, Schmidt OG. Engineering microrobots for targeted cancer therapies from a medical perspective. Nat Commun. 2020;11:5618.
2. Stanton MM, et al. Magnetotactic bacteria powered biohybrids target *E. coli* biofilms. ACS Nano. 2017;11:9968–78.
3. Yan X, et al. Multifunctional biohybrid magnetite microrobots for imaging-guided therapy. Sci Robot. 2017;2:eaaq1155.
4. Xu H, et al. Sperm-hybrid micromotor for targeted drug delivery. ACS Nano. 2018;12:327–37.

5. Xie L, Altindal T, Chattopadhyay S, Wu X-L. Bacterial flagellum as a propeller and as a rudder for efficient chemotaxis. Proc Natl Acad Sci. 2011;108:2246–51.
6. Arrieta J, Barreira A, Chioccioli M, Polin M, Tuval I. Phototaxis beyond turning: persistent accumulation and response acclimation of the microalga *Chlamydomonas reinhardtii*. Sci Rep. 2017;7:3447.
7. Sourjik V, Wingreen NS. Responding to chemical gradients: bacterial chemotaxis. Curr Opin Cell Biol. 2012;24:262–8.
8. Sun L, et al. Biohybrid robotics with living cell actuation. Chem Soc Rev. 2020;49:4043–69.
9. Yasa O, Erkoc P, Alapan Y, Sitti M. Microalga-powered microswimmers toward active cargo delivery. Adv Mater. 2018;30:1804130.
10. Pan T, et al. Bio-micromotor tweezers for noninvasive bio-cargo delivery and precise therapy. Adv Func Mater. 2022;32:2111038.
11. Xin H, et al. Optically controlled living micromotors for the manipulation and disruption of biological targets. Nano Lett. 2020;20:7177–85.
12. Xiong J, et al. Light-controlled soft bio-microrobot. Light Sci Appl. 2024;13:55.
13. Tang Y, et al. Pentapeptide AYP from Isochrysis zhanjiangensis exhibits antiangiogenic activity in HT1080 cells and HUVECs by suppressing migration and invasion in vitro. J Agric Food Chem. 2022;70:8481–91.
14. Zhou J, et al. The impact of liquid-pressurized extracts of Spirulina, Chlorella and *Phaedactylum tricornutum* on in vitro antioxidant, antiinflammatory and bacterial growth effects and gut microbiota modulation. Food Chem. 2023;401:134083.
15. Geyer VF, Jülicher F, Howard J, Friedrich BM. Cell-body rocking is a dominant mechanism for flagellar synchronization in a swimming alga. Proc Natl Acad Sci. 2013;110:18058–63.
16. Drescher K, Goldstein RE, Michel N, Polin M, Tuval I. Direct measurement of the flow field around swimming microorganisms. Phys Rev Lett. 2010;105:168101.
17. Ehrenfeld C, et al. Microalgae share key features with human erythrocytes and can safely circulate through the vascular system in mice. Appl Microbiol Biotechnol. 2023;107:4621–33.
18. Chen H, et al. Dissolved oxygen from microalgae-gel patch promotes chronic wound healing in diabetes. Sci Adv. 2020;6:eaba4311.
19. Ashkin A, Dziedzic JM, Bjorkholm JE, Chu S. Observation of a single-beam gradient force optical trap for dielectric particles. Opt Lett. 1986;11:288–90.
20. Xu B, et al. Power micromachines with light. Laser Photon Rev. 2024;19:2400791.
21. Sasaki K, Koshioka M, Misawa H, Kitamura N, Masuhara H. Optical trapping of a metal particle and a water droplet by a scanning laser beam. Appl Phys Lett. 1992;60:807–9.
22. Xin H, et al. Optical forces: from fundamental to biological applications. Adv Mater. 2020;32:2001994.
23. Ueki N, et al. Eyespot-dependent determination of the phototactic sign in *Chlamydomonas reinhardtii*. Proc Natl Acad Sci. 2016;113:5299–304.
24. Hegemann P. Algal sensory photoreceptors. Annu Rev Plant Biol. 2008;59:167–89.
25. Akolpoglu MB, et al. High-yield production of biohybrid microalgae for on-demand cargo delivery. Adv Sci. 2020;7:2001256.
26. Xiong J, et al. Wake-riding effect-inspired opto-hydrodynamic diatombot for non-invasive trapping and removal of nano-biothreats. Adv Sci. 2023;10:2301365.
27. Ozcelik A, et al. Acoustic tweezers for the life sciences. Nat Methods. 2018;15:1021–8.
28. Ashkin A, Dziedzic JM. Optical trapping and manipulation of viruses and bacteria. Science. 1987;235:1517–20.

Optical Manipulation and Assembly of Nanoparticles

14

Tianli Wu, Chenchen Liu, Zeran Li, and Zhongpeng Huang

Abstract

Periodic arrays of nanoparticles (NPs) have garnered significant interest in the realm of nanodevice fabrication, owing to their distinctive photonic characteristics. A multitude of methodologies have been documented for the orchestrated assembly of nanoparticles into intricate superstructures, encompassing self-assembly, acoustic, optical, electrical, magnetic, and DNA ligation approaches. Among these, optical-based assembly techniques have emerged as a promising avenue, boasting the benefits of high resolution, adaptable manipulation, and the capacity to achieve arbitrary configurations of assembled structures. This chapter presents a comprehensive review of the recent advancements in the optical assembly of NPs and their associated applications. An in-depth discussion is provided on a variety of techniques predicated on disparate physical mechanisms, including optical force, photomechanical effects, optoelectronic interactions, photochemical reactions, laser-induced bubble formation, and photothermal effects. Given its versatility and adaptability, optical assembly technology is poised to have extensive application potential across numerous domains, including nanophotonics, energy, and nanomedicine.

Keywords

Periodic arrays · Optical trapping and assembly · Nanoparticles · Light control

T. Wu (✉)
Institute of Physics, Henan Academy of Science, Zhengzhou, China
e-mail: wutianli@hnas.ac.cn

T. Wu · C. Liu · Z. Li · Z. Huang
Guangdong Provincial Key Laboratory of Nanophotonic Manipulation, Institute of Nanophotonics, College of Physics & Optoelectronic Engineering, Jinan University, Guangzhou, China

14.1 Introduction

The unique surface effects, size effects, and macroscopic quantum tunneling effects inherent to NPs endow microscale assemblies with distinct functional and property profiles, significantly different from their bulk counterparts [1]. This has prompted extensive research into the construction of ordered superstructures [2, 3]. Traditional top-down methods, such as photolithography, are capable of producing high-precision nanostructures [4–6]. However, the cumbersome processes and complex instruments make it difficult to fabricate reconfigurable complex superstructures.

Recently, bottom-up techniques have gained prominence for their potential in fabricating superstructures. These methods offer the advantage of low cost, simple equipment, and single-step processing through controllably assembled ordered single NPs [6]. NPs utilized in assembly processes can be synthesized chemically, yielding particles with precisely tunable dimensions, morphology, and composition [7]. Self-assembly presents a cost-effective approach for the large-scale assembly of NPs; however, it is constrained to thermodynamically stable colloidal configurations, which can result in limited stability and controllability [8–10]. Template-assisted self-assembly methods can produce more complex nanostructures with precise configurations but are restricted by the shapes of the templates used [11, 12]. Alternatively, the regioselective modification of NPs with organic or biomolecules (copolymers or DNA) enables programmable self-assembly of different structures through directional bonding between single particles [13]. In addition, researchers have proposed external field-assisted methods to assemble NPs, such as electric and magnetic to manipulate the directional motion of particles to assemble one-dimensional chains or two-dimensional thin films [14–16]. However, these methods apply to particles that respond to electric and magnetic fields.

B. Li et al. (eds.), *Biophotonic Manipulation*, Advances in Optics and Optoelectronics,
https://doi.org/10.1007/978-981-96-4982-2_14

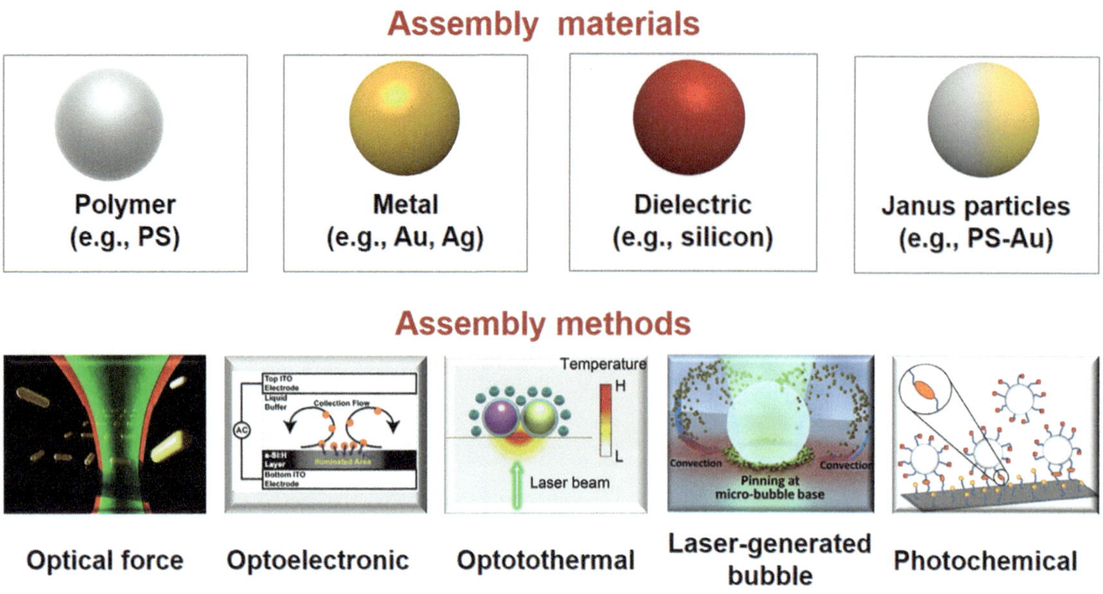

Fig. 14.1 Schematic illustration of optical assembly methods for NPs with different materials

The assembly of NPs into superstructures necessitates precise spatial manipulation of individual particles. Optical methods offer the flexibility to manipulate and position NPs with precision, enabling the trapping and placement of individual NPs into arbitrary structures, particularly in three dimensions [17]. Optical assembly-assisted methods are very promising in the fabrication of superstructures. This chapter reviews the research progress of the assembly of NPs based on different optical physical mechanisms. First, the optical force generated by optical tweezers was directly used to capture and assemble NPs onto the substrate with high precision [18]. Assembly of NPs can also be achieved under the indirect optical coupling mechanism under an electric or temperature field [19, 20]. For example, laser irradiation of a photosensitive substrate causes local conductivity changes, resulting in electrophoresis to drive NPs to move to specific regions; or laser irradiation of thermally responsive materials. The temperature gradient field was used to manipulate and assemble the NPs. In addition, the particles were assembled using thermal Marangoni water flow to drive the NPs by generating bubbles under laser irradiation [21]; chemically functionalized NPs are also driven and assembled through photochemical reactions [22]. Finally, this chapter summarizes the application prospects and challenges associated with assembling NPs based on optical methods.

14.2 Optical Assembly Techniques

In this section, we will discuss the physical mechanism underlying the assembly of NPs of different materials (Fig. 14.1). Initially, the optical force trapping and assembly method is introduced; Subsequently, we describe methods for collecting NPs using light-controlled electric fields, opto-thermoelectric effects, and laser heating to generate microbubbles; Finally, we describe the photochemical reactions-based NPs assembly and patterning method.

14.2.1 Optical Force Assembly

Optical tweezers utilize highly focused laser beams to precisely capture and manipulate nanoscale objects, such as single molecules, NPs, and biological cells [23–25]. Under incident light irradiation, metal NPs can be trapped to the center of the beam waist by optical gradient force and radiation pressure (Fig. 14.2a) [26]. The high focusability and intensity of the laser beam allow for the control of single particles within an exceedingly minute irradiation area, enabling the theoretical realization of various periodic NP arrangements. In 2015, N. Huang et al. realized an optical epitaxy growth experiment of gold particles with a size of 200 nm (Fig. 14.2b). The optical force and beam

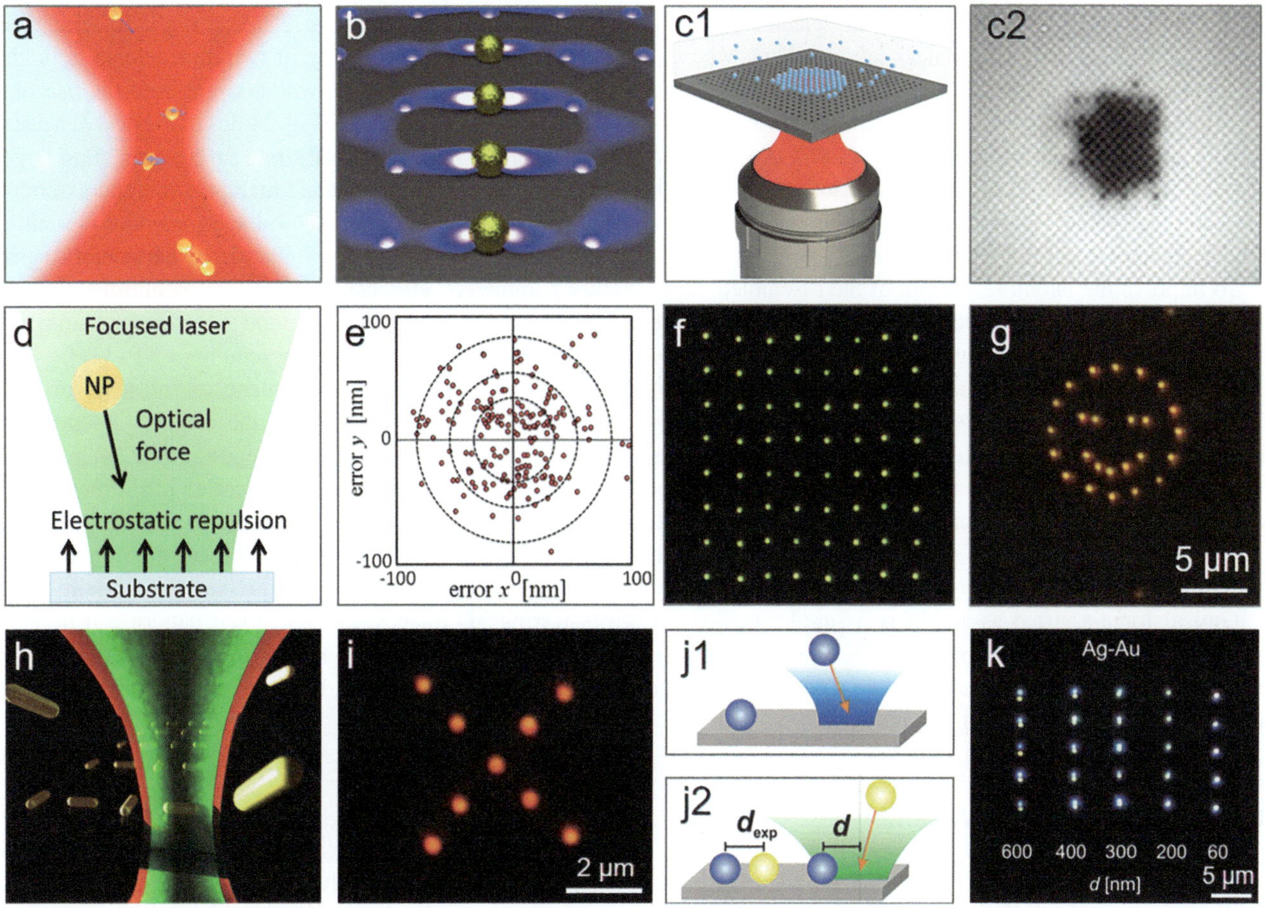

Fig. 14.2 NPs Assembly via optical force. **a** Schematic illustration of trapping NPs using optical force. **b** Optical epitaxial growth of AuNPs arrays. **c** Light-assisted and templated self-assembly of PS on a photonic crystal slab. **d** Schematic of the optical force stamping process. **e** 2D scatter plot of the assembling error of the Au NPs. **f** Dark image of 8 × 8 array of optically assembled Au NPs. **g** A smile pattern composed of 80 nm Au NPs. **h** Schematic illustration of two-color laser assembling metal NPs. **i** Dark field image of the "X" pattern. **j** Schematic of the trapping Ag (**j1**) and assembly process of Ag-Au heterodimer (**j2**). **k** Dark-field image of Ag–Au heterodimer. **a** Reprinted with permission from Lehmuskero et al. [26], Copyright 2015, Springer Nature. **b** Reprinted with permission from Huang et al. [27], Copyright 2015, American Chemical Society. **c** Reprinted with permission from Jaquay et al. [28], Copyright 2013, American Chemical Society. **d–f** Reprinted with permission from Gargiulo et al. [29], Copyright 2017, American Chemical Society. **g** Reprinted with permission from Nedev et al. [30], Copyright 2011, American Chemical Society. **h, i** Reprinted with permission from Do et al. [32], Copyright 2013, American Chemical Society. **j, k** Reprinted with permission from J. Gargiulo et al. [33], Copyright 2016, American Chemical Society

confinement facilitated the attachment of NPs to a template, achieving the periodic arrangement of the NPs [27]. Similarly, Jaquacy et al. placed polystyrene nanoparticles (PS) with a diameter of 260 nm on a photonic crystal template (Fig. 14.2c1). Since the photonic crystal resonates with the incident light to create an array of optical potential wells, the near-field effect of the photonic crystal can trap PS particles. Upon deactivation of the laser beam, the PS particles aggregate in the local area to complete the assembly process (Fig. 14.2c2), thereby achieving light-assisted self-assembly of NPs [28].

Template-assisted assembly of NPs is often limited by shape, precision, etc. Optical beams can precisely and directly manipulate the nanoparticle assembly process (Fig. 14.2d). When the localized surface plasmon resonance occurs between the wavelength of the incident beam and the manipulated metal particles, the metal particles can be trapped with high precision due to the enhanced plasmon-enhanced optical force. As visualized in Fig. 14.2e, the scatter plot of the motion trajectories of the 2D trapped NPs was shown. The beam can stably capture 90% of the gold particles with a diameter of 60 nm within the precision range of 200 nm. Therefore, the beam manipulates the localization of the metal single particles in space and arranges them into 8 × 8 regular arrays (Fig. 14.2f) [29], or forms arbitrary patterns (Fig. 14.2g) [30]. The plasmonic-enhanced optical force not only captures and assembles NPs, but also provides directional control over the orientation of NPs.

Scherer and co-workers demonstrated the selective trapping of bipyramidal AuNPs using a near-infrared laser, whereas non-resonant spherical NPs could not be trapped [31]. Thus, two different wavelengths of lasers can be used to directionally assemble gold nanorods (Fig. 14.2h). Au NPs can be trapped and oriented using the non-resonant wavelength [32], and then rapidly immobilize on substrates at resonant wavelengths, such as "X" pattern (Fig. 14.2i).

To improve the lateral resolution of the assembled structure, Stefani, et al. proposed to reduce the temperature gradient caused by the absorption of light energy by the metal particles [33]. One strategy to avoid light absorption by the trapped metal NPs was to choose a laser wavelength that minimizes light interactions with the trapped NPs (Fig. 14.2j1–2). In this way, the NPs can be continuously fixed and the distance between the particles can be reduced, such as Au–Ag heterodimers with well-defined orientation, a minimum distance of 60 nm can be achieved (Fig. 14.2k).

Optical force-assisted assembly is a simple method for manipulating NPs. The use of spatial light modulators can efficiently assemble NPs, which provides a better method for large-scale fabrication of superstructures.

14.2.2 Optoelectronic Assembly

In this section, we will discuss particle assembly under dielectrophoretic force mediated in a light-controlled electronic field. Optoelectronic tweezers utilize light and alternating current (AC) bias to manipulate the motion, aggregation, and patterning of NPs on photosensitive substrates [34]. Since the photosensitive substrate generates electron–hole pairs under the irradiation of the light beam, the conductivity of the region is increased, thereby generating a non-uniform electric field. Dielectrophoretic (DEP) force F_{DEP} is applied to the particle through the interaction of the dipole moment within the nanoparticle with the surrounding medium [35]:

$$F_{\text{DEP}} = 2\pi r^3 \varepsilon_m \text{Re}(f_{\text{CM}}) \nabla |E|^2$$

The force depends on the electric field gradient and the polarizability of the particle, which is related to the dielectric properties of the particle and the surrounding medium. In 2005, P. Y. Chiou et al. proposed the projection of an incoherent beam onto photosensitive hydrogenated amorphous silicon (a-Si:H) to generate a non-uniform electric field under an AC bias (Fig. 14.3a), multiple particles can be captured and manipulated simultaneously due to the dielectric force [36], and 15,000 single particles can be transported in parallel and in opposite directions (Fig. 14.3b1, 2). NPs and biological cells can be successfully separated due to their dielectric differences (Fig. 14.3b3). Using an incoherent light source, the cost of the laser is relatively high. To achieve an efficient and simple assembly of two-dimensional colloidal particles, H. Hwang and co-workers proposed the use of a photoelectric control device (Fig. 14.3c1). The heat generates electrothermal flow on the substrate surface, enabling large-area, long-distance nanoparticle manipulation and assembly (Fig. 14.3c2). In addition, AC voltage parameters, such as peak voltage and frequency, light source parameters, operating parameters, etc., can be adjusted to control the speed of assembling particles. This method provides a versatile tool for large-area, high-throughput assembly of NPs [37].

To assemble particles on a plane to form 3D ordered arrays, S. Zhang et al. reported the ordered assembly of NPs (PS, metal particles, and graphene nanosheets) into arbitrary shapes using optoelectronic tweezers [38], freeze-dried and transferred onto flexible substrates such as PDMS films (Fig. 14.3d), or transfer 3D micro-images onto double-sided tape substrates. Since this technique has good fidelity after transferring 3D images, the assembled pattern can be used as a security/identification mark. Figure 14.3e depicts the assembly of red fluorescent PS with optoelectronic tweezers to form an Einstein head. And transferred to banknotes, proving that the technology can be used for anti-counterfeiting encryption.

Besides using AC electric field-induced dielectrophoresis, DC electric field-based optoelectronic tweezers can also be used for the ordered assembly of NPs [39]. The optoelectronic assembly of plasmonic NPs on substrates using laser and direct current was recently reported (Fig. 14.3f). The laser was focused on the ITO substrate to form a heating zone, and under a uniform electric field, the NPs were trapped and formed an Archimedes spiral along the laser direction (Fig. 14.3g). This method does not require a high-frequency AC electric field, and directly uses a DC electric field to achieve directional assembly of metal particles [40]. This technology has broad application prospects in the field of micro-nano scale and biosensing.

14.2.3 Opto-thermophoretic Assembly

The opto-thermophoretic assembly technique involves irradiating a plasmonic substrate with a light beam to induce a temperature gradient, facilitating the capture, manipulation, and assembly of NPs [41, 42]. To achieve orderly arrangement of NPs, Y. Zheng et al. introduced cationic surfactant cetyltrimethylammonium chloride (CTAC) into the solution, thereby generating a thermoelectric field [43]. Under local laser heating, the temperature gradient prompts CTAC micelles and negatively charged chloride ions to migrate towards the cooler region, establishing a photo-directed thermoelectric field. Consequently, positively charged NPs are

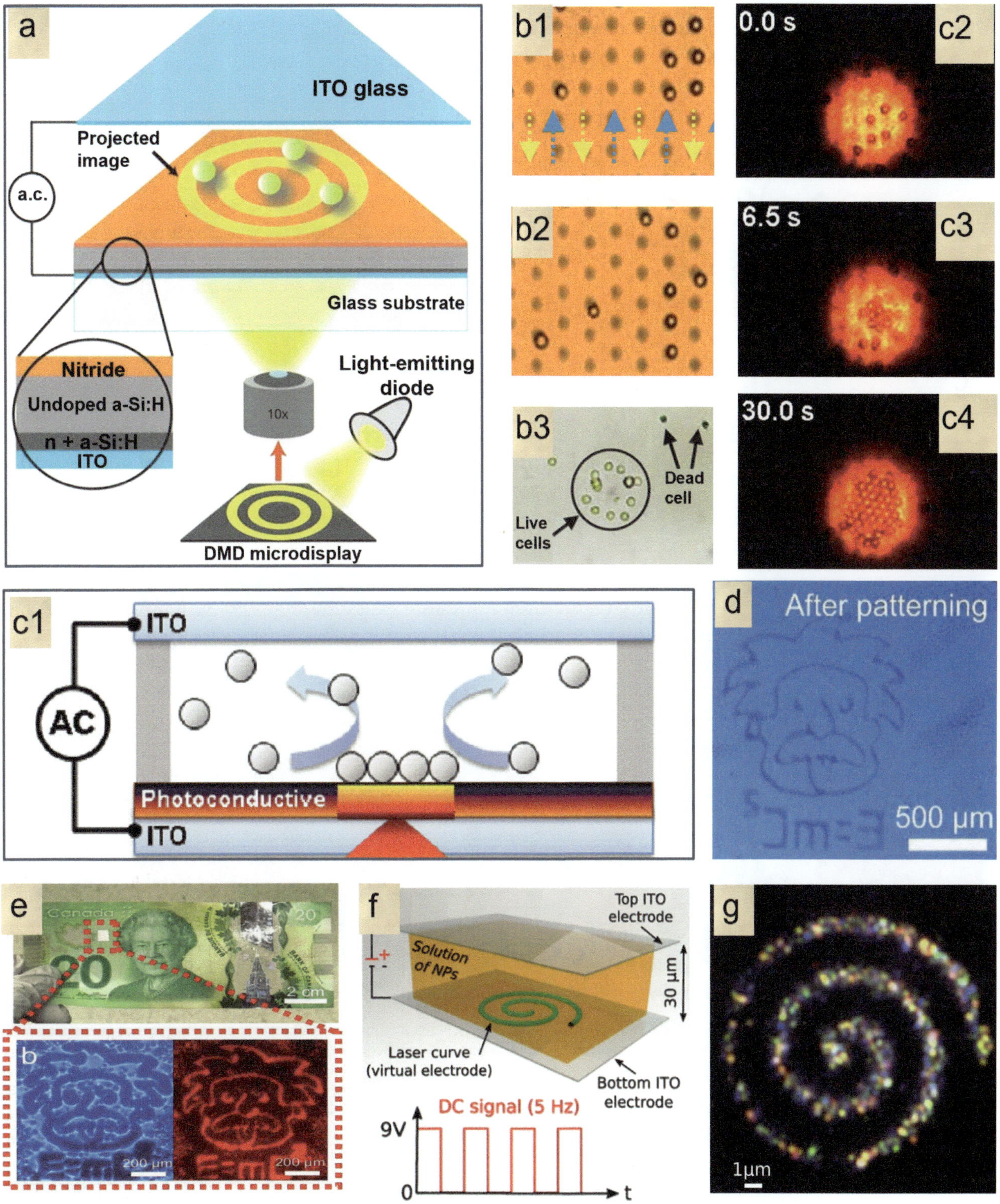

Fig. 14.3 Optoelectronic assembly of NPs. **a** Schematic illustration of manipulating of NPs using optoelectronic tweezers. **b** Parallel transportation of PS with optoelectronic tweezers. **b3** Selective sorting and assembly of living cells from a mixture of live and dead cells. **c** Schematic diagram and experimental setup of assembly NPs using optoelectrofluidic (**c1**), microscopic picture of assembled PS at 0 s (**c2**), 6.5 s (**c3**) and 30 s (**c4**). **d** Assembly of PS into topographical micropatterns with optoelectronic tweezers. **e** Image of a Canadian bill modified with a red-fluorescent assembly structure. The enlarged display of bright-field and fluorescence microscopy images of an Albert Einstein caricature assembly structure on the bill. **f**, **g** Schematic showing the setup for optoelectronic assembly (**f**), and AgNP assemblies created along an Archimedean spiral circuit (**g**). **a**, **b** Reprinted with permission from Chiou et al. [36], Copyright 2005, Springer Nature. **c** Reprinted with permission from Hwang et al. [37], Copyright 2009, American Chemical Society. **d**, **e** Reprinted with permission from Zhang et al. [38], Copyright 2019, John Wiley and Sons Inc. **f**, **g** Reprinted with permission from Rodrigo [40], Copyright 2017, Springer Nature

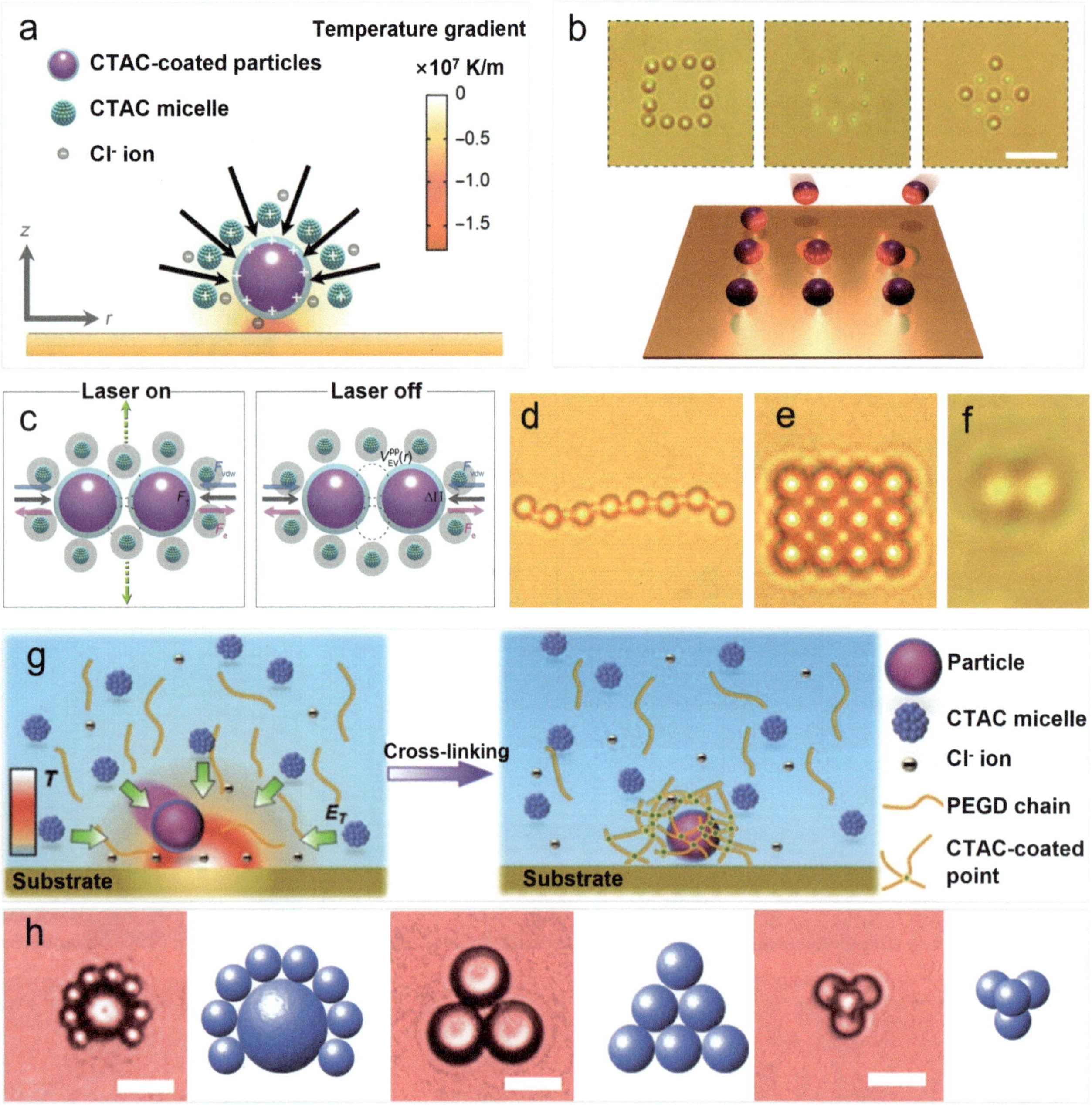

Fig. 14.4 Opto-thermoelectric assembly of NPs. **a** Diagram illustrating the opto-hermoelectric trapping of NPs. **b** Experimental micrographs (top) and corresponding schematics (bottom) depicting the trapping and assembly of NPs. **c** Schematic diagram of depletion gravity between two colloidal particles assembled. **d–f** Brightfield micrographs of colloidal particles assembled into 1D chain (**d**), 2D (**e**), and 3D structures (**f**). **g** Schematic illustration of opto-thermoelectric trapping of colloidal particles in a hydrogel solution and immobilization of the trapped particles by UV crosslinking. Scale bar: 5 μm. **a–f** Reprinted with permission from Lin et al. [43], Copyright 2017, American Chemical Society. **g** Reprinted with permission from Peng et al. [44], Copyright 2017, American Chemical Society

thermoelectrically trapped within the laser beam (Fig. 14.4a). Through optoelectronic tweezers and a digital micromirror device, multiple NPs can be dynamically manipulated and assembled into arbitrary patterns in parallel (Fig. 14.4b). The inclusion of CTAC serves a dual purpose: generating a thermoelectric field and providing inter-particle bonding. When two NPs are lased-trapped, thermophoresis induces the detachment of CTAC micelles, creating a depletion force between the particles (Fig. 14.4c). Given the constant CTAC concentration, the assembled NPs maintain their configuration even after the temperature field is removed and the laser is turned off. This method allows for the assembly of NPs from various materials and sizes into one-dimensional chains (Fig. 14.4d), 2D lattices (Fig. 14.4e), and 3D hybrid assemblies (Fig. 14.4f) using this method.

To fix the assembled structure on the substrate, X. Peng et al. suggested utilizing a photocurable hydrogel solution for colloidal assembly, with CTAC micelles acting as a depletion agent that generates inter-particle depletion forces. Controlled assembly of NPs. This enables the controlled assembly of NPs (Fig. 14.4g) [44]. NPs are trapped and arranged into desired configurations using thermoelectric fields and gravitational forces, with the assembled patterns retaining their integrity even after rinsing and drying. Additionally, the structures can be attached to the substrate through depletion attraction between the substrate and the particles. The laser captures the target particles, and the CTAC micelles create repulsive electrostatic forces, allowing the particles to be released into the solution and facilitating the ordered arrangement of NPs, enabling transformations between different configurations, such as from a "sad face" to a "smiley face." [45].

14.2.4 Microbubble Assembly

Microbubbles growing at the liquid–solid interface promote the directional assembly of mesoscopic soft matter and colloidal particles, a phenomenon reported under various terminologies such as laser-induced microbubbles [46, 47], thermo-optically manipulated laser-induced microbubbles [48], bubble pen and so on [49]. Essentially, they use photothermally generated microbubbles to assemble NPs in solution. The laser is focused on the substrate to heat, generate local "hot spots", establish a temperature gradient, and thus induce natural convection. The Marangoni convection on the microbubble surface and its in-plane resistance pull the nanoparticles toward the microbubble [50, 51], completing the assembly process as visualized in (Fig. 14.5a, b). Q. Zhang et al. employed finite element simulation to elucidate the details of thermocapillary convection surrounding surface bubbles, including the heated regions, direction of convection, and temperature distribution. The simulation results, which align closely with experimental observations, confirm the critical role of volumetric heating in inducing specific directions of thermocapillary convection, providing a theoretical explanation for the experimental phenomena (Fig. 14.5c) [52]. Additionally, the influence of different heat source positions on the direction of convection is illustrated in Fig. 14.5d.

The generation of microbubbles is related to the absorption rate of NPs or substrates and the surrounding medium, and the radius of microbubbles is [51]:

$$r = \frac{2\gamma}{\delta p}$$

γ is the surface tension of the fluid at a particular temperature, and δp is the pressure difference between the fluid inside the bubble and the fluid outside. According to a report by Lin et al. in 2015, when the temperature gradient reaches 60°, bubbles with a diameter of 1 μm are generated. To obtain larger diameter bubbles, the temperature gradient should reach 250 °C [49].

The temperature gradient, in turn, creates a density gradient in the medium, resulting in natural and Marangoni convection, and the flow lines around the bubble [52]. Microbubbles appear during the movement of the hot spot, and the NPs are driven by convection (capillary force) and anchored at the gas/liquid/solid interface (Fig. 14.5e) [53]. The size of the nanoparticle assembly structure is related to the size of the microbubble, which can be controlled by adjusting the laser power. As shown in Fig. 14.5f, the ability of microbubbles to assemble PS particles was experimentally demonstrated [54] and remained stable even when the laser was removed. Except for 2D assembly structures (Fig. 14.5h), 3D structures can also be assembled due to the reduced surface tension gradient of the trapped particles. L. Lin et al. proposed the use of a scanning laser to continuously write particles, enabling the assembly of NPs into arbitrary patterned structures (Fig. 14.5i), such as the "Mona Lisa" fluorescent pattern formed by assembling quantum dots [55].

Compared with other assembly methods, the microbubble-assisted assembly method exhibits additional versatility in assembling low-dimensional objects and materials (dielectric particles, semiconductors, and metal particles). However heating the plasma requires high temperatures, which can cause damage to delicate biological objects such as cells. In addition, this method remains a big challenge in the manipulation and assembly of single particles.

14.2.5 Photochemical Reaction Assembly

Photocatalytic or photolytic materials initiate chemical reactions by light, consuming reactants to produce ions or molecules in solution [56, 57]. A chemical gradient field established in the solution induces the assembly of NPs following the resulting diffusive motion of the generated ions or molecules. By depositing AuNPs functionalized with a self-assembled monolayer (SAM) of 11-mercaptoundecane (MUD) onto a hydrogen-terminated silicon substrate and irradiating with visible light, the assembly of AuNPs was efficiently facilitated through a photochemical reaction aided by the localized surface plasmon resonance (LSPR) effect of the AuNPs [58]. This process resulted in the formation of Si–C covalent bonds between the AuNPs and the

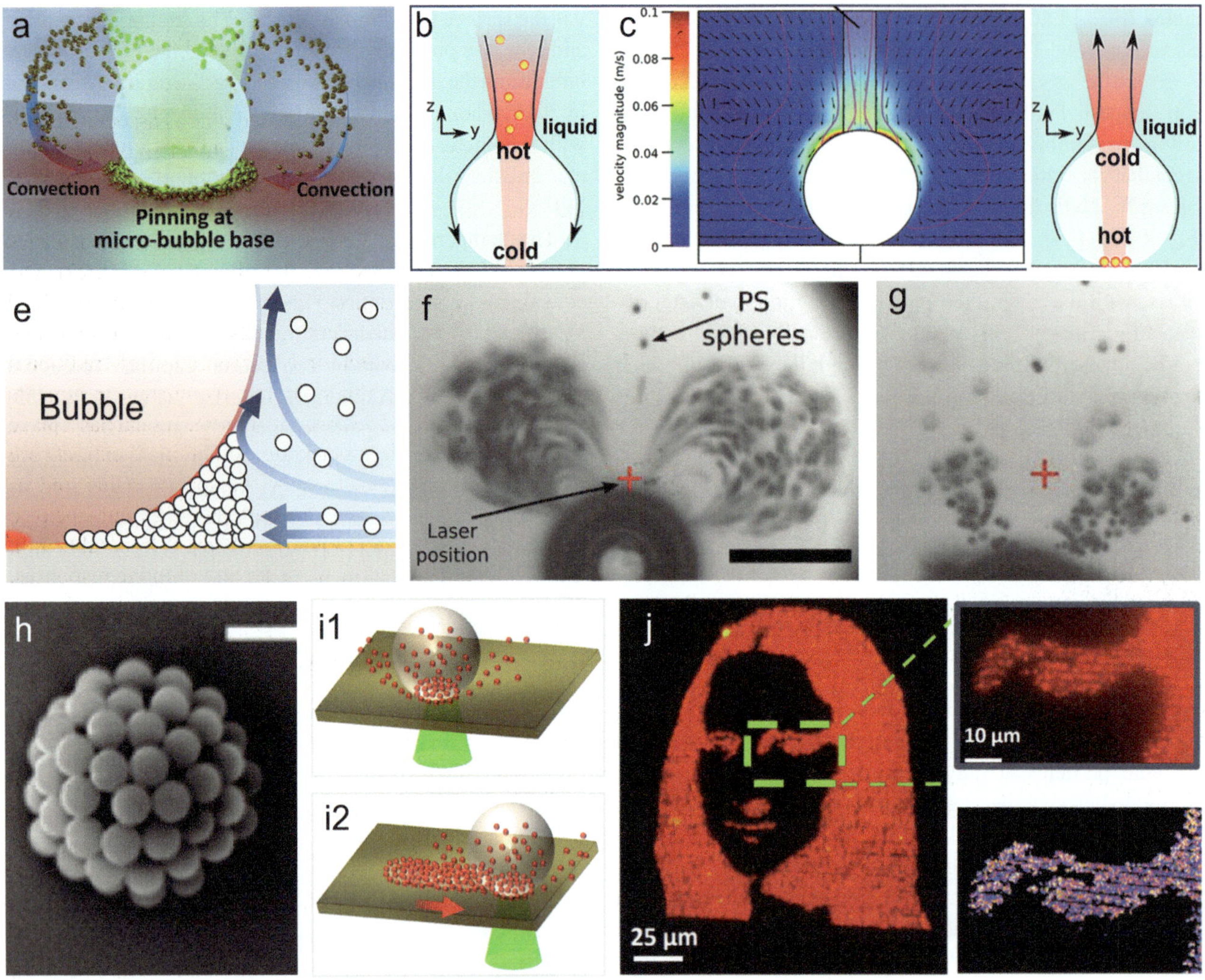

Fig. 14.5 Microbubble assembly NPs. **a**, **b** Schematic illustration of laser heating to generate microbubbles and assemble particles. **c** Simulation of thermo-capillary flow around a surface bubble, with the shaded area indicating the laser-heated region. Black arrows denote flow direction, and magenta lines show isothermal contours. **d** Schematic of thermo-capillary flow with the heat source at the substrate surface. **e** Cross-sectional view of NPs at the gas/liquid/solid interface. **f**, **g** Experimental micrographs of microbubbles assembling and releasing NPs. **h** Micrograph of microbubbles assembling NPs to form 3D structures. **i** Schematic representation of particle deposition at the bottom of microbubbles with and moving in front of the laser (**i1** and **i2**, respectively). **j** Fluorescent image of Mona Lisa assembled with red quantum dots using raster scanning method. Top right: magnified view of fluorescence; bottom right: fluorescence lifetime map. **a** Reprinted with permission from Armon et al. [50], Copyright 2017, Royal Society of Chemistry. **b–d** Reprinted with permission from Zhang et al. [52], Copyright 2019, American Chemical Society. **e** Reprinted with permission from Yamamoto et al. [53], Copyright 2019, American Chemical Society. **f**, **g** Reprinted with permission from Namura et al. [54], Copyright 2015, American Institute of Physics. **h** Reprinted with permission from Lin et al. [49], Copyright 2016, American Chemical Society. **i**, **j** Reprinted with permission from Namura et al. [55], Copyright 2017, American Chemical Society

silicon substrate, thereby achieving high-precision and efficient assembly of gold nanoparticles (Fig. 14.6a, b). Or establish a local electric field combined with a photochemical reaction to induce the stable assembly of colloidal polymethyl methacrylate (PMMA) in the light-exposed region (Fig. 14.6c) [59]. Without the help of an electric field, for example, D. P. Singh et al. reported that under blue light irradiation, H_2O_2 was decomposed under light exposure (Fig. 14.6d1) to form a chemical gradient that causes colloidal particles to tightly aggregate (Fig. 14.6d2). Turning off the light, the structure assemblies disperse due to electrostatic repulsive interactions (Fig. 14.6d3) [60]. To fix the NPs on the substrate, the coordination of the NPs to the substrate surface can be controlled by forming and breaking covalent bonds by photochemical methods [61, 62]. For example, under UV irradiation, surfactant-modified gold NPs and magnetic NPs are directionally immobilized on the substrate [63, 64]. Fischer et al. reported the use of UV light to trigger the dynamic

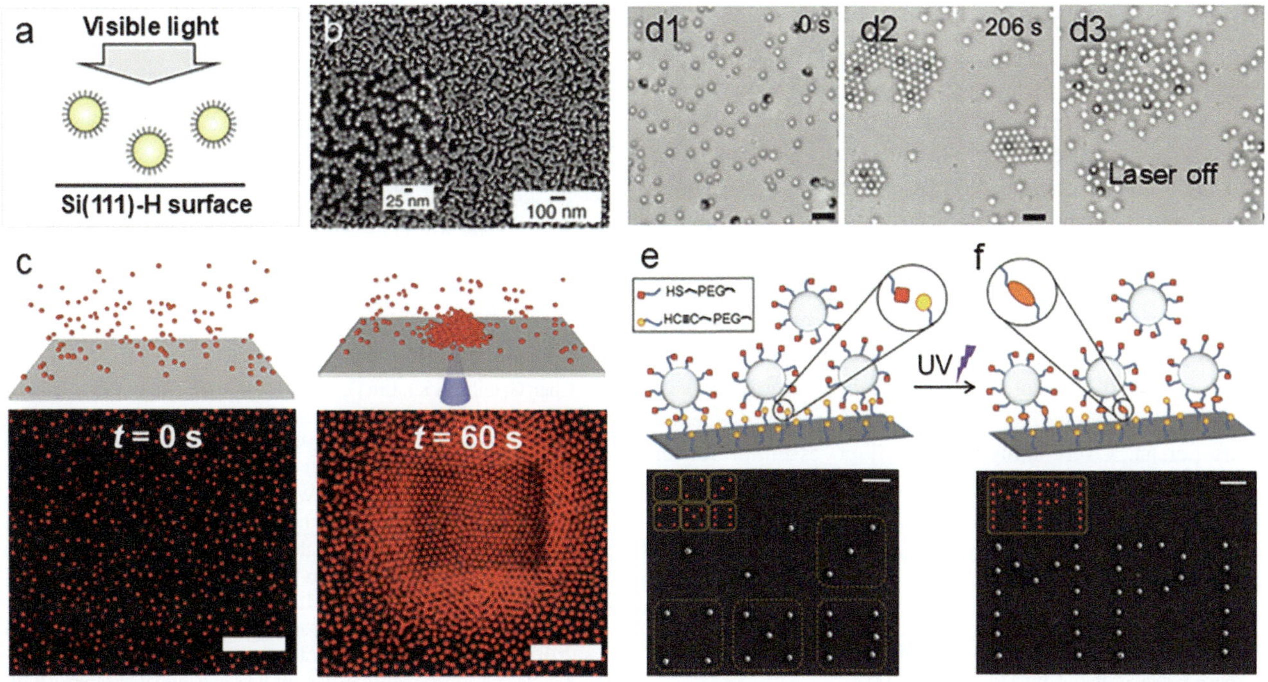

Fig. 14.6 Photochemical reaction-triggered assembly of CPs. **a**, **b** Diagram of AuNPs assembled under visible light irradiation (**a**), and arranged into a 2-D array (**b**). **c** Colloidal poly (methyl methacrylate) (PMMA) particles concentrate in the region of light irradiation. **d1**–**d3** Light-controlled assembly of TiO_2–SiO_2 Janus micromotors and passive NPs at 0 s (**d1**), 206 s (**d2**) and light off (**d3**) during the mixture process. **e** Schematic representation of the UV-induced thiol-one reaction and its use to immobilize particles on surfaces, obtained by immobilizing 1 μm PEG-SH-functionalized silica particle assemblies by UV irradiation. **a**, **b** Reprinted with permission from Sugimura et al. [58], Copyright 2015, American Chemical Society. **c** Reprinted with permission from Kim et al. [59], Copyright 2014, Springer Nature. **d** Reprinted with permission from Singh et al. [60], Copyright 2017, Wiley-Blackwell. **e** Reprinted with permission from Walker et al. [22], Copyright 2016, Wiley-Blackwell

assembly of colloidal particles, and the organic reaction of covalent bond formation between thiols and alkynes occurred in solution to fix the colloidal particles on the substrate (Fig. 14.6e) [22].

14.3 Applications of Assemblies

In nature, groups of organisms can perform complex tasks, such as ants swarming together to carry food, wolves working together to hunt large prey and so on. Assemblies formed from nanostructures also have a variety of applications, such as drug delivery [65, 66], targeted therapy [67], and magnetic resonance imaging [68]. The advantages of optical are remote controllability, high spatiotemporal precision, and non-invasiveness. Therefore, the use of optical methods to control nanoparticle assemblies has broad applications.

Firstly, the assembly can tune the optical properties of nanoparticle assemblies. Therefore, it has broad application prospects in optically responsive devices and photothermal agents [69, 70]. Gao et al. reported that light-triggered assemblies of gold NPs in mice shifted surface plasmon resonance wavelength from the visible to the near-infrared region, enhancing the efficacy of photothermal tumor ablation [71]. Secondly, assemblies of NPs can be used to load drugs or nanoreactors [72, 73]. For example, DNA and NPs are assembled into core–shell structures, which serve as carriers for therapeutic drugs and then encapsulate the drugs [74]. In addition to the free assemblies in the solution, the assemblies immobilized on the substrate can be used as functionalized micro/nanodevices [75, 76]. Metal structures tightly assembled on substrates by photothermal methods, with plasmonic nanostructures with strong electric field enhancement, enable surface-enhanced Raman spectroscopy (SERS) to detect molecules at low concentrations [77, 78]. Finally, the assembly of fluorescent particles (quantum dots, fluorescent proteins, etc.) can be used for micro-display applications, or the assembly of different emission wavelengths can be used to form multicolor displays, showing the potential to fabricate full-color displays [79–81].

14.4 Conclusions

Significant strides in nanotechnology have been marked by advancements in assembling NPs into complex structures using photodynamic, photothermal, photochemical, and microbubble-assisted methods. These innovative techniques allow for the precise construction of nanostructures at the single-particle level, holding the potential to revolutionize nanofabrication and broaden its impact across microelectronics, nanophotonics, colloidal sciences, and biology. Despite these advancements, challenges such as the specific requirements of optical tweezers-based assembly and the sensitivity of photochemical processes to ion concentrations persist. Overcoming these will be essential for the field' progress. Improving these assembly methods will drive further breakthroughs in nanofabrication, offering unprecedented potential for the manipulation and applications of nanoparticles.

Acknowledgements This work was supported by the National Natural Science Foundation of China (No. 62305132).

References

1. Klimov VI, Mikhailovsky AA, Xu S, Malko A, Hollingsworth JA, Leatherdale CA, Eisler H-J, Bawendi MG. Optical gain and stimulated emission in nanocrystal quantum dots. Science. 2000;290(5490):314–7. https://doi.org/10.1126/science.290.5490.314.
2. Qiu L, Zhang H, Bick T, Martin J, Wendler P, Böker A, Glebe U, Xing C. Construction of highly ordered glycol-inside nanoassemblies through RAFT dispersion polymerization of galactose-decorated monomer. Angew Chem. 2021;133(20):11198–203. https://doi.org/10.1002/ange.202015692.
3. Gupta A. Highly ordered supramolecular nanoassemblies of paramagnetic amphiphilic chelates as potential MRI contrast agents. Aust J Chem. 2017;71(3):195–6. https://doi.org/10.1071/CH17523.
4. Subramani C, Ofir Y, Patra D, Jordan BJ, Moran IW, Park MH, Carter KR, Rotello VM. Nanoimprinted polyethyleneimine: a multimodal template for nanoparticle assembly and immobilization. Adv Funct Mater. 2009;19(18):2937–42. https://doi.org/10.1002/adfm.200900805.
5. Mendes PM, Jacke S, Critchley K, Plaza J, Chen Y, Nikitin K, Palmer RE, Preece JA, Evans SD, Fitzmaurice D. Gold nanoparticle patterning of silicon wafers using chemical e-beam lithography. Langmuir. 2004;20(9):3766–8. https://doi.org/10.1021/la049803g.
6. Liu Y, Liu J, Sohn S, Li Y, Cha JJ, Schroers J. Metallic glass nanostructures of tunable shape and composition. Nature Commun. 2015;6:7043. https://doi.org/10.1038/ncomms8043.
7. Matijevic E. Production of monodispersed colloidal particles. Annu Rev Mater Sci. 1985;15(1):483–516. https://doi.org/10.1146/annurev.ms.15.0801.
8. Grzelczak M, Vermant J, Furst EM, Liz-Marzán LM. Directed self-assembly of nanoparticles. ACS Nano 2010;4(7):3591–605. https://doi.org/10.1021/nn100869j.
9. Zhang H, Edwards EW, Wang D, Möhwald H. Directing the self-assembly of nanocrystals beyond colloidal crystallization. Phys Chem Chem Phys. 2006;8(28):3288–99. https://doi.org/10.1039/b604745a.
10. Zhou Y, Yan D. Supramolecular self-assembly of amphiphilic hyperbranched polymers at all scales and dimensions: progress, characteristics and perspectives. Chem Commun. 2009;10:1172–88. https://doi.org/10.1039/B814560C.
11. Rycenga M, Camargo PH, Xia Y. Template-assisted self-assembly: a versatile approach to complex micro-and nanostructures. Soft Matter. 2009;5(6):1129–36. https://doi.org/10.1039/B811021B.
12. Confalonieri GB, Vega V, Ebbing A, Mishra D, Szary P, Prida VM, Petracic O, Zabel H. Template-assisted self-assembly of individual and clusters of magnetic nanoparticles. Nanotechnology. 2011;22(28): 285608. https://doi.org/10.1088/0957-4484/22/28/285608.
13. Chen G, Gibson KJ, Liu D, Rees HC, Lee JH, Xia W, Lin R, Xin HL, Gang O, Weizmann Y. Regioselective surface encoding of nanoparticles for programmable self-assembly. Nat Mater. 2019;18:169–74. https://doi.org/10.1038/s41563-018-0231-1.
14. Jiang C, Leung CW, Pong PW. Magnetic-field-assisted assembly of anisotropic superstructures by iron oxide nanoparticles and their enhanced magnetism. Nanoscale Res Lett. 2016;11:189. https://doi.org/10.1186/s11671-016-1406-9.
15. Dey S, Mohanta K, Pal AJ. Magnetic-field-assisted layer-by-layer electrostatic assembly of ferromagnetic nanoparticles. Langmuir. 2010;26(12):9627–31. https://doi.org/10.1021/la101132z.
16. Kotov NA. Nanoparticle assemblies and superstructures. CRC Press; 2016. https://doi.org/10.1201/9781420027884.
17. Grier DG. A revolution in optical manipulation. Nature. 2003;424(6950):810–6. https://doi.org/10.1038/nature01935.
18. Swartzlander GA, Peterson TJ, Artusio-Glimpse AB, Raisanen AD. Stable optical lift. Nat Photonics. 2011;5:48–51. https://doi.org/10.1364/fio.2010.jwa11.
19. Li J, Lin L, Inoue Y, Zheng Y. Opto-thermophoretic tweezers and assembly. J Micro Nano Manuf. 2018;6(4): 040801. https://doi.org/10.1115/1.4041615.
20. Wu MC. Optoelectronic tweezers. Nat Photonics. 2011;5(6):322–4. https://doi.org/10.1038/nphoton.2011.98.
21. Xu X, Luo J. Marangoni flow in an evaporating water droplet. Appl Phys Lett. 2007;91(12): 124102. https://doi.org/10.1063/1.2789402.
22. Walker D, Singh DP, Fischer P. Capture of 2D microparticle arrays via a UV-triggered thiol-une "click" reaction. Adv Mater. 2016;28(44):9846–50. https://doi.org/10.1002/adma.201603586.
23. Ashkin A, Dziedzic JM, Yamane T. Optical trapping and manipulation of single cells using infrared laser beams. Nature. 1987;330(6150):769–71. https://doi.org/10.1038/330769a0.
24. Ashkin A, Dziedzic JM. Optical trapping and manipulation of viruses and bacteria. Science. 1987;235(4795):1517–20. https://doi.org/10.1126/science.3547653.
25. Ashkin A. Optical trapping and manipulation of neutral particles using lasers. Proc Natl Sci. 1997;94(10):4853–60. https://doi.org/10.1142/9789812774897_0024.
26. Lehmuskero A, Johansson P, Rubinsztein-Dunlop H, Tong L, Kall M. Laser trapping of colloidal metal nanoparticles. ACS Nano. 2015;9(4):3453–69. https://doi.org/10.1016/b978-0-12-813357-6.00003-6.
27. Huang N, Martínez LJ, Jaquay E, Nakano A, Povinelli ML. Optical epitaxial growth of gold nanoparticle arrays. Nano Lett. 2015;15(9):5841–5. https://doi.org/10.1021/acs.nanolett.5b01929.
28. Jaquay E, Martínez LJ, Mejia CA, Povinelli ML. Light-assisted, templated self-assembly using a photonic-crystal slab. Nano Lett. 2013;13(5):2290–4. https://doi.org/10.1021/nl400918x.
29. Gargiulo J, Violi IL, Cerrota O, Chvátal L, Cortés E, Perassi OM, Diaz F, Zemánek P, Stefani FD. Accuracy and mechanistic details of optical printing of single Au and Ag nanoparticles. ACS Nano. 2017;11(10):9678–88. https://doi.org/10.1021/acsnano.8b00573.

30. Nedev S, Urban AS, Lutich AA, Feldmann J. Optical force stamping lithography. Nano Lett. 2011;11(11):5066–70. https://doi.org/10.1021/nl203214n.
31. Guffey MJ, Miller RL, Gray SK, Scherer NF. Plasmon-driven selective deposition of au bipyramidal nanoparticles. Nano Lett. 2011;11(10):4058–66. https://doi.org/10.1021/nl201020g.
32. Do J, Fedoruk M, Jäckel F, Feldmann J. Two-color laser printing of individual gold nanorods. Nano Lett. 2013;13(9):4164–8. https://doi.org/10.1021/nl401788w.
33. Gargiulo J, Cerrota S, Cortés E, Violi IL, Stefani FD. Connecting metallic nanoparticles by optical printing. Nano Lett. 2016;16(2):1224–9. https://doi.org/10.1021/acs.nanolett.5b04542.
34. Chiou PY, Ohta AT, Wu MC. Massively parallel manipulation of single cells and microparticles using optical images. Nature. 2005;436(7049):370–2. https://doi.org/10.3410/f.1013743.327623.
35. Pethig R. Dielectrophoresis: status of the theory, technology, and applications. Biomicrofluidics. 2010;4(2): 022811. https://doi.org/10.1063/1.3456626.
36. Chiou PY, Ohta AT, Wu MC. Massively parallel manipulation of single cells and microparticles using optical images. Nature. 2005;436(7049):370–2. https://doi.org/10.1038/nature03831.
37. Hwang H, Park YH, Park JK. Optoelectrofluidic control of colloidal assembly in an optically induced electric field. Langmuir. 2009;25(11):6010–4. https://doi.org/10.1021/la9005604.
38. Zhang S, Zhai Y, Peng R, Shayegannia M, Flood AG, Qu J, Liu X, Kherani N, Wheeler AR. Assembly of topographical micropatterns with optoelectronic tweezers. Adv Opt Mater. 2019;7(20):1900669. https://doi.org/10.1002/adom.201900669.
39. Hayward RC, Saville DA, Aksay IA. Electrophoretic assembly of colloidal crystals with optically tunable micropatterns. Nature. 2000;404(6773):56–9. https://doi.org/10.1038/35003530.
40. Rodrigo JA. Fast optoelectric printing of plasmonic nanoparticles into tailored circuits. Sci Rep. 2017;7:46506. https://doi.org/10.1038/srep46506.
41. Li J, Zheng Y. Optothermally assembled nanostructures. Acc. Mater Res. 2021;2(5):352–63. https://doi.org/10.1021/accountsmr.1c00033.
42. Lin L, Peng X, Wei X, Mao Z, Xie C, Zheng Y. Thermophoretic tweezers for low-power and versatile manipulation of biological cells. ACS Nano. 2017;11(3):3147–54. https://doi.org/10.1021/acsnano.7b00207.
43. Lin L, Zhang J, Peng X, Wu Z, Coughlan AC, Mao Z, Bevan MA, Zheng Y. Opto-thermophoretic assembly of colloidal matter. Sci Adv. 2017;3(9): e1700458. https://doi.org/10.1126/sciadv.1700458.
44. Peng X, Li J, Lin L, Liu Y, Zheng Y. Opto-thermophoretic manipulation and construction of colloidal superstructures in photocurable hydrogels. ACS Appl Nano Mater. 2018;1(8):3998–4004. https://doi.org/10.1021/acsanm.8b00766.
45. Lin L, Peng X, Zheng Y. Reconfigurable opto-thermoelectric printing of colloidal particles. Chem Commun. 2017;53(53):7357–60. https://doi.org/10.1039/c7cc03530f.
46. Armon N, Greenberg E, Layani M, Rosen YS, Magdassi O, Shpaisman O. Continuous nanoparticle assembly by a modulated photo-induced microbubble for fabrication of micrometric conductive patterns. ACS Appl Mater Inter. 2017;9(50):44214–21. https://doi.org/10.1021/acsami.7b14614.
47. Edri E, Armon N, Greenberg E, Hadad E, Bockstaller MR, Shpaisman H. Assembly of conductive polyaniline microstructures by a laser-induced microbubble. ACS Appl Mater Inter. 2020;12(19):22278–86. https://doi.org/10.1021/acsami.0c00904.
48. Fujii S, Kanaizuka K, Toyabe S, Kobayashi K, Muneyuki E, Haga M. Fabrication and placement of a ring structure of nanoparticles by a laser-induced micronanobubble on a gold surface. Langmuir. 2011;27(14):8605–10. https://doi.org/10.1021/la201616s.
49. Lin L, Peng X, Mao Z, Li W, Yogeesh MN, Rajeeva BB, Perillo EP, Dunn AK, Akinwande D, Zheng Y. Bubble-pen lithography. Nano Lett. 2016;16(1):701–8. https://doi.org/10.1021/acs.nanolett.5b04524.
50. Armon N, Greenberg E, Edri E, Nagler-Avramovitz O, Elias Y, Shpaisman H. Laser-based printing: from Liquids to microstructures. Adv Funct Mater. 2021;31(13):2008547. https://doi.org/10.1002/adfm.202008547.
51. Ghosh S, Biswas A, Roy B, Banerjee A. Self-assembly and complex manipulation of colloidal mesoscopic particles by active thermocapillary stress. Soft Matter. 2019;15(23):4703–13. https://doi.org/10.1039/C9SM00721K.
52. Zhang Q, Pang Y, Schiffbauer J, Jemcov A, Chang H, Lee E, Luo T. Light-guided surface plasmonic bubble movement via contact line de-pinning by in-situ deposited plasmonic nanoparticle heating. ACS Appl Mater Interfaces. 2019;11(51):48525–32. https://doi.org/10.1021/acsami.9b16067.
53. Yamamoto Y, Tokonami S, Iida T. Surfactant-controlled photothermal assembly of nanoparticles and microparticles for rapid concentration measurement of microbes. ACS Appl Bio Mater. 2019;2(4):1561–8. https://doi.org/10.1021/acsabm.8b00838.
54. Namura K, Nakajima K, Kimura K, Suzuki M. Photothermally controlled marangoni flow around a micro bubble. Appl Phys Lett. 2015;106(4): 043101. https://doi.org/10.1063/1.4906929.
55. Rajeeva BB, Lin L, Perillo EP, Peng X, Yu WW, Dunn OK, Zheng Y. High-resolution bubble printing of quantum dots. ACS Appl Mater Interfaces. 2017;9(19):16725–33. https://doi.org/10.1021/acsami.7b04881.
56. Yamamoto D, Shioi A. Self-propelled nano/micromotors with a chemical reaction: underlying physics and strategies of motion control. KONA Powder Part J 2015;32:2015005. https://doi.org/10.14356/kona.2015005.
57. Kline TR, Iwata J, Lammert PE, Mallouk TE, Sen A, Velegol D. Catalytically driven colloidal patterning and transport. J Phys Chem B. 2006;110(48):24513–21. https://doi.org/10.1021/jp0643931.
58. Sugimura H, Mo S, Yamashiro K, Ichii T, Murase K. Photochemical assembly of gold nanoparticle arrays covalently attached to silicon surface assisted by localized plasmon in the nanoparticles. J Phys Chem C. 2013;117(6):2480–5. https://doi.org/10.1021/jp306290w.
59. Kim Y, Shah AA, Solomon MJ. Spatially and temporally reconfigurable assembly of colloidal crystals. Nat Commun. 2014;5:3637. https://doi.org/10.1038/ncomms4676.
60. Singh DP, Choudhury U, Fischer P, Mark AG. Non-equilibrium assembly of light-activated colloidal mixtures. Adv Mater. 2017;29(32):1701328. https://doi.org/10.1002/adma.201701328.
61. Huang X, Zhou X, Wu S, Wei Y, Qi X, Zhang J, Boey F, Zhang H. Reduced graphene oxide-templated photochemical synthesis and in situ assembly of Au nanodots to orderly patterned Au nanodot chains. Small. 2010;6(4):513–6. https://doi.org/10.1002/smll.200902001.
62. Xu BB, Wang L, Ma Z-C, Zhang R, Chen Q-D, Lv C, Han B, Xiao X-Z, Zhang X-L, Zhang Y-L, Ueno K, Misawa H, Sun H-B. Surface-plasmon-mediated programmable optical nanofabrication of an oriented silver nanoplate. ACS Nano 2014;8(7):6682–92. https://doi.org/10.1021/nn5029345.
63. Fanizza E, Cozzoli PD, Curri ML, Striccoli M, Sardella E, Agostiano A. UV-light-driven immobilization of surface-functionalized oxide nanocrystals onto Silicon. Adv Funct Mater. 2007;17(2):201–11. https://doi.org/10.1002/adfm.200600288.
64. van Geel R, Pruijn GJ, van Delft FL, Boelens WC. Preventing thiol-yne addition improves the specificity of strain-promoted azide-alkyne cycloaddition. Bioconjugate Chem. 2012;23(3):392–8. https://doi.org/10.1021/bc200365k.

65. Ling D, Park W, Park S-j, Lu Y, Kim KS, Hackett MJ, Kim BH, Yim H, Jeon, Na K, Hyeon T. Multifunctional tumor pH-sensitive self-assembled nanoparticles for bimodal imaging and treatment of resistant heterogeneous tumors. J Am Chem Soc 2014;136(15):5647–55. https://doi.org/10.1021/ja4108287.
66. Wang W, Cheng D, Gong F, Miao X, Shuai X. Design of multifunctional micelle for tumor-targeted intracellular drug release and fluorescent imaging. Adv Mater. 2012;24(1):115–20. https://doi.org/10.1002/adma.201104066.
67. Gallo J, Kamaly N, Lavdas I, Stevens E, Nguyen Q-D, Wylezinska-Arridge M, Aboagye EO, Long NJ. CXCR4-targeted and MMP-responsive iron oxide nanoparticles for enhanced magnetic resonance imaging. Angew Chem Int Edit 2014:53(36):9550–54. https://doi.org/10.1002/ange.201405442.
68. Ito A, Shinkai M, Honda H, Kobayashi T. Medical application of functionalized magnetic nanoparticles. J Biosci Bioeng. 2005;100(1):1–11. https://doi.org/10.1263/jbb.100.1.
69. Klajn R, Wesson PJ, Bishop KJ, Grzybowski BA. Writing self-erasing images using metastable nanoparticle "inks." Angew Chem Int Edit. 2009;48(38):7035–9. https://doi.org/10.1002/ange.200901119.
70. Kundu PK, Samanta D, Leizrowice R, Margulis B, Zhao H, Börner M, Udayabhaskararao T, Manna D, Klajn R. Light-controlled self-assembly of non-photoresponsive nanoparticles. Nat Chem. 2015;7(8):646–52. https://doi.org/10.1038/nchem.2303.
71. Cheng X, Sun R, Yin L, Chai Z, Shi H, Gao M. Light-triggered assembly of gold nanoparticles for photothermal therapy and photoacoustic imaging of tumors in vivo. Adv Mater. 2017;29(6):1604894. https://doi.org/10.1002/adma.201604894.
72. Li F, Lu J, Kong X, Hyeon T, Ling D. Dynamic nanoparticle assemblies for biomedical applications. Adv Mater. 2017;29(14):1605897. https://doi.org/10.1002/adma.201605897.
73. Zhao H, Sen S, Udayabhaskararao T, Sawczyk M, Kučanda K, Manna D, Kundu PK, Lee J-W, Král P, Klajn R. Reversible trapping and reaction acceleration within dynamically self-assembling nanoflasks. Nat Nanotechnol. 2016;11:82–8. https://doi.org/10.1038/nnano.2015.256.
74. Raeesi V, Chou LY, Chan WC. Tuning the drug loading and release of DNA-assembled gold-nanorod superstructures. Adv Mater. 2016;28(38):8511–8. https://doi.org/10.1002/adma.201600773.
75. Vieu C, Carcenac F, Pepin A, Mejias M, Lebib A, Manin-Ferlazzo L, Couraud L, Launois H. Electron beam lithography: resolution limits and applications. Appl Surf Sci. 2000;164(1):111–7. https://doi.org/10.1016/S0169-4332(00)00352-4.
76. Feldheim D, Keating C. Self-assembly of single electron transistors and related devices. Chem Soc Rev. 1998;27:1–12. https://doi.org/10.1039/A827001Z.
77. Ou FS, Hu M, Naumov I, Kim A, Wu W, Bratkovsky AM, Li X, Williams RS, Li Z. Hot-spot engineering in polygonal nanofinger assemblies for surface enhanced Raman spectroscopy. Nano Lett. 2011;11(6):2538–42. https://doi.org/10.1021/nl201212n.
78. Thacker VV, Herrmann LO, Sigle DO, Zhang T, Liedl T, Baumberg JJ, Keyser UF. DNA origami based assembly of gold nanoparticle dimers for surface-enhanced Raman scattering. Nat Commun. 2014;5:3448. https://doi.org/10.1038/ncomms4448.
79. Bae J, Lee S, Ahn J, Kim JH, Wajahat M, Chang WS, Yoon S-Y, Kim JT, Seol SK, Pyo J. 3D-printed quantum dot nanopixels. ACS Nano. 2020;14(9):10993–1001. https://doi.org/10.1021/acsnano.0c04075.
80. Yu D, Cao F, Gao Y, Xiong Y, Zeng H. Room-temperature ion-exchange-mediated self-assembly toward formamidinium perovskite nanoplates with finely tunable, ultrapure green emissions for achieving rec. 2020 displays. Adv Funct Mater. 2018;28(19):1800248. https://doi.org/10.1002/adfm.201800248.
81. Huang Y, Li W, Qin M, Zhou H, Zhang X, Li F, Song Y. Printable functional chips based on nanoparticle assembly. Small. 2017;13(4):1503339. https://doi.org/10.1002/smll.201503339.

15 Optical Nanotweezers for Biological Applications

Xufeng Zhang, Tong He, and Yuchao Li

Abstract

Ashkin et al. proved that laser radiation pressure can be used to manipulate bacteria and viruses in bioscience, and officially named "optical tweezers". Nowadays, the technology of optical tweezers has become one of the most powerful tools in the field of biology, but with the reduction of the scale of capturing objects, due to the existence of Brownian motion and laser diffraction limit, traditional optical tweezers are difficult to achieve stable capture. The emergence of nano-optical tweezers brings the possibility of capturing and manipulating nanoscale objects. Nano-optical tweezers have great application value and development potential in biomedical, quantum science, material science and other fields. In biomedical research, it can capture and manipulate tiny particles such as biological cells, DNA molecules, bacteria and viruses. In the research of quantum science, optical tweezers array, as a platform of quantum system, affects the subfields of quantum optics, quantum networks, cold quantum computing and atomic computing. In the research of material science, the emergence of nano-optical tweezers makes it possible to analyze individual nanoparticles by fluorescence spectrum, Raman spectrum and absorption spectrum. This paper introduces the basic principle of nano-optical tweezers, summarizes four kinds of nano-optical tweezers: fiber optical tweezers, waveguide optical tweezers, photonic crystal optical tweezers and plasmon optical tweezers, and introduces their main applications in manipulating nanoparticles and biomolecules.

Keywords

Optical force · Optical tweezers · Plasmonics · Nanoparticles · Biological molecule

15.1 Introduction

The famous astronomer Kepler found that the comet's tail deviates from the sun because of the radiation pressure of the sun, which exerts a steady pressure so that the comet's tail always points away from the sun. Maxwell speculated the existence of radiation pressure from his electromagnetic theory, but no experiments could prove this theory, until Soviet physicists Lebedev, Nicholas and Hull proved the existence of radiation pressure by experiments in 1900. Lebedev confirmed Maxwell's prediction with arc lamp and extremely light suspensions, but the light produced by arc lights was so weak that the radiation pressure was so small that it was difficult to have practical applications. Until 1960, the ruby laser, the first laser in the world, was successfully invented by Theodore Mailer. As a high-intensity light source, its invention provided great help for the research of optics and promoted the application and development of optical manipulation. In 1986, Professor Ashkin first reported the optical trapping of dielectric silica particles by "single beam gradient force trap" [1]. A year later, Ashkin et al. proved that laser radiation pressure can be used to manipulate bacteria and viruses, offically named "optical tweezer" [2], and steadily captured a microsphere with a diameter of 2.68 μm [3] using radiation pressure of continuous laser in 1970. Professor Ashkin made these outstanding contributions and was awarded the Nobel Prize in Physics in 2018. In view of the fact that optical tweezers have the advantages of non-contact, non-injury and high precision, optical tweezers have a good prospect and application value in biomedical, material science and other fields.

X. Zhang · T. He · Y. Li (✉)
Guangdong Provincial Key Laboratory of Nanophotonic Manipulation, Institute of Nanophotonics, College of Physics and Optoelectronic Engineering, Jinan University, Guangzhou, China
e-mail: liyuchao@jnu.edu.cn

B. Li et al. (eds.), *Biophotonic Manipulation*, Advances in Optics and Optoelectronics,
https://doi.org/10.1007/978-981-96-4982-2_15

The precise manipulation of nanoparticles plays an important role in biotechnology [4], medicine, quantum science and material physics. However, with the smaller the volume of the captured object (Fig. 15.1), the capture force will decrease sharply, such as Rayleigh nanoparticles and biomolecules. Due to the existence of Brownian motion and the diffraction limit of laser, it is difficult for traditional optical tweezers to achieve stable trapping, and it's more difficult to manipulate them. In order to capture and manipulate smaller biological cells and particles, it is necessary to increase the capture force of optical tweezers to overcome the Brownian motion of particles in liquid, and to solve this problem by introducing additional capture force, or by increasing the gradient and intensity of the capture field. Therefore, scientists began research on nano-optical tweezers. In 1986, Ashkin used a 1.5 W laser to capture nanoparticles with a diameter of 9–14 nm, but if the laser power was too high, it was easy to damage the sample. The emergence of nano-optical tweezers brings the possibility of capturing and manipulating nano-scale objects under lower laser power, and expands the research scope of optical tweezers to nano-scale particles or single molecules. This paper introduces the basic principle of nano-optical tweezers, summarizes four kinds of nano-optical tweezers: fiber optical tweezers, waveguide optical tweezers, photonic crystal optical tweezers and plasmon optical tweezers, and introduces their main applications in manipulating nanoparticles and biomolecules.

15.2 Nano-optical Tweezer

15.2.1 Basic Principles of Nano-optical Tweezer

In the 1970s, Ashkin observed the optical force produced by focused laser beams in the course of experiments, which can accelerate and capture micron particles [3]. Optical force comes from the interaction between light and matter. The two most common optical forces in optical tweezers are scattering force and gradient force. The light scattering force is caused by the light momentum transmitted to the particles, and the direction is the same as the propagation direction (Fig. 15.2a-I), the light momentum changes when the direction of light scattering is different from that of incident light. The optical gradient force depends on the electric field intensity gradient of the focused beam, and the direction always points to the maximum of the electric field intensity. If the object is transparent and satisfies that the size of the object is much larger than the wavelength, the gradient force can be regarded as the superposition of the scattering force in Fig. 15.2a-II.

For nanoparticles, because the linear size of nanoparticles is much smaller than the wavelengths of visible and nearinfrared light, Rayleigh approximation can be used to regard nanoparticles as dipoles moving in non-uniform electromagnetic fields [7]. The F (the gradient force) of the dipole:

$$F = \frac{1}{2}|\alpha|\nabla\left\langle E^2\right\rangle \tag{15.1}$$

where α is the polarizability of the particle, E is the electromagnetic field, and parentheses indicate the average time. $|E^2|$ is proportional to the field strength. Thus, as shown in Fig. 15.2b-I, in a three-dimensional focused Gaussian beam, a dipole is captured in focus. In addition to the gradient force, when the photon is scattered or absorbed on the surface of the incident particle, there is also the radiation pressure F_{rad} of the incident particle [8]:

$$F_{\text{rad}} = \frac{n\langle P\rangle}{c}C_{ext} \tag{15.2}$$

where n is the refractive index of the surrounding liquid, c is the speed of light, and $\langle P\rangle$ is the time-average Poynting vector:

$$\langle P\rangle = \frac{1}{2}\text{Re}\left(E\times H^*\right) \tag{15.3}$$

σ is the extinction cross section, which is the sum of the absorption cross section σ_{scat} and the scattering cross section σ_{abs}:

$$\sigma = \sigma_{scat} + \sigma_{abs} = \frac{k^4|\alpha|^2}{4\pi} + k\alpha'' \tag{15.4}$$

So, the time average force acting on the dipole is $\langle F\rangle$ [9]:

$$\langle F\rangle = \frac{1}{4}\text{Re}(\alpha)\nabla|E|^2 + \frac{\sigma}{2c}\text{Re}\left(E\times H^*\right) + \frac{\sigma c\varepsilon_0}{4\omega i}\nabla\times E\times E^* \tag{15.5}$$

where E is the electric field, H is the magnetic field, and ω is the angular frequency of the light field. The first term in formula (15.5) is the optical gradient force, the second term is the radiation pressure, and the third term is the force caused by the existence of spatial polarization gradient [10]. The radiation pressure plays a role in destroying the trapping potential in the axial direction, but as long as the first term is dominated by the second and third terms, the gradient force can overcome the radiation pressure, in other words, when the optical gradient force is equal to the light scattering force, the nanoparticles can be trapped stably in three dimensions, and this stable equilibrium position can be regarded as a simple harmonic spring oscillator in Fig. 15.2b-II.

Another key factor for the realization of light energy capture is the materials of nanoparticles. Their polarizability α has a certain influence on the stable capture, and α is

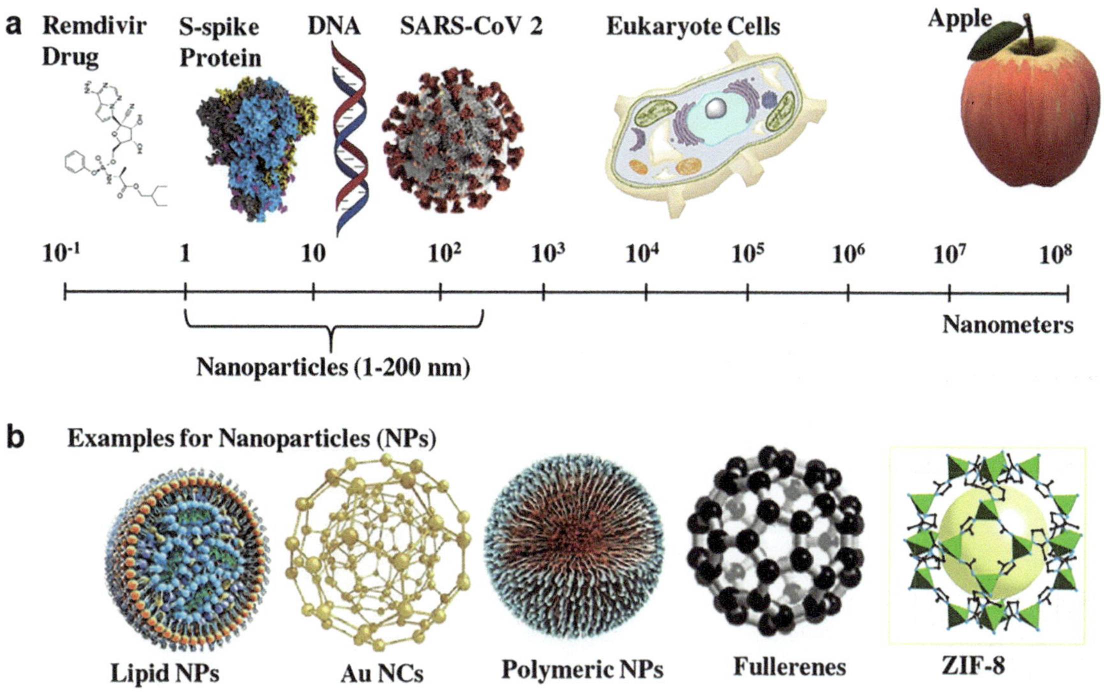

Fig. 15.1 Scale of objects and some examples of nanoparticles. Reprinted from Ashkin et al. [5], Copyright (1986), with permission from The Optical Society

proportional to the polarization volume V of nanoparticles:

$$\alpha = 3V\frac{\varepsilon_P - \varepsilon_m}{\varepsilon_p + 2\varepsilon_m} \tag{15.6}$$

Here ε_m is the dielectric constant of the medium, and ε_p refers to the dielectric constant of the particle at the appropriate wavelength [10]. Formula (15.6) is only applicable to very small nanoparticles ($d \ll \lambda$). If the particles have a coating, the coating will also affect the total polarizability of the particles. For capturing larger reflective particles, such as metallic particles, the trapping potential will become unstable, and the particles can easily get out of the trapping potential and can not be trapped stably.

15.2.2 Nano-optical Tweezers Application

Nano-optical tweezers are widely used in biomedicine, quantum science, material science and other fields. In biomedical research, optical tweezers have been a core technology in this field. Nano-optical tweezers can capture and manipulate tiny particles such as biological cells, DNA molecules, bacteria and viruses [11]. Because the particles in the liquid have Brownian motion, they will move irregularly all the time, which is not good for researchers to observe with a microscope. If we can steadily capture the particles, it will be convenient for researchers to observe and detect them. Moreover, optical tweezers are also a powerful tool for arrangement and assembly. Researchers can arrange and assemble particles or cells according to a specific location or order to study some characteristics and interactions between particles and cells. With the development of nano-optical tweezers, nano-optical tweezers are gradually combined with other technologies. In the research of quantum science, optical tweezers array, as a platform of quantum system, affects quantum optics, quantum networks, cold quantum computing and atomic computing and other sub-fields. In the study of various spectroscopy in material science, the emergence of nano-optical tweezers makes it possible to analyze the fluorescence spectrum, Raman spectrum and absorption spectrum of a single nanoparticle [12].

In addition to steadily capturing objects, nano-optical tweezers can also manipulate the movement and rotation of objects. We know that light carries not only linear momentum, but also angular momentum (AM). The optical force generated by linear momentum transfer can move an object, and angular momentum can be transmitted in optical tweezers to achieve the effect of rotating an object. As shown in Fig. 15.2c, this is called optical wrench [13], which is another application of optical tweezers. AM has two forms: spin angular momentum (SAM) and orbital angular momentum (OAM) [14]. The difference between SAM and OAM is that SAM depends on the polarization of the beam, and the SAM transfer

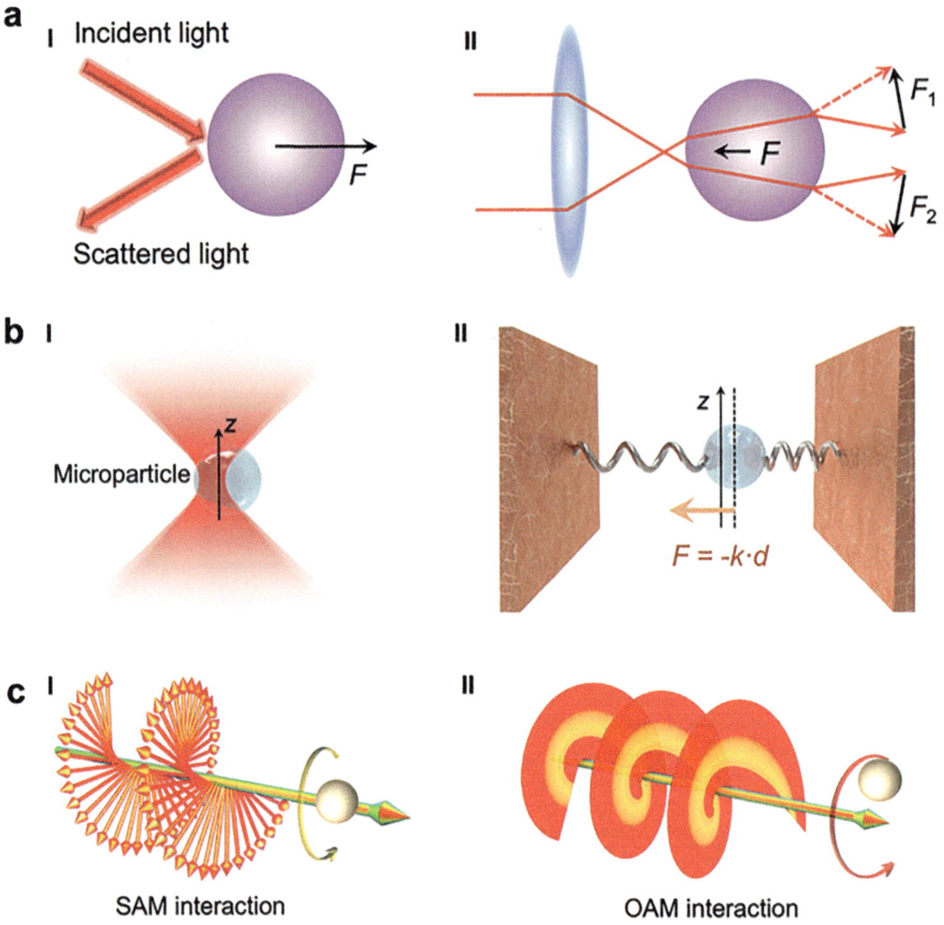

Fig. 15.2 The model of optical force [6]. **a** Principle of optical force. (I) Scattering force. (II) Gradient force. **b** Conventional optical tweezers. (I) Particles at the focus of a Gaussian beam. (II) COTs are conceptually analogous to a harmonic spring oscillator. **c** Angular momentum of light. (I) SAM. (II) OAM. Reprinted from Xin et al. [6], Copyright (2020), with permission from Wiley-Blackwell

causes the particles to rotate around their own axis (Fig. 15.2c-I), however, The OAM transfer will cause the particles to orbit around the optical axis of the incident beam (Fig. 15.2c-II). The mechanical effect of AM has turned optical tweezers into optical wrenches [13], which has a variety of potential applications. For example, the rapid rotation of gold nanoparticles in liquid is realized under the irradiation of circularly polarized light [15].

Based on Professor Askin's pioneering work on optical tweezers, optical tweezers have promoted the development of various fields, but the traditional optical tweezers have a short working distance and can not capture too small particles, researchers have begun to study nano-optical tweezers that can break through the diffraction limit. Based on the review of four typical nano-optical tweezers, the applications of nano-optical tweezers in capturing nanoparticles and biomolecules are introduced below. The four kinds of nano-optical tweezers are fiber optical tweezers, waveguide optical tweezers, photonic crystal optical tweezers and plasmon optical tweezers [16].

15.2.2.1 Fiber Optical Tweezers

Traditional optical tweezers are not flexible and difficult to integrate and apply in biological systems and in vivo. In order to solve this problem, researchers focus on optical fiber. The process of capturing particles by optical fiber tweezers is as follows: the laser is emitted and focused from the front end of the optical fiber to form the gradient distribution of the light field, and the longitudinal gradient force captures the particles near the focus to the optical axis, and then moves to the focus under the action of the transverse scattering force. Finally, the transverse gradient force captures the particles stably at the focus. In 1995, Ikeda et al. proved that nanoparticles can be successfully captured based on single-lens optical fiber [17]. For optical fiber tweezers, the shape of the front end of the fiber basically determines the distribution of the light field. Different shapes of the front end of the fiber can be prepared by heating stretching, physical polishing and other methods, usually spherical or paraboloid structure. These fiber optical tweezers already have the ability to capture particles, but because of the diffraction limit, it is difficult to capture when the size of the target object is submicron. The suitable shape and size of the optical fiber front end can be prepared by

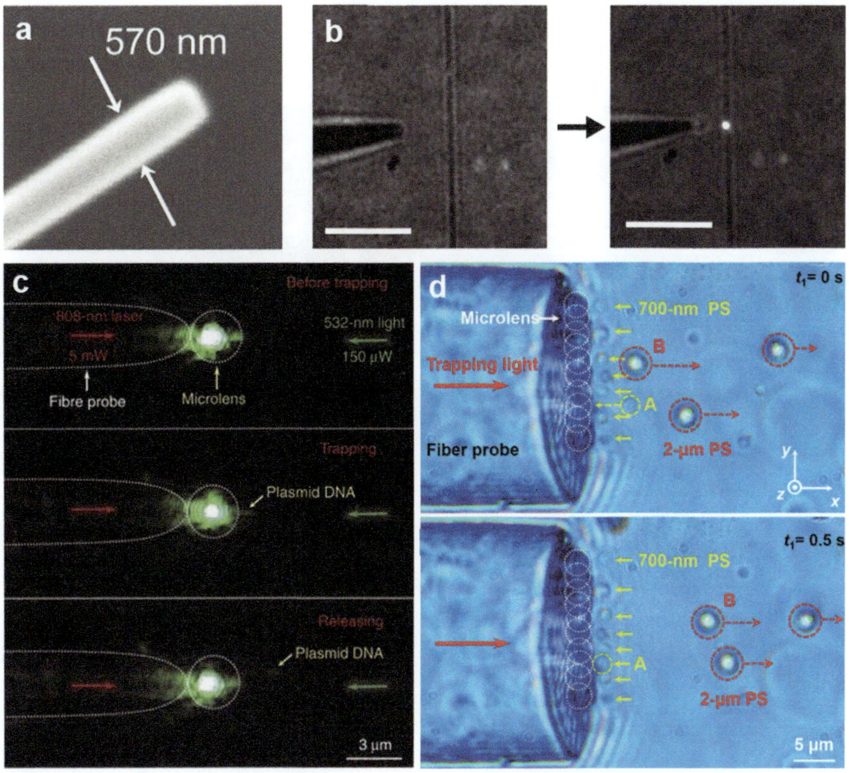

Fig. 15.3 Application of nano-fiber optical. **a** Optical fiber tip of diameter 570 nm. **b** The tip grasping the nanodiamond approaches the nanofiber and the nanodiamond is placed on the nanofiber and shows fluorescence. Reprinted from Fujiwara et al. [18], Copyright (2016), with permission from IOP Publishing Ltd. **c** Capture the dark-field light-scattering image of a single DNA molecule. Reprinted from Li et al. [20], Copyright (2016), with permission from Springer Nature. **d** 700 nm PS nanoparticles were selectively captured from a mixture of 700 nm PS and 2 µm PS particles. Reprinted from Li et al. [21], Copyright (2016), with permission from American Chemical Society

controlling the physical parameters of the preparation method, such as drawing temperature, speed and time.

As shown in Fig. 15.3 a, the fiber is first drawn into a needle shape with a diameter of about 1–2 µm under the excitation of a carbon dioxide laser, and then the tip is immersed in 5% HF solution to further reduce the size of the tip, resulting in a tip with a diameter of 400–600 nm. Manipulating the fiber with a submicron tip can pick up and place a single nano-diamond, in Fig. 15.3b [18], coupling a single nitrogen vacancy center to the ultrathin fiber-taper nano fibers. Fujiwara et al. coupled photons emitted by a single CdSe/ZnS nanocrystal in a tapered fiber with a diameter of 300 nm [19].

In addition to preparing finer optical fiber probes, Li et al. demonstrated a new method of assembling the dielectric microlens with the fiber tip to capture the nanoparticles near the microlens in a non-contact way. The nanojet generated by the dielectric microlens enhances the trapping ability of the potential well. This method successfully captures 85 nm PS particles and manipulates the optical fiber to achieve three-dimensional manipulation. Compared with dielectric nanoparticles, a single biomolecule is more difficult to manipulate, such as DNA molecules in a partially extended state, but this method can successfully capture a single plasmid DNA molecule (Fig. 15.3c) [20]. In addition to trapping a single molecule, people want to be able to selectively capture objects of multiple nanometer sizes. Li et al. reported a method to capture nanoparticles and sub-wavelength cells by using parallel photon nanojet arrays produced by assembling microlenses on optical fiber probes to capture multiple 190 nm fluorescent nanoparticles at the same time. The method also has the function of selective capture in Fig. 15.3d, particles A and B are affected by optical gradient force and scattering force, 700 nm PS particles (particle A) are captured, 2 µm PS particles (particle B) are pushed away from [21], and the optical separation of nanoparticles is realized.

15.2.2.2 Waveguide Optical Tweezers

The structures of waveguides can be divided into four types: optical fiber waveguides, SOI-based straight waveguides, resonant waveguides and SPR waveguides. Nano-fiber has a sensor tip with a topological structure, which can be integrated into microfluidic channels to capture and transport particles. For example, it was reported in 2012 that PS microspheres with a diameter of about 700 nm were transmitted along nanofibers with arbitrary curved diameters of 600 nm [22]. However, from the perspective of silicon-based manufacturing, optical fiber waveguides can't solve the problem of commercial chips. Therefore, scientists have proposed a variety of SOI optical waveguides, such as slot, trench, and groove. The evanescent wave on the surface of the waveguide has a non-uniform field strength, which can also excite the scattering force and gradient force. Because the optical force is a non-conservative force, which depends on the size of the particles, it is difficult for far-field optical tweezers to capture smaller particles. In contrast, near-field optical technology can overcome the diffraction limit and manipulate smaller

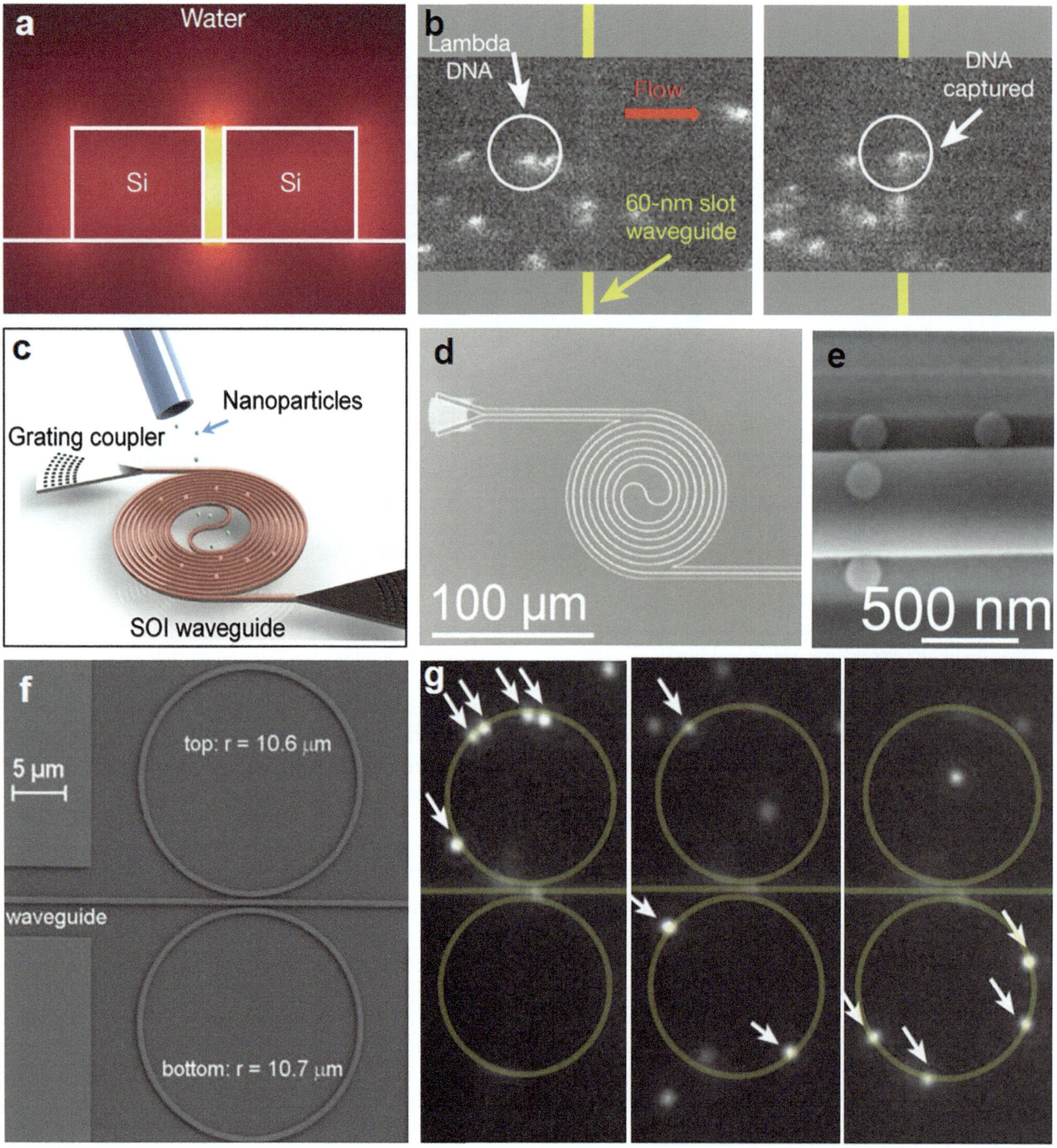

Fig. 15.4 Waveguide optical tweezers and their applications. **a** Mode profile for a silicon-on-insulator 40-nm slot waveguide immersed in water. **b** Capture and transmission of L-DNA. Reprinted from Yang et al. [25], Copyright (2009), with permission from Springer Nature. **c** Spiral nano waveguide sensor. **d** SEM of spiral nano waveguide sensor. Reprinted from Tang et al. [27], Copyright (2018), with permission from Springer Nature. **e** PS particles of 100 nm are deposited on the surface of waveguide. **f** SEM of cascaded waveguide. **g** Different nano particles are captured at different laser wavelengths. Reprinted from Lin et al. [28], Copyright (2011), with permission from Royal Society of Chemistry

particles and even nanoparticles by enhancing the local light intensity of the optical potential trap [23].

Almeida et al. developed a nano-photonic structure called slit waveguide [24] in 2004. The slot waveguide is composed of nano slot sandwiched between two high refractive index materials (Fig. 15.4a), Almeida et al. successfully captured and transmitted PS nanoparticles and L-DNA molecules of 75 nm using 100 and 60 nm slot waveguides in Fig. 15.4b [25]. Compared with slit waveguides, the optical modes of trench waveguides are deeper and wider, and can make full use of optical power. V-groove waveguides based on SOI have higher ability to capture nanoparticles [26]. In order to further improve the particle trapping ability and shorten the target analysis time, some special structure waveguides have been gradually studied and invented. For example, a spiral nano-waveguide was reported in 2018, which increased the sensing ability by two orders of magnitude in Fig. 15.4c [27].

Although both optical fibers and SOI waveguides can capture and transmit particles, the particles can only move in one direction. WGM waveguides effectively solve this problem. For example, Lin et al. proposed a cascade waveguide, such as Fig. 15.4f, which tunes the laser wavelength to the resonant wavelengths of different rings to realize the transmission of captured particles back and forth between rings. In addition, there is a panda micro-ring resonator composed of large-size microring and two small microrings [28]. WGM waveguides have great potential in drug targeted transport.

Various inventions and improvements of waveguide tweezers have promoted the development of waveguide structures, such as photonic crystal (PC), surface plasmon resonance (SPR) and whispering gallery mode (WGM) to improve the intensity of evanescent waves. Ta_2O_5 with high refractive index can be fabricated into submicron optical waveguides to realize the optical capture and transmission of nanoparticles [29]. In terms of nanostructure mode, two-dimensional periodic PC cavity [30] can obtain tunable optical properties through self-assembly of gold nanoparticles. Therefore, the near-field optical fluidic chip based on near-field waveguide technology applied to microfluidic channels or containers has the advantages of low cost and good stability, and is also suitable for a variety of biological and medical research and applications.

15.2.2.3 Photonic Crystal Optical Tweezers

Slot waveguide optical tweezers and fiber waveguide optical tweezers can only move transport particles along the waveguide surface and can't capture nanoparticles stably. The emergence of photonic crystal optical tweezers solves this problem. In 1987, E. Yablonovitch found that in the dielectric materials with periodic permittivity, the electromagnetic wave intensity of some bands will decay exponentially and can't be transmitted in the system, which is equivalent to the formation of energy gap [31] in the spectrum, that is, the so-called photonic band structure. The dielectric material with photonic band structure is called photonic crystal. According to the spatial distribution characteristics of dielectric materials, photonic crystals can be divided into one-dimensional photonic crystals, two-dimensional photonic crystals and three-dimensional photonic crystals. The dielectric constant of photonic crystals has spatial periodicity. This special structure makes photonic crystals have physical properties such as photonic bandgap and photon localization, which can be used to enhance the interaction between light and matter.

Photonic crystal optical tweezers can be designed as a local resonance mode, and the energy of the light field will be strongly concentrated, which will greatly enhance the optical gradient force. As in Fig. 15.5a, in 2010, it was reported that the optical resonance in one-dimensional photonic crystals was used to form a nano-scale optical potential well, and taking advantage of the stability of light in the resonator and the propagation characteristics of light in the waveguide, the PS particles of 48 nm and 62 nm and the PS particles transporting 500 nm were successfully captured in Fig. 15.5b [32]. In addition to linear transport, photonic crystal nano-optical tweezers can also control the angular orientation and rotation of microtubules and carbon nanotubes in Fig. 15.5d, which shows that short-range nano-optical manipulation can be achieved by photonic crystal optical tweezers [33]. In biological applications, traditional optical tweezers can cause photothermal damage to organisms and proteins. Later, researchers found that switching from Si devices to Si_3N_4 devices and using laser wavelengths from 1064 to 1550 nm can greatly reduce the local light and heat of the resonator, realizing the capture of individual bacteria [34]. Chen et al. proved that photonic crystal resonators can be used to capture and release Wilson disease proteins, and the temperature is below about 0.3 K in Fig. 15.5f [35]. With the continuous development of the excited photonic crystal cavity, the stiffness of the potential well is also increasing. Wu et al. [36] proposed an optical Brownian ratchet of the near-field traps based on the asymmetric pattern photonic crystal. The stiffness of the potential well is more than 25 times that of the traditional optical tweezers.

15.2.2.4 Plasmon Optical Tweezers

Since Professor Askin invented optical tweezers, optical capture technology has been developing continuously, and many applications have been reported, especially in biological systems. In order to overcome diffraction limit, near-field optics technology is developing rapidly, for example, several kinds of nano-optical tweezers introduced above, although near-field optical methods can capture and manipulate sub-wavelength objects, there generally exists the problem of weak scattering field, and microscopic metal objects have

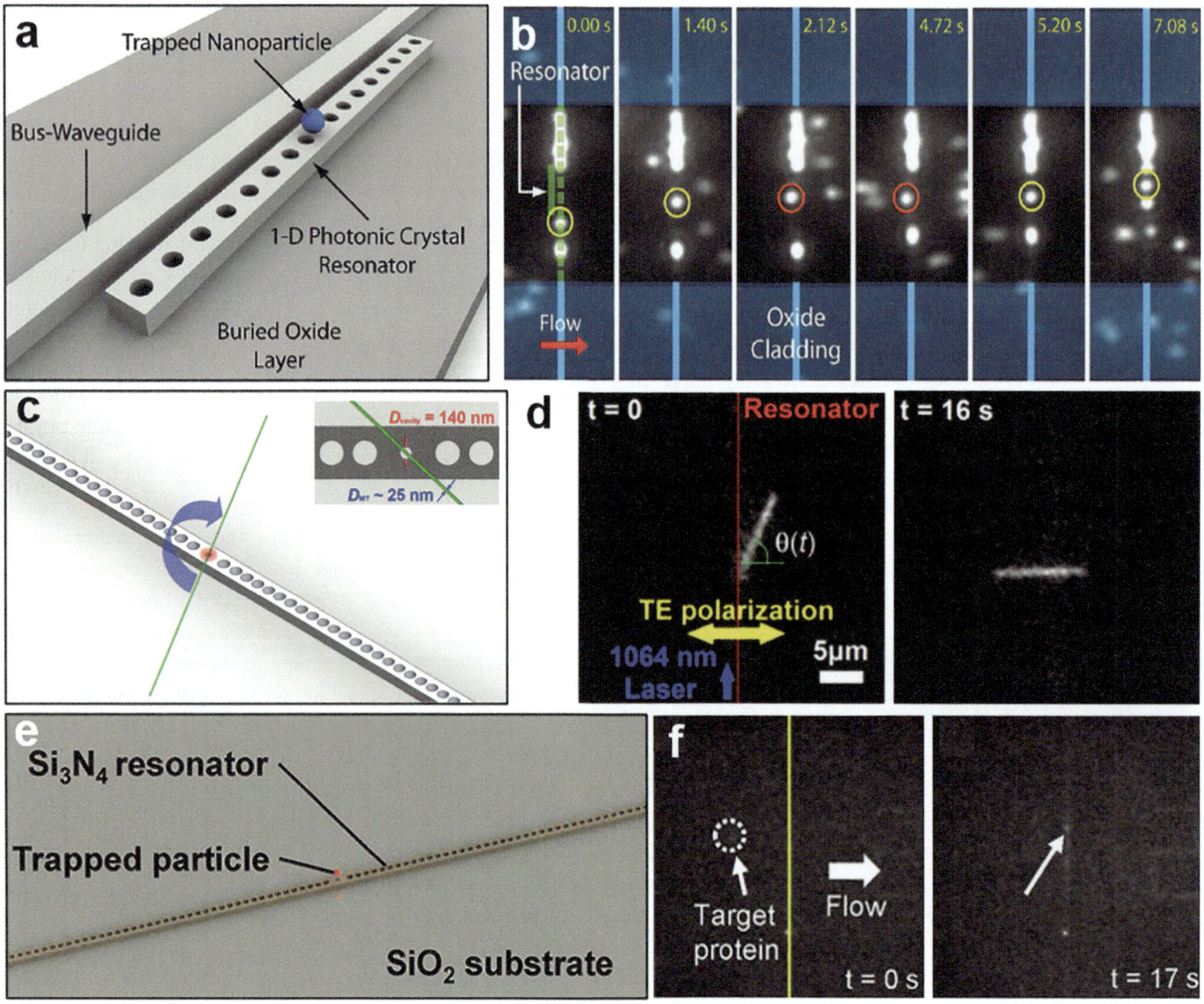

Fig. 15.5 Photonic crystal optical tweezers and application. **a** Schematic of photonic crystal resonator. **b** PS particles of 500 nm are captured and transported. Reprinted from Mandal et al. [32], Copyright (2010), with permission from American Chemical Society. **c** Schematic of microtubule rotation on photonic crystal resonator. **d** Microtubule rotation image under fluorescence microscope. Reprinted from Kang et al. [33], Copyright (2012), with permission from American Chemical Society. **e** Schematic of capturing nanoparticles on silicon nitride photonic crystal resonator. **f** Capture Wilson disease proteins on silicon nitride photonic crystal resonator. Reprinted from Chen et al. [35], Copyright (2012), with permission from American Chemical Society

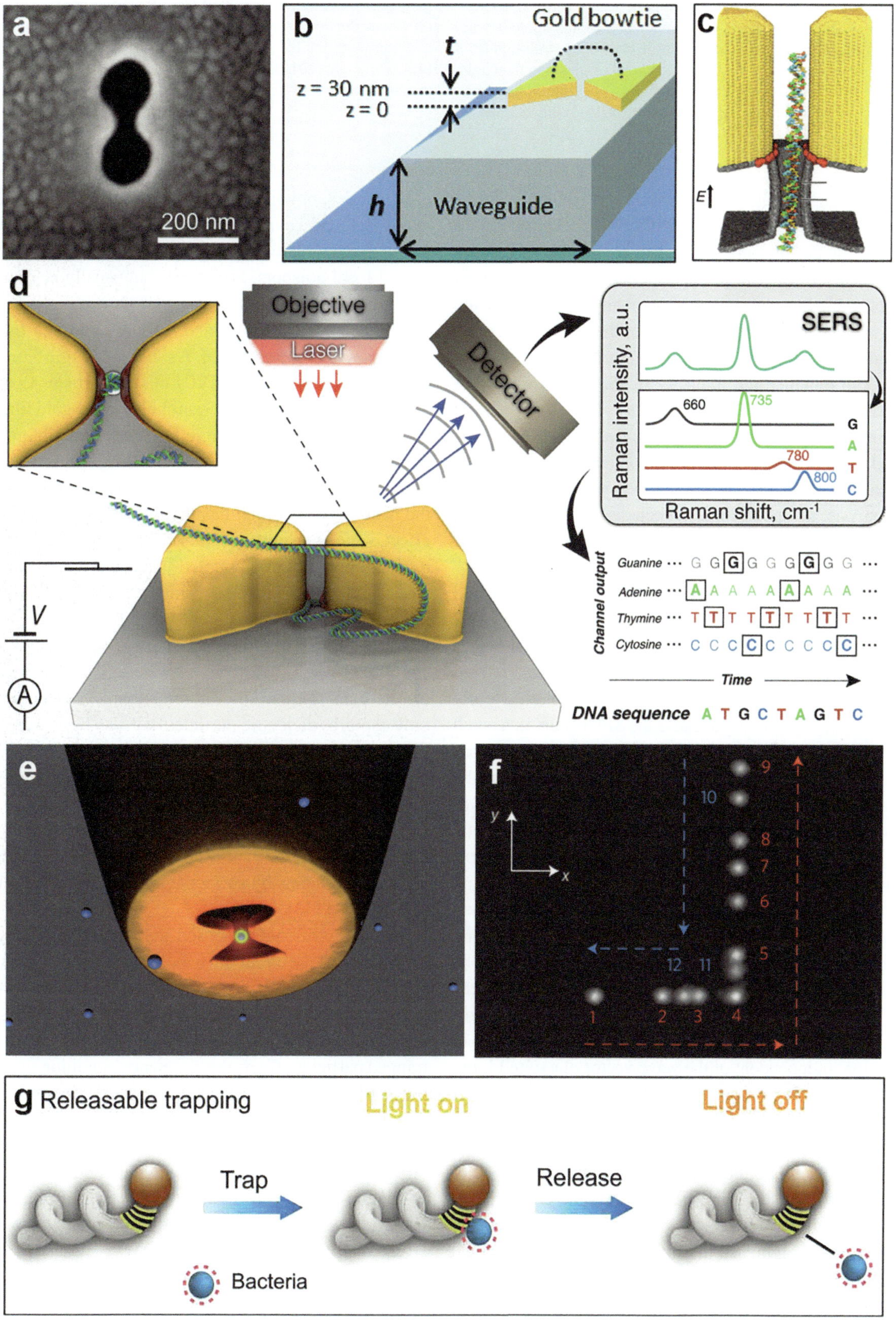
a
200 nm
b
Gold bowtie
t
z = 30 nm
z = 0
h
Waveguide
c
E
d
Objective
Laser
Detector
SERS
Raman intensity, a.u.
660
735
780
800
G
A
T
C
Raman shift, cm-1
Channel output
Guanine ··· G G G G G G G G G ···
Adenine ··· A A A A A A A A A ···
Thymine ··· T T T T T T T T T ···
Cytosine ··· C C C C C C C C C ···
Time
DNA sequence A T G C T A G T C
V
A
e
f
y
x
1 2 3 4 5 6 7 8 9 10 11 12
g Releasable trapping
Light on
Light off
Trap
Release
Bacteria

◄**Fig. 15.6** Plasmon optical tweezers and their applications. **a** Double-nanohole with tip spacing of 15 nm. Reprinted from Pang et al. [38], Copyright (2011), with permission from American Chemical Society. **b** Capturing schematic of optical waveguide-coupled nano gold bowtie. Reprinted from Lin et al. [42], Copyright (2014), with permission from Royal Society of Chemistry. **c** All-atomic model of plasmon nanohole. **d** Schematic of DNA for capture and sequencing of plasmon nanohole. Reprinted from Belkin et al. [43], Copyright (2015), with permission from American Chemical Society. **e** A three-dimensional manipulation diagram of a single 50 nm PS particle. **f** Composite diagram of particle displacement. Reprinted from Berthelot et al. [45], Copyright (2014), with permission from Springer Nature. **g** Schematic of capture and transportation bacteria of under light. Reprinted from Ghosh et al. [46], Copyright (2018), with permission from American Association for the Advancement of Science

strong characteristics of light absorption and scattering. Ordinary nano-optical tweezers are difficult to capture metal particles [37], but plasmon optical tweezers can overcome these obstacles. In various branches of near-field nano-optics, surface plasmon has great potential in manipulating nanoscale objects, which can be excited by the coupling of free electrons and incident photons on the material surface. The surface plasmon can enhance the intensity of the local evanescent field, and the enhancement coefficient is as high as 103 orders of magnitude, which provides a stable trap for 12 nm particles in diameter [38] without damaging the heat-sensitive biological samples, and can also improve the performance of bioanalysis in spectroscopy, such as surface-enhanced Raman scattering. Therefore, plasmon optical tweezers have unlimited biomedical possibilities.

Surface plasmons are electromagnetic waves are excited on the surface of a conductor. It has two types: the first is surface plasmon polariton (SPP), which propagates on a smooth dielectric-metal interface, and the second is localized surface plasmon (LSP), which exists in finite geometric structures such as nanoparticles. Because the SPP occurs in a much smaller range than the light wavelength, the energy is limited to the sub-wavelength region near the metal surface, where the light field is strongly enhanced. The surface plasmon waves can be excited and propagated on the smooth dielectric-metal interface. In addition, the surfaces of small geometric patterns can effectively restore the moving electrons, thus causing resonance [39]. Localized plasmon resonance (LSPR) refers to a collective oscillation of electrons at the interface of metal structures. Unlike the propagating SPP, SPP usually can't be directly excited by light incident from free space and needs the enhancement of wave vector to meet the matching condition of excitation, while LSP can be excited by direct light illumination, independent of the wave vector of excited light.

The metal geometric structure whose size is equal to or less than the incident wavelength can excite the enhanced local surface electromagnetic field, so most of the plasmon tweezers will integrate the nanostructure into the surface of the potential well and compress the energy into a very small nanostructure, as a result, the potential well field produced by the surface plasmon breaks through the diffraction limit and improves the strength and efficiency of the potential well. The development of nanomanufacturing technology has promoted the proposal and research of various nanostructures, such as nanoholes, bowties, rings, waveguides, square nanoplates, nanowires and cavities [40]. These structures can enhance the capture intensity of plasmons by resonance or the formation of local hotspots. Structures can usually be divided into two types: embossed patterns that generate electric potential around the substrate and hollow apertures that excite hotspots within the structure. As shown in Fig. 15.6a, Gordon et al. successfully captured 20 nm's PS particles and 12 nm's silica spheres using double-nanoholes in the gold film [38], in addition to capturing a single bovine serum albumin (BSA) molecule with a radius of 3.4 nm [41]. In addition to the double-nanohole, there is a very common structure, that is the nano-bowtie, which has the tip of lightning rod and gap effect, which can enhance the LSPR around the gap. In 2014, a structure combining coupled waveguides and nano bowtie was proposed in Fig. 15.6b, which can transport, capture and release particles with a diameter of 20 nm. This structure has strong capture force, high efficiency, and relatively simple equipment, only 350 nm, which is helpful to the circuit design on the chip [42]. Like double-nanoholes, nano bowtie not only captures dielectric particles, but also has a wide range of biomedical applications. For example, DNA is an important biomolecule in biology. In order to develop a method of DNA sequencing, the team studied the exposure of DNA fragments by using plasmon optical tweezers combined with gold nano bowtie and nanohole, and studied the nucleotide composition of DNA fragments by surface-enhanced Raman scattering in Fig. 15.6c–d [43].

In 2009, Quidant proposed and proved the self-induced reaction (SIBA) effect in nanohole potential wells, which is a particularly useful method in single molecule trapping experiments [44]. The SIBA potential well makes use of the quadratic potential well effect, and the motion of the trapped particles affects the resonance condition of the potential well itself, thus affecting the local electric field. The end effect refers to the fact that the trapped particles leave the trap center resulting in an increase in the optical force on the particles towards the trap center, an effect that helps to capture the particles, which was experimentally achieved by Quidant et al., for PS particles smaller than 100 nm at incident power less than 1 mW [44], and reduce the thermal effect to deal with the damage of particles. In addition, the general plasmon optical tweezers have to rely on the substrate to design the structure, which can not achieve three-dimensional optical manipulation. In 2014, their team reported that the design of a nano bowtie at the end of the optical fiber to achieve three-dimensional manipulation of a single 50 nm dielectric particles in Fig. 15.6e–f, the whole system is very portable, there are no other clunky optical elements [45], SIBA effect is expected to be widely used in the future. In 2018, it was also reported that a mobile nano-optical tweezer, which combines plasmon nanostructures with magnetically driven micro-robots, successfully manipulates one or more nano-level bacteria and fluorescent nano-diamonds in a three-dimensional way in Fig. 15.6g [46].

In addition to the above applications, plasmon optical tweezers have many applications in various fields because of their high precision and field enhancement. In biomedicine, the size of the virus is also within the range of plasmon capture. For example, Durmanov et al. developed a substrate that can capture virus particles and detect their Raman signals [47]. Vesicles in vivo can be used as a drug carrier. Oh et al. reported an integrated NanoGap plasmon optical tweezers that can capture vesicles below 100 nm and be used for real-time Raman imaging [48]. The signal of an object needs to be used not only in imaging, but also in sensing. In 2017, Gordon et al. reported a highly sensitive plasmon nanohole system, which can detect and capture the changes of 1 nm particles and improve the sensing efficiency of double nanoholes [49]. The rapid development of plasmon optical tweezers provides new opportunities and applications for nano-science and technology, although there are still some shortcomings and difficulties to overcome, but the development and application of plasmon optical tweezers will never stop.

15.3 Conclusion

In conclusion, we introduce the basic principle of nano-optical tweezers, and start with four common nano-optical tweezers, including fiber optical tweezers, waveguide optical tweezers, photonic crystal optical tweezers and plasmon optical tweezers, and briefly introduce the applications of different nano-optical tweezers in various fields. Nano-optical tweezers have the advantages of strong capture ability, wide control range, high precision, no damage and no contact, and are widely used to capture and manipulate nano-sized particles or materials in different fields of science. It has become a powerful tool for us to research the micro world. The rapid development of optical capture technology shows that nano-optical tweezers have great potential in capturing smaller particles and manipulating substances of various shapes, and have great application value and development potential in biomedical, quantum information, material science and other fields. However, the thermal effect caused by the interaction between light and matter can not be completely avoided, the thermal effect may affect the precise manipulation of particles, and it is easy to cause damage to cells and biomolecules, so that nano-optical tweezers are difficult to be used in vivo. The application of special nanostructures can reduce the influence of thermal effect, in addition, the appearance of femtosecond optical tweezers [50] and photonic crystal optical tweezers also reduces the photothermal damage to cells and biomolecules. However, the limitation of penetration depth also affects the application in vivo. Optical fiber tweezers have no limitation of manipulation depth and can be combined with biochips. With the development of optical tweezers technology, it is believed that the precise manipulation of cells in vivo will be realized in the future. Nano-optical tweezers still have some other shortcomings, such as accuracy and efficiency, there is room for improvement, and the simplification of the structure of nano-optical tweezers also needs to be studied.

Acknowledgements This work was supported by the National Key Research and Development Program of China (No. 2022YFA1206300), and National Natural Science Foundation of China (No. 62422505). X.Z. and T.H. contributed equally to this work.

References

1. Ashkin A, Dziedzic JM, Bjorkholm JE, Chu S. Observation of a single-beam gradient force optical trap for dielectric particles. Opt Lett. 1986;11(5):288–90. https://doi.org/10.1364/OL.11.000288.
2. Ashkin A, Dziedzic JM. Optical trapping and manipulation of viruses and bacteria. Science. 1987;235(4795):1517–20. https://doi.org/10.1126/science.3547653.
3. Ashkin A. Acceleration and trapping of particles by radiation pressure. Phys Rev Lett. 1970;24:156. https://doi.org/10.1103/PhysRevLett.24.156.
4. Bustamante CJ, Chemla YR, Liu S, Wang MD. Optical tweezers in single-molecule biophysics. Nat Rev Methods Primers. 2021;1:25. https://doi.org/10.1038/s43586-021-00021-6.
5. Abdelhamid HN, Badr G. Nanobiotechnology as a platform for the diagnosis of COVID-19: a review. Nanotechnol Environ Eng. 2021;6:19. https://doi.org/10.1007/s41204-021-00109-0.
6. Xin H, Li Y, Liu Y-C, Xiao Y-F, Li B. Optical forces: from fundamental to biological applications. Adv Mater. 2020;32(37):2001994. https://doi.org/10.1002/adma.202001994.
7. Bendix PM, Jauffred L, Norregaard K, Oddershede LB. Optical trapping of nanoparticles and quantum dots. IEEE J Sel Top Quantum Electron. 2013;20(3):15–26. https://doi.org/10.1109/JSTQE.2013.2287094.
8. Harada Y, Asakura T. Radiation forces on a dielectric sphere in the Rayleigh scattering regime. Opt Commun. 1996;124(5–6):529–41. https://doi.org/10.1016/0030-4018(95)00753-9.
9. Albaladejo S, Marqués MI, Laroche M, Sáenz JJ. Scattering forces from the curl of the spin angular momentum of a light field. Phys Rev Lett. 2009;102: 113602. https://doi.org/10.1103/PhysRevLett.102.113602.
10. Johnson PB, Christy RW. Optical constants of the noble metals. Phys Rev B. 1972;6:4370. https://doi.org/10.1103/PhysRevB.6.4370.

11. Vickers NJ. Animal communication: when I'm calling you, will you answer too? Curr Biol. 2017;27(14):R713–5. https://doi.org/10.1016/j.cub.2017.05.064.
12. Jones S, Al Balushi AA, Gordon R. Raman spectroscopy of single nanoparticles in a double-nanohole optical tweezer system. J Opt. 2015;17(10):102001. https://doi.org/10.1016/j.cub.2017.05.064.
13. Simpson NB, Dholakia K, Allen L, Padgett MJ. Mechanical equivalence of spin and orbital angular momentum of light: an optical spanner. Opt Lett. 1997;22(1):52–4. https://doi.org/10.1364/OL.22.000052.
14. He H, Friese MEJ, Heckenberg NR, RubinszteinDunlop H. Direct observation of transfer of angular momentum to absorptive particles from a laser beam with a phase singularity. Phys Rev Lett. 1995;75:826. https://doi.org/10.1103/PhysRevLett.75.826.
15. Lehmuskero A, Ogier R, Gschneidtner T, Johansson P, Käll M. Ultrafast spinning of gold nanoparticles in water using circularly polarized light. Nano Lett. 2013;13(7):3129–34. https://doi.org/10.1021/nl4010817.
16. Kolbow JD, Lindquist NC, Ertsgaard CT, Yoo D, Oh SH. Nano-optical tweezers: methods and applications for trapping single molecules and nanoparticles. ChemPhysChem. 2021;22(14):1409–20. https://doi.org/10.1002/cphc.202100004.
17. Ikeda M, Optical trapping using optical fibers. In: Proceedings of the fifth Sino-Japanese joint meeting; 1995.
18. Fujiwara M, Yoshida K, Noda T, Takashima H, Schell AW, Mizuochi N, Takeuchi S. Manipulation of single nanodiamonds to ultrathin fiber-taper nanofibers and control of NV-spin states toward fiber-integrated λ-systems. Nanotechnology 2016;27(45):455202. https://iopscience.iop.org/article/10.1088/0957-4484/27/45/455202/meta.
19. Fujiwara M, Toubaru K, Noda T, Zhao H-Q, Takeuch S. Highly efficient coupling of photons from nanoemitters into single-mode optical fibers. Nano Lett. 2011;11(10):4362–5. https://doi.org/10.1021/nl2024867.
20. Li YC, Xin HB, Lei HX, Lin LL, Li YZ, Zhang Y, Li BJ. Manipulation and detection of single nanoparticles and biomolecules by a photonic nanojet. Light: Sci Appl.; 2016;5(12):e16176–e16176. https://doi.org/10.1038/lsa.2016.176.
21. Li YC, Xin HB, Liu XS, Zhang Y, Lei HX, Li BJ. Trapping and detection of nanoparticles and cells using a parallel photonic nanojet array. ACS Nano. 2016;10(6):5800–8. https://doi.org/10.1021/acsnano.5b08081.
22. Li YC, Xu LL, Li BJ. Optical delivery of nanospheres using arbitrary bending nanofibers. J Nanopart Res. 2012;14(4):1–7. https://doi.org/10.1007/s11051-012-0799-3.
23. Kotsifaki DG, Chormaic SN. Plasmonic optical tweezers based on nanostructures: fundamentals, advances and prospects. Nanophotonics. 2019;8(7):1227–45. https://doi.org/10.1515/nanoph-2019-0151.
24. Almeida VR, Xu QF, Barrios CA, Lipson M. Guiding and confining light in void nanostructure. Opt Lett. 2004;29(11):1209–11. https://doi.org/10.1364/OL.29.001209.
25. Yang AHJ, Moore SD, Schmidt B, Klug M, Lipson M, Erickson D. Optical manipulation of nanoparticles and biomolecules in subwavelength slot waveguides. Nature. 2009;457:71–5. https://doi.org/10.1038/nature07593.
26. Sahafi M, Habibzadeh-Sharif A. Robust increase of the optical forces in waveguide-based optical tweezers using V-groove structure. JOSA B. 2018;35(8):1905–9. https://doi.org/10.1364/JOSAB.35.001905.
27. Tang S-J, Liu S, Yu X-C, Song Q, Gong Q, Xiao Y-F. On-chip spiral waveguides for ultrasensitive and rapid detection of nanoscale objects. Adv Mater. 2018;30(25):1800262. https://doi.org/10.1002/adma.201800262.
28. Lin S, Crozier KB. Planar silicon microrings as wavelength-multiplexed optical traps for storing and sensing particles. Lab Chip. 2011;11:4047–51. https://doi.org/10.1039/C1LC20574A.
29. Ahluwalia BS, McCourt P, Huser T, Hellesø OG. Optical trapping and propulsion of red blood cells on waveguide surfaces. Opt Express. 2010;18(20):21053–61. https://doi.org/10.1364/OE.18.021053.
30. Huang N, Martínez LJ, Jaquay E, Nakano A, Povinelli ML. Optical epitaxial growth of gold nanoparticle arrays. Nano Lett. 2015;15(9):5841–5. https://doi.org/10.1021/acs.nanolett.5b01929.
31. Yablonovitch E. Inhibited spontaneous emission in solid-state physics and electronics. Phys Rev Lett. 1987;58(20):2059. https://doi.org/10.1103/PhysRevLett.58.2059.
32. Mandal S, Serey X, Erickson D. Nanomanipulation using silicon photonic crystal resonators. Nano Lett. 2010;10(1):99–104. https://doi.org/10.1021/nl9029225.
33. Kang P, Serey X, Chen Y-F, Erickson D. Angular orientation of nanorods using nanophotonic tweezers. Nano Lett. 2012;12(12):6400–7. https://doi.org/10.1021/nl303747n.
34. Leest T, Caro J. Cavity-enhanced optical trapping of bacteria using a silicon photonic crystal. Lab Chip. 2013;13:4358–65. https://doi.org/10.1039/C3LC50879J.
35. Chen YF, Serey X, Sarkar R, Chen P, Erickson D. Controlled photonic manipulation of proteins and other nanomaterials. Nano Lett. 2012;12(3):1633–7. https://doi.org/10.1021/nl204561r.
36. Wu S-H, Huang N, Jaquay E, Povinelli ML. Near-field, on-chip optical Brownian ratchets. Nano Lett. 2016;16(8):5261–6. https://doi.org/10.1021/acs.nanolett.6b02426.
37. Hansen PM, Bhatia VK, Harrit N, Oddershede L. Expanding the optical trapping range of gold nanoparticles. Nano Lett. 2005;5(10):1937–42. https://doi.org/10.1021/nl051289r.
38. Pang Y, Gordon R. Optical trapping of 12 nm dielectric spheres using double-nanoholes in a gold film. Nano Lett. 2011;11(9):3763–7. https://doi.org/10.1021/nl201807z.
39. Maier SA. Plasmonics: fundamentals and applications. New York: Springer; 2007. https://link.springer.com/book/10.1007/0-387-37825-1.
40. P. Jones, O. Maragó, G. Volpe. Optical tweezers. Cambridge University Press; 2015. https://doi.org/10.1017/CBO9781107279711.
41. Pang Y, Gordon R. Optical trapping of a single protein. Nano Lett. 2012;12(1):402–6. https://doi.org/10.1021/nl203719v.
42. Lin PT, Chu HY, Lu TW, Lee P-T. Trapping particles using waveguide-coupled gold bowtie plasmonic tweezers. Lab Chip. 2014;14:4647–52. https://doi.org/10.1039/C4LC00731J.
43. Belkin M, Chao SH, Jonsson MP, Dekker C, Aksimentiev A. Plasmonic nanopores for trapping, controlling displacement, and sequencing of DNA. ACS Nano. 2015;9(11):10598–611. https://doi.org/10.1021/acsnano.5b04173.
44. Juan ML, Gordon R, Pang Y, Eftekhari F, Quidant R. Self-induced back-action optical trapping of dielectric nanoparticles. Nat Phys. 2009;5:915–9. https://doi.org/10.1038/nphys1422.
45. Berthelot J, Aćimović SS, Juan ML, Kreuzer MP, Renger J, Quidant R. Three-dimensional manipulation with scanning near-field optical nanotweezers. Nat Nanotechnol. 2014;9:295–9. https://doi.org/10.1038/nnano.2014.24.
46. Ghosh S, Ghosh A. Mobile nanotweezers for active colloidal manipulation. Sci Robot. 2018;3(14):eaaq0076. https://doi.org/10.1126/scirobotics.aaq0076.
47. Durmanov NN, Guliev RR, Eremenko AV, Boginskaya IA, Ryzhikov IA, Trifonova EA, Putlyaev EV, Mukhin AN, Kalnov SL, Balandina MV, Tkachuk AP, Gushchin VA, Sarychev AK, Lagarkov AN, Rodionov IA, Gabidullin AR, Kurochkin IN. Non-labeled selective virus detection with novel SERS-active porous silver nanofilms fabricated by electron beam physical vapor deposition. Sens Actuators, B Chem. 2018;257:37–47. https://doi.org/10.1016/j.snb.2017.10.022.

48. Ertsgaard CT, Wittenberg NJ, Klemme DJ, Barik A, Shih W-C, Oh S-H. Integrated nanogap platform for sub-volt dielectrophoretic trapping and real-time Raman imaging of biological nanoparticles. Nano Lett. 2018;18(9):5946–53. https://doi.org/10.1021/acs.nanolett.8b02654.
49. Ghorbanzadeh M, Jones S, Moravvej-Farshi MK, Gordon R. Improvement of sensing and trapping efficiency of double nanohole apertures via enhancing the wedge plasmon polariton modes with tapered cusps. ACS Photonics. 2017;4(5):1108–13. https://doi.org/10.1021/acsphotonics.6b00923.
50. Jing P, Wu J, Liu GW, Keeler EG, Pun SH, Lin LY. Photonic crystal optical tweezers with high efficiency for live biological samples and viability characterization. Sci Rep. 2016;6:1–7. https://doi.org/10.1038/srep19924.

GPSR Compliance

The European Union's (EU) General Product Safety Regulation (GPSR) is a set of rules that requires consumer products to be safe and our obligations to ensure this.

If you have any concerns about our products, you can contact us on ProductSafety@springernature.com

In case Publisher is established outside the EU, the EU authorized representative is:

Springer Nature Customer Service Center GmbH
Europaplatz 3
69115 Heidelberg, Germany

Batch number: 10705949

Printed by Printforce, the Netherlands